OXFORD SERIES ON MATERIALS MODELLING

OXFORD SERIES ON MATERIALS MODELLING

Materials modelling is one of the fastest growing areas in the science and engineering of materials, both in academe and in industry. It is a very wide field covering materials phenomena and processes that span ten orders of magnitude in length and more than twenty in time. A broad range of models and computational techniques has been developed to model separately atomistic, microstructural and continuum processes. A new field of multi-scale modelling has also emerged in which two or more length scales are modelled sequentially or concurrently. The aim of this series is to provide a pedagogical set of texts spanning the atomistic and microstructural scales of materials modelling, written by acknowledged experts. Each book will assume at most a rudimentary knowledge of the field it covers and it will bring the reader to the frontiers of current research. It is hoped that the series will be useful for teaching materials modelling at the postgraduate level.

APS, London
RER, Livermore, California

1. M.W. Finnis: *Interatomic forces in condensed matter*
2. K. Bhattacharya: *Microstructure of martensite - Why it forms and how it gives rise to the shape-memory effects*
3. V.V. Bulatov, W. Cai: *Computer simulations of dislocations*
4. A.S. Argon: *Strengthening mechanisms in crystal plasticity*
5. L.P. Kubin: *Dislocations, mesoscale simulations and plastic flow*
6. A. P. Sutton: *Physics of elasticity and crystal defects, Second edition*
7. D. Steigmann: *A course on plasticity theory*
8. J.A. Moriarty: *Theory and application of quantum-based interatomic potentials in metals and alloys*

Forthcoming:
D.N. Theodorou, V. Mavrantzas: *Multiscale modelling of polymers*

From reviews of the first edition

This is a concise, modern and refreshing introduction to a topic that is key to all of materials science and engineering. The writing is lucid and clear. It not only conveys the concepts and mathematical details but also gives a historical perspective. Sutton has bridged the gap between the atomistic and continuum pictures and, more importantly, makes the leap across the seemingly unbridgeable divide, easy, appealing and enjoyable. The steps in the mathematical derivations are shown clearly ... It is a great textbook to teach from as well as for those who want to learn the material on their own.

Reviewed in Contemporary Physics by Mogadalai P. Gururajan.

Sutton is a giant in the field... I am certain this book will be a classic.

Craig Carter, MIT.

Superb ... and written in an excellent, engaging style ... Sutton is an internationally respected expert in structural materials science and condensed matter physics, one of very few people to have such status in these two domains simultaneously.

Tom Swinburne, CNRS, Aix-Marseille Université.

Sutton emphasizes the physical meaning behind the mathematical models he clearly introduces. The style is simple, didactic, and effective. The coverage of some of the Open Questions in Chapter 10 (e.g. electroplasticity) is entirely unique to this book.

Beñat Gurrutxaga-Lerma, University of Cambridge.

Although there are other relevant texts in this field, this book includes connections to atomic treatments of defects. These are timely additions, and provide new physical insights. Although the book contains much mathematics, it is essentially readable, and stimulating.

Sir Peter Hirsch, University of Oxford.

This is an outstanding book. Students will appreciate the clarity of the arguments, including careful derivations.

Robert Rudd, Series Editor, Oxford Series on Materials Modelling.

The book is highly accessible, and provides the level of insight into the subject that you would rarely find in academic literature... It is particularly significant that the author has made a clear connection between Physics and Elasticity and Defects in this book. There is an established element of tradition here, where L.D. Landau and E.M. Lifshitz included Theory of Elasticity in their famous Course in Theoretical Physics. This new book by Adrian Sutton matches the Landau-Lifshitz book extremely well, providing new, modern insights into the phenomena, and matching the needs of contemporary generations of students and researchers.

Sergei Dudarev, University of Oxford.

It is quite obvious that the majority of the content is material that the author has worked through from scratch, much of it original, and this is especially reflected in the problems, which are detailed and novel.

Tony Paxton, King's College London.

About Adrian P Sutton

Educated in materials science at the Universities of Oxford and Pennsylvania, Adrian Sutton has held professorships at Oxford University, Aalto University and Imperial College London. He takes a physicist's approach to address fundamental aspects of the science of materials. He has taught across the undergraduate curriculum of materials science. He has consulted for companies in the United States, Japan and the UK. He is a founder of the Thomas Young Centre, the London Centre for Theory and Simulation of Materials. He was also the founding director in 2009 of the Centre for Doctoral Training (CDT) on Theory and Simulation of Materials at Imperial College, which attracted into materials science more than a hundred graduates from the UK and overseas with first class honours degrees or equivalent in physics and engineering. His book *Rethinking the PhD* is the extraordinary story of this acclaimed CDT. In 2012 he was awarded the Rector's Medal for Outstanding Innovation in Teaching by Imperial College. He was elected to a Fellowship of the Royal Society in 2003. In 2018 he left paid employment to focus on research and scholarship. He is a visiting professor at the Department of Physics at Imperial College London where he is also an emeritus professor, and he is a supernumerary fellow of Linacre College Oxford. He lives in Oxford with his wife Pat White.

Also by Adrian P Sutton

Electronic structure of materials, Oxford University Press (1993), 260 pages + xv. Published in German translation as *Elektronische Struktur in Materialen*, by VCH, Weinheim (1996).

Interfaces in crystalline materials, with R W Balluffi, Oxford University Press (1995), 819 pages + xxxi. Reissued in 2006 in the series *Oxford Classic Texts in the Physical Sciences*. Published in Chinese by Higher Education Press Limited, Beijing (2015).

Rethinking the PhD, published by Amazon.com (2020), 94 pages.

Concepts of materials science, Oxford University Press (2021), 137 pages + xi.

Physics of Elasticity and Crystal Defects

2nd Edition

Adrian P. Sutton

Emeritus Professor, Department of Physics, Imperial College London, UK

OXFORD
UNIVERSITY PRESS

OXFORD
UNIVERSITY PRESS

Great Clarendon Street, Oxford, OX2 6DP,
United Kingdom

Oxford University Press is a department of the University of Oxford.
It furthers the University's objective of excellence in research, scholarship,
and education by publishing worldwide. Oxford is a registered trade mark of
Oxford University Press in the UK and in certain other countries

First Edition published in 2020
Second Edition published in 2024

Published in the United States of America by Oxford University Press
198 Madison Avenue, New York, NY 10016, United States of America

British Library Cataloguing in Publication Data
Data available

Library of Congress Control Number: 2023949595

ISBN 9780198908081

DOI: 10.1093/oso/9780198908081.001.0001

Printed and bound by

CPI Group (UK) Ltd, Croydon, CR0 4YY

Foreword

All crystalline materials are imperfect. Common defects are vacancies, dislocations and cracks. These affect mechanical properties in important ways—some beneficial, others causing problems. Dislocations confer ductility to crystalline materials, cracks facilitate fracture. Dislocations enable metals and alloys to be mechanically worked, rolled, pressed, drawn etc., to change their shape without cracking, thus increasing immensely their range of use, and enriching our everyday lives. The behaviour of dislocations and cracks is greatly affected by stress, and in order to optimise their effect on mechanical properties, it is important to understand their interaction with stress. This is an important objective of this book. The first part deals with the fundamentals of strain and stress, and discusses the important role of Green's Function in linear elasticity, and introduces Hooke's Law. This is followed by discussion of properties of defects in crystals and their interaction with stress and with each other.

Dislocations are particularly complicated defects, firstly because they are flexible line defects characterised by the atomic displacements (Burgers vector) these cause when moving through the crystal lattice, and whose modes of motion are affected by their directions in the lattice. Secondly, the displacement at the centre of a dislocation results in a long-range elastic stress field which interacts with other dislocations over long distances, while at the centre of the atomic line defect linear elasticity does not apply, but large forces exist which lead to strong interactions with point defects, e.g. solute atoms, causing solution hardening. These short-range interactions require atomic scale quantum mechanical treatments. While such treatments are outside the scope of this book, the author identifies the situations where such treatments are necessary. A further complication is that depending on the crystal lattice and the Burgers vector, such dislocations can dissociate into partial dislocations, with smaller Burgers vector, bounding a stacking fault and lowering their energy.

The mobility of dislocations is important, as it affects the strain-rate at which materials can be deformed. As the dislocation moves from one line in the atomic lattice to the next, bonds are broken and reformed. The mobility is therefore determined by the interatomic forces at the centre (the core) of the dislocation. The fact that diamond is hard and aluminium relatively soft is due to the different interatomic bond strengths in these materials. The classical treatments are discussed carefully by the author. But just as the macroscopic shear of a crystal occurs by the motion of individual dislocations which cause one atomic step displacement at a time, so the individual dislocations do not advance simultaneously along their entire line, but the motion occurs by the creation and movement of atomic scale lengths, called kinks, on the dislocations. So it is these kinks, which are point-type defects, which control mobility. The author includes an example in his book. Sometimes the mobility of a dislocation causing a particular atomic displacement can differ greatly depending on the direction of the dislocation line in the lattice. An extreme example occurs in the body-centred cubic lattice, where transmission electron microscopy observations showed that edge dislocations (Burgers vector normal to the dislocation line) move much more rapidly than screw dislocations (same Burgers vector but parallel to the dislocation line, and to the triad symmetry axis in the crystal). Such screw dislocations can spread out their cores on three planes intersecting along the dislocation. They have to be constricted before movement can take place on a particular atomic slip plane. Such a process is assisted by thermal activation, and results in a sharp rise in yield stress at low temperature, which can cause structural integrity problems under such conditions.

The book includes an excellent discussion of the interaction of dislocations with sharp cracks in the presence of stress. The stress field of a stressed crack is conveniently modelled in terms of arrays of virtual dislocations with infinitesimal Burgers vectors. In the absence of crack tip plasticity, cracks cause brittle fracture at critical stresses, as described in the famous Griffith formula. Such behaviour is found in ceramic crystals in which dislocation mobility is low. But in metals and alloys the plastic deformation generated by the stress concentrations ahead of a stressed sharp crack can shield and blunt the cracks, resulting in stable crack extension and large increases in failure stress. This property is of great importance in the structural integrity of stressed components such as nuclear reactor pressure vessels, or bridges, etc. The variation with temperature of dislocation mobility, and correspondingly the plastic yield stress, gives rise to a brittle–ductile transition with temperature, which affects critically the structural integrity of pressure vessels in nuclear reactors where the yield stress can be impaired by radiation damage.

Much research has been carried out on problems which can be treated by the interactions of individual dislocations with each other, and with other defects. This approach has provided much understanding of the processes occurring in the interaction of dislocations with different Burgers vectors intersecting each other (leading to 'forest hardening'), for which there is much corroborating experimental evidence by transmission electron microscopy. The interaction of dislocations with non-deformable obstacles (e.g. a dispersion of oxide particles in copper) can be solved in a similar way and leads to a simple formula (the Orowan equation) relating yield stress with inter-particle separation (as explained in the text), which can be used to predict the macroscopic yield stress in terms of particle separation.

But other macroscopic mechanical properties, such as workhardening, involve the mutual interactions of many dislocations through their long-range elastic stress field, a many-bodied problem which cannot be treated in this way. Early transmission electron microscope observations showed that in heavily deformed polycrystals (such as aluminium, copper, nickel, etc.) the dislocations tend to form cell structures; in relatively soft materials such as aluminium, the cells are subgrains separated by low-energy low-angle boundaries on the scale of microns. It appears therefore that a mechanism of self-organisation into low-energy structures is operating, very likely facilitated by the ability of screw dislocations to cross-slip from one plane to another and annihilate. The scale at which these substructures occur is not properly understood.

The last chapter of the book on unsolved problems, includes an excellent account by the author of recent theories of workhardening of single crystals. Instead of treating the deformation in terms of the movement of individual dislocations in this many-bodied problem, the unit of deformation is a slip band formed by avalanches of dislocations generated by dislocation sources on individual glide planes. These slip bands are modelled as ellipsoids in which the stress inside the ellipsoid does not affect the primary dislocations forming the slip bands, and producing the shear, but generates dislocations on intersecting slip systems, which stabilise the slip bands and cause 'forest hardening'. It has been shown that such a model gives rise to a linear increase in hardening with increasing strain, as observed experimentally. But, as the author points out, there remain some unanswered questions.

This book provides materials scientists, physicists, and engineers with interest in mechanical properties of crystalline materials with an excellent account of the basic armoury needed to address the many unsolved problems in the mechanical properties of materials. It is beautifully written and the author emphasises the physical basis underlying the mathematical treatment of the various topics, providing insight and understanding. This book should be essential reading for graduate students, and in fact for anyone working in this field.

P.B. Hirsch
22 September 2019

Preface to the first edition

My aim in writing this book is to provide mathematically inclined physicists and engineers with an introduction to elasticity, and to take them relatively quickly to advanced concepts and ideas about its application to defects in crystals. The first six chapters are based on a course I gave for ten years in the EPSRC Centre for Doctoral Training (CDT) on Theory and Simulation of Materials at Imperial College London. The last four chapters cover material I would have included had there been more time. The CDT was created in 2009 to attract first class physicists and engineers with a taste for theory into materials science. Some of their research appears in this book.

Some justification is needed for another book in an area that already benefits from excellent texts. By discussing the connections to treatments of defects at the atomic scale I have tried to make the text more appealing to physicists. This approach is also quite novel for engineers, and I suspect that even the treatment of stress in the second chapter contains sections that are new to them. Having given this course to physics graduates from the UK and continental Europe I know that most of the material covered in this course, in many cases all of it, was new to them too. I hope this book treats the subject in a way that will appeal to students with backgrounds in physics and engineering. I hope it will also appeal to materials scientists who would like to see a more mathematical approach to the subject.

Until the late 1960s, in some of the strongest departments of physics around the World, *metal physics* was focused on the physics of crystal defects and their interactions. But today such metal physics has become unfashionable and it has vanished almost completely from physics departments. The study of metals and alloys by physicists has morphed into functional properties such as superconductivity, plasmonics and magnetism. Undergraduate courses of physics in the UK rarely include anything on defects in crystals, other than dopants in semiconductors. Physics students could be forgiven for thinking the world is made of perfect single crystals. However, the industrial need for an understanding of defect-related mechanisms of deformation and mechanical failures has never been greater. Some major manufacturing companies have recognised they cannot make reliable assessments of the life-times of certain components critical for safety unless they understand the defect-related mechanisms that limit their service life. On the other hand, university engineering departments do not have a tradition of thinking about materials at the atomic scale. For example, hydrogen embrittlement is a problem that cuts across swathes of current and proposed technologies, but it has fallen between physics and engineering, and very little progress has been made into the fundamental mechanisms after a century and a half of research. The physics of defects will remain important as long as metals and alloys are used to make things like jet engines, cars, ships, rail track, bridges, skyscrapers, wind turbines and nuclear power stations.

Such a short book as this about such a huge subject cannot be self-contained. I have included a list of recommended books which should be consulted. I have also included references to research papers in footnotes, and a list of them all at the end of the book, with web addresses when I have been able to find them online. Clicking on the url will take the reader directly to where the paper is located on the web, although most of them are behind pay-walls. Exercises and problem sets are included in each chapter, except the last. Some of the problems extend the material covered in the text into more advanced areas, and the reader is guided through them.

The theory of elasticity has a long history going back to Euler in the 18th century. Its evolution through the 19th and 20th centuries involved some of the most familiar names in the development of mathematical physics. Brief information is included about these people where it is available. I regret that for some prominent contributors to the subject I could find no information. It will be seen that many of them were Fellows of the Royal Society (FRS), with two Presidents (PRS) and six Foreign Members (ForMemRS)[1]. Four recipients of Nobel prizes in physical sciences also appear in this book. I hope this information will persuade readers that the subject has attracted some of the finest minds in theoretical and mathematical physics, and that it will help to inspire them to make their own contributions to advancing the subject. The subject is still evolving and the final chapter introduces four areas for further research with suggestions for challenging PhD projects.

Since the Second World War the global number of published science papers has doubled approximately every 9 years[2]. In 1665 The Royal Society published the first scientific journal – *The Philosophical Transactions of the Royal Society*. In 2009 it was estimated[3] there were more than 50 million published scholarly papers. Today in 2019 that figure is likely to be more than 100 million, published in tens of thousands of journals. Modern science has become so specialised and fragmented it has become very difficult to raise our eyes above our own narrow furrows of research. I hope this book will provide some satisfaction to readers who feel the need to broaden their horizons.

---

[1]The Royal Society is the National Academy of Sciences for the UK and the Commonwealth. It was founded in 1660.

[2]`tinyurl.com/m8sc884p`

[3]Jinha, Arif E, Learned Publishing **23**, 258-263 (2010).

I am very grateful to Professor Sir Peter Hirsch FRS for writing the Foreword. I thank Luca Cimbaro, Luca Reali, Tchavdar Todorov, Michele Valsecchi and Kang Wang for spotting errors in earlier drafts of the manuscript, and Professor Bob Balluffi for encouragement throughout the writing of this book. Professor Stan Lynch provided helpful comments on an earlier draft of section 11.5. Professors Mick Brown FRS, Tony Paxton and Vasek Vitek read the whole manuscript and provided many helpful suggestions and comments. I am also grateful for useful suggestions from an anonymous reviewer. Remaining errors are entirely my responsibility.

I am grateful to my colleagues in the Department of Physics at Imperial College London who in 2004 took the bold decision to hire a materials scientist, thereby allowing a trojan horse into their midst to recruit some of their best students into materials science. Working with these students has been the most fulfilling period of my academic career. I dedicate this book to them.

Imperial College London
August 2019

Preface to the second edition

In view of its exceptionally wide-ranging applications in materials science and earth sciences I have expanded the discussion of the Eshelby ellipsoidal inclusion into a chapter of its own. I have gone to considerable lengths to make the treatment self-contained. Expressions are derived for the fields outside the inclusion, which I have applied to a slip band enabling it to be treated as a single entity, rather than as a collection of discrete dislocations, in coarse-grained models of plasticity.

I have included solutions to all the exercises and problems. I hope this will benefit all readers, particularly those working through the book by themselves.

Following suggestions from several readers I have included references in the main text to some of the results in the problem sets, which may otherwise be overlooked. I thank Kevin Knowles for pointing out an error in the discussion of Poisson's ratio in cubic crystals in the first edition. I am especially grateful to Stan Lynch for lengthy correspondence which has convinced me there is much not understood about the ductile-brittle transition in metals. This has led to the introduction of an extended discussion of the ductile-brittle transition in the final chapter on open questions, and I am grateful to Sir Peter Hirsch FRS and Luca Reali for their comments on that section. The section on electroplasticity in the chapter on open questions has been rewritten in the light of recent theoretical advances.

While working on this second edition I have reflected on its relevance to a modern postgraduate education in materials science. This book is about one of the most richly developed areas of *theory* in materials science spanning more than two centuries of thought by some of the greatest scientists of all time. The final chapter describes a few of the many remaining problems of great scientific and technological interest concerning defects and their collective behaviour in crystalline materials. They require fresh ideas and physical insight to *develop models* of the processes involved.

This second edition has been typeset by the author using the Legrand Orange Book Template.

Imperial College London
July 2023

Contents

1. Strain

1.1 The continuum approximation

Condensed matter is lumpy. It comprises very dense atomic nuclei packed a few angstroms apart, with electrons in much less dense clouds between them. The continuum approximation smears out this lumpiness into a uniform, structureless jelly with the same density as a macroscopic lump of the matter it approximates. In addition to its density the continuum is given elastic properties, which characterise how easy it is to deform it in a reversible manner. The elastic properties of the continuum are equated to those of the material it approximates.

What physics do we leave out by approximating the discrete atomic structure of a material with a continuum model? Whenever the discrete atomic structure of the material becomes essential to the physics we can expect the continuum model to be a poor approximation. For example, when we consider structural defects inside the material we can expect the continuum model to become increasingly unreliable as we get closer to the centre of the defect because there the discrete atomic structure of the defect can no longer be ignored. But once we get beyond a few nanometres, in many cases just one nanometre, from the centre of a defect the continuum approximation becomes an accurate description of the distortion the defect generates.

Whereas the smallest separation of atoms in a material provides a natural length-scale there is no natural length-scale associated with the continuum. This has significant consequences for dynamical properties, such as the propagation of atomic vibrations. There are just three waves that can propagate in the continuum with speeds that vary in general with the direction of propagation. In contrast to vibrations in a crystal the elastic waves are dispersionless, i.e. their speed of propagation does not depend on their wavelength. The dispersion of vibrations in a crystal is a direct consequence of its discrete atomic structure. Waves in the crystal and in its continuum representation coincide only in the limit of long wavelengths compared with the spacing of atoms, where the discreteness of the atomic structure no longer plays a significant role in wave propagation.

1.2 What is deformation?

When a body changes shape or volume in response to external or internal forces it is *deformed.* In contrast to a rigid rotation or translation, deformation alters distances between points within the body. Materials may undergo changes of volume and shape in response to changes of temperature, applied electric and magnetic fields and other fields including gravitational. In this book we will be concerned principally with deformation created by mechanical forces applied to bodies and by defects within them. The focus of this chapter is the mathematical description of deformation and strain.

The simplest deformation is a homogeneous expansion or contraction where the vector between two points does not change direction but its length changes by an amount proportional to their separation in the undeformed state. The word 'homogeneous' here means 'the same everywhere'. When a body is deformed inhomogeneously the deformation depends on position in the undeformed state. The deformation is then a *field.*

Relative to an arbitrary origin, let $\mathbf{X}$ be the position vector of a point in the body before any deformation occurs. Let $\mathbf{X}+\mathrm{d}\mathbf{X}$ be the position vector of a point in the undeformed body infinitesimally close to $\mathbf{X}$. These two points are separated by $|\mathrm{d}\mathbf{X}| = \sqrt{\mathrm{d}X_i\,\mathrm{d}X_i}$, where X_i is cartesian component i of $\mathbf{X}$. Summation is implied throughout this book whenever subscripts are repeated, except where stated otherwise. Thus, $\mathrm{d}X_i\,\mathrm{d}X_i$ is shorthand for $\mathrm{d}X_1^2+\mathrm{d}X_2^2+\mathrm{d}X_3^2$.

Suppose $\mathbf{X}$ and $\mathbf{X}+\mathrm{d}\mathbf{X}$ become $\mathbf{x}$ and $\mathbf{x}+\mathrm{d}\mathbf{x}$ in the deformed state. In general $\mathbf{x}$ is a function of $\mathbf{X}$. The chain rule enables us to write down the components of $\mathrm{d}\mathbf{x}$ in terms of the components of $\mathrm{d}\mathbf{X}$:

$$\mathrm{d}x_i = \frac{\partial x_i}{\partial X_j}\mathrm{d}X_j \tag{1.1}$$

The 3×3 matrix $F_{ij} = \partial x_i/\partial X_j$ is called the *deformation tensor*[1]. It follows that the change in the squared separation of the points is given by:

$$\mathrm{d}x_i\mathrm{d}x_i - \mathrm{d}X_j\mathrm{d}X_j = \left(\frac{\partial x_i}{\partial X_j}\frac{\partial x_i}{\partial X_k} - \delta_{jk}\right)\mathrm{d}X_j\mathrm{d}X_k, \tag{1.2}$$

where δ_{jk} is the Kronecker delta: $\delta_{jk} = 1$ if $j = k$, and $\delta_{jk} = 0$ if $j \neq k$.

[1]You may be wondering why we do not distinguish between contravariant and covariant components of vectors and tensors. In almost all of this book we use rectangular cartesian coordinates, so the distinction is unnecessary. Exceptionally, in section 5.2 care has to be taken with the use of spherical polar rather than rectangular cartesian coordinates to describe the elastic displacement field.

Exercise 1.1 (i) Prove that the deformation tensor F_{ij} satisfies the tensor transformation law under a rotation of the coordinate system. This is what makes F_{ij} a tensor.
(ii) Show that in matrix notation eq 1.2 becomes:

$$(\mathbf{dx}^T \cdot \mathbf{dx}) - (\mathbf{dX}^T \cdot \mathbf{dX}) = \mathbf{dX}^T \cdot \left(\mathbf{F}^T\mathbf{F} - \mathbf{I}\right) \cdot \mathbf{dX}$$

where the T superscript denotes transpose, and $\mathbf{I}$ is the identity matrix.
(iii) Hence prove that if the deformation tensor is a rotation the change in the separation of points is zero. ■

Solution (i) We begin with how the components of a vector transform under a rotation of the coordinate axes. If you are familiar with this already skip to the paragraph beginning 'The transformation law for the second rank tensor . . . '

Let the unit vectors $\hat{\mathbf{e}}_1, \hat{\mathbf{e}}_2, \hat{\mathbf{e}}_3$ be along the axes x_1, x_2, x_3 of a right-handed Cartesian coordinate system. Let the unit vectors $\hat{\mathbf{e}}'_1, \hat{\mathbf{e}}'_2, \hat{\mathbf{e}}'_3$ be along the axes x'_1, x'_2, x'_3 of a second right-handed Cartesian coordinate system, obtained by rotating the first coordinate system. Consider a vector $\mathbf{X}$ in three dimensional space. It may be expressed in terms of its components (X_1, X_2, X_3) in the first coordinate system and (X'_1, X'_2, X'_3) in the rotated coordinate system:

$$\mathbf{X} = X_1\hat{\mathbf{e}}_1 + X_2\hat{\mathbf{e}}_2 + X_3\hat{\mathbf{e}}_3 = X'_1\hat{\mathbf{e}}'_1 + X'_2\hat{\mathbf{e}}'_2 + X'_3\hat{\mathbf{e}}'_3, \tag{1.3}$$

which in suffix notation is as follows

$$\mathbf{X} = X_i\hat{\mathbf{e}}_i = X'_j\hat{\mathbf{e}}'_j.$$

By taking the scalar product of this equation with respect to $\hat{\mathbf{e}}'_1$, $\hat{\mathbf{e}}'_2$, $\hat{\mathbf{e}}'_3$ in turn we may express the coordinates (X'_1, X'_2, X'_3) in the rotated coordinate system in terms of the coordinates (X_1, X_2, X_3) in the unrotated coordinate system:

$$\begin{aligned}
X'_1 &= (\hat{\mathbf{e}}'_1 \cdot \hat{\mathbf{e}}_1)X_1 + (\hat{\mathbf{e}}'_1 \cdot \hat{\mathbf{e}}_2)X_2 + (\hat{\mathbf{e}}'_1 \cdot \hat{\mathbf{e}}_3)X_3 \\
X'_2 &= (\hat{\mathbf{e}}'_2 \cdot \hat{\mathbf{e}}_1)X_1 + (\hat{\mathbf{e}}'_2 \cdot \hat{\mathbf{e}}_2)X_2 + (\hat{\mathbf{e}}'_2 \cdot \hat{\mathbf{e}}_3)X_3 \\
X'_3 &= (\hat{\mathbf{e}}'_3 \cdot \hat{\mathbf{e}}_1)X_1 + (\hat{\mathbf{e}}'_3 \cdot \hat{\mathbf{e}}_2)X_2 + (\hat{\mathbf{e}}'_3 \cdot \hat{\mathbf{e}}_3)X_3
\end{aligned}$$

In suffix notation these equations are expressed succinctly as follows:

$$X'_i = (\hat{\mathbf{e}}'_i \cdot \hat{\mathbf{e}}_j)X_j$$

The 3×3 matrix of direction cosines $(\hat{\mathbf{e}}'_i \cdot \hat{\mathbf{e}}_j)$ is the rotation matrix $R_{ij} = \hat{\mathbf{e}}'_i \cdot \hat{\mathbf{e}}_j$ that rotates the unprimed coordinate system into the primed coordinate system[a]. Thus,

$$X'_i = R_{ij} X_j. \tag{1.4}$$

A physical quantity represented by a vector, such as a displacement or velocity, does not depend on the choice of coordinate system. In a given Cartesian coordinate system a vector is uniquely defined by its components. But in a rotated Cartesian coordinate system those components change. A quantity is defined as a vector if its components change under a rotation of the Cartesian coordinate system according to the transformation law of eq 1.4.

The transformation law for the second rank tensor T_{ij} is as follows[b]:

$$T'_{ij} = R_{ik} R_{jl} T_{kl}. \tag{1.5}$$

We are asked to show that the deformation tensor F_{ij} satisfies this transformation law. The deformation tensor relates the vectors $\mathbf{dx}$ and $\mathbf{dX}$:

$$\mathrm{d}x_i = F_{ij} \mathrm{d}X_j$$

In the rotated coordinate system this relation becomes:

$$\mathrm{d}x'_i = F'_{ij} \mathrm{d}X'_j.$$

The components of the vectors $\mathbf{dx}'$ and $\mathbf{dX}'$ are determined by the transformation law for vectors we have just derived: $\mathrm{d}x'_i = R_{ik} \mathrm{d}x_k$ and $\mathrm{d}X'_j = R_{jl} \mathrm{d}X_l$. Therefore,

$$\mathrm{d}x'_i = R_{ik} \mathrm{d}x_k = R_{ik} F_{km} \mathrm{d}X_m = R_{ik} F_{km} R_{pm} \mathrm{d}X'_p.$$

The last equality uses the inverse of the rotation matrix is its transpose to write $dX_m = R_{pm} \mathrm{d}X'_p$. Therefore,

$$F'_{ip} = R_{ik} R_{pm} F_{km}$$

which proves the deformation $\mathbf{F}$ satisfies the transformation law of second rank tensors. This is the condition for $\mathbf{F}$ to be a tensor. In matrix notation the transformation law is as follows:

$$\mathbf{F}' = \mathbf{R}\mathbf{F}\mathbf{R}^T.$$

(ii) The scalar product $\mathrm{d}x_i \mathrm{d}x_i$ in matrix notation is $\mathbf{dx}^T \cdot \mathbf{dx}$ where $\mathbf{dx}^T$ is a row matrix and $\mathbf{dx}$ is a column matrix. The partial derivative $\partial x_i/\partial X_j$ is the element F_{ij} of the deformation tensor $\mathbf{F}$. Therefore the product $(\partial x_i/\partial X_j)(\partial x_i/\partial X_k) = F_{ij}F_{ik}$. But $F_{ij}F_{ik} = F^T_{ji}F_{ik} = (\mathbf{F}^T\mathbf{F})_{jk}$. Therefore,

$$\left(\frac{\partial x_i}{\partial X_j}\frac{\partial x_i}{\partial X_k} - \delta_{jk}\right)\mathrm{d}X_j\mathrm{d}X_k = \mathbf{dX}^T \cdot (\mathbf{F}^T\mathbf{F} - \mathbf{I}) \cdot \mathbf{dX},$$

and the matrix equation of part (ii) follows.

(iii) If the deformation tensor is a rotation then $\mathbf{F}^T\mathbf{F} = \mathbf{I}$ and hence $\mathbf{dx}^T \cdot \mathbf{dx} - \mathbf{dX}^T \cdot \mathbf{dX} = 0$.

[a] Strictly speaking, we are discussing the transformation of the components of a *polar* vector, not an *axial* vector, under a rotation of the coordinate system. Whereas the components of a polar vector change sign under an inversion of the coordinate system, i.e. $X_i' = -X_i$, the components of an axial vector do not change sign. An example of an axial vector is a cross product of two polar vectors, such as angular velocity.

[b] The *rank* of a tensor is the total number of indices (which are subscripts only in this book) associated with it. In an N-dimensional space a scalar quantity T has rank 0; a first rank tensor T_i has rank 1 and N components, and it is more commonly called a vector; a second rank tensor T_{ij} has rank 2, and its N^2 components may be represented by the elements of an $N \times N$ matrix; a third rank tensor T_{ijk} has rank 3 and N^3 components, and so on. Note that the rank of a tensor is independent of N.

You may recognise $\mathbf{F}^T\mathbf{F}$ as the metric tensor $\mathbf{g}$ which is used to measure the distance between two points in the deformed state. Consider the distance between two points in the *undeformed* state. If $\mathbf{X} = \mathbf{X}(\lambda)$ is the parametric equation of a path in the undeformed state between two points A and B, where $\lambda = \lambda_A$ and $\lambda = \lambda_B$ respectively, then the distance between these points along the path in the undeformed state is given by:

$$\begin{aligned} s &= \int_{\lambda_A}^{\lambda_B} \sqrt{\left(\frac{\mathrm{d}X_1}{\mathrm{d}\lambda}\right)^2 + \left(\frac{\mathrm{d}X_2}{\mathrm{d}\lambda}\right)^2 + \left(\frac{\mathrm{d}X_3}{\mathrm{d}\lambda}\right)^2}\,\mathrm{d}\lambda \\ &= \int_{\lambda_A}^{\lambda_B} \sqrt{\frac{\mathrm{d}X_i}{\mathrm{d}\lambda} g_{ij} \frac{\mathrm{d}X_j}{\mathrm{d}\lambda}}\,\mathrm{d}\lambda \end{aligned}$$

and we see the metric tensor in the undeformed state is the identity matrix δ_{ij}.

In the *deformed* state the path becomes $\mathbf{x} = \mathbf{x}(\mathbf{X}(\lambda))$ and points A and B are moved to new positions. The distance between points A and B along the deformed path is given by

$$s = \int_{\lambda_A}^{\lambda_B} \sqrt{dx_k(\lambda)dx_k(\lambda)}d\lambda = \int_{\lambda_A}^{\lambda_B} \sqrt{\frac{dX_i}{d\lambda}\frac{\partial x_k}{\partial X_i}\frac{\partial x_k}{\partial X_j}\frac{dX_j}{d\lambda}}d\lambda, \tag{1.6}$$

and the metric tensor is $g_{ij} = (\partial x_k/\partial X_i)(\partial x_k/\partial X_j)$, or $\mathbf{g} = \mathbf{F}^T\mathbf{F}$ in matrix notation.

1.3 The displacement vector and the strain tensor

We may always express $\mathbf{x}(\mathbf{X})$ as $\mathbf{X} + \mathbf{u}(\mathbf{X})$, where $\mathbf{u}(\mathbf{X})$ is the displacement undergone by a point at $\mathbf{X}$ in the undeformed state when the body is deformed. The gradient of the displacement vector is related to the deformation tensor as follows:

$$F_{ki} = \frac{\partial x_k}{\partial X_i} = \delta_{ki} + \frac{\partial u_k}{\partial X_i}. \tag{1.7}$$

Exercise 1.2 Show that the squared separation of points that were at $\mathbf{X}$ and $\mathbf{X}+d\mathbf{X}$ in the undeformed state becomes the following in the deformed state:

$$(ds)^2 = \left(\delta_{kj} + \frac{\partial u_k}{\partial X_j} + \frac{\partial u_j}{\partial X_k} + \frac{\partial u_i}{\partial X_k}\frac{\partial u_i}{\partial X_j}\right)dX_k dX_j \tag{1.8}$$

■

Solution Since

$$dx_i = \left(\delta_{ij} + \frac{\partial u_i}{\partial X_j}\right)dX_j$$

then

$$\begin{aligned}(ds)^2 &= dx_i dx_i = \left(\delta_{ij} + \frac{\partial u_i}{\partial X_j}\right)\left(\delta_{ik} + \frac{\partial u_i}{\partial X_k}\right)dX_j dX_k \\ &= \left(\delta_{ij}\delta_{ik} + \delta_{ij}\frac{\partial u_i}{\partial X_k} + \delta_{ik}\frac{\partial u_i}{\partial X_j} + \frac{\partial u_i}{\partial X_j}\frac{\partial u_i}{\partial X_k}\right)dX_j dX_k \\ &= \left(\delta_{kj} + \frac{\partial u_j}{\partial X_k} + \frac{\partial u_k}{\partial X_j} + \frac{\partial u_i}{\partial X_j}\frac{\partial u_i}{\partial X_k}\right)dX_j dX_k\end{aligned}$$

Equation 1.8 is exact provided $|\mathbf{dX}|$ is infinitesimal, and it leads to non-linear theories of elasticity. To obtain a linear approximation the assumption is made that the *gradients* of the displacement vector are small in comparison to unity. It is important to recognise that this assumption does not require the displacements themselves to be small. Taking the square root of both sides of eq 1.8, and dividing both sides by $|\mathbf{dX}|$, we obtain:

$$\begin{aligned}\frac{ds}{|\mathbf{dX}|} &= \sqrt{1+\hat{l}_k\left(\frac{\partial u_k}{\partial X_j}+\frac{\partial u_j}{\partial X_k}+\frac{\partial u_i}{\partial X_k}\frac{\partial u_i}{\partial X_j}\right)\hat{l}_j} \\ &\approx 1+\hat{l}_k e_{kj}\hat{l}_j, \end{aligned} \tag{1.9}$$

where the unit vector $\hat{\mathbf{l}}$ is parallel to $\mathbf{dX}$ and e_{kj} is the *strain tensor*:

$$e_{kj} = \frac{1}{2}\left(\frac{\partial u_k}{\partial X_j}+\frac{\partial u_j}{\partial X_k}+\frac{\partial u_i}{\partial X_k}\frac{\partial u_i}{\partial X_j}\right) \tag{1.10}$$

Since the displacement gradients are assumed to be small the last term in eq 1.10 is neglected and thus we obtain the strain tensor used in linear elasticity, and in the rest of this book:

$$e_{kj} = \frac{1}{2}\left(\frac{\partial u_k}{\partial X_j}+\frac{\partial u_j}{\partial X_k}\right). \tag{1.11}$$

We see the strain tensor is symmetric. The displacement gradient may always be written as the sum of a symmetric and an a symmetric part:

$$\frac{\partial u_k}{\partial X_j} = e_{kj}+\omega_{kj}$$

where $\omega_{kj} = \frac{1}{2}\left(\partial u_k/\partial X_j - \partial u_j/\partial X_k\right)$ is asymmetric because $\omega_{kj} = -\omega_{jk}$.

Exercise 1.3 Given the following general expression for the rotation matrix describing a rotation by θ about an axis $\hat{\boldsymbol{\rho}}$:

$$R_{ij} = \cos\theta\,\delta_{ij} + \hat{\rho}_i\hat{\rho}_j(1-\cos\theta) - \varepsilon_{ijk}\hat{\rho}_k\sin\theta,$$

where ε_{ijk} is the permutation tensor[a] show that in the limit $\theta \to 0$ the rotation matrix becomes:

$$R_{ij} = \delta_{ij} - \varepsilon_{ijk}\hat{\rho}_k\theta.$$

Hence show that the axis of the rotation described by the three independent components of ω_{kj} is parallel to curl $\mathbf{u}$. ■

[a] The permutation tensor equals 1 if i, j, k are 1,2,3 or 2,3,1 or 3,1,2 respectively, it equals −1 if i, j, k are 1,3,2 or 2,1,3 or 3,2,1 respectively, and it equals 0 otherwise.

Solution In the limit $\theta \to 0$, $\cos\theta \to 1$ and $\sin\theta \to \theta$ to first order in θ. Therefore, $R_{ij} \to \delta_{ij} - \varepsilon_{ijk}\hat{\rho}_k\theta$, and hence

$$\mathbf{R} \to \begin{pmatrix} 1 & -\hat{\rho}_3\theta & \hat{\rho}_2\theta \\ \hat{\rho}_3\theta & 1 & -\hat{\rho}_1\theta \\ -\hat{\rho}_2\theta & \hat{\rho}_1\theta & 1 \end{pmatrix}.$$

Note $\mathbf{R}$ is an orthogonal matrix only to first order in θ. The asymmetric matrix $\omega_{ij} = (1/2)(\partial u_i/\partial X_j - \partial u_j/\partial X_i)$. Written out in full it is:

$$\omega = \begin{pmatrix} 0 & \frac{1}{2}\left(\frac{\partial u_1}{\partial X_2} - \frac{\partial u_2}{\partial X_1}\right) & \frac{1}{2}\left(\frac{\partial u_1}{\partial X_3} - \frac{\partial u_3}{\partial X_1}\right) \\ \frac{1}{2}\left(\frac{\partial u_2}{\partial X_1} - \frac{\partial u_1}{\partial X_2}\right) & 0 & \frac{1}{2}\left(\frac{\partial u_2}{\partial X_3} - \frac{\partial u_3}{\partial X_2}\right) \\ \frac{1}{2}\left(\frac{\partial u_3}{\partial X_1} - \frac{\partial u_1}{\partial X_3}\right) & \frac{1}{2}\left(\frac{\partial u_3}{\partial X_2} - \frac{\partial u_2}{\partial X_3}\right) & 0 \end{pmatrix}$$

Equating ω to $\mathbf{R} - \mathbf{I}$ we find:

$$\hat{\rho}_1 \text{ is parallel to } \left(\frac{\partial u_3}{\partial X_2} - \frac{\partial u_2}{\partial X_3}\right) = (\text{curl}\,\mathbf{u})_1$$

$$\hat{\rho}_2 \text{ is parallel to } \left(\frac{\partial u_1}{\partial X_3} - \frac{\partial u_3}{\partial X_1}\right) = (\text{curl}\,\mathbf{u})_2$$

$$\hat{\rho}_3 \text{ is parallel to } \left(\frac{\partial u_2}{\partial X_1} - \frac{\partial u_1}{\partial X_2}\right) = (\text{curl}\,\mathbf{u})_3$$

This proves the axis of rotation, $\hat{\boldsymbol{\rho}}$, is parallel to curl $\mathbf{u}$.

1.3.1 Normal strain and shear strain

The diagonal and off-diagonal components of the strain tensor of eq 1.11 are called *normal* and *shear* strains respectively. An example of a normal strain in two dimensions is illustrated in Fig 1.1(a), for which the deformation tensor is:

$$\mathbf{F} = \begin{pmatrix} 1+\varepsilon & 0 \\ 0 & 1 \end{pmatrix}$$

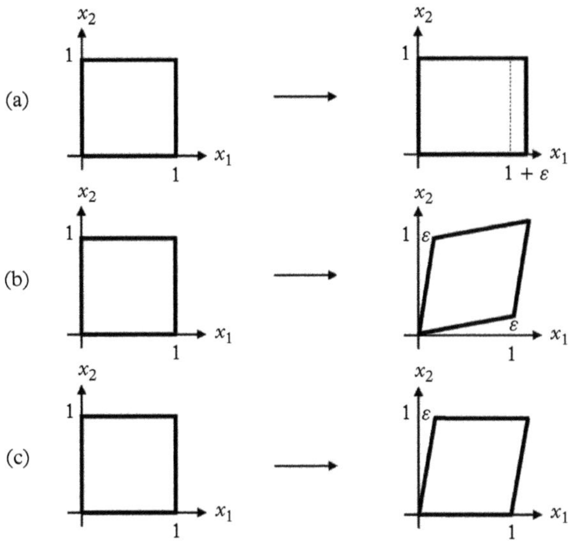

Figure 1.1: Illustrations in two dimensions of (a) normal strain along x_1 of magnitude ε, (b) pure shear strain of magnitude ε and (c) simple shear strain of magnitude ε. In each case the unit square on the left is deformed into the shape on the right by the corresponding strain.

and the corresponding strain tensor is

$$\mathbf{e} = \begin{pmatrix} \varepsilon & 0 \\ 0 & 0 \end{pmatrix}.$$

A shear strain arises when the displacement in a particular direction varies with a perpendicular distance. Two common types of shear strain are a *pure shear* and a *simple shear*. Fig 1.1(b) illustrates the pure shear described by the following deformation tensor:

$$\mathbf{F} = \begin{pmatrix} 1 & \varepsilon \\ \varepsilon & 1 \end{pmatrix}$$

for which the corresponding strain tensor is

$$\mathbf{e} = \begin{pmatrix} 0 & \varepsilon \\ \varepsilon & 0 \end{pmatrix}.$$

The symmetric nature of the deformation tensor ensures it is a pure shear because no rotations are involved.

A simple shear involves a pure shear and a rotation. An example of a simple shear is illustrated in Fig 1.1(c) for which the deformation tensor is:

$$\mathbf{F} = \begin{pmatrix} 1 & \varepsilon \\ 0 & 1 \end{pmatrix}.$$

This may be decomposed into a pure shear strain of $\varepsilon/2$ and a rotation by $\theta = \varepsilon/2$ about an axis normal to the page:

$$\mathbf{F} = \begin{pmatrix} 1 & 0 \\ 0 & 1 \end{pmatrix} + \begin{pmatrix} 0 & \varepsilon/2 \\ \varepsilon/2 & 0 \end{pmatrix} + \begin{pmatrix} 0 & \varepsilon/2 \\ -\varepsilon/2 & 0 \end{pmatrix}.$$

Simple shears occur in *mechanical twinning*, which is a mechanism of deformation of many crystalline materials, and in slip bands.

1.4 Closing remarks

The normal strain illustrated in Fig 1.1(a) changes the volume of the material. The shear strains illustrated in Figs.1.1(b) and (c) do not change the volume, but unless the material is a liquid[2] there will be an elastic resistance to these deformations. Young[3] was the first to consider shear as an elastic strain in 1807. He noticed that the elastic resistance of a body to shear is different from its resistance to extension or compression, but he did not introduce a separate elastic modulus to characterise the rigidity to shear[4]. The famous *Young's modulus* refers only to the rigidity of the material to elastic extension or compression.

Love[5] commenting[6] on Young's introduction of his modulus wrote:

This introduction of a definite physical concept, associated with the coefficient of elasticity which descends, as it were from a clear sky, on the reader of mathematical

[2]*Real* liquids do resist shear deformations in a time-dependent manner through their viscosities. We are thinking here of the response of the material after a long period of time when any time-dependent relaxation processes have finished. An *ideal* liquid has no viscosity and displays no resistance to shear stresses.

[3]Thomas Young FRS, 1773-1829, British physicist, physician, optometrist, linguist, Egyptologist, musician, and actuary for life insurance. For an engrossing biography of this exceptional polymath see *The last man who knew everything*, Andrew Robinson, One World: Oxford (2006). ISBN 978-0452288058.

[4]Navier appears to be the first to introduce the shear modulus in 1833 on p.96 of the second edition of his Navier, C-L, *Leçons... (1833)* (in French). Claude-Louis Marie Henri Navier 1785-1836, French mathematician, engineer and physicist.

[5]Augustus Edward Hough Love FRS, 1863-1940, British physicist.

[6]Love, A E H, *A Treatise on the Mathematical Theory of Elasticity*, Dover: New York (1944), p.4. ISBN 0-486-60174-9.

memoirs, marks an epoch in the history of the science.

We will come back to the moduli of elasticity in Chapter 3.

It should be noted that the concepts of the deformation tensor and the strain tensor rest on the existence of a reference state of the material where the deformation is sensibly regarded as zero. In a crystalline material the perfect crystal itself is a natural choice for the reference state. But in a glass or amorphous material there is no obvious choice of a reference state, and the concept of strain is much less useful. However, as we shall see in the next chapter, the concept of stress is just as applicable in an amorphous material as it is in a crystal.

1.5 Problems

Problem 1.1 Prove that the strain tensor of eq 1.11 satisfies the tensor transformation law under a rotation of the coordinate system.

Solution In the rotated coordinate system the strain tensor is as follows:

$$e'_{pq} = \frac{1}{2}\left(\frac{\partial u'_p}{\partial X'_q} + \frac{\partial u'_q}{\partial X'_p}\right).$$

We are asked to prove that the relation between the components e'_{pq} and e_{kj} is as follows:

$$e'_{pq} = R_{pk}R_{qj}e_{kj}.$$

This is the transformation law of second rank tensors.

Since the displacement $\mathbf{u}$ is a vector it satisfies the transformation law of eq 1.4:

$$u'_i = R_{ij}u_j.$$

The partial derivative $\partial/\partial X'_m$ may expressed in terms of the partial derivatives $\partial/\partial X_k$ using the chain rule:

$$\frac{\partial}{\partial X'_m} = \frac{\partial X_k}{\partial X'_m}\frac{\partial}{\partial X_k}$$

By taking the scalar product of eq 1.3 with $\hat{\mathbf{e}}_k$ we obtain:

$$X_k = X'_i(\hat{\mathbf{e}}'_i \cdot \hat{\mathbf{e}}_k).$$

Therefore,

$$\frac{\partial X_k}{\partial X'_m} = \hat{\mathbf{e}}'_m \cdot \hat{\mathbf{e}}_k = R_{mk}.$$

It follows that

$$\frac{\partial}{\partial X'_m} = R_{mk}\frac{\partial}{\partial X_k},$$

and the partial derivative $\partial u'_p/\partial X'_q$ transforms as follows:

$$\frac{\partial u'_p}{\partial X'_q} = R_{pk}R_{qj}\frac{\partial u_k}{\partial X_j}$$

and the partial derivative $\partial u'_q/\partial X'_p$ transforms as follows:

$$\frac{\partial u'_q}{\partial X'_p} = R_{qj}R_{pk}\frac{\partial u_j}{\partial X_k}.$$

We arrive at the transformation law for the strain tensor:

$$e'_{pq} = \frac{1}{2}\left(\frac{\partial u'_p}{\partial X'_q} + \frac{\partial u'_q}{\partial X'_p}\right) = R_{pk}R_{qj}\frac{1}{2}\left(\frac{\partial u_k}{\partial X_j} + \frac{\partial u_j}{\partial X_k}\right) = R_{pk}R_{qj}e_{kj}.$$

Problem 1.2 Under a homogeneous strain the displacement of a point at $\mathbf{X}$ in the undeformed body is given by $\mathbf{u}(\mathbf{X}) = \mathbf{e}\cdot\mathbf{X}$, where $\mathbf{e}$ is a constant symmetric matrix, in which the elements are small compared to unity. Consider two points $\mathbf{X}^{(1)}$ and $\mathbf{X}^{(2)}$ in the undeformed body. Show that the change in the separation of the two points in the homogeneously deformed state to first order in the strain is given by $\left|\mathbf{X}^{(2)} - \mathbf{X}^{(1)}\right| l_i e_{ij} l_j$, where l_i is the unit vector $(X_i^{(2)} - X_i^{(1)})/\left|\mathbf{X}^{(2)} - \mathbf{X}^{(1)}\right|$.

Solution The homogeneous deformation changes the coordinates of $\mathbf{X}^{(1)}$ and $\mathbf{X}^{(2)}$ as follows:

$$\begin{aligned} X_i^{(1)} &\rightarrow X_i^{(1)} + e_{ij}X_j^{(1)} \\ X_i^{(2)} &\rightarrow X_i^{(2)} + e_{ij}X_j^{(2)} \end{aligned}$$

Therefore,

$$\begin{aligned}
|\mathbf{X}^{(2)} - \mathbf{X}^{(1)}|^2 \quad &\rightarrow \quad |\mathbf{X}^{(2)} - \mathbf{X}^{(1)}|^2 + 2(X_i^{(2)} - X_i^{(1)})e_{ij}(X_j^{(2)} - X_j^{(1)}) + O(e^2) \\
&\approx \quad |\mathbf{X}^{(2)} - \mathbf{X}^{(1)}|^2 \left\{1 + 2\frac{(X_i^{(2)} - X_i^{(1)})}{|\mathbf{X}^{(2)} - \mathbf{X}^{(1)}|} e_{ij} \frac{(X_j^{(2)} - X_j^{(1)})}{|\mathbf{X}^{(2)} - \mathbf{X}^{(1)}|}\right\}.
\end{aligned}$$

In the second line the term second order in the strain has been dropped because the components of the strain tensor are all assumed to be much less than 1 in magnitude. The change in separation between $\mathbf{X}^{(2)}$ and $\mathbf{X}^{(1)}$ is then as follows:

$$\begin{aligned}
&|\mathbf{X}^{(2)} - \mathbf{X}^{(1)}| \left\{1 + 2\frac{(X_i^{(2)} - X_i^{(1)})}{|\mathbf{X}^{(2)} - \mathbf{X}^{(1)}|} e_{ij} \frac{(X_j^{(2)} - X_j^{(1)})}{|\mathbf{X}^{(2)} - \mathbf{X}^{(1)}|}\right\}^{1/2} - |\mathbf{X}^{(2)} - \mathbf{X}^{(1)}| \\
&= \quad |\mathbf{X}^{(2)} - \mathbf{X}^{(1)}| \left\{1 + \frac{1}{2} 2 l_i e_{ij} l_j\right\} - |\mathbf{X}^{(2)} - \mathbf{X}^{(1)}| + O(e^2) \\
&= \quad |\mathbf{X}^{(2)} - \mathbf{X}^{(1)}| l_i e_{ij} l_j.
\end{aligned}$$

where l_i is the unit vector $(X_i^{(2)} - X_i^{(1)})/|\mathbf{X}^{(2)} - \mathbf{X}^{(1)}|$, and terms only first order in strain components have been retained.

We see that under a homogeneous strain the change in separation between two points depends on the magnitude and direction of the vector separating them in the unstrained state and the full strain tensor.

Problem 1.3 Prove that the trace of the strain tensor, e_{kk}, is invariant with respect to rotations of the coordinate system.

Solution The trace of the strain tensor is $e_{11} + e_{22} + e_{33} = e_{ii}$. It becomes the following in the rotated coordinate system:

$$e'_{ii} = R_{ij} R_{ik} e_{jk}$$

But $R_{ij} R_{ik} = R^T_{ji} R_{ik} = \delta_{jk}$. The final equality follows because a rotation matrix is orthogonal, so that its inverse is its transpose. Therefore,

$$e'_{ii} = \delta_{jk} e_{jk} = e_{jj}.$$

This proves the invariance of the trace of the strain tensor under a rotation of the coordinate system.

Comments

The orthogonality of the rotation matrix may be proved using $R_{ij} = \hat{\mathbf{e}}'_i \cdot \hat{\mathbf{e}}_j$, which was derived in Exercise 1.1. We have:

$$R_{ij}R_{ik} = (\hat{\mathbf{e}}'_i \cdot \hat{\mathbf{e}}_j)(\hat{\mathbf{e}}'_i \cdot \hat{\mathbf{e}}_k) = (\hat{\mathbf{e}}_j \cdot \hat{\mathbf{e}}'_i)(\hat{\mathbf{e}}'_i \cdot \hat{\mathbf{e}}_k) = \hat{\mathbf{e}}_j \cdot (\hat{\mathbf{e}}'_i\, \hat{\mathbf{e}}'_i) \cdot \hat{\mathbf{e}}_k.$$

The part involving the implied sum on i is $(\hat{\mathbf{e}}'_i\, \hat{\mathbf{e}}'_i)$. It is a sum of 'outer products'. The symbol for an outer product between two vectors is $\otimes$. That is to say, $\hat{\mathbf{e}}'_i\, \hat{\mathbf{e}}'_i \equiv \hat{\mathbf{e}}'_i \otimes \hat{\mathbf{e}}'_i = \hat{\mathbf{e}}'_1 \otimes \hat{\mathbf{e}}'_1 + \hat{\mathbf{e}}'_2 \otimes \hat{\mathbf{e}}'_2 + \hat{\mathbf{e}}'_3 \otimes \hat{\mathbf{e}}'_3$. Therefore we may rewrite $R_{ij}R_{ik}$ as follows:

$$R_{ij}R_{ik} = \hat{\mathbf{e}}_j \cdot \hat{\mathbf{e}}'_i \otimes \hat{\mathbf{e}}'_i \cdot \hat{\mathbf{e}}_k.$$

In this rewriting we see that $\hat{\mathbf{e}}'_i \otimes \hat{\mathbf{e}}'_i$ is an operator that acts on $\hat{\mathbf{e}}_j$ on the left and $\hat{\mathbf{e}}_k$ on the right. Consider the action of this operator on the vector $\mathbf{a} = a'_j \hat{\mathbf{e}}'_j$:

$$\hat{\mathbf{e}}'_i \otimes \hat{\mathbf{e}}'_i \cdot \mathbf{a} = \hat{\mathbf{e}}'_i \otimes \hat{\mathbf{e}}'_i \cdot a'_j \hat{\mathbf{e}}'_j = \hat{\mathbf{e}}'_i (\hat{\mathbf{e}}'_i \cdot \hat{\mathbf{e}}'_j) a'_j = \hat{\mathbf{e}}'_i \delta_{ij} a'_j = a'_i \hat{\mathbf{e}}'_i = \mathbf{a}.$$

Similarly,

$$\mathbf{a} \cdot \hat{\mathbf{e}}'_i \otimes \hat{\mathbf{e}}'_i = a'_j \hat{\mathbf{e}}'_j \cdot \hat{\mathbf{e}}'_i \otimes \hat{\mathbf{e}}'_i = a'_j (\hat{\mathbf{e}}'_j \cdot \hat{\mathbf{e}}'_i) \hat{\mathbf{e}}'_i = a'_j \delta_{ij} \hat{\mathbf{e}}'_i = a'_i \hat{\mathbf{e}}'_i = \mathbf{a}.$$

It follows that the operator $\hat{\mathbf{e}}'_i \otimes \hat{\mathbf{e}}'_i$ is the identity[a] (remember summation on i is implied). Therefore,

$$R_{ij}R_{ik} = \hat{\mathbf{e}}_j \cdot \hat{\mathbf{e}}'_i \otimes \hat{\mathbf{e}}'_i \cdot \hat{\mathbf{e}}_k = \hat{\mathbf{e}}_j \cdot \hat{\mathbf{e}}_k = \delta_{jk}.$$

The trace of the strain tensor is one of three invariants of the strain tensor. Physically, it is the local dilation. To see this we observe that $e_{ii} = \nabla \cdot \mathbf{u}$. Consider the volume integral $(1/V) \int_R \nabla \cdot \mathbf{u} \mathrm{d}V$ where V is the volume of the region R over which the integral is taken. Using the divergence theorem:

$$\frac{1}{V} \int_R \nabla \cdot \mathbf{u} \mathrm{d}V = \frac{1}{V} \int_S \mathbf{u} \cdot \hat{\mathbf{n}} \mathrm{d}S = \frac{\Delta V}{V},$$

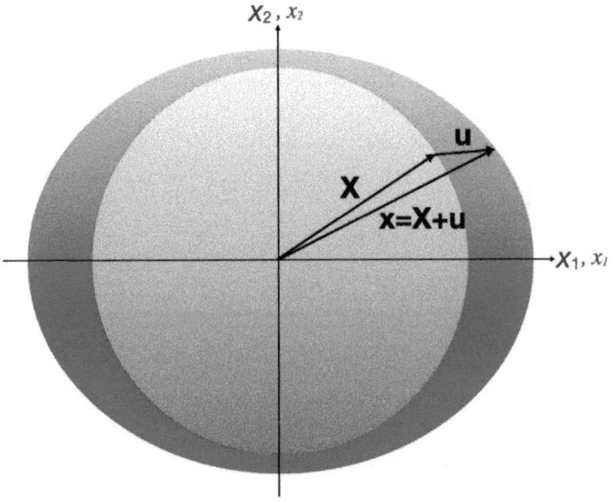

Figure 1.2: The unit sphere $X_1^2+X_2^2+X_3^2=1$ shown in blue is deformed into the strain ellipsoid $x_1^2/(1+\lambda_1)^2+x_2^2/(1+\lambda_2)^2+x_3^2/(1+\lambda_3)^2=1$ shown in red. The X_3,x_3 axis is normal to the page, towards the reader. The vector $\mathbf{X}$ is deformed into the vector $\mathbf{x}=\mathbf{X}+\mathbf{u}$, where $\mathbf{u}$ is the displacement vector.

where $\hat{\mathbf{n}}$ is the local unit normal vector to the surface S enclosing R. ΔV is the change in the volume of R to first order in the displacement field $\mathbf{u}$. As $V\to 0$, $\Delta V/V$ becomes the dilation at a point in the body.

[a] There is an equivalent relation in quantum mechanics for a complete set of orthonormal basis states $|j\rangle$, i.e. $\Sigma_j |j\rangle\langle j| = \mathbf{I}$, where $\mathbf{I}$ is the identity operator.

Problem 1.4 Recall the eigenvectors of a symmetric matrix are orthogonal and the eigenvalues are real numbers. Consider a unit sphere $\mathbf{X}^T\cdot\mathbf{X}=1$ embedded in a body before it is deformed. The body is subjected to a *homogeneous* strain $\mathbf{e}$. Show that the sphere becomes an ellipsoid with its axes aligned along the eigenvectors of the strain tensor. This ellipsoid is called the *strain ellipsoid*. If the eigenvalues of the strain tensor are λ_1, λ_2 and λ_3 determine the equation of the ellipsoid. These eigenvalues are called principal strains. To first order in the strain show that the ratio of the change in the volume of the sphere to its original volume is equal to the trace of the strain tensor. The trace of the strain tensor is called the dilation. Sketch the sphere and the ellipsoid, and for an arbitrarily chosen $\mathbf{X}$, show $\mathbf{x}$ and $\mathbf{u}$ in your sketch. The strain ellipsoid is a helpful visualisation of the distortion caused by the strain.

Solution In the coordinate system defined by its three orthonormal eigenvectors the strain tensor is represented by the diagonal matrix:

$$e = \begin{pmatrix} \lambda_1 & 0 & 0 \\ 0 & \lambda_2 & 0 \\ 0 & 0 & \lambda_3 \end{pmatrix}.$$

Then, X_i becomes $x_i = (1+\lambda_i)X_i$ and the unit sphere $X_1X_1 + X_2X_2 + X_3X_3 = 1$ becomes the strain ellipsoid:

$$\frac{x_1^2}{(1+\lambda_1)^2} + \frac{x_2^2}{(1+\lambda_2)^2} + \frac{x_3^2}{(1+\lambda_3)^2} = 1.$$

The semi-axes of the strain ellipsoid are $(1+\lambda_1)$, $(1+\lambda_2)$ and $(1+\lambda_3)$. Therefore, the volume of the strain ellipsoid is:

$$\frac{4\pi}{3}(1+\lambda_1)(1+\lambda_2)(1+\lambda_3) = \frac{4\pi}{3}(1+\lambda_1+\lambda_2+\lambda_3+O(\lambda^2)).$$

To first order in the strain the dilation is as follows:

$$\frac{\Delta V}{V} = \frac{\frac{4\pi}{3}(1+\lambda_1+\lambda_2+\lambda_3) - \frac{4\pi}{3}}{\frac{4\pi}{3}} = \lambda_1+\lambda_2+\lambda_3.$$

We see that the dilation is the trace of the strain tensor, as shown in the previous problem.

Fig 1.2 illustrates the unit sphere and the strain ellipsoid, in which a vector **X** on the surface of the unit sphere is deformed by the strain into the vector $\mathbf{x} = \mathbf{X} + \mathbf{u}$ on the surface of the strain ellipsoid.

2. Stress

2.1 What is stress?

Imagine you are stretching an elastic band between your fingers by applying an equal and opposite tensile force F at either end. The elastic band has reached a new stable length and you keep your hands in a fixed position. There is no net force acting on any element of the elastic band. If that were not true the elastic band would not be in equilibrium and some further displacement would take place. But if you release the elastic band from one end it immediately shrinks back to its natural length.

We say "the elastic band is under tension" to describe its state when it is stretched. If the stretched elastic band were cut anywhere between your fingers the tension would be released. The tension is transmitted across every transverse plane within the elastic band. The atoms on the left side of every transverse plane in the elastic band are exerting forces on the atoms of the right hand side. Conversely the atoms on the right hand side of every transverse plane are exerting forces on the left hand side. The resultant forces exerted by atoms on each side on the other side are equal and opposite when the elastic band is in equilibrium.

Suppose we stretch a thicker elastic band of the same material with the same force F. Obviously it will not stretch as much as the first elastic band. The task of transmitting the tension F across every transverse plane is shared by more atoms on either side of the plane. It follows that the area of the transverse plane is just as significant as the tension F in characterising the internal mechanical state of the elastic band. The concept of stress brings together the force F and the area of the plane on which it acts:

The stress acting on an element of area of a plane within a body is defined as the resultant force exerted by atoms on one side of the plane on atoms on the other side of the plane, where lines connecting those atoms pass through the element of area. The resultant force is divided by the area of the element through which it acts to yield the stress.

This is illustrated in Fig 2.1. This atomic-level definition of stress was developed

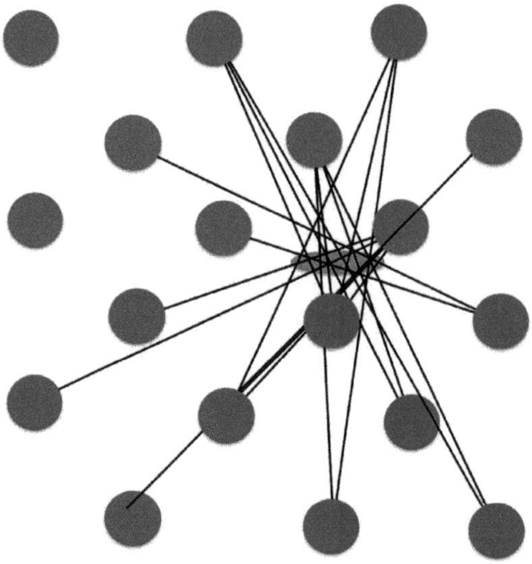

Figure 2.1: To illustrate the Cauchy-Saint Venant definition of stress at the atomic scale. The green shaded area is a circle viewed in perspective. The stress acting on it is defined by those forces exerted by blue atoms on red atoms, where the line between the centres of each pair of blue and red atoms passes through the green shaded area. The total force exerted through the shaded area becomes proportional to the area as the area increases. This is the continuum limit.

by Cauchy[1] and Saint-Venant[2] in the 19th century before the existence of atoms was universally accepted, and before any understanding of interatomic forces was developed[3]. It assumes the total force on an atom can be decomposed into a sum of forces directed along the line to each surrounding atom. This assumption is made in some models of interatomic forces, but it is not true in general.

Nevertheless, it is always true that the stress acting on a plane within a body is the resultant force exerted by atoms on one side of the plane on atoms on the other side, divided by the area of the plane. The resultant force $\mathbf{F}$ can be evaluated by the following procedure based on the principle of virtual work. If the body on one side of the plane is displaced rigidly by a small amount $\delta\mathbf{r}$ relative to the other side, the change

[1]Baron Augustin-Louis Cauchy ForMemRS, 1789-1857. French mathematician, engineer and physicist.

[2]Adhémar Jean Claude Barré de Saint-Venant, 1797-1886. French mathematician, engineer and physicist.

[3]see Timoshenko, S P, *History of strength of materials*, Dover Publications: New York (1983), p.108. ISBN 0-486-61187-6. Stephen Prokofievitch Timoshenko ForMemRS 1878-1972. Ukrainian born US engineer.

in the total energy δE is equal to $-\mathbf{F}\cdot\delta\mathbf{r}$ to first order in $|\delta\mathbf{r}|$. The three components of $\mathbf{F}$ can be found by repeating the calculation for $\delta\mathbf{r}$ along the plane normal and two non-collinear directions in the plane. This procedure yields the average stress acting on the entire plane. However, it is difficult to see how to modify this procedure to calculate the stress in a small region of the plane, which is necessary to reach the continuum limit where stress is defined at a point. An alternative approach is to calculate the stress tensor at atomic sites in terms of interatomic forces, as illustrated in section 2.4 with a simple model of atomic interactions.

Physicists have continued to develop descriptions of stress at a microscopic level[4]. But given the lack of knowledge about interatomic forces in the 1820s Cauchy developed a pragmatic definition of stress in a continuum, which is the way most engineers think about stress today. In this chapter we will discuss the continuum and atomistic approaches to defining stress.

The concept of pressure is closely related to stress: they both have dimensions of force per unit area, with units Newtons per square metre, which are called Pascals. The key difference is that with pressure the force acting on the plane is always normal to the plane, whereas with stress the force can be inclined to the plane normal and even in the plane. The tensorial nature of stress then becomes apparent because it depends on the direction of the resultant force *and* the direction of the normal to the plane on which it acts.

2.2 Cauchy's stress tensor in a continuum

Cauchy's definition of stress is simple but brilliant. He considered the force $\mathbf{f}$ per unit area acting on a plane in a continuum with unit normal $\hat{\mathbf{n}}$. Not knowing about the spatial range and nature of atomic interactions this force was assumed to have a negligible range and to exist only when the regions of the continuum on either side of the plane were in direct contact. If the plane is not flat its normal will vary with position. Apart from corners, where the normal changes discontinuously, it is always possible to define an infinitesimal element of area where the normal is constant. The concept of the continuum makes all this possible.

Consider a body in mechanical equilibrium but loaded in some arbitrary way, for example by forces applied to its surfaces. How do we calculate the infinitesimal force $\mathbf{df}$ per unit area acting on an infinitesimal area $\mathrm{d}S$ of a plane with normal $\hat{\mathbf{n}}$ at any point inside the body? At first sight this might appear to be a hopeless task because there is an infinite number of directions of plane normals. Cauchy showed it can be done by

[4]For example, a quantum mechanical formulation of stress at any point in a body was presented by Nielsen, O H, and Martin, R M, Phys. Rev. B **32**, 3780 (1985). Ole Holm Nielsen, Danish physicist. Richard M Martin, US physicist.

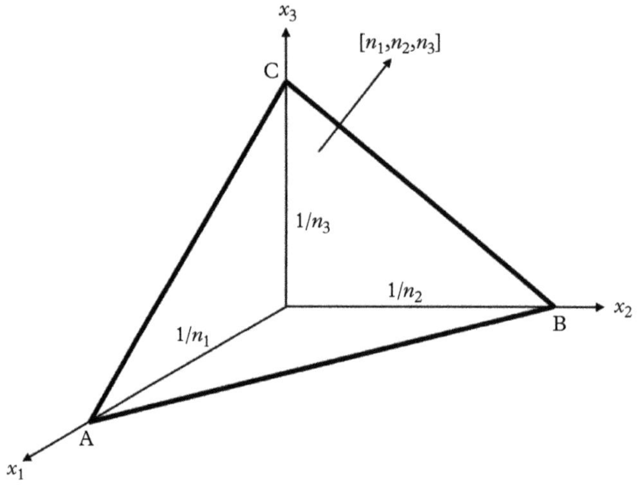

Figure 2.2: The triangle ABC has unit normal $\hat{\mathbf{n}} = [n_1, n_2, n_3]$ and area $1/(2n_1n_2n_3)$.

defining a tensor field of just six independent components, the stress field, inside the body.

Define a right-handed global cartesian coordinate system in the body, with axes x_1, x_2, x_3 and an arbitrary origin.

Consider the plane through the point $\mathbf{X}$ *with normal along the positive* x_j *direction. Let* $\mathrm{d}S$ *be an infinitesimal element of area of this plane at* $\mathbf{X}$*. Then component i of the force per unit area acting on* $\mathrm{d}S$ *is the component* $\sigma_{ij}(\mathbf{X})$*, of the Cauchy stress tensor* $\boldsymbol{\sigma}(\mathbf{X})$*.*

We will show that the force per unit area acting on an infinitesimal element of area with an *arbitrary* outward unit normal $\hat{\mathbf{n}}$ at $\mathbf{X}$ can be expressed in terms of the stress tensor components $\sigma_{ij}(\mathbf{X})$ and $\hat{\mathbf{n}}$. Let $\hat{\mathbf{n}} = [n_1, n_2, n_3]$ in this coordinate system. A plane with this normal has intercepts $1/n_1, 1/n_2, 1/n_3$ along the x_1, x_2, x_3 axes respectively, which is the plane ABC in Fig 2.2. The area of the triangle ABC is $1/(2n_1n_2n_3)$.

The plane ABC is parallel to the infinitesimal area element at $\mathbf{X}$. At equilibrium the net force acting on the body OABC must be zero. Component i of the force acting on the face OBC is $-\sigma_{i1} \times 1/(2n_2n_3)$, where the negative sign is because the outward normal to the face OBC is along $-x_1$, and $1/(2n_2n_3)$ is the area of OBC. Similarly, component i of the forces acting on the faces OAC and OAB are $-\sigma_{i2} \times 1/(2n_1n_3)$ and $-\sigma_{i3} \times 1/(2n_1n_2)$ respectively. If $\mathbf{f}$ is the force per unit area acting on the plane ABC then equilibrium of OABC requires:

$$\frac{f_i}{2n_1n_2n_3} - \frac{\sigma_{i1}}{2n_2n_3} - \frac{\sigma_{i2}}{2n_1n_3} - \frac{\sigma_{i3}}{2n_1n_2} = 0,$$

from which it follows that:

$$f_i = \sigma_{i1}n_1 + \sigma_{i2}n_2 + \sigma_{i3}n_3,$$

or,

$$f_i = \sigma_{ij}n_j. \tag{2.1}$$

This equation gives the force per unit area acting on a plane with normal $\hat{\mathbf{n}}$ in terms of the components of the stress tensor. In Exercise 2.2 the requirement that there is no torque acting on a volume element in the body leads to the condition $\sigma_{ij} = \sigma_{ji}$, i.e. the stress tensor is symmetric. Thus, there are only six independent components of the stress tensor. Equation 2.1 also shows that at equilibrium the forces per unit area acting on either side of a plane are equal and opposite because the sense of the plane normal reverses on either side. Finally, if σ_{ij} varies with position it becomes a stress field and the force per unit area acting on planes with a given normal also varies with position.

Exercise 2.1 Using eq 2.1 prove that σ_{ij} satisfies the tensor transformation law under a rotation of the coordinate system. ■

Solution The components of the force per unit area, and the components of the normal to the plane on which it acts, satisfy the transformation law for vectors. In the rotated coordinate system the components of the force become $f'_k = R_{ki}f_i$ and the components of the plane normal become $n'_p = R_{pj}n_j$. Therefore, in the rotated coordinate system the relation between the force per unit area and the plane normal is $f'_k = \sigma'_{kp}n'_p = R_{ki}f_i = R_{ki}\sigma_{ij}n_j = R_{ki}\sigma_{ij}R_{pj}n'_p$, where we have used the orthogonality of the rotation matrix to equate the inverse of the rotation matrix to its transpose: $n_j = (R^{-1})_{jp}n'_p = R_{pj}n'_p$. Therefore, $\sigma'_{kp} = R_{ki}R_{pj}\sigma_{ij}$, which proves σ_{ij} satisfies the tensor transformation law for second rank tensors.

Comment

In matrix notation the transformation law becomes $\boldsymbol{\sigma}' = \mathbf{R}\boldsymbol{\sigma}\mathbf{R}^T$, where $\mathbf{R}^T$ is the transpose of $\mathbf{R}$.

Exercise 2.2 (a) Sketch a cube of side length equal to unity with edges parallel to the axes Ox_1, Ox_2, Ox_3. Each face has unit area. As a result of a stress field each face experiences a force. On your sketch show the force acting on each face as an arrow and label it in terms of the components of the stress tensor. For example, on the face with normal parallel to the positive x_1 direction the three components of the force are $(\sigma_{11}, \sigma_{21}, \sigma_{31})$ with respect to the x_1, x_2 and x_3 axes respectively. But on the face with normal parallel to the negative x_1 axis the three components of the force are $(-\sigma_{11}, -\sigma_{21}, -\sigma_{31})$, so their arrows point in the opposite directions to those on the first face.
(b) By taking moments of the forces about the centre of the cube show that the condition for there to be no torque on a volume element is $\sigma_{ij} = \sigma_{ji}$. A more rigorous derivation of this relation is outlined in Exercise 2.5. ■

Solution (a) See Fig 2.3.
(b) The area of each face is 1. Taking moments about a line parallel to the X_1-axis passing through the centre of the cube:

$$2 \times \frac{1}{2}\sigma_{23} - 2 \times \frac{1}{2}\sigma_{32} = 0$$

Therefore, $\sigma_{23} = \sigma_{32}$. Similarly, by taking moments about a line parallel to the X_2-axis passing through the centre of the cube we find $\sigma_{13} = \sigma_{31}$, and by taking moments about a line parallel to the X_3-axis passing through the centre of the cube we find $\sigma_{12} = \sigma_{21}$. Therefore, $\sigma_{ij} = \sigma_{ji}$ ensures there is no torque acting on the cube.

2.3 Normal stresses and shear stresses

We have seen that the force per unit area acting on an element of area of a plane may have components in the plane as well as normal to it. The components parallel to the plane give rise to *shear* stresses, and they correspond to off-diagonal components of the stress tensor. The components normal to the plane are called *normal* stresses. Normal stresses are often called *tensile* or *compressive* stresses depending on whether they are positive or negative respectively. However, the designation of normal or shear stress may change under a rotation of the coordinate system, as shown in the next Exercise. Shear stresses arise in frictional sliding and they play a central role in plastic deformation of crystalline materials.

Exercise 2.3 A stress σ has the following representation in a cartesian coordinate system $Ox_1x_2x_3$:

$$\sigma = \begin{bmatrix} -s & 0 & 0 \\ 0 & s & 0 \\ 0 & 0 & 0 \end{bmatrix}.$$

Although this matrix has only diagonal elements it must correspond to a state of shear stress because the trace of the matrix is zero. When the same stress is represented in a rotated coordinate system $Ox_1'x_2'x_3'$, obtained by rotating Ox_1 and Ox_2 in a positive sense by $\pi/4$ about Ox_3, show that its matrix representation

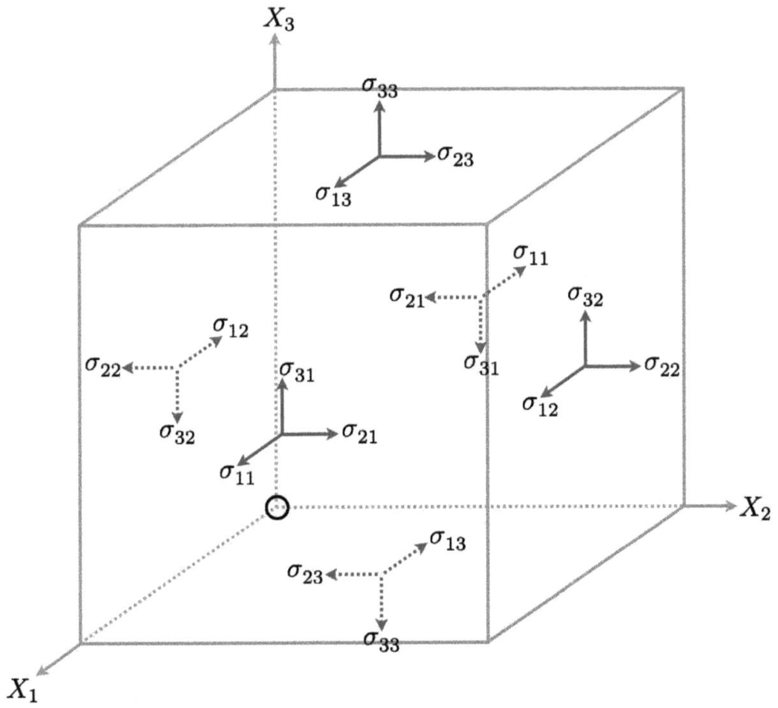

Figure 2.3: A cube of side unity showing the components of the stress tensor acting on each of its six faces. A right-hand Cartesian coordinate system is parallel to the cube edges.

becomes:

$$\sigma' = \begin{bmatrix} 0 & s & 0 \\ s & 0 & 0 \\ 0 & 0 & 0 \end{bmatrix}.$$

■

Solution The rotation matrix has elements $R_{ij} = \hat{\mathbf{e}}'_i \cdot \hat{\mathbf{e}}_j$, where $\hat{\mathbf{e}}'_i$ and $\hat{\mathbf{e}}_j$ are unit vectors along the positive x'_i and x_j axes:

$$\mathbf{R} = \begin{pmatrix} 1/\sqrt{2} & 1/\sqrt{2} & 0 \\ -1/\sqrt{2} & 1/\sqrt{2} & 0 \\ 0 & 0 & 1 \end{pmatrix}.$$

In the rotated coordinate system the stress tensor $\boldsymbol{\sigma}'$ has components $\sigma'_{ij} = R_{ik}R_{jl}\sigma_{kl} = R_{ik}\sigma_{kl}R^T_{lj}$. Therefore, $\boldsymbol{\sigma}' = \mathbf{R}\boldsymbol{\sigma}\mathbf{R}^T$:

$$\begin{aligned} \mathbf{R}\sigma\mathbf{R}^T &= \begin{pmatrix} 1/\sqrt{2} & 1/\sqrt{2} & 0 \\ -1/\sqrt{2} & 1/\sqrt{2} & 0 \\ 0 & 0 & 1 \end{pmatrix} \begin{pmatrix} -s & 0 & 0 \\ 0 & s & 0 \\ 0 & 0 & 0 \end{pmatrix} \begin{pmatrix} 1/\sqrt{2} & -1/\sqrt{2} & 0 \\ 1/\sqrt{2} & 1/\sqrt{2} & 0 \\ 0 & 0 & 1 \end{pmatrix} \\ &= \begin{pmatrix} 0 & s & 0 \\ s & 0 & 0 \\ 0 & 0 & 0 \end{pmatrix}. \end{aligned}$$

The physical equivalence of these two descriptions of the state of stress is illustrated in Fig 2.4.

2.4 Stress at the atomic scale

After studying Cauchy's analysis of stress in terms of forces between elastic continua in contact it may be surprising that the concept of stress can be developed at the atomic scale involving interactions between discrete atoms. An early treatment of stress at the atomic scale may be found in Note B, p.616 of Love's treatise of 1927, albeit for a perfect crystal. The treatment we shall give here is more general, but it does assume the total force on an atom may be expressed as a sum of contributions from surrounding atoms.

Consider a cluster of atoms of any size, from molecules to macroscopic components. The potential energy of the cluster is defined by the energy of interaction between all

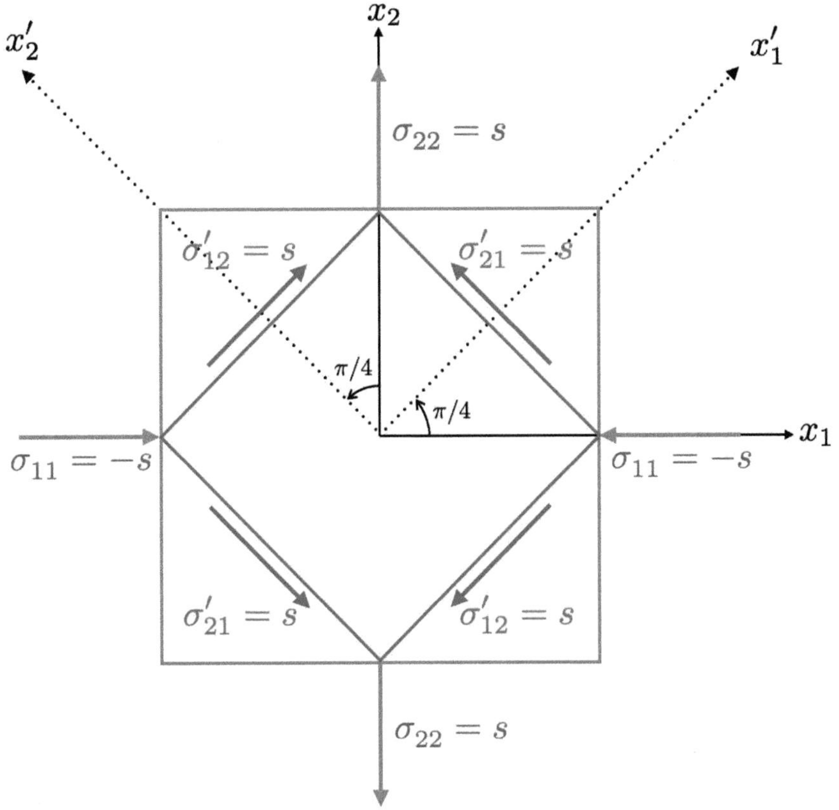

Figure 2.4: To illustrate the solution of Exercise 2.3. The axes x_1, x_2 are rotated clockwise by $\pi/4$ to new positions x'_1, x'_2 about the positive $x_3 = x'_3$ axis. The rotation appears anti-clockwise because the positive x_3 axis is normal to the page towards the reader. Equal and opposite normal stresses $\mp s$ (blue) along the x_1, x_2 axes are equivalent to shear stresses (red) of the same magnitude, s, on planes at $\pi/4$ to the x_1, x_2 axes.

atoms in the cluster. Atoms in the cluster may be in a perfect crystal configuration, or a defective crystal or an amorphous state. Let $\mathbf{f}^{(a)}$ be the total force acting on atom a in the cluster. We assume this force comprises a sum of forces exerted by other atoms in the cluster. Let $\mathbf{f}^{(m/a)}$ be the force exerted by atom m upon atom a. Then $\mathbf{f}^{(a)} = \sum_{m \neq a} \mathbf{f}^{(m/a)}$.

Let the position of atom a be $\mathbf{X}^{(a)}$. Imagine the positions of all atoms are displaced by infinitesimal amounts in arbitrary directions. The change in the total potential energy of the cluster is:

$$\begin{aligned}\delta V &= -\sum_a \mathbf{f}^{(a)} \cdot \delta \mathbf{X}^{(a)} \\ &= -\frac{1}{2}\sum_a \sum_{m \neq a} \mathbf{f}^{(m/a)} \cdot \delta\left(\mathbf{X}^{(a)} - \mathbf{X}^{(m)}\right),\end{aligned} \tag{2.2}$$

where $\delta \mathbf{X}^{(j)}$ is the displacement of atom j. The factor of $\frac{1}{2}$ takes into account the sharing of the interaction between atoms a and m.

Suppose the infinitesimal displacements of atoms are a result of the application of a homogeneous, infinitesimal strain δe_{ij}. Since the applied strain is homogeneous and infinitesimal the change of $(X_i^{(a)} - X_i^{(m)})$ is $\delta e_{ij}(X_j^{(a)} - X_j^{(m)})$. Therefore the change in potential energy of the cluster in eq 2.2 becomes the following:

$$\delta V = -\frac{1}{2}\sum_a \sum_{m \neq a} f_i^{(m/a)} \delta e_{ij} \left(X_j^{(a)} - X_j^{(m)}\right). \tag{2.3}$$

We may use eq 2.3 to *define* atomic level stresses:

$$\delta V = \sum_a \Omega^{(a)} \sigma_{ij}^{(a)} \delta e_{ij}, \tag{2.4}$$

where $\Omega^{(a)}$ and $\sigma_{ij}^{(a)}$ are the volume associated with atomic site a and the stress tensor associated with atomic site a inside the cluster. Comparing with eq 2.3 we obtain:

$$\sigma_{ij}^{(a)} = \frac{1}{2\Omega^{(a)}} \sum_{m \neq a} f_i^{(m/a)} \left(X_j^{(m)} - X_j^{(a)}\right) \tag{2.5}$$

where the sum over m is taken over all atoms in the cluster. The atomic volume $\Omega^{(a)}$ may be defined by a Voronoi construction.

In eq 2.5 it is clear that even if the net force on an atom is zero it may still be in a state of stress. For example, the atoms in the cluster may be in a perfect crystal environment where each atom experiences zero net force, but it may still be in a state of compression or tension.

Provided no heat is exchanged between the cluster and its surroundings as a result of the application of the strain the change in the potential energy of the cluster is equal to the change in its internal energy, δE. Then in a continuum approximation eq 2.4 becomes the following volume integral over the cluster:

$$\delta E = \sum_n \Omega^{(n)} \sigma_{ij}^{(n)} \delta e_{ij} \approx \int \mathrm{d}^3 X \, \sigma_{ij}(\mathbf{X}) \delta e_{ij}(\mathbf{X})$$

which leads to a new definition of stress in a continuum as the following functional derivative:

$$\sigma_{ij}(\mathbf{X}) = \frac{\delta E}{\delta e_{ij}(\mathbf{X})}, \tag{2.6}$$

where the variation is carried out adiabatically, i.e. at constant entropy. The definition of stress in eq 2.6 is based on the existence of a strain energy function describing the potential energy of the body as a function of a homogeneous elastic strain applied to it. The notion of a strain energy function was introduced by Green[5] in 1837 and put on a rigorous thermodynamic foundation by Thomson[6] in 1855. This definition of stress appears to be quite different from Cauchy's definition, but they are equivalent when the body is in mechanical equilibrium, as we shall see in Exercise 2.6.

As we have already noted, not all models of atomic interactions enable the total force on an atom to be expressed as a sum of contributions from surrounding atoms. In quantum mechanics the Ehrenfest-Hellmann-Feynman force on an atomic nucleus depends on the self-consistent electronic charge density at the nucleus. The self-consistent charge density at the nucleus depends on the positions of surrounding atoms in a way that cannot be broken down into a sum of separate contributions from each surrounding atom. However, the definition of stress in eq 2.6 may be applied to all atomic interactions.

2.5 Invariants of the stress tensor

Let $\hat{\mathbf{e}}_1, \hat{\mathbf{e}}_2, \hat{\mathbf{e}}_3$ be unit vectors along the right-handed cartesian coordinate system x_1, x_2, x_3. Let $\hat{\mathbf{e}}'_1, \hat{\mathbf{e}}'_2, \hat{\mathbf{e}}'_3$ be unit vectors along the right-handed cartesian coordinate system x'_1, x'_2, x'_3. The rotation matrix which rotates $\hat{\mathbf{e}}_1, \hat{\mathbf{e}}_2, \hat{\mathbf{e}}_3$ into $\hat{\mathbf{e}}'_1, \hat{\mathbf{e}}'_2, \hat{\mathbf{e}}'_3$ has components $R_{ij} = (\hat{\mathbf{e}}'_i \cdot \hat{\mathbf{e}}_j)$, that is $\hat{\mathbf{e}}'_i = R_{ij}\hat{\mathbf{e}}_j$. Since stress is a second rank tensor it satisfies the following transformation law:

$$\sigma'_{jk} = R_{ji} R_{kp} \sigma_{ip}, \tag{2.7}$$

where $\boldsymbol{\sigma}'$ and $\boldsymbol{\sigma}$ are the matrix representations of the stress tensor in the primed and unprimed coordinate systems respectively.

Since the components of the matrix representing the stress tensor change under a coordinate transformation, individual stress tensor components have limited physical significance. However, there are three quantities that remain invariant under a rotation

[5] George Green 1793-1841, British mathematical physicist and miller.

[6] Sir William Thomson PRS OM 1824-1907, Scots-Irish mathematical physicist and engineer, who became Lord Kelvin in 1892.

of the coordinate system, and they can be used to construct more physically significant quantities. The key to identifying invariants of the stress tensor is to recall that the eigenvalues do not depend on the choice of coordinate system. It follows that the cubic polynomial that defines the eigenvalues must have the same coefficients in all coordinate systems. Let the eigenvalues be s_1, s_2, s_3. These eigenvalues are called *principal* stresses. When the coordinate system is aligned with the three eigenvectors of the matrix representing the stress tensor, the matrix becomes diagonal with s_1, s_2, s_3 along the leading diagonal. The cubic polynomial defining these eigenvalues is:

$$(s - s_1)(s - s_2)(s - s_3) = 0,$$

or

$$s^3 - (s_1 + s_2 + s_3)s^2 + (s_1 s_2 + s_2 s_3 + s_3 s_1)s - s_1 s_2 s_3 = 0.$$

Therefore, the following three quantities are invariants:

$$\begin{aligned} I_1 &= s_1 + s_2 + s_3 \\ I_2 &= s_1 s_2 + s_2 s_3 + s_3 s_1 \\ I_3 &= s_1 s_2 s_3. \end{aligned} \tag{2.8}$$

Any quantity that may be expressed in terms of these invariants is also invariant. I_1 is the trace of the stress tensor, $\mathrm{Tr}\,\sigma$. The hydrostatic stress is defined as the average normal stress, which is $I_1/3$. The hydrostatic pressure, p, is the negative of the hydrostatic stress:

$$p = -\frac{1}{3}\mathrm{Tr}\,\sigma. \tag{2.9}$$

The second and third invariants may be expressed in any coordinate system as follows:

$$I_2 = \frac{1}{2}\left[(\mathrm{Tr}\,\sigma)^2 - \mathrm{Tr}\,\sigma^2\right] \tag{2.10}$$

$$I_3 = \frac{1}{6}\left[(\mathrm{Tr}\,\sigma)^3 + 2\mathrm{Tr}\,\sigma^3 - 3\,(\mathrm{Tr}\,\sigma)\left(\mathrm{Tr}\,\sigma^2\right)\right] \tag{2.11}$$

We note also that I_3 is the determinant of the stress tensor.

Stress invariants are useful for characterising the stress fields of defects in crystals, e.g. grain boundaries, because they are independent of the coordinate system. Strain energy functions are also often expressed in terms of stress invariants for similar reasons. The strain tensor has equivalent invariants.

2.6 Shear stress on a plane and the Von Mises stress invariant

In the previous section we saw the hydrostatic stress is an invariant of the stress tensor. As the average of the three normal stresses in any coordinate system it is a scalar quantity which indicates the degree of compression or tension. It is useful to have another invariant quantity that measures the shear content of the stress tensor. This is somewhat more difficult because shear stresses, unlike hydrostatic stresses, depend on the normal of the plane where they act. Therefore, we begin this section by evaluating the magnitude of the shear stress acting on any plane for an arbitrary stress tensor.

Consider a stress tensor σ_{ij} with eigenvalues s_1, s_2, s_3 and corresponding unit eigenvectors $\hat{\mathbf{e}}_1, \hat{\mathbf{e}}_2, \hat{\mathbf{e}}_3$. The eigenvectors form an orthonormal set, which defines the cartesian coordinate system we shall use. Consider a plane with unit normal $\hat{\mathbf{n}} = n_i \hat{\mathbf{e}}_i$. The force per unit area acting on this plane is $\mathbf{f} = s_1 n_1 \hat{\mathbf{e}}_1 + s_2 n_2 \hat{\mathbf{e}}_2 + s_3 n_3 \hat{\mathbf{e}}_3$. The magnitude of the component of $\mathbf{f}$ along the normal $\hat{\mathbf{n}}$ is $f_n = \mathbf{f} \cdot \hat{\mathbf{n}} = s_1 n_1^2 + s_2 n_2^2 + s_3 n_3^2$. Therefore, the force per unit area normal to the plane is $\mathbf{f_n} = f_n \hat{\mathbf{n}}$. The force per unit area parallel to the plane is $\mathbf{f_p} = \mathbf{f} - (\mathbf{f} \cdot \hat{\mathbf{n}})\, \hat{\mathbf{n}}$. Thus,

$$\mathbf{f_p} = [s_1 n_1 - f_n n_1, s_2 n_2 - f_n n_2, s_3 n_3 - f_n n_3] \, .$$

We may obtain a useful expression for the square of the magnitude of $\mathbf{f_p}$ as follows (the summation convention is temporarily suspended to derive eq 2.12):

$$\begin{aligned}
f_p^2 &= \sum_i (s_i n_i - f_n n_i)^2 \\
&= \sum_i n_i^2 s_i^2 - 2 \sum_i s_i n_i^2 \sum_j s_j n_j^2 + \Big(\sum_j s_j n_j^2\Big)^2 \sum_i n_i^2 \\
&= \sum_i n_i^2 s_i^2 - \sum_i s_i n_i^2 \sum_j s_j n_j^2 \\
&= \sum_i n_i^2 s_i^2 \sum_j n_j^2 - \sum_i s_i n_i^2 \sum_j s_j n_j^2 \\
&= \sum_i \sum_j n_i^2 n_j^2 s_i^2 - n_i^2 n_j^2 s_i s_j \\
&= \frac{1}{2} \sum_i \sum_j n_i^2 n_j^2 (s_i^2 + s_j^2 - 2 s_i s_j) \\
&= \sum_{i<j} \sum_j n_i^2 n_j^2 (s_i - s_j)^2 \\
&= (s_1 - s_2)^2 n_1^2 n_2^2 + (s_2 - s_3)^2 n_2^2 n_3^2 + (s_3 - s_1)^2 n_3^2 n_1^2 \qquad (2.12)
\end{aligned}$$

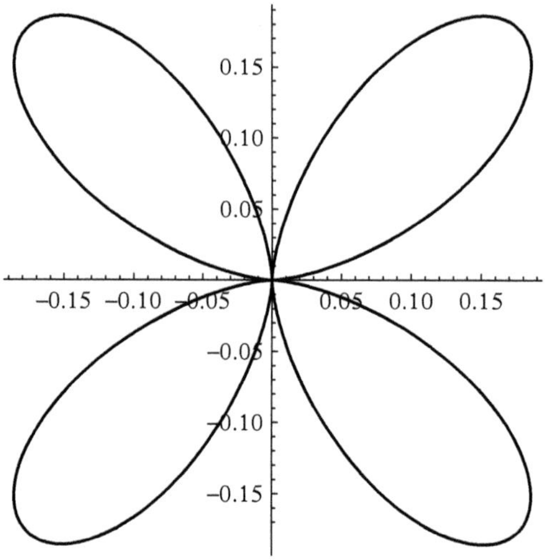

Figure 2.5: Plot of $f_p^2 = n_1^2 n_2^2 (s_1 - s_2)^2$ in the plane $n_3 = 0$ with $s_1 - s_2 = \pm 1$. The axes are aligned with the principal stress directions $\hat{\mathbf{e}}_1$ and $\hat{\mathbf{e}}_2$.

If the stress is purely hydrostatic then, as expected, eq 2.12 shows that the shear stress on all planes is zero. In the plane $n_3 = 0$ we have $f_p^2 = (s_1 - s_2)^2 n_1^2 n_2^2$. This function is plotted in Fig 2.5 for $s_1 - s_2 = \pm 1$. It is seen that the maximum value of f_p^2 is 0.25 and it occurs on planes at 45° to the principal stress directions $\hat{\mathbf{e}}_1$ and $\hat{\mathbf{e}}_2$.

Exercise 2.4 (a) Show that $f_p^2 = \hat{\mathbf{n}}^T \sigma^2 \hat{\mathbf{n}} - \left(\hat{\mathbf{n}}^T \sigma \hat{\mathbf{n}}\right)^2$, where $\hat{\mathbf{n}}^T$ is the transpose of $\hat{\mathbf{n}}$.
(b) Show that $\mathbf{f_p}$ is unaffected if σ_{ij} is replaced by $\sigma_{ij}^{(D)} = \sigma_{ij} - \frac{1}{3}(\mathrm{Tr}\,\sigma)\delta_{ij}$, where $\sigma^{(D)}$ is called the *deviatoric* stress.
(c) Using eq 2.12 show that the average value of f_p^2, where the averaging is over all orientations of $\hat{\mathbf{n}}$ on the surface of a unit sphere, is:

$$\begin{aligned} \left\langle f_p^2 \right\rangle &= \frac{1}{15}\left((s_1 - s_2)^2 + (s_2 - s_3)^2 + (s_3 - s_1)^2\right) \\ &= \frac{1}{15}\left(3\mathrm{Tr}\,\sigma^2 - (\mathrm{Tr}\,\sigma)^2\right) \qquad (2.13) \end{aligned}$$

■

Solution (a) The force per unit area, $\mathbf{f}_p$, parallel to the plane with normal $\hat{\mathbf{n}}$ is $\mathbf{f} - (\mathbf{f}^T \cdot \hat{\mathbf{n}})\hat{\mathbf{n}} = \sigma\hat{\mathbf{n}} - \left(\hat{\mathbf{n}}^T \sigma \hat{\mathbf{n}}\right)\hat{\mathbf{n}}$, where $\mathbf{f}$ is the total force per unit area acting on the plane with normal $\hat{\mathbf{n}}$. We are using matrix notation to write vectors as column matrices and their transposes as row matrices. Taking the transpose of $\mathbf{f}_p$ we get $\mathbf{f}_p^T = \hat{\mathbf{n}}^T\sigma - (\hat{\mathbf{n}}^T\sigma\hat{\mathbf{n}})\hat{\mathbf{n}}^T$. The squared magnitude of $\mathbf{f}_p$ is then:

$$\begin{aligned} f_p^2 &= \mathbf{f}_p^T \cdot \mathbf{f}_p = \left(\hat{\mathbf{n}}^T\sigma - \left(\hat{\mathbf{n}}^T\sigma\hat{\mathbf{n}}\right)\hat{\mathbf{n}}^T\right)\cdot\left(\sigma\hat{\mathbf{n}} - \left(\hat{\mathbf{n}}^T\sigma\hat{\mathbf{n}}\right)\hat{\mathbf{n}}\right) \\ &= \hat{\mathbf{n}}^T\sigma^2\hat{\mathbf{n}} - \left(\hat{\mathbf{n}}^T\sigma\hat{\mathbf{n}}\right)^2. \end{aligned}$$

(b) If σ is replaced by the deviatoric stress, $\sigma^{(D)} = \sigma - (\mathrm{Tr}\sigma/3)\mathbf{I}$, where $\mathbf{I}$ is the identity matrix, then $\mathbf{f}_p$ becomes:

$$\begin{aligned} \mathbf{f}_p &\rightarrow \left(\sigma - \frac{\mathrm{Tr}\sigma}{3}\mathbf{I}\right)\hat{\mathbf{n}} - \left\{\hat{\mathbf{n}}^T\left(\sigma - \frac{\mathrm{Tr}\sigma}{3}\mathbf{I}\right)\hat{\mathbf{n}}\right\}\hat{\mathbf{n}} \\ &= \sigma\hat{\mathbf{n}} - \frac{\mathrm{Tr}\sigma}{3}\hat{\mathbf{n}} - \left(\hat{\mathbf{n}}^T\sigma\hat{\mathbf{n}}\right)\hat{\mathbf{n}} + \frac{\mathrm{Tr}\sigma}{3}\hat{\mathbf{n}} \\ &= \sigma\hat{\mathbf{n}} - \left(\hat{\mathbf{n}}^T\sigma\hat{\mathbf{n}}\right)\hat{\mathbf{n}}. \end{aligned}$$

We see that $\mathbf{f}_p$ is unaffected by the replacement, as we would expect since $\mathbf{f}_p$ measures the shear stress on a plane, which is independent of the hydrostatic stress, $\mathrm{Tr}\sigma/3$.

(c) From the symmetry of the sphere $\langle n_1^2 n_2^2\rangle = \langle n_2^2 n_3^2\rangle = \langle n_3^2 n_1^2\rangle$, where $\langle\dots\rangle$ denotes the average taken over all orientations in a unit sphere. Consider $\langle n_3^2 n_1^2\rangle$. In spherical polar coordinates $n_3 = \cos\theta$ and $n_1 = \sin\theta\cos\phi$, and an element of area of the sphere at (θ,ϕ) is $\mathrm{d}S = \sin\theta\mathrm{d}\theta\mathrm{d}\phi$. Therefore,

$$\begin{aligned} \langle n_3^2 n_1^2\rangle &= \frac{1}{4\pi}\int_0^\pi \mathrm{d}\theta\sin\theta\int_0^{2\pi}\mathrm{d}\phi\cos^2\theta\sin^2\theta\cos^2\phi \\ &= \frac{1}{4\pi}\int_{-1}^{1}\mathrm{d}(\cos\theta)\left(1-\cos^2\theta\right)\cos^2\theta\int_0^{2\pi}\mathrm{d}\phi\left(\frac{1+\cos(2\phi)}{2}\right) \\ &= \frac{1}{4\pi}\left[\frac{\cos^3\theta}{3} - \frac{\cos^5\theta}{5}\right]_{-1}^{1}\times\pi \\ &= \frac{1}{4\pi}\frac{4}{15}\pi = \frac{1}{15}. \end{aligned}$$

and hence,

$$
\begin{aligned}
\langle f_p^2 \rangle &= \frac{1}{15}\left((s_1 - s_2)^2 + (s_2 - s_3)^2 + (s_3 - s_1)^2\right) \\
&= \frac{2}{15}\left(s_1^2 + s_2^2 + s_3^2 - s_1 s_2 - s_2 s_3 - s_3 s_1\right) \\
&= \frac{2}{15}\left\{\frac{3}{2}\left(s_1^2 + s_2^2 + s_3^2\right) - \frac{1}{2}\left(s_1^2 + s_2^2 + s_3^2 + 2\left(s_1 s_2 + s_2 s_3 + s_3 s_1\right)\right)\right\} \\
&= \frac{1}{15}\left(3\mathrm{Tr}\,\sigma^2 - (\mathrm{Tr}\,\sigma)^2\right)
\end{aligned}
$$

The von Mises[7] stress is an invariant used to characterise the degree of shear in a stress tensor. It is used for example as a yield criterion to decide whether the stress in a body enables plastic deformation to take place. It is defined by:

$$
\sigma^{vM} = \frac{1}{\sqrt{2}}\sqrt{(s_1 - s_2)^2 + (s_2 - s_3)^2 + (s_3 - s_1)^2} \tag{2.14}
$$

The von Mises shear stress is $\sigma^{vM} = \sqrt{15/2} \times \sqrt{\langle f_p^2 \rangle}$, and is therefore directly related to the root mean square shear stress on a plane.

2.7 Mechanical equilibrium

We come now to one of the most important ideas in the continuum theory of elasticity, that of mechanical equilibrium. We have seen already that the stress tensor has to be symmetric if there are no torques acting. We consider now the equilibrium of a region $\mathcal{R}$ that experiences a force per unit volume $\mathbf{f}(\mathbf{X})$ within a body. This force may be gravitational for example, and it is called a *body force*. The continuum surrounding $\mathcal{R}$ distorts generating stresses that balance the net body force acting on $\mathcal{R}$. These stresses are transmitted to $\mathcal{R}$ through the surface $\mathcal{S}$ surrounding it.

Mechanical equilibrium requires that the total force acting on $\mathcal{R}$ is zero:

$$
\int_{\mathcal{R}} f_i(\mathbf{X})\mathrm{d}V - \int_{\mathcal{S}} \sigma_{ij} n_j \mathrm{d}S = 0,
$$

where n_j is component j of the *inward* pointing normal $\hat{\mathbf{n}}$ at the surface $\mathcal{S}$ of the region $\mathcal{R}$. We have chosen the inward pointing normal because we are considering the force

[7]Richard Edler von Mises 1883-1953, US mathematician, engineer and philosopher, born in what is now Ukraine.

that the surrounding medium exerts on the region $\mathcal{R}$. But we normally use the *outward* pointing normal in which case we must change the minus sign to a plus sign:

$$\int_{\mathcal{R}} f_i(\mathbf{X})\mathrm{d}V + \int_{\mathcal{S}} \sigma_{ij} n_j \mathrm{d}S = 0,$$

where n_j is now component j of the *outward* pointing normal $\hat{\mathbf{n}}$ at the surface $\mathcal{S}$ of the region $\mathcal{R}$. Using the divergence theorem this may be rewritten as:

$$\int_{\mathcal{R}} \left(f_i(\mathbf{X}) + \sigma_{ij,j} \right) \mathrm{d}V = 0,$$

where the comma denotes differentiation ($Z_{,j}$ means $\partial Z / \partial X_j$) and $\sigma_{ij,j}$ is the divergence of the stress tensor. Since this balance of forces must hold for all regions $\mathcal{R}$ within the body we arrive at the following differential equations for mechanical equilibrium:

$$\sigma_{ij,j} + f_i = 0. \tag{2.15}$$

There are three equations here, one for each component i of the body force, and $\sigma_{ij,j}$ consists of three derivatives for each value of i. These equations are analogous to Poisson's equation in electrostatics where the divergence of the electric displacement vector is the electric charge density, and electric charges are the sources of electric fields. In eq 2.15 body forces are sources of stress fields originating within the body. However, they are not the only sources of stress fields. There may also be stress fields generated by forces applied to the surface of the body, but their divergence is zero within the body.

How does eq 2.15 relate to the atomic level stresses of eq 2.5? Those stresses are defined only at atomic nuclei. If the stress field could be reliably interpolated between the nuclei the integral $\int \sigma_{ij} n_j \mathrm{d}S$ taken over a closed surface surrounding a single atom, with n_j pointing inwards, should equal component i of the force acting on the atom. The difficulty is that atomic level stresses vary much too rapidly in the core of a defect, where they are most useful, to interpolate confidently the stress field between the nuclei[8]. In problem 2.4 it is shown that eq 2.15 is satisfied by the atomic level stress tensor if the atoms are replaced by a continuum in which volume elements exert forces on each other that are directly related to the force of interaction between discrete atoms.

[8]However, the stress fields calculated quantum mechanically (see footnote 4 of this chapter) are defined throughout the body, and they do satisfy eq 2.15.

Exercise 2.5 Consider the resultant torque T_i acting on a region $\mathcal{R}$ with surface $\mathcal{S}$ within a body. It is acted on by a distribution of body forces $f_i(\mathbf{x})$ within $\mathcal{R}$ and surface tractions[a] $t_i(\mathbf{x})$ on $\mathcal{S}$. If the region $\mathcal{R}$ is in mechanical equilibrium prove that the condition for the torque T_i to be zero is that the stress tensor is symmetric.
Hint
With respect to an arbitrary origin the torque acting on the region $\mathcal{R}$ is given by

$$\mathbf{T} = \int_{\mathcal{R}} \mathbf{x} \times \mathbf{f} \, \mathrm{d}V + \int_{\mathcal{S}} \mathbf{x} \times \mathbf{t} \, \mathrm{d}S.$$

Rewrite this equation in component form using suffix notation and use the divergence theorem to convert the surface integral into a volume integral. Then use the equilibrium condition $\sigma_{ij,j} + f_i = 0$ to simplify the terms and deduce the symmetry of the stress tensor as the condition for $T_i = 0$.
Why is the expression for the torque $\mathbf{T}$ independent of the choice of origin? ▪

[a] A *surface traction* is a force per unit area acting on a surface.

Solution Since the traction $\mathbf{t}$ has to equal $\boldsymbol{\sigma}\hat{\mathbf{n}}$, where $\hat{\mathbf{n}}$ is the outward pointing normal of the surface $\mathcal{S}$ we may write the following expression for the torque in component form:

$$T_i = \int_{\mathcal{R}} \varepsilon_{ijk} x_j f_k \mathrm{d}V + \int_{\mathcal{S}} \varepsilon_{ijk} x_j \sigma_{kl} n_l \mathrm{d}S.$$

The surface integral is converted into a volume integral by the divergence theorem:

$$\begin{aligned} T_i &= \int_{\mathcal{R}} \varepsilon_{ijk} x_j f_k \mathrm{d}V + \int_{\mathcal{R}} \varepsilon_{ijk} \frac{\partial}{\partial x_l} \left(x_j \sigma_{kl} \right) \mathrm{d}V \\ &= \int_{\mathcal{R}} \varepsilon_{ijk} x_j f_k \mathrm{d}V + \int_{\mathcal{R}} \varepsilon_{ijk} \left(\delta_{jl} \sigma_{kl} + x_j \sigma_{kl,l} \right) \mathrm{d}V \\ &= \int_{\mathcal{R}} \varepsilon_{ijk} x_j \underbrace{\left(f_k + \sigma_{kl,l} \right)}_{=0} \mathrm{d}V + \int_{\mathcal{R}} \varepsilon_{ijk} \sigma_{kj} \mathrm{d}V \end{aligned}$$

$$= \int_{\mathcal{R}} \varepsilon_{ijk}\sigma_{kj}\mathrm{d}V = \frac{1}{2}\int_{\mathcal{R}} \varepsilon_{ijk}\sigma_{kj} + \varepsilon_{ikj}\sigma_{jk}\mathrm{d}V = \frac{1}{2}\int_{\mathcal{R}} \varepsilon_{ijk}\left(\sigma_{kj} - \sigma_{jk}\right)\mathrm{d}V,$$

where we have used the condition for mechanical equilibrium. Therefore, $T_i = 0$ if and only if $\sigma_{kj} = \sigma_{jk}$.
Suppose the origin is at $\mathbf{x}_0$. Then the expression for the torque vector becomes:

$$\mathbf{T} \rightarrow \int_{\mathcal{R}} (\mathbf{x} - \mathbf{x}_0)\times\mathbf{f}\,\mathrm{d}V + \int_{\mathcal{S}} (\mathbf{x} - \mathbf{x}_0)\times\mathbf{t}\,\mathrm{d}S$$

$$= \underbrace{\int_{\mathcal{R}} \mathbf{x}\times\mathbf{f}\,\mathrm{d}V + \int_{\mathcal{S}} \mathbf{x}\times\mathbf{t}\,\mathrm{d}S}_{\text{original expression for torque}} - \mathbf{x}_0 \times \underbrace{\left(\int_{\mathcal{R}} \mathbf{f}\,\mathrm{d}V + \int_{\mathcal{S}} \mathbf{t}\,\mathrm{d}S\right)}_{\text{total force on body}}$$

The torque is independent of $\mathbf{x}_0$ provided the total force on the body is zero.

Exercise 2.6 In this exercise we use the equilibrium condition $\sigma_{ij,j} + f_i = 0$ to prove $\sigma_{ij} = \delta E/\delta e_{ij}(\mathbf{x})$.
Consider a region $\mathcal{R}$ with surface $\mathcal{S}$ within a body in which there is a distribution of body forces $f_i(\mathbf{x})$ within $\mathcal{R}$ and surface tractions $t_i(\mathbf{x})$ on $\mathcal{S}$. There is no net force acting on $\mathcal{R}$. Suppose an infinitesimal displacement field $\delta u_i(\mathbf{x})$ is applied to points within $\mathcal{R}$ and on $\mathcal{S}$, which does not disturb the equilibrium of the body. The work done δW by the body forces in $\mathcal{R}$ and surface tractions on $\mathcal{S}$ is:

$$\delta W = -\left\{\int_{\mathcal{R}} f_i(\mathbf{x})\delta u_i(\mathbf{x})\,\mathrm{d}V + \int_{\mathcal{S}} \sigma_{ij}(\mathbf{x})\delta u_i(\mathbf{x}) n_j(\mathbf{x})\,\mathrm{d}S\right\},$$

where the normal vector in the surface integral points outwards. The corresponding change in the internal energy of the region $\mathcal{R}$ is $\delta E = -\delta W$. Thus,

$$\delta E = \int_{\mathcal{R}} f_i(\mathbf{x})\delta u_i(\mathbf{x})\,\mathrm{d}V + \int_{\mathcal{S}} \sigma_{ij}(\mathbf{x})\delta u_i(\mathbf{x}) n_j(\mathbf{x})\,\mathrm{d}S.$$

Use the divergence theorem to express the surface integral as a volume integral and simplify the resulting terms using the equilibrium condition to show that:

$$\delta E = \int_{\mathcal{R}} \sigma_{ij}\delta u_{i,j}\mathrm{d}V.$$

Using the symmetry of the stress tensor show that this expression is equivalent to:

$$\delta E = \int_{\mathcal{R}} \sigma_{ij}\delta e_{ij}\mathrm{d}V.$$

Hence deduce:

$$\sigma_{ij}(\mathbf{x}) = \delta E/\delta e_{ij}(\mathbf{x}). \tag{2.16}$$

■

Solution Using the divergence theorem we have

$$\begin{aligned}
\delta E &= \int_{\mathcal{R}} f_i(\mathbf{x})\delta u_i(\mathbf{x})\mathrm{d}V + \int_{\mathcal{R}} \left(\sigma_{ij}(\mathbf{x})\delta u_i(\mathbf{x})\right)_{,j}\mathrm{d}V \\
&= \int_{\mathcal{R}} \underbrace{\left(f_i(\mathbf{x}) + \sigma_{ij,j}(\mathbf{x})\right)}_{=0}\delta u_i(\mathbf{x}) + \sigma_{ij}(\mathbf{x})\delta u_{i,j}(\mathbf{x})\mathrm{d}V \\
&= \int_{\mathcal{R}} \sigma_{ij}(\mathbf{x})\delta u_{i,j}(\mathbf{x})\mathrm{d}V \\
&= \int_{\mathcal{R}} \sigma_{ij}(\mathbf{x})\Bigg(\underbrace{\frac{1}{2}\left(\delta u_{i,j}(\mathbf{x}) + \delta u_{j,i}(\mathbf{x})\right)}_{=\delta e_{ij}(\mathbf{x})} + \frac{1}{2}\left(\delta u_{i,j}(\mathbf{x}) - \delta u_{j,i}(\mathbf{x})\right)\Bigg)\mathrm{d}V
\end{aligned}$$

Using the symmetry of the stress tensor we may write $\sigma_{ij}\delta u_{j,i} = \sigma_{ji}\delta u_{j,i}$. Therefore,

$$\sigma_{ij}(\delta u_{i,j} - \delta u_{j,i}) = \sigma_{ij}\delta u_{i,j} - \sigma_{ji}\delta u_{j,i} = 0.$$

Hence

$$\delta E = \int_{\mathcal{R}} \sigma_{ij}(\mathbf{x})\delta e_{ij}(\mathbf{x})\mathrm{d}V$$

Therefore, the stress at $\mathbf{x}$ is given by the following functional derivative:

$$\sigma_{ij}(\mathbf{x}) = \frac{\delta E}{\delta e_{ij}(\mathbf{x})}.$$

Note that to obtain this relationship we had to assume the body is in mechanical equilibrium.

2.8 Adiabatic and isothermal stresses

We will show in this section that the stress in eq 2.16 is the adiabatic (or isentropic) stress because it assumes there is no exchange of heat. It is not the same as a stress calculated when heat flows to maintain a constant temperature, which is called the isothermal stress. Both adiabatic and isothermal stresses may arise depending on how the strain is applied. If a material is deformed so rapidly that there is insufficient time for heat to flow the immediate stress generated will be adiabatic. But with the passage of time heat will flow, and the stress will evolve to the isothermal limit.

In a continuum model there is no distinction between isothermal and adiabatic stresses unless the continuum model also displays thermal strain, which is strain caused by a change of temperature at constant stress. In a continuum model the influence of the rate of elastic deformation is usually limited to distinguishing between adiabatic and isothermal elastic constants, which we shall return to in the next chapter.

In an atomistic model the isothermal and adiabatic stresses are not the same in general. But the definition of an isothermal stress atomistically has been controversial, with some investigators claiming the momenta of atoms contribute to the Cauchy stress, and others arguing the Cauchy stress arises only from forces acting between atoms. The controversy appears to originate from a failure to distinguish conceptually between the pressure exerted by an ideal gas on the walls of a container and the Cauchy stress.

In an ideal gas there are no forces acting between atoms except when they collide. The gas exerts a pressure on the wall of a container through the exchange of momentum when gas atoms bounce off it. Therefore, the pressure exerted by the gas on the wall of a container is determined by the distribution of atomic momenta in the gas, as in the kinetic theory of gases.

In contrast, we see in Fig 2.1 that the Cauchy stress does not arise from an exchange of momentum: it arises *exclusively* from interatomic forces acting across a plane. For example, if the plane is within the solid it is only notional because no atoms are bouncing off it and no impulses are imparted to it. The role of the plane is merely to separate the medium into two distinct parts so that the total force per unit area acting on one part due to the other can be calculated. However, *interatomic forces vary with temperature owing to their anharmonicity*[9]. It is this temperature dependence that gives rise to thermal strain. If atomic interactions were strictly harmonic there would be no thermal strain no matter how much atoms vibrate about their equilibrium positions: the solid would get hot but its shape and volume would not change because no internal stresses would be generated to drive those changes.

The combined first and second laws of thermodynamics for a solid may be expressed as follows:

$$\mathrm{d}E = T\mathrm{d}S + \int \sigma^S_{ij}(\mathbf{x})\,\mathrm{d}e_{ij}(\mathbf{x})\,\mathrm{d}V, \tag{2.17}$$

where E is the internal energy of the solid and S is the entropy. The integral $\int \sigma^S_{ij}(\mathbf{x})\,\mathrm{d}e_{ij}(\mathbf{x})\,\mathrm{d}V$ is the work done *on* the solid when an infinitesimal strain field $\mathrm{d}e_{ij}(\mathbf{x})$ is applied to the solid and there is a pre-existing stress field $\sigma^S_{ij}(\mathbf{x})$. The position dependence of the stress tensor allows for the possibility that the body is not homogeneous. The local stress in eq 2.16 is obtained from eq 2.17 by an adiabatic variation of the internal energy with respect to a local strain, that is at constant entropy. The superscript S on the stress tensor is to remind us it is obtained by a variation of the internal energy at constant entropy:

$$\sigma^S_{ij}(\mathbf{x}) = \left(\frac{\delta E}{\delta e_{ij}(\mathbf{x})}\right)_S. \tag{2.18}$$

The Helmholtz free energy of the solid is defined by $A = E - TS$, so that

$$\mathrm{d}A = -S\mathrm{d}T + \int \sigma^T_{ij}(\mathbf{x})\,\mathrm{d}e_{ij}(\mathbf{x})\,\mathrm{d}V. \tag{2.19}$$

It follows that

$$\sigma^T_{ij}(\mathbf{x}) = \left(\frac{\delta A}{\delta e_{ij}(\mathbf{x})}\right)_T. \tag{2.20}$$

[9]There is a related discussion of the temperature dependence of interatomic forces in section 3.9 of Sutton, A P and Balluffi, R W, *Interfaces in crystalline materials*, Oxford classic texts in the physical sciences, Clarendon Press: Oxford (2006). ISBN 978-0-19-921106-7. Robert Weierter Balluffi (1924-2022), US materials physicist.

In contrast to the local adiabatic stress in eq 2.18, the local isothermal stress is obtained by a variation of the Helmholtz free energy with respect to the local strain tensor at constant temperature, which is indicated by the superscript T on the stress tensor. The two stress tensors σ^S_{ij} and σ^T_{ij} are equal at absolute zero[10]. But in general they differ as the temperature becomes finite.

As an illustration of eq 2.20 consider a perfect crystal with one atom in each unit cell. The crystal is initially in equilibrium at absolute zero with a volume V_0, so that the stress tensor is zero throughout the crystal. The temperature of the crystal is then raised to T, less than the melting point, while its shape and volume V_0 are constrained to remain as they were at absolute zero. We will show that a stress is generated within the crystal if and only if atomic interactions are anharmonic. It is this stress which drives the thermal strain of the crystal when the constraints on its shape and volume are relaxed.

The homogeneity of the crystal enables eq 2.19 to be rewritten as follows:

$$\mathrm{d}A = -S\mathrm{d}T + \sigma^T_{ij}\mathrm{d}e_{ij}V_0. \tag{2.21}$$

Since the Helmholtz free energy is a state function the following Maxwell relation must apply:

$$-\left(\frac{\partial \sigma^T_{ij}}{\partial T}\right)_e = \frac{1}{V_0}\left(\frac{\partial S}{\partial e_{ij}}\right)_{T,e'}. \tag{2.22}$$

$-(\partial\sigma^T_{ij}/\partial T)_e = \beta_{ij}$ is known as the thermal stress tensor. It is evaluated with *all* strain components held constant, which is indicated by the e outside the bracket. Similarly, the thermal strain tensor is defined as $\alpha_{ij} = (\partial e_{ij}/\partial T)_\sigma$, where the σ outside the bracket indicates that *all* stress components are held constant, and usually that constant is zero. The partial derivative on the right is evaluated at constant temperature and all strain components *except* e_{ij} and e_{ji}, which is indicated by the prime on the e outside the bracket . When this Maxwell relation is applied to a crystal it shows that the temperature dependence of a stress component, at a constant crystal configuration, is determined by the dependence of the entropy of the crystal on the same component of the strain tensor when all other strain components and the temperature are held constant. This immediately tells us that anharmonicity is involved in the temperature dependence of the stress, because in a harmonic crystal the elastic stiffness matrix is independent of strain. We will now show this explicitly.

Let the sum of the atomic interaction energies at absolute zero be the potential energy E_P. Then since the crystal is in equilibrium at absolute zero we have $\partial E_P/\partial e_{kl} = 0$

[10]This statement assumes the zero point energy is included in both the internal energy E and the free energy A.

because there are no internal stresses. When the temperature is raised to T there is a free energy associated with the thermal vibrations. In the harmonic approximation the Helmholtz free energy of the crystal becomes:

$$A = E_P + k_B T \sum_n \ln\left[2\sinh\left(\frac{h\omega_n}{4\pi k_B T}\right)\right] \tag{2.23}$$

where h is Planck's constant, k_B is Boltzmann's constant, and ω_n is the angular frequency of normal mode n. The normal modes are obtained by solving the equations of motion:

$$m\ddot{u}_{Ai} = -\sum_{Bj} S_{AiBj} u_{Bj}, \tag{2.24}$$

where u_{Ai} is the displacement of atom A along the x_i direction, m is the atomic mass, and S is the elastic stiffness matrix consisting of the second derivatives of the potential energy[11] with elements $S_{AiBj} = \partial^2 E_P/\partial u_{Ai}\partial u_{Bj}$. Since we are assuming each atom is performing harmonic vibrations we write $u_{Ai} = U_{Ai}e^{i\omega t}$. Then the angular frequency of normal mode n satisfies the following equation:

$$\omega_n^2 = \frac{1}{m}\sum_{Ai}\sum_{Bj} U_{Ai}^{(n)} U_{Bj}^{(n)} S_{BjAi}, \tag{2.25}$$

where we have used the symmetry of the stiffness matrix to write $S_{AiBj} = S_{BjAi}$, the orthonormality of its eigenvectors $U^{(n)}$, and the eigenvectors of a symmetric matrix can always be expressed as real numbers.

Differentiating the free energy in eq 2.23 with respect to a homogeneous strain at constant temperature, and remembering $\partial E_P/\partial e_{kl} = 0$, we obtain:

$$\begin{aligned} \sigma_{kl}^T &= \frac{1}{2}\sum_n \coth\left(\frac{h\omega_n}{4\pi k_B T}\right)\frac{h}{4\pi\omega_n}\frac{\partial\omega_n^2}{\partial e_{kl}} \\ &= \frac{1}{2}\sum_n \coth\left(\frac{h\omega_n}{4\pi k_B T}\right)\frac{h}{4\pi\omega_n}\sum_{Ai}\sum_{Bj} U_{Ai}^{(n)} U_{Bj}^{(n)}\frac{\partial S_{BjAi}}{\partial e_{kl}} \\ &= \frac{1}{2}\sum_{Ai}\sum_{Bj}\langle U_{Ai}U_{Bj}\rangle \frac{\partial S_{BjAi}}{\partial e_{kl}}. \end{aligned} \tag{2.26}$$

[11]These second derivatives are sometimes called force constants. Since they are constant only if E_P is a sum of harmonic interactions (i.e. a sum of quadratic functions of the atomic separations) we prefer to call them stiffnesses. This terminology allows for their variation when separations between atoms change owing to the existence of higher order derivatives in E_P.

$\langle U_{Ai}U_{Bj}\rangle$ is the equal time displacement-displacement correlation function. We see that the stress is non-zero if and only if there are non-zero derivatives of the elastic stiffness matrix with respect to strain. If E_P consists of only harmonic interactions each S_{BjAi} is constant and the strain derivatives are all zero. In that case the stress is independent of temperature and $\sigma^T_{ij} = \sigma^S_{ij}$ at all temperatures. But in a real crystal atomic interactions are never purely harmonic and there are higher order derivatives of E_P. The stress is then dependent on temperature and $\sigma^T_{ij} \neq \sigma^S_{ij}$. By taking the limit $T \to 0$ in eq 2.26 it is seen that there is a contribution to σ^T_{ij} even at absolute zero which arises from the zero point motion.

When the constraints on the surface of the crystal to maintain its shape and volume are removed it undergoes a spontaneous strain to relieve the stress σ^T of eq 2.26. This is the origin of thermal strain. Once this strain has occurred the time-average separations of atoms in the crystal change slightly, and they have new mean positions, as determined by the anharmonicity of the interatomic forces. The model of atomic interactions is then called quasi-harmonic because the potential energy E_P is still expanded only to second order in the displacements of atoms from their mean positions, but the stiffnesses S_{AiBj}, which are evaluated at the new mean atomic positions, change owing to the existence of higher order derivatives in E_P.

To evaluate the isothermal Cauchy stress of eq 2.20 in a molecular dynamics simulation we calculate the time average of the expression in eq 2.5, with the vectors defining the periodic supercell and the temperature held constant. When the supercell vectors are allowed to relax thermal stresses create thermal strains changing the volume and/or shape of the supercell. But it should be remembered that at temperatures below the Debye temperature quantum effects become significant and classical molecular dynamics does not capture them.

To summarise, isothermal and adiabatic stresses differ only because interatomic forces are anharmonic. Increasing the kinetic energies of atoms enables them to experience forces that are increasingly anharmonic, but atomic momenta do not appear explicitly in the Cauchy stress at a finite temperature.

2.9 Problems

Problem 2.1 With respect to cartesian axes x_1, x_2, x_3 a stress tensor σ is represented by the matrix

$$\sigma = \begin{bmatrix} 2 & 1 & 3 \\ 1 & 0 & -1 \\ 3 & -1 & 1 \end{bmatrix}$$

(a) Show that the force per unit area on the plane $2x_1 + x_2 - 2x_3 = 0$ is $\mathbf{f} = 1/3[-1,4,3]$.

(b) Show that the normal stress on this plane is $-4/9$.
(c) Show that the shear stress on this plane is $\sqrt{1962}/27$ and that it acts along the $[-1, 40, 19]$ direction.
(d) Show that the stress tensor when referred to a new set of cartesian axes obtained by rotating the x_1 and x_3 axes by $-45°$ about the positive x_2 axis (i.e. [010]) is given by:

$$\sigma = \frac{1}{2}\begin{bmatrix} 9 & 0 & -1 \\ 0 & 0 & -2\sqrt{2} \\ -1 & -2\sqrt{2} & -3 \end{bmatrix}$$

Solution a) The normal to the plane is $\hat{\mathbf{n}} = (1/3)[2, 1, -2]$. The force per unit area acting on this plane is:

$$\mathbf{f} = \sigma\hat{\mathbf{n}} = \begin{pmatrix} 2 & 1 & 3 \\ 1 & 0 & -1 \\ 3 & -1 & 1 \end{pmatrix} \frac{1}{3}\begin{pmatrix} 2 \\ 1 \\ -2 \end{pmatrix} = \frac{1}{3}\begin{pmatrix} -1 \\ 4 \\ 3 \end{pmatrix}$$

(b) The normal stress is the component of the force per unit area normal to the plane:

$$f_n = \mathbf{f} \cdot \hat{\mathbf{n}} = \frac{1}{3}[-1, 4, 3] \cdot \frac{1}{3}[2, 1, -2]] = \frac{-4}{9}.$$

(c) The shear stress is the component of the force per unit area parallel to the plane:

$$f_p = |\mathbf{f} - f_n\hat{\mathbf{n}}| = \left|\frac{1}{3}[-1, 4, 3] + \frac{4}{27}[2, 1, -2]\right| = \left|\frac{1}{27}[-1, 40, 19]\right| = \frac{\sqrt{1962}}{27}.$$

Therefore, the shear stress has magnitude $\sqrt{1962}/27$ and it is parallel to $[-1, 40, 19]$.

(d) A rotation by $-45°$ about the positive x_2 axis is an anticlockwise rotation about the positive x_2 axis, as shown in Fig 2.6. The rotation matrix $\mathbf{R}$, with components $R_{ij} = \hat{\mathbf{e}}'_i \cdot \hat{\mathbf{e}}_j$, that accomplishes this rotation of the coordinate system is:

$$\mathbf{R} = \begin{pmatrix} 1/\sqrt{2} & 0 & 1/\sqrt{2} \\ 0 & 1 & 0 \\ -1/\sqrt{2} & 0 & 1/\sqrt{2} \end{pmatrix}$$

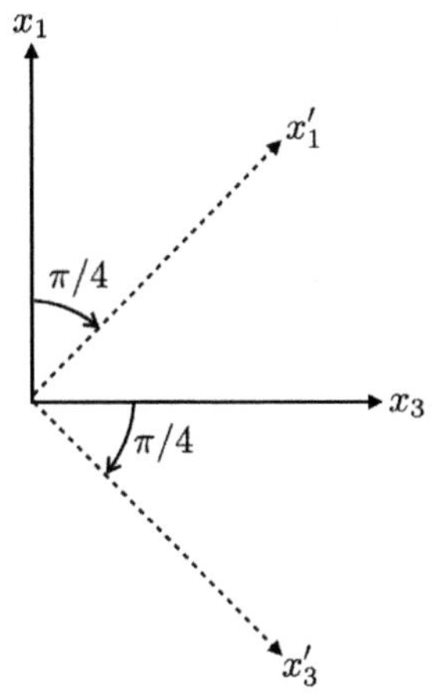

Figure 2.6: To illustrate the rotation of a right-handed Cartesian coordinate system x_1, x_2, x_3 by $-45°$ about the positive x_2 axis, to a new coordinate system x'_1, x'_2, x'_3. The positive x_2, x'_2 axis is normal to the page and points towards the reader.

The stress tensor expressed in the rotated coordinate system is then

$$\begin{aligned}\sigma &\to \mathbf{R}\sigma\mathbf{R}^T \\ &= \begin{pmatrix} 1/\sqrt{2} & 0 & 1/\sqrt{2} \\ 0 & 1 & 0 \\ -1/\sqrt{2} & 0 & 1/\sqrt{2} \end{pmatrix} \begin{pmatrix} 2 & 1 & 3 \\ 1 & 0 & -1 \\ 3 & -1 & 1 \end{pmatrix} \begin{pmatrix} 1/\sqrt{2} & 0 & -1/\sqrt{2} \\ 0 & 1 & 0 \\ 1/\sqrt{2} & 0 & 1/\sqrt{2} \end{pmatrix} \\ &= \frac{1}{2}\begin{pmatrix} 9 & 0 & -1 \\ 0 & 0 & -2\sqrt{2} \\ -1 & -2\sqrt{2} & -3 \end{pmatrix}\end{aligned}$$

Problem 2.2 Consider an atomistic model in which the total potential energy is described by a sum of pairwise interactions:

$$E = \frac{1}{2}\sum_m \sum_{n \neq m} V(X^{(mn)})$$

where $X^{(mn)}$ is the separation $\left|\mathbf{X}^{(m)} - \mathbf{X}^{(n)}\right|$ between atoms m and n and $V(X)$ is a function of the separation between pairs of atoms, e.g. a Lennard-Jones potential. The factor of one half is to correct for the double counting in the sum over m and n. Show that the atomic level stress tensor at atom k is given by:

$$\sigma_{ij}^{(k)} = \frac{1}{2\Omega^{(k)}} \sum_{n \neq k} \frac{\left(X_i^{(n)} - X_i^{(k)}\right)\left(X_j^{(n)} - X_j^{(k)}\right)}{X^{(nk)}} \left(\frac{\mathrm{d}V}{\mathrm{d}X}\right)_{X=X^{(nk)}}.$$

Solution To first order the change in the potential energy δE due to the application of a homogeneous virtual strain δe_{ij} is:

$$\begin{aligned}\delta E &= \frac{1}{2}\sum_m \sum_{n \neq m} \delta X^{(mn)} \left(\frac{\mathrm{d}V}{\mathrm{d}X}\right)_{X=X^{(mn)}} \\ &= \frac{1}{2}\sum_m \sum_{n \neq m} \frac{\left(X_i^{(m)} - X_i^{(n)}\right)\delta e_{ij}\left(X_j^{(m)} - X_j^{(n)}\right)}{X^{(mn)}} \left(\frac{\mathrm{d}V}{\mathrm{d}X}\right)_{X=X^{(mn)}},\end{aligned}$$

where we have used the result of problem 1.2 for $\delta X^{(mn)}$. This is equated to:

$$\delta E = \sum_m \sigma_{ij}^{(m)} \delta e_{ij} \Omega^{(m)},$$

where $\Omega^{(m)}$ is the volume associated with atom m. Note the absence of a factor of $1/2$ in front of the last sum. That is because the strain δe_{ij} is applied to a configuration in which there may already exist local stresses $\sigma_{ij}^{(m)}$. Equating these two expressions for δE we obtain:

$$\sigma_{ij}^{(k)} = \frac{1}{2\Omega^{(k)}} \sum_{n \neq k} \frac{\left(X_i^{(n)} - X_i^{(k)}\right)\left(X_j^{(n)} - X_j^{(k)}\right)}{X^{(nk)}} \left(\frac{\mathrm{d}V}{\mathrm{d}X}\right)_{X=X^{(nk)}}$$

Problem 2.3 Using eqs 2.14 and 2.13 show that

$$\sigma^{vM} = \frac{1}{\sqrt{2}}\sqrt{(\sigma_{11} - \sigma_{22})^2 + (\sigma_{22} - \sigma_{33})^2 + (\sigma_{33} - \sigma_{11})^2 + 6(\sigma_{12}^2 + \sigma_{23}^2 + \sigma_{31}^2)}. \quad (2.27)$$

Show that the second invariant I_2 (see eq 2.10) of the deviatoric stress tensor $(\sigma_{ij} - \delta_{ij}\sigma_{kk}/3)$ is directly proportional to the square of the von Mises stress σ^{vM}, eq 2.14.

Solution This is just algebra. $\mathrm{Tr}\sigma^2 = \sigma_{ik}\sigma_{ki}$ and $\mathrm{Tr}\sigma = \sigma_{jj}$. Therefore,

$$\begin{aligned}
3\mathrm{Tr}\,\sigma^2 - (\mathrm{Tr}\,\sigma)^2 &= 3\left(\sigma_{11}^2 + \sigma_{22}^2 + \sigma_{33}^2 + 2\sigma_{12}^2 + 2\sigma_{23}^2 + 2\sigma_{31}^2\right) \\
&\quad - \left(\sigma_{11}^2 + \sigma_{22}^2 + \sigma_{33}^2 + 2\sigma_{11}\sigma_{22} + 2\sigma_{22}\sigma_{33} + 2\sigma_{33}\sigma_{11}\right) \\
&= 2\left(\sigma_{11}^2 + \sigma_{22}^2 + \sigma_{33}^2 - (\sigma_{11}\sigma_{22} + \sigma_{22}\sigma_{33} + \sigma_{33}\sigma_{11})\right) \\
&\quad +6\left(\sigma_{12}^2 + \sigma_{23}^2 + \sigma_{31}^2\right) \\
&= (\sigma_{11} - \sigma_{22})^2 + (\sigma_{22} - \sigma_{33})^2 + (\sigma_{33} - \sigma_{11})^2 \\
&\quad +6\left(\sigma_{12}^2 + \sigma_{23}^2 + \sigma_{31}^2\right).
\end{aligned}$$

Therefore,

$$\begin{aligned}
\sigma^{vM} &= \frac{1}{\sqrt{2}}\sqrt{3\mathrm{Tr}\,\sigma^2 - (\mathrm{Tr}\,\sigma)^2} \\
&= \frac{1}{\sqrt{2}}\sqrt{(\sigma_{11} - \sigma_{22})^2 + (\sigma_{22} - \sigma_{33})^2 + (\sigma_{33} - \sigma_{11})^2 + 6(\sigma_{12}^2 + \sigma_{23}^2 + \sigma_{31}^2)}.
\end{aligned}$$

As seen in eq 2.10, the second invariant of the stress tensor σ is $(1/2)\left((\mathrm{Tr}\sigma)^2 - \mathrm{Tr}\sigma^2\right)$. The trace of the deviatoric stress is zero. The trace of the square of the deviatoric stress is:

$$\left(\sigma_{ij} - \frac{1}{3}\sigma_{kk}\delta_{ij}\right)\left(\sigma_{ji} - \frac{1}{3}\sigma_{ll}\delta_{ij}\right) = \sigma_{ij}\sigma_{ji} - \frac{2}{3}\sigma_{ii}\sigma_{kk} + \frac{1}{3}\sigma_{kk}\sigma_{ll} = \mathrm{Tr}\sigma^2 - \frac{1}{3}(\mathrm{Tr}\sigma)^2.$$

Therefore, the second invariant of the deviatoric stress is $-(1/3)(\sigma^{vM})^2$.

Problem 2.4 *This question is more advanced. It provides insight into the relationship between the equilibrium condition $\sigma_{ij,j} + f_i$ in a continuum and the Cauchy-Saint Venant definition of stress in terms of interatomic forces illustrated in Fig 2.1. It is based on the work of Noll*[12].

We return to the definition of stress in terms of interatomic potentials. Atoms are discrete objects and this poses a mathematical difficulty in applying the condition for mechanical equilibrium in a continuum embodied in eq 2.15. Noll's analysis overcomes this difficulty by replacing the discrete force that one atom exerts on another with a

[12]Walter Noll 1925-2017, US mathematician, born in Germany, see https://www.math.cmu.edu/ wn0g/. On this web-site there are many fascinating articles, including a very thoughtful short essay entitled *The role of the professor*.

continuous force density $\mathbf{f}(\mathbf{x}',\mathbf{x})$ with units of force per unit volume squared (N·m^{-6}). Then $\mathbf{f}(\mathbf{x}',\mathbf{x})\mathrm{d}V_{\mathbf{x}'}\mathrm{d}V_{\mathbf{x}}$ is the force that a volume element $\mathrm{d}V_{\mathbf{x}'}$ at $\mathbf{x}'$ exerts on a volume element $\mathrm{d}V_{\mathbf{x}}$ at $\mathbf{x}$. Notice that $\mathbf{f}(\mathbf{x}',\mathbf{x}) = -\mathbf{f}(\mathbf{x},\mathbf{x}')$. If the stress tensor at $\mathbf{x}$ is $\sigma_{ij}(\mathbf{x})$ then its divergence must equal the net force per unit volume exerted by the surrounding medium exerts at $\mathbf{x}$:

$$\sigma_{ij,j}(\mathbf{x}) = \int_V f_i(\mathbf{x}',\mathbf{x})\mathrm{d}V_{\mathbf{x}'} \tag{2.28}$$

It is important to recognise that although $\mathbf{f}(\mathbf{x}',\mathbf{x})$ depends only on $\mathbf{x}'$ and $\mathbf{x}$ this does not amount to an assumption of pairwise interactions that depend only on the separation of $\mathbf{x}'$ and $\mathbf{x}$. In other words $\mathbf{f}(\mathbf{x}',\mathbf{x})$ may depend on the *environments* of $\mathbf{x}'$ and $\mathbf{x}$. It is also *not* necessarily the case that the force $\mathbf{f}(\mathbf{x}',\mathbf{x})$ is parallel to $\mathbf{x}-\mathbf{x}'$.

Following Noll we will show that:

$$\sigma_{ij}(\mathbf{x}) = \frac{1}{2}\int_S \mathrm{d}\Omega_m \int_{r=0}^{\infty} \mathrm{d}r\, r^2 \int_{\alpha=0}^{1} \mathrm{d}\alpha\, f_i(\mathbf{x}+\alpha r\hat{\mathbf{m}}, \mathbf{x}-(1-\alpha)r\hat{\mathbf{m}})rm_j \tag{2.29}$$

where $\mathrm{d}\Omega_m$ is an element of solid angle centred on the direction $\hat{\mathbf{m}}$, and the integral over S is over the unit sphere centred at $\mathbf{x}$. The magnitude of the vector $\mathbf{r}$ is r. The points $\mathbf{x}+\alpha r\hat{\mathbf{m}}$ and $\mathbf{x}-(1-\alpha)r\hat{\mathbf{m}}$ are separated by r at all values of $0 \le \alpha \le 1$. As α varies between 0 and 1 in the third integral the forces of interaction, passing through $\mathbf{x}$ between all points separated by r along the direction $\hat{\mathbf{m}}$, are included. In the second integral r ranges over all possible separations, and in the first integral all possible directions $\hat{\mathbf{m}}$ are considered. In this way all forces of interaction that pass through $\mathbf{x}$ between points on either side of $\mathbf{x}$ contribute to the stress, in accord with the definition of stress due to Cauchy and Saint-Venant in section 2.1. Each interaction is counted twice, and this is corrected by the factor of one half. The reason for the final factor rm_j will become clear shortly.

To prove eq 2.29 we show that it satisfies eq 2.28.
Let $\mathbf{u} = \mathbf{x}+\alpha r\hat{\mathbf{m}}$ and $\mathbf{v} = \mathbf{x}-(1-\alpha)r\hat{\mathbf{m}}$.
Show that:

$$\frac{\partial f_i}{\partial x_j} = \frac{\partial f_i}{\partial u_j} + \frac{\partial f_i}{\partial v_j}.$$

Using the chain rule show that:

$$\frac{\partial f_i}{\partial \alpha} = \left(\frac{\partial f_i}{\partial u_j} + \frac{\partial f_i}{\partial v_j}\right) rm_j = \frac{\partial f_i}{\partial x_j} rm_j$$

Hence obtain eq 2.28.

We observe the following:

- The stress in eq 2.29 has the correct units, i.e. force per unit area.
- The stress at **x** is attributed to forces that act not only on **x** but also through **x**, in accord with the Cauchy-Saint Venant conception of stress at the atomic scale.
- Equation 2.29 is a continuum version of the discrete atomic-level stress tensor of eq 2.5.
- In eq 2.28 the force flux is inward towards the point **x**: it is the resultant force per unit volume the surrounding medium exerts on the point **x**. If this resultant force per unit volume is not zero mechanical equilibrium requires there is an equal and opposite force per unit volume exerted at **x** on the surrounding medium. This is the body force at **x**. The right hand side of eq 2.28 is therefore equal and opposite to the body force at **x**. Therefore, eq 2.28 is equivalent to the equation of mechanical equilibrium in a continuum, eq 2.15.

Solution Using the chain rule we have:

$$\frac{\partial f_i}{\partial x_j} = \frac{\partial f_i}{\partial u_k}\frac{\partial u_k}{\partial x_j} + \frac{\partial f_i}{\partial v_k}\frac{\partial v_k}{\partial x_j}.$$

But

$$\frac{\partial u_k}{\partial x_j} = \delta_{kj} \text{ and } \frac{\partial v_k}{\partial x_j} = \delta_{kj}.$$

Therefore,

$$\frac{\partial f_i}{\partial x_j} = \frac{\partial f_i}{\partial u_j} + \frac{\partial f_i}{\partial v_j}.$$

Using the chain rule again:

$$\frac{\partial f_i}{\partial \alpha} = \frac{\partial f_i}{\partial u_k}\frac{\partial u_k}{\partial \alpha} + \frac{\partial f_i}{\partial v_k}\frac{\partial v_k}{\partial \alpha} = \frac{\partial f_i}{\partial u_k} r m_k + \frac{\partial f_i}{\partial v_k} r m_k = r m_k \frac{\partial f_i}{\partial x_k}.$$

These relations are needed in the final part of the question. The divergence of the stress tensor in eq 2.29 is as follows:

$$\begin{aligned}
\sigma_{ij,j}(\mathbf{x}) &= \frac{1}{2}\int_{S} \mathrm{d}\Omega_m \int_{r=0}^{\infty} \mathrm{d}r\, r^2 \int_{\alpha=0}^{1} \mathrm{d}\alpha \frac{\partial}{\partial x_j} f_i(\mathbf{x}+\alpha r\hat{\mathbf{m}}, \mathbf{x}-(1-\alpha)r\hat{\mathbf{m}}) r m_j \\
&= \frac{1}{2}\int_{S} \mathrm{d}\Omega_m \int_{r=0}^{\infty} \mathrm{d}r\, r^2 \int_{\alpha=0}^{1} \mathrm{d}\alpha \frac{\partial}{\partial \alpha} f_i(\mathbf{x}+\alpha r\hat{\mathbf{m}}, \mathbf{x}-(1-\alpha)r\hat{\mathbf{m}}) \\
&= \frac{1}{2}\int_{S} \mathrm{d}\Omega_m \int_{r=0}^{\infty} \mathrm{d}r\, r^2 \left(f_i(\mathbf{x}+r\hat{\mathbf{m}}, \mathbf{x}) - f_i(\mathbf{x}, \mathbf{x}-r\hat{\mathbf{m}})\right) \\
&= \frac{1}{2}\int_{S} \mathrm{d}\Omega_m \int_{r=0}^{\infty} \mathrm{d}r\, r^2 f_i(\mathbf{x}+r\hat{\mathbf{m}}, \mathbf{x}) + \frac{1}{2}\int_{S} \mathrm{d}\Omega_m \int_{r=0}^{\infty} \mathrm{d}r\, r^2 f_i(\mathbf{x}-r\hat{\mathbf{m}}, \mathbf{x}) \\
&= \int_{S} \mathrm{d}\Omega_m \int_{r=0}^{\infty} \mathrm{d}r\, r^2 f_i(\mathbf{x}+r\hat{\mathbf{m}}, \mathbf{x}).
\end{aligned}$$

In the penultimate line each of the two terms is equal to half the total force per unit volume the surrounding medium exerts at $\mathbf{x}$, which leads to the last line.

It follows that $\sigma_{ij,j}(\mathbf{x})$ is the net force per unit volume exerted by the surrounding medium at $\mathbf{x}$, which validates Noll's expression for the stress tensor in eq 2.29. For mechanical equilibrium this net force per unit volume at $\mathbf{x}$ must either be zero or it must be balanced by a body force at $\mathbf{x}$. Conversely, if there is a body force acting at $\mathbf{x}$, then it must be equal and opposite to $\sigma_{ij,j}(\mathbf{x})$, which is the usual condition for mechanical equilibrium in continuum theory.

Noll's analysis is a revelation because it provides a connection between the Cauchy-Saint Venant atomistic conception of stress and the condition for mechanical equilibrium in a continuum involving the divergence of the stress tensor and a body force.

3. Hooke's law and elastic constants

3.1 Generalised Hooke's law: elastic constants and compliances

Robert Hooke 1635-1703 was one of the most versatile and accomplished experimentalists of all time[1]. He was appointed *Curator of Experiments* in the Royal Society in 1662, two years after the Society was formed, a post he held for 40 years until his death.

The modern form of the law which takes his name is that the stress tensor is proportional to the strain tensor, and conversely the strain tensor is proportional to the stress tensor. Since both stress and strain are second rank tensors the proportionality constants are fourth rank tensors:

$$\sigma_{ij} = c_{ijkl}\, e_{kl} \tag{3.1}$$

$$e_{ij} = s_{ijkl}\, \sigma_{kl}, \tag{3.2}$$

where c_{ijkl} are components of the elastic constant tensor, and s_{ijkl} are components of the elastic compliance tensor. By substituting eq 3.2 into eq 3.1, and noting that the stress and strain tensors are symmetric, it is found that the elastic constant and compliance tensors are related as follows:

$$c_{ijkl}\, s_{klmn} = \frac{1}{2}\left(\delta_{im}\delta_{jn} + \delta_{in}\delta_{jm}\right). \tag{3.3}$$

Exercise 3.1 Verify eq 3.3. ■

Solution Using Hooke's law first to write stress as a function of strain and then to express strain as a function of stress we have $\sigma_{ij} = c_{ijkl}e_{kl} = c_{ijkl}s_{klmn}\sigma_{mn}$. We show that this equation is satisfied provided eq 3.3 holds. Consider $\sigma_{11} = c_{11kl}s_{klmn}\sigma_{mn}$. The condition for this equation to be satisfied is $c_{11kl}s_{klmn} =$

[1] see Jardine, L, *The curious life of Robert Hooke*, Harper Collins: London (2003). ISBN 978-0007151752

$\delta_{1m}\delta_{1n}$. This condition is satisfied by eq 3.3: $c_{11kl}s_{klmn} = (1/2)(\delta_{1m}\delta_{1n} + \delta_{1n}\delta_{1m}) = \delta_{1m}\delta_{1n}$. Consider $\sigma_{12} = c_{12kl}s_{klmn}\sigma_{mn}$. This equation is satisfied by eq 3.3 because $c_{12kl}s_{klmn}\sigma_{mn} = (1/2)(\delta_{1m}\delta_{2n} + \delta_{2m}\delta_{1n})\sigma_{mn} = (1/2)(\sigma_{12} + \sigma_{21}) = \sigma_{12} = \sigma_{21}$. Therefore, eq 3.3 ensures $\sigma_{ij} = c_{ijkl}s_{klmn}\sigma_{mn}$ is true for diagonal and off-diagonal stress components. Equation 3.3 also ensures $e_{ij} = s_{ijkl}c_{klmn}e_{mn}$ is satisfied.

The direct proportionality between stress and strain is the basis of *linear* elasticity. Hooke's 'law' is an approximation because nonlinear terms become significant as the magnitude of the strain increases, but are neglected in the linear theory. Physically, stiffer bonds between atoms lead to larger elastic constants and smaller elastic compliances. Thus, in diamond, which is a very stiff insulator, c_{1111} is 1,079 GPa, while in lead, which is a soft metal, it is 49.66 GPa. In most metals the elastic constants are of order 10^{11} Pa. One GPa (gigapascal) is 10^9 Pa, and 1 Pa = 1 N·m^{-2} = 1 J·m^{-3}; 1 GPa $\approx 6.24\times10^{-3}$ eV·Å^{-3}.

3.2 Maximum number of independent elastic constants in a crystal

Since the elastic constant tensor is a fourth rank tensor it appears at first sight that there are $3^4 = 81$ independent elastic constants. If that were true the theory of elasticity would be much less useful because it would require the measurement of 81 parameters for each material. Symmetry enables the number of independent elastic constants to be reduced to a much more manageable number. The smallest number of independent elastic constants is just two, and this is the case in an elastically isotropic material like rubber. In cubic crystals there are just three independent elastic constants and in hexagonal crystals five. Since these restrictions are determined by symmetry they apply to the elastic compliance tensor in the same way as they do to the elastic constant tensor.

The largest number of independent elastic constants in any material is 21. This is the case in a triclinic crystal where there are no rotational symmetries in the point group. The first reduction is achieved by enforcing the symmetry of the stress and strain tensors: $\sigma_{ij} = \sigma_{ji}$, $e_{kl} = e_{lk}$. Therefore we must have $c_{ijkl} = c_{jikl} = c_{ijlk} = c_{jilk}$. The second reduction is more subtle and was first shown by George Green when he introduced the strain energy function, or elastic energy density.

3.2.1 The elastic energy density

In linear elasticity the elastic energy density is given by:

$$E = \frac{1}{2}\sigma_{ij}e_{ij}$$

$$= \frac{1}{2} c_{ijkl} e_{ij} e_{kl}. \tag{3.4}$$

This expression comes from integrating $dE = \sigma_{ij} de_{ij}$ with respect to strain from $e_{ij} = 0$ to the final strain and using Hooke's law to express stress in terms of strain. The factor of one half is a consequence of the linear relationship between stress and strain. The elastic energy density has units of $J \cdot m^{-3}$, the same as the elastic constants. The total elastic energy is then the integral of the elastic energy density over the volume of the body.

The elastic constants are second derivatives of the elastic energy density with respect to strain. For example, consider the terms involving the product $e_{12}e_{32}$. Since $e_{12} = e_{21}$ and $e_{32} = e_{23}$ we cannot vary e_{12} and e_{32} without also varying e_{21} and e_{23}. Therefore, there are four terms in the elastic energy density to consider: $\frac{1}{2}[c_{1232}e_{12}e_{32} + c_{2132}e_{21}e_{32} + c_{1223}e_{12}e_{23} + c_{2123}e_{21}e_{23}]$. We have already seen that $c_{1232} = c_{2132} = c_{1223} = c_{2123}$. Therefore these 4 terms amount to $2c_{1232}\, e_{12}\, e_{32}$, and

$$c_{1232} = \frac{1}{2} \frac{\partial^2 E}{\partial e_{12} \partial e_{32}}.$$

But there are also four terms in the elastic energy density involving the product $e_{32}e_{12}$. They amount to $2c_{3212}\, e_{32}\, e_{12}$, and

$$c_{3212} = \frac{1}{2} \frac{\partial^2 E}{\partial e_{32} \partial e_{12}}.$$

Green argued that the order of differentiation in these two second derivatives cannot matter. It follows that $c_{1232} = c_{3212}$. More generally,

$$c_{ijkl} = c_{klij}. \tag{3.5}$$

Thus the elastic constant tensor displays the following symmetries in all materials:

$$c_{ijkl} = c_{jikl} = c_{ijlk} = c_{jilk} = c_{klij} = c_{lkij} = c_{klji} = c_{lkji}. \tag{3.6}$$

There are six independent $\{ij\}$ combinations: 11, 22, 33, 23, 13 and 12. The symmetry embodied in eq 3.5 reduces the number of independent elastic constants from $6 \times 6 = 36$ to $6+5+4+3+2+1 = 21$. This was first demonstrated by Green in 1837[2]. Any further reduction in the number of independent elastic constants depends on the point group symmetry of the material.

[2]Green, G, Transactions of the Cambridge Philosophical Society, **7**, 1 (1839).

3.2.2 Matrix notation

Green's analysis above suggests that Hooke's law can be expressed in a convenient matrix form where each index signifies two indices in the tensor form of the equation:

$$11 \to 1,\ 22 \to 2,\ 33 \to 3,\ 23 \text{ and } 32 \to 4,\ 13 \text{ and } 31 \to 5,\ 12 \text{ and } 21 \to 6.$$

For example, $\sigma_{31} \to \sigma_5$ and $c_{1232} \to C_{64}$. Hooke's law may then be written in the following matrix form:

$$\begin{pmatrix} \sigma_1 \\ \sigma_2 \\ \sigma_3 \\ \sigma_4 \\ \sigma_5 \\ \sigma_6 \end{pmatrix} = \begin{pmatrix} C_{11} & C_{12} & C_{13} & C_{14} & C_{15} & C_{16} \\ C_{21} & C_{22} & C_{23} & C_{24} & C_{25} & C_{26} \\ C_{31} & C_{32} & C_{33} & C_{34} & C_{35} & C_{36} \\ C_{41} & C_{42} & C_{43} & C_{44} & C_{45} & C_{46} \\ C_{51} & C_{52} & C_{53} & C_{54} & C_{55} & C_{56} \\ C_{61} & C_{62} & C_{63} & C_{64} & C_{65} & C_{66} \end{pmatrix} \begin{pmatrix} e_1 \\ e_2 \\ e_3 \\ e_4 \\ e_5 \\ e_6 \end{pmatrix} \tag{3.7}$$

Now we see explicitly that there are just 21 independent components of the matrix **C**. Let us compare this equation with the tensor form of Hooke's law, eq 3.1. For example, consider σ_{11} in eq 3.1:

$$\sigma_{11} = c_{1111}e_{11} + c_{1122}e_{22} + c_{1133}e_{33} + 2c_{1123}e_{23} + 2c_{1113}e_{13} + 2c_{1112}e_{12}.$$

This has to be equivalent to σ_1 in eq 3.7:

$$\sigma_1 = C_{11}e_1 + C_{12}e_2 + C_{13}e_3 + C_{14}e_4 + C_{15}e_5 + C_{16}e_6.$$

For these two expressions to be equivalent we must have:

$$\begin{pmatrix} e_1 \\ e_2 \\ e_3 \\ e_4 \\ e_5 \\ e_6 \end{pmatrix} = \begin{pmatrix} e_{11} \\ e_{22} \\ e_{33} \\ 2e_{23} \\ 2e_{13} \\ 2e_{12} \end{pmatrix}$$

Note the factors of 2 for the off-diagonal elements of the strain tensor.

It is important to recognise that $\boldsymbol{\sigma}, \mathbf{C}$ and $\mathbf{e}$ in eq 3.7 are *not* tensors because they do not transform according to the tensor transformation law under a rotation. They are merely a convenient way of writing the tensor relationship in eq 3.1 as a matrix equation.

Physical properties of a material and physical fields are independent of the choice of cartesian coordinate system. That is why they are described mathematically by tensors because tensors obey a strict law of transformation when the cartesian coordinate system is rotated. The transformation law ensures the invariance of the physical entity to rotations of the coordinate system.

An example of a triclinic crystal for which all 21 elastic constants have been determined experimentally is low albite ($NaAlSi_3O_8$), which is a plagioclase feldspar mineral[3]. With the x_2-axis parallel to the crystal b-axis, the x_1-axis perpendicular to crystal b and c axes, and the x_3-axis completing a right-handed cartesian coordinate system, the matrix **C** is as follows:

$$\mathbf{C} = \begin{pmatrix} 69.1 & 34.0 & 30.8 & 5.1 & -2.4 & -0.9 \\ 34.0 & 183.5 & 5.5 & -3.9 & -7.7 & -5.8 \\ 30.8 & 5.5 & 179.5 & -8.7 & 7.1 & -9.8 \\ 5.1 & -3.9 & -8.7 & 24.9 & -2.4 & -7.2 \\ -2.4 & -7.7 & 7.1 & -2.4 & 26.8 & 0.5 \\ -0.9 & -5.8 & -9.8 & -7.2 & 0.5 & 33.5 \end{pmatrix} \text{GPa.}$$

3.3 Transformation of the elastic constant tensor under a rotation

We have seen that since stress and strain are second rank tensors they are related in Hooke's law by a fourth rank tensor, which is either the elastic constant tensor or the elastic compliance tensor. For a rotation of the cartesian coordinate system defined as in eq 2.7 the elastic constant tensor transforms as follows:

$$c'_{ijkl} = R_{im}R_{jn}R_{kp}R_{lq}c_{mnpq} \tag{3.8}$$

We shall use this transformation to reduce the number of independent elastic constants to less than 21 when **R** represents a rotational symmetry of the material.

We note a further useful transformation property. If (x'_1, x'_2, x'_3) and (x_1, x_2, x_3) are the coordinates of a point in the rotated and unrotated coordinate systems respectively then

$$x'_i x'_j x'_k x'_l = R_{im}R_{jn}R_{kp}R_{lq}x_m x_n x_p x_q. \tag{3.9}$$

This equation shows that the elastic constant tensor c_{mnpq} transforms under a rotation in exactly the same way as the product of coordinates $x_m x_n x_p x_q$. We shall make use of this observation extensively below.

[3]Brown, J M, Abramson, E H and Angel, R J Phys. Chem. Minerals **33**, 256-265 (2006).

3.3.1 Neumann's principle

This is a fundamental principle that relates the symmetry displayed by a physical property of a crystal to the point group symmetry of the crystal. It is arguably the most fundamental structure-property relationship in materials science. It was formulated by Neumann[4] and first appeared in print in 1885[5]. Here is how the International Union of Crystallography states the principle:

The symmetry elements of any physical property of a crystal must include all the symmetry elements of the point group of the crystal.

Neumann's principle is based on the requirement that any physical property is invariant with respect to every symmetry operation of the crystal. This means that when we transform the elastic constant tensor according to eq 3.8, with the rotation $\mathbf{R}$ being one of the symmetry rotations of the crystal, we must obtain an elastic constant tensor that is equivalent to the elastic constant tensor before the rotation was applied.

Note the word *include* in Neumann's principle: the physical property may display more symmetry than the point group of the crystal. For example, the diffusivity tensor in a cubic crystal is isotropic, so that it displays the symmetry of a sphere in 3D, i.e. the rotation group SO(3), which has infinitely more rotational symmetries than a cube or octahedron or tetrahedron. The elastic constant tensor always displays inversion symmetry because if a homogeneous stress and strain were inverted through any centre no change would be apparent in the elastic properties since a state of homogeneous stress or strain is centrosymmetrical[6]. This remains true even in a crystal that does not display inversion symmetry in its point group. All point group operations are either rotations or rotations combined with an inversion (e.g. mirror planes are 2-fold rotations followed (or preceded) by an inversion). Since the elastic constant tensor already displays inversion symmetry it is necessary to ask how it transforms under only the rotational symmetries of the point group. Eleven of the 32 point groups contain only rotational symmetries, and they are known as the proper groups, or enantiomorphous groups. They are the point groups that determine the numbers of independent elastic constants in all 32 point groups.

---

[4]Franz Ernst Neumann ForMemRS 1798-1895, German mineralogist, physicist and mathematician.

[5]Neumann, F E, *Vorlesungen über die Theorie der Elastizität der festen Körper und des Lichtäthers*, (1885), ed. O E Meyer. Leipzig, B G Teubner-Verlag

[6]Nye, J F, *Physical properties of crystals*, Oxford University Press (1957), p.21. ISBN 0-19-851165-5. John Frederick Nye FRS 1923 - 2019, British physicist.

3.4 Isotropic materials

An elastically isotropic material is one in which the elastic constants do not depend on direction in the material: they have the symmetry of SO(3). Examples of isotropic materials are rubber, glass and amorphous materials.

If c_{ijkl} is the same in all directions then $c_{ijkl} = \langle c_{ijkl} \rangle$ where $\langle \ldots \rangle$ means an average taken over all radial directions within a sphere. It follows from eq 3.9 that $\langle c_{ijkl} \rangle$ is proportional to $\langle x_i x_j x_k x_l \rangle$, where x_i are the coordinates of a point on the surface of the unit sphere, with respect to an origin at its centre. We find:

$$\begin{aligned}
\langle x_i x_j x_k x_l \rangle &= \langle x_1^2 x_2^2 \rangle \delta_{ij}\delta_{kl}(1-\delta_{jk}) \\
&\quad + \langle x_1^2 x_2^2 \rangle \delta_{ik}\delta_{jl}(1-\delta_{kj}) + \langle x_1^2 x_2^2 \rangle \delta_{il}\delta_{jk}(1-\delta_{jl}) + \langle x_1^4 \rangle \delta_{ij}\delta_{jk}\delta_{kl} \\
&= \frac{1}{15}(\delta_{ij}\delta_{kl} + \delta_{ik}\delta_{jl} + \delta_{il}\delta_{jk}) \\
&\quad + \frac{1}{15}(3\delta_{ij}\delta_{jk}\delta_{kl} - \delta_{ij}\delta_{kl}\delta_{jk} - \delta_{ik}\delta_{jl}\delta_{ij} - \delta_{il}\delta_{jk}\delta_{ik}) \\
&= \frac{1}{15}(\delta_{ij}\delta_{kl} + \delta_{ik}\delta_{jl} + \delta_{il}\delta_{jk}) \qquad (3.10)
\end{aligned}$$

where $\langle x_1^2 x_2^2 \rangle = \langle x_2^2 x_3^2 \rangle = \langle x_3^2 x_1^2 \rangle = \frac{1}{15}$ and $\langle x_1^4 \rangle = \langle x_2^4 \rangle = \langle x_3^4 \rangle = \frac{1}{5}$ have been used. It follows that an isotropic elastic constant tensor has the following form:

$$c_{ijkl} = \lambda\delta_{ij}\delta_{kl} + \mu\delta_{ik}\delta_{jl} + \mu'\delta_{il}\delta_{jk},$$

where λ, μ and μ' are constants.

Since $c_{ijij} = c_{ijji}$ (no summation) we must have $\mu = \mu'$. Therefore there are just two independent elastic constants in an isotropic material:

$$c_{ijkl} = \lambda\delta_{ij}\delta_{kl} + \mu\left(\delta_{ik}\delta_{jl} + \delta_{il}\delta_{jk}\right). \qquad (3.11)$$

λ is called Lamé's first constant[7], and μ is sometimes called Lamé's second constant but more commonly the shear modulus.

Exercise 3.2 Derive eq 3.10 in detail. ■

[7]named after Gabriel Lamé 1795-1870, French mathematician, engineer and physicist.

Solution The only non-zero terms are $< x_1^2 x_2^2 >=< x_2^2 x_3^2 >=< x_3^2 x_1^2 >$ and $< x_1^4 >=$ $< x_2^4 >=< x_3^4 >$. In exercise 2.4(c) it was shown that $< x_3^2 x_1^2 >= 1/15$. Consider $< x_3^4 >$:

$$
\begin{aligned}
< x_3^4 > &= \frac{1}{4\pi}\int_0^{2\pi} \mathrm{d}\phi \int_0^{\pi} \mathrm{d}\theta \sin\theta \cos^4\theta \\
&= \frac{1}{4\pi} 2\pi \int_{-1}^{1} \mathrm{d}(\cos\theta) \cos^4\theta \\
&= \frac{1}{5}.
\end{aligned}
$$

Therefore,

$$
\begin{aligned}
< x_i x_j x_k x_l > &= \frac{1}{15}\left[\delta_{ij}\delta_{kl}(1-\delta_{jk}) + \delta_{ik}\delta_{jl}(1-\delta_{ij}) + \delta_{il}\delta_{jk}(1-\delta_{ik})\right] \\
&\quad + \frac{1}{5}\delta_{ij}\delta_{jk}\delta_{kl} \\
&= \frac{1}{15}\left(\delta_{ij}\delta_{kl} + \delta_{ik}\delta_{jl} + \delta_{il}\delta_{jk}\right).
\end{aligned}
$$

Exercise 3.3 Verify that $c_{ijkl} = \lambda\delta_{ij}\delta_{kl} + \mu\left(\delta_{ik}\delta_{jl} + \delta_{il}\delta_{jk}\right)$ is invariant when it substituted into eq 3.8 for any rotation **R**.

■

Solution The key to this question is that the rows and columns of a rotation matrix are orthonormal vectors. Under a rotation **R** of the coordinate system, components of the elastic constant tensor change according to the usual transformation law of fourth rank tensors:

$$
c'_{ijkl} = R_{im} R_{jn} R_{kp} R_{lq} c_{mnpq}
$$

In the isotropic elastic case $c_{mnpq} = \lambda\delta_{mn}\delta_{pq} + \mu(\delta_{mp}\delta_{nq} + \delta_{mq}\delta_{np})$. Therefore,

$$\begin{aligned} c'_{ijkl} &= R_{im}R_{jn}R_{kp}R_{lq}\left(\lambda\delta_{mn}\delta_{pq} + \mu(\delta_{mp}\delta_{nq} + \delta_{mq}\delta_{np})\right) \\ &= \lambda R_{im}R_{jm}R_{kp}R_{lp} + \mu(R_{im}R_{km}R_{jn}R_{ln} + R_{im}R_{lm}R_{jn}R_{kn}) \\ &= \lambda\delta_{ij}\delta_{kl} + \mu(\delta_{ik}\delta_{jl} + \delta_{il}\delta_{jk}) \\ &= c_{ijkl} \end{aligned}$$

Therefore the isotropic elastic constant tensor is invariant with respect to a rotation of the coordinate system.

When we substitute the isotropic elastic constants, eq 3.11, into Hooke's law, eq 3.1, we obtain the following equations:

$$\begin{aligned} \sigma_{11} &= 2\mu e_{11} + \lambda(e_{11} + e_{22} + e_{33}) \\ \sigma_{22} &= 2\mu e_{22} + \lambda(e_{11} + e_{22} + e_{33}) \\ \sigma_{33} &= 2\mu e_{33} + \lambda(e_{11} + e_{22} + e_{33}) \\ \sigma_{23} &= 2\mu e_{23} \\ \sigma_{13} &= 2\mu e_{13} \\ \sigma_{12} &= 2\mu e_{12}, \end{aligned} \tag{3.12}$$

where we recognise $e_{kk} = e_{11} + e_{22} + e_{33}$ as the dilation $\Delta V/V$. Thus $C_{11} = 2\mu + \lambda$, $C_{12} = \lambda$ and $C_{44} = \mu$. Therefore in an isotropic material we have

$$A = \frac{2C_{44}}{C_{11} - C_{12}} = 1 \tag{3.13}$$

This is called the anisotropy ratio, about which we will say more in the context of cubic crystals where $A \neq 1$.

To relate λ and μ to Young's modulus Y consider a tensile test where a sample is loaded in tension along the x_3 axis and no constraints or loads are applied along x_1 and x_2. There are no shear strains and eqs 3.12 become:

$$\begin{aligned} 0 &= 2\mu e_{11} + \lambda(e_{11} + e_{22} + e_{33}) \\ 0 &= 2\mu e_{22} + \lambda(e_{11} + e_{22} + e_{33}) \\ \sigma_{33} &= 2\mu e_{33} + \lambda(e_{11} + e_{22} + e_{33}). \end{aligned}$$

Solving these equations for e_{11}, e_{22} and e_{33} we find $e_{11} = e_{22} = -\lambda\sigma_{33}/\{2\mu(2\mu + 3\lambda)\}$

and $e_{33} = (\mu+\lambda)\sigma_{33}/\{\mu(2\mu+3\lambda)\}$. From these relations we deduce the following:

$$Y = \frac{\mu(2\mu+3\lambda)}{\mu+\lambda} \tag{3.14}$$

$$\nu = -\frac{e_{11}}{e_{33}} = \frac{\lambda}{2(\mu+\lambda)} \tag{3.15}$$

$$\mu = \frac{Y}{2(1+\nu)} \tag{3.16}$$

$$\lambda = \frac{2\mu\nu}{1-2\nu}, \tag{3.17}$$

where ν is called Poisson's[8] ratio. Poisson's ratio is the ratio of the contraction in the lateral x_1 and x_2 directions to the tensile strain along x_3. Most materials contract along the lateral directions when they are stretched, and expand along the lateral directions when they are compressed. Materials that do the opposite are called *auxetic*, and they have negative Poisson's ratios[9]. In terms of the Young's modulus and Poisson's ratio the strains may be expressed in terms of the stresses as follows:

$$\begin{aligned}
e_{11} &= \frac{\sigma_{11}}{Y} - \frac{\nu\sigma_{22}}{Y} - \frac{\nu\sigma_{33}}{Y} \\
e_{22} &= \frac{\sigma_{22}}{Y} - \frac{\nu\sigma_{11}}{Y} - \frac{\nu\sigma_{33}}{Y} \\
e_{33} &= \frac{\sigma_{33}}{Y} - \frac{\nu\sigma_{11}}{Y} - \frac{\nu\sigma_{22}}{Y} \\
e_{23} &= \frac{1+\nu}{Y}\sigma_{23} = \frac{\sigma_{23}}{2\mu} \\
e_{13} &= \frac{1+\nu}{Y}\sigma_{13} = \frac{\sigma_{13}}{2\mu} \\
e_{12} &= \frac{1+\nu}{Y}\sigma_{12} = \frac{\sigma_{12}}{2\mu}.
\end{aligned} \tag{3.18}$$

Another commonly used elastic constant is the bulk modulus, B. This relates the

[8]Siméon Denis Poisson 1781-1840. French mathematician, engineer and physicist.

[9]Most auxetic materials are cellular solids such as honeycombs and foams. But they also occur naturally, e.g. human artery walls and skin, and a form of silica (SiO_2) known as α-cristobalite. The Poisson ratio of cork is almost zero, which makes it ideal for sealing wine in bottles.

hydrostatic pressure $p = -\mathrm{Tr}\sigma/3$ to the dilation $\Delta V/V = \mathrm{Tr}e$:

$$p = -B\frac{\Delta V}{V} = -B\,\mathrm{Tr}e. \tag{3.19}$$

Using eqs 3.12 it is deduced that:

$$B = \frac{1}{3}(2\mu + 3\lambda) = \frac{2\mu(1+\nu)}{3(1-2\nu)} \tag{3.20}$$

It is stressed that in isotropic elasticity only two of the Young's modulus Y, the shear modulus μ, Poisson's ratio ν, the bulk modulus B and Lamé's first constant λ are independent.

Exercise 3.4 (a) Why must the value of ν always be between -1 and $\frac{1}{2}$?
(b) What do these two limits correspond to physically? ▪

Solution (a) It is because in isotropic elasticity the bulk modulus is $B = 2\mu(1+\nu)/\{3(1-2\nu)\}$. Therefore as $\nu \longrightarrow -1$ we find $B \longrightarrow 0$, and as $\nu \longrightarrow 1/2$ we find $B \longrightarrow \infty$. Thus, the range $-1 \leq \nu \leq 1/2$ encompasses the entire range of possible values of the bulk modulus.

(b)When $\nu \longrightarrow -1$ the bulk modulus tends to 0. This is the limit of infinite compressibility. When $\nu \longrightarrow 1/2$ the bulk modulus tends to infinity. This is the limit of zero compressibility.

Comment

These limiting values of Poisson's ratio apply only when the material is elastically isotropic.

Exercise 3.5 Show that in an isotropic medium Hooke's law may be expressed in the following equivalent ways:

$$\begin{aligned} \sigma_{ij} &= 2\mu e_{ij}^{(d)} + \delta_{ij} B e_{kk} \\ e_{ij} &= \frac{1}{2\mu}\sigma_{ij}^{(d)} + \delta_{ij}\frac{\sigma_{kk}}{9B} \end{aligned}$$

where $e_{ij}^{(d)}$ and $\sigma_{ij}^{(d)}$ are the deviatoric strain and stress tensors. By introducing the deviatoric stress and strain tensors we see a clear separation between shear and dilational contributions, involving the shear modulus and bulk modulus respectively, to the total stress and strain tensors. ■

Solution In isotropic elasticity Hooke's law is:

$$
\begin{aligned}
\sigma_{ij} &= 2\mu e_{ij} + \lambda \delta_{ij} e_{mm} \\
&= 2\mu \left(e_{ij}^{(d)} + \frac{1}{3} e_{nn} \delta_{ij} \right) + \lambda \delta_{ij} e_{mm} \\
&= 2\mu e_{ij}^{(d)} + \delta_{ij} e_{kk} \left(\lambda + \frac{2\mu}{3} \right) \\
&= 2\mu e_{ij}^{(d)} + B \delta_{ij} e_{mm}
\end{aligned}
$$

Starting from Hooke's law again we have:

$$
\sigma_{ij} = \sigma_{ij}^{(d)} + \delta_{ij} \frac{\sigma_{pp}}{3} = 2\mu e_{ij} + \lambda \delta_{ij} \frac{\sigma_{pp}}{3B}.
$$

where we have used $e_{mm} = \sigma_{pp}/(3B)$. Therefore,

$$
2\mu e_{ij} = \sigma_{ij}^{(d)} + \frac{1}{3} \delta_{ij} \left(1 - \frac{\lambda}{B} \right) \sigma_{pp}
$$

Since $1 - (\lambda/B) = 2\mu/(3B)$ we have:

$$
e_{ij} = \frac{\sigma_{ij}^{(d)}}{2\mu} + \delta_{ij} \frac{\sigma_{pp}}{9B}.
$$

Exercise 3.6 (a) By orienting the axes along the eigenvectors of the stress tensor show that the elastic energy density in an isotropic medium may be expressed as follows:

$$
E = \frac{1}{2Y} \left(s_1^2 + s_2^2 + s_3^2 - 2\nu \left(s_1 s_2 + s_2 s_3 + s_3 s_1 \right) \right),
$$

where s_i are the eigenvalues of the stress tensor.

(b) Show that the elastic energy density may be expressed as:

$$E = \frac{I_1^2}{18B} + \frac{I_1^2 - 3I_2}{6\mu} = \frac{p^2}{2B} + \frac{\left(\sigma^{vM}\right)^2}{6\mu}$$

where I_1 and I_2 are the first and second invariants of the stress tensor, p is the hydrostatic pressure and σ^{vM} is the von Mises shear stress given by eq 2.14. We see here that the elastic energy density in an isotropic medium also separates into dilational and shear contributions. ▪

Solution (a) In the coordinate system defined by the eigenvectors of the stress (and strain) tensor we have:

$$\begin{aligned} e_1 &= [s_1 - \nu(s_2 + s_3)]/Y \\ e_2 &= [s_2 - \nu(s_1 + s_3)]/Y \\ e_3 &= [s_3 - \nu(s_1 + s_2)]/Y \end{aligned}$$

The elastic energy density is as follows:

$$E = \frac{1}{2}(s_1 e_1 + s_2 e_2 + s_3 e_3) = \frac{1}{2Y}\left[s_1^2 + s_2^2 + s_3^2 - 2\nu(s_1 s_2 + s_2 s_3 + s_3 s_1)\right]$$

(b) First we rewrite the elastic energy density in terms of the stress invariants I_1 and I_2. The first invariant is $I_1 = \mathrm{Tr}\sigma = s_1 + s_2 + s_3$ and the second is $I_2 = s_1 s_2 + s_2 s_3 + s_3 s_1$. Therefore $s_1^2 + s_2^2 + s_3^2 = I_1^2 - 2I_2$. The elastic energy density is therefore,

$$E = \frac{1}{2Y}[I_1^2 - 2(1+\nu)I_2] = \frac{I_1^2}{2Y} - \frac{(1+\nu)I_2}{Y}$$

Using $Y = 9\mu B/(\mu + 3B)$ and $(1+\nu)/Y = 1/(2\mu)$ we obtain:

$$E = \left(\frac{3B+\mu}{18\mu B}\right) I_1^2 - \frac{I2}{2\mu} = \frac{I_1^2}{18B} + \frac{I_1^2 - 3I_2}{6\mu}.$$

Since $p = -I_1/3$ and $\left(\sigma^{vM}\right)^2 = I_1^2 - 3I_2$ (as shown in problem 2.3) we have:

$$E = \frac{p^2}{2B} + \frac{\left(\sigma^{vM}\right)^2}{6\mu}.$$

3.5 Anisotropic materials

There are no crystalline materials that are exactly elastically isotropic, but tungsten is almost isotropic. In this section we will illustrate how point group symmetry is used to reduce the number of independent elastic constants from 21 in a crystal. As an example we will show there are 3 independent elastic constants in cubic crystals.

3.5.1 Cubic crystals

In this section we will make use of the observation in eq 3.9 that the elastic constant tensor c_{mnpq} transforms under a rotation in the same way as the product of coordinates $x_m x_n x_p x_q$.

Cubic crystals are defined by four 3-fold rotational symmetry axes along $\langle 111 \rangle$ directions. These rotational symmetries generate a further three 2-fold rotation axes along $\langle 100 \rangle$. In this way we obtain the cubic point group *23* in Hermann-Mauguin notation or T in Schönflies notation.

Rotating the coordinate axes by π about [100] results in $x_1' = x_1, x_2' = -x_2, x_3' = -x_3$. Therefore the following eight elastic constants must be zero because they are equal to their own negative under this rotation: $c_{1112} = C_{16}$, $c_{1113} = C_{15}$, $c_{2212} = C_{26}$, $c_{2213} = C_{25}$, $c_{3312} = C_{36}$, $c_{3313} = C_{35}$, $c_{2312} = C_{46}$, $c_{2313} = C_{45}$, where we are specifying the 4-index tensor component and its corresponding element of the 6×6 matrix in eq 3.7. Similarly rotating the coordinate axes by π about [010] results in $x_1' = -x_1, x_2' = x_2, x_3' = -x_3$, and four additional elastic constants are found to be zero: $c_{1123} = C_{14}$, $c_{2223} = C_{24}$, $c_{3323} = C_{34}$, $c_{1312} = C_{56}$.

No additional information is obtained by rotating by π about [001]. Rotating by $2\pi/3$ anti-clockwise about [111] results in $x_1' \to x_2, x_2' \to x_3, x_3' \to x_1$. Therefore the following elastic constants must be equal: $c_{1111} = c_{2222} = c_{3333}$; $c_{1122} = c_{2233} = c_{3311}$; $c_{2323} = c_{3131} = c_{1212}$, which in matrix notation are $C_{11} = C_{22} = C_{33}$; $C_{12} = C_{23} = C_{31}$; $C_{44} = C_{55} = C_{66}$. No additional information is obtained by invoking any of the other symmetry operations.

The conclusion is that there are three independent elastic constants in a cubic crystal: C_{11}, C_{12}, C_{44}:

$$\mathbf{C} = \begin{pmatrix} C_{11} & C_{12} & C_{12} & 0 & 0 & 0 \\ C_{12} & C_{11} & C_{12} & 0 & 0 & 0 \\ C_{12} & C_{12} & C_{11} & 0 & 0 & 0 \\ 0 & 0 & 0 & C_{44} & 0 & 0 \\ 0 & 0 & 0 & 0 & C_{44} & 0 \\ 0 & 0 & 0 & 0 & 0 & C_{44} \end{pmatrix}. \tag{3.21}$$

This conclusion remains the same with the cubic point group *432* in Hermann-Mauguin notation or O in Schönflies notation. Therefore, all cubic point groups have an elastic constant matrix of the same form as that shown in eq 3.21. This is conveniently summarised in the following formula for the elastic constants in cubic crystals:

$$c_{ijkl} = C_{12}\delta_{ij}\delta_{kl} + C_{44}\left(\delta_{ik}\delta_{jl} + \delta_{il}\delta_{jk}\right) + (C_{11} - C_{12} - 2C_{44})\,\delta_{ij}\delta_{jk}\delta_{kl}. \quad (3.22)$$

The elements of the elastic constant matrix with value zero in a cubic crystal are the same as those in an isotropic medium. The only difference between the cubic and isotropic cases is that the anisotropy ratio, eq 3.13, in a cubic crystal is not unity. Let us look at this more closely. A pure shear strain e_{23} in a cubic crystal is on (010) and (001) planes, and it is created by the shear stress $\sigma_{23} = 2C_{44}e_{23}$. Therefore, C_{44} measures the resistance to shear on {100} planes in the cubic crystal. If we rotate the coordinate system by $\pi/4$ about [100] then e'_{23} is a pure shear on (011) and $(01\bar{1})$ planes[10]. After transforming the elastic constant tensor it is found that $c'_{2323} = (C_{11} - C_{12})/2$. Therefore, $(C_{11} - C_{12})/2$ measures the resistance to shear on {110} planes in the cubic crystal, and it is called C' (pronounced "C prime"). It follows that the anisotropy ratio in a cubic crystal is the ratio of the shear resistance on {100} planes to the shear resistance on {110} planes. In an isotropic crystal the resistances are the same. The anisotropy ratio can have a strong influence on the elastic fields of defects and modes of plastic deformation in cubic crystals.

The reduction in the number of independent elastic constants in other crystal systems is considered in problems 3.7, 3.8 and 3.9.

3.5.2 The directional dependence of the elastic constants in anisotropic media

In an anisotropic medium the elastic constants vary with direction. For a chosen elastic constant this variation can be depicted graphically by plotting a surface $r(\theta, \phi)$ where r is the magnitude of the elastic constant along the direction (θ, ϕ) in spherical coordinates. In an isotropic medium this surface is a sphere.

As a first example consider the variation of C_{11} with direction in a cubic crystal. Orienting the cartesian axes along the sides of the cubic unit cell the variation of $C_{11} = c_{1111}$ as the coordinate system is rotated is given by eq 3.8:

$$c'_{1111} = R_{1i}R_{1j}R_{1k}R_{1l}c_{ijkl}. \quad (3.23)$$

Let $R_{1i} = \hat{\eta}_i$. This vector is parallel to the x'_1 axis. Equation 3.23 provides the value of C_{11} along the direction $\hat{\eta}$ with respect to the cube axes. We obtain:

$$C'_{11} = c'_{1111} = C_{11} + 2(C_{11} - C_{12})\left(\frac{2C_{44}}{C_{11} - C_{12}} - 1\right)\left(\hat{\eta}_1^2\hat{\eta}_2^2 + \hat{\eta}_2^2\hat{\eta}_3^2 + \hat{\eta}_3^2\hat{\eta}_1^2\right), \quad (3.24)$$

[10] A line over a number signifies the negative of the number. Thus $\bar{1}$, which is pronounced "bar one", means -1. This is standard crystallographic notation.

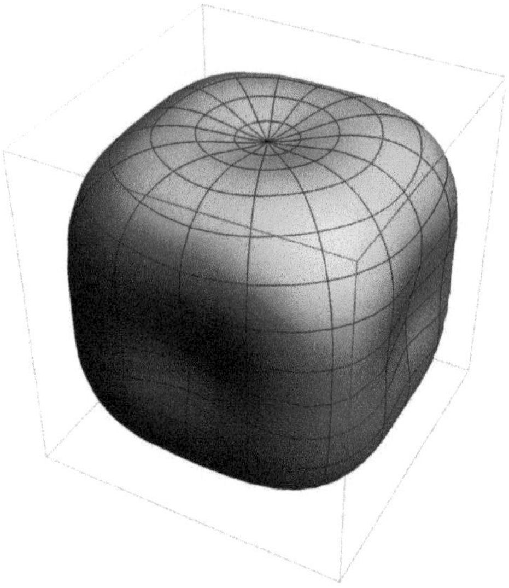

Figure 3.1: Polar plot of C'_{11} given by eq 3.24 for copper inside a bounding cube aligned with the $\langle 100 \rangle$ directions of the fcc crystal.

where we see that the directional dependence is proportional to the deviation of the anisotropy ratio A (eq 3.13) from unity. For $A > 1$ the maximum value of C'_{11} is along $\langle 111 \rangle$ directions. A plot of C'_{11} is shown in Fig 3.1 for copper, where $C_{11} = 168.4$ GPa, $C_{12} = 121.4$ GPa, $C_{44} = 75.4$ GPa and the anisotropy ratio is $A = 3.21$. The average value of C'_{11}, where the averaging is over all directions, is $C_{11} + \frac{2}{5}(A-1)(C_{11} - C_{12})$, and in copper this is 210 GPa.

The variation of the shear elastic constant C_{44} with direction is more complicated because it depends on two directions: the plane normal and the direction of shear. Invoking the transformation law:

$$c'_{1212} = R_{1i}R_{2j}R_{1k}R_{2l}c_{ijkl} = \hat{\eta}_i\hat{\xi}_j\hat{\eta}_k\hat{\xi}_l c_{ijkl} \tag{3.25}$$

where $\hat{\xi}_j = R_{2j}$ is any unit vector perpendicular to $\hat{\eta}$. In this equation $\hat{\eta}$ may be interpreted as the normal to the plane where c'_{1212} is evaluated and $\hat{\xi}$ as the direction of shear in that plane. Thus, c'_{1212} is a function of 3 independent variables. After some algebraic manipulations we obtain:

$$C'_{44} = C'_{66} = c'_{1212} = C_{44} - (C_{11} - C_{12})\left(\frac{2C_{44}}{C_{11} - C_{12}} - 1\right)\left(\hat{\eta}_1^2\hat{\xi}_1^2 + \hat{\eta}_2^2\hat{\xi}_2^2 + \hat{\eta}_3^2\hat{\xi}_3^2\right). \tag{3.26}$$

For example, when $\hat{\eta} = [110]/\sqrt{2}$ and $\hat{\xi} = [1\bar{1}0]/\sqrt{2}$ we obtain $C'_{44} = \frac{1}{2}(C_{11} - C_{12})$, which is C'. For completeness we also state the variation of the elastic constant C_{12}

with two orthonormal directions $\hat{\eta}$ and $\hat{\xi}$:

$$C'_{12} = c'_{1122} = C_{12} - (C_{11} - C_{12})\left(\frac{2C_{44}}{C_{11} - C_{12}} - 1\right)\left(\hat{\eta}_1^2\hat{\xi}_1^2 + \hat{\eta}_2^2\hat{\xi}_2^2 + \hat{\eta}_3^2\hat{\xi}_3^2\right). \tag{3.27}$$

3.6 Further restrictions on the elastic constants

For structural stability the elastic energy density must be positive definite. Otherwise the material will spontaneously distort to a lower energy structure. The elastic energy density may be written in matrix notation as $\frac{1}{2}\sigma_i e_i = \frac{1}{2}C_{ij}e_i e_j$. For the quadratic form $C_{ij}e_i e_j$ to be positive definite all 6 of the leading principal minors of the 6×6 matrix **C** must be positive definite.

In an isotropic medium this condition leads to $\lambda + \frac{2}{3}\mu > 0$ and $\mu > 0$. Since the bulk modulus is given by $B = \lambda + \frac{2}{3}\mu$ (see eq 3.20) the first condition is equivalent to requiring the bulk modulus is positive. In Exercise 3.6 it was also shown that the elastic energy density is positive definite provided $B > 0$ and $\mu > 0$.

In a cubic crystal the elastic energy density is positive definite provided $C_{11} - C_{12} > 0$, $C_{44} > 0$ and $C_{11} + 2C_{12} > 0$. Thus, the elastic energy density is positive definite provided the two shear elastic constants and the bulk modulus are all positive. C_{12} has to lie between $-C_{11}/2$ and C_{11}.

3.7 Elastic constants and atomic interactions

Elastic constants may be calculated for a crystal if we have a description of atomic interactions. Consider a crystal with one atom at each lattice site. Since all atoms are at centres of inversion there is no net force acting on any atom. Let $u_i^{(n)}$ be a small arbitrary displacement of atom n. Then the change in the energy of the crystal to second order in the displacements is:

$$E = \sum_n \frac{\partial E}{\partial u_i^{(n)}} u_i^{(n)} + \frac{1}{2}\sum_n \sum_p \frac{\partial^2 E}{\partial u_i^{(n)} \partial u_j^{(p)}} u_i^{(n)} u_j^{(p)} \tag{3.28}$$

This is the usual harmonic expansion of the energy of the crystal, where the derivatives are evaluated in the perfect crystal configuration. The first term on the right is zero because the forces on all atoms are zero at equilibrium.

If $u_i^{(n)} = t_i$ for all n, where t_i is a small constant vector, then the energy E in eq 3.28 should be invariant, because the crystal has undergone a rigid body translation. This is achieved if the following equation is satisfied:

$$\frac{\partial^2 E}{\partial u_i^{(n)} \partial u_j^{(n)}} = -\sum_{p \neq n} \frac{\partial^2 E}{\partial u_i^{(n)} \partial u_j^{(p)}},$$

for each site n. When this is substituted into eq 3.28 the second order term becomes:

$$\frac{1}{2}\sum_n\sum_p \frac{\partial^2 E}{\partial u_i^{(n)}\partial u_j^{(p)}} u_i^{(n)} u_j^{(p)} = \frac{1}{2}\sum_n\sum_{p\neq n} \frac{\partial^2 E}{\partial u_i^{(n)}\partial u_j^{(p)}} u_i^{(n)} (u_j^{(p)} - u_j^{(n)})$$

$$= -\frac{1}{2}\cdot\frac{1}{2}\cdot\sum_n\sum_{p\neq n} \frac{\partial^2 E}{\partial u_i^{(n)}\partial u_j^{(p)}} (u_i^{(p)} - u_i^{(n)})(u_j^{(p)} - u_j^{(n)}) \tag{3.29}$$

Let the displacements be created by a small homogeneous strain e_{kl}, such that $u_i^{(p)} - u_i^{(n)} = e_{ik}\left(X_k^{(p)} - X_k^{(n)}\right)$, where $\mathbf{X}^{(p)}$ is the position of atom p in the unstrained crystal. All atoms remain at centres of inversion during this operation and therefore the net force on any atom remains zero. Since the strain is homogeneous and since all atoms in the crystal remain equivalent we need to consider the change in the energy of just atom n, which we call δE_n:

$$\delta E_n = -\frac{1}{4}\sum_{p\neq n} e_{ik}\left(X_k^{(p)} - X_k^{(n)}\right) S_{ij}^{(np)} e_{jl}\left(X_l^{(p)} - X_l^{(n)}\right), \tag{3.30}$$

where $S_{ij}^{(np)} = \partial^2 E/\partial u_i^{(n)}\partial u_j^{(p)}$. We may rewrite this equation in terms of the elastic constants:

$$\delta E_n = \frac{1}{2}\Omega c_{ikjl}\, e_{ik}\, e_{jl}$$

where Ω is the volume of a primitive unit cell of the crystal. Comparing this with eq 3.30 we obtain:

$$c_{ikjl} = -\frac{1}{2\Omega}\sum_{p\neq n}\left(X_k^{(p)} - X_k^{(n)}\right) S_{ij}^{(np)}\left(X_l^{(p)} - X_l^{(n)}\right). \tag{3.31}$$

It is evident that this expression satisfies the symmetry $c_{ikjl} = c_{jlik}$. To satisfy the other symmetries of eqs 3.6 we set $c_{ikjl} = \frac{1}{4}\left(c_{ikjl} + c_{kijl} + c_{iklj} + c_{kilj}\right)$:

$$\begin{aligned} c_{ikjl} \quad = \quad & -\frac{1}{8\Omega}\left\{\sum_{p\neq n}\left(X_k^{(p)} - X_k^{(n)}\right) S_{ij}^{(np)}\left(X_l^{(p)} - X_l^{(n)}\right)\right. \\ & + \sum_{p\neq n}\left(X_i^{(p)} - X_i^{(n)}\right) S_{kj}^{(np)}\left(X_l^{(p)} - X_l^{(n)}\right) \\ & + \sum_{p\neq n}\left(X_k^{(p)} - X_k^{(n)}\right) S_{il}^{(np)}\left(X_j^{(p)} - X_j^{(n)}\right) \\ & \left. + \sum_{p\neq n}\left(X_i^{(p)} - X_i^{(n)}\right) S_{kl}^{(np)}\left(X_j^{(p)} - X_j^{(n)}\right)\right\}. \end{aligned}$$

If there is more than one atom associated with each lattice site, those atoms not on lattice sites may undergo small displacements in addition to those prescribed by a homogeneous strain. These additional displacements are sometimes called the *internal strain*. Although the strain is still imposed by displacing atoms at lattice sites, atoms between lattice sites will experience net forces as a result of the strain. Relaxation of those forces reduces the energy of the homogeneously strained crystal, and therefore it affects the calculated elastic constants.

Problem 3.1 considers the three elastic constants in a face-centred cubic crystal in which bonding between nearest neighbours is described by linear springs. It is found that $C_{12} = C_{44}$. This is an example of a *Cauchy relation* between elastic constants. In problem 3.3 this result is found in all models of cubic crystals at their equilibrium volumes if atomic interactions are described by potentials between pairs of atoms.

3.8 Isothermal and adiabatic elastic moduli

So far the elastic moduli[11] we have considered are those obtained by an adiabatic variation of the internal energy. In this section we follow the treatment[12] by Wallace[13] to show the relationships between elastic moduli obtained adiabatically and isothermally. Insightful relationships for the temperature dependencies of the isothermal elastic constant tensor and the isothermal elastic compliance tensor are also derived. Whether adiabatic or isothermal moduli should be used in any given thermoelastic process depends on the rate of elastic deformation. For example, ultrasonic pulse experiments measure adiabatic elastic moduli, whereas isothermal elastic moduli are measured in tensile and torsion tests at room temperature and small strain rates.

In a homogeneous crystal the adiabatic and isothermal elastic moduli differ because there is always a finite thermal strain or stress tensor owing to the anharmonicity of atomic interactions. However the difference between them tends to zero as the temperature approaches absolute zero[14]. The elastic moduli normally assigned to a continuum are either adiabatic or isothermal. But if a continuum model is required to display either adiabatic or isothermal moduli over a range of thermoelastic conditions[15] the continuum must also be assigned a thermal stress tensor and/or a thermal strain tensor, and specific heats at constant strain and/or constant stress spanning the range of temperatures required.

[11] i.e. elastic constants and elastic compliances

[12] Wallace, D C, *Thermodynamics of crystals*, section 2, John Wiley & Sons Inc.: New York (1972). ISBN 9780471918554.

[13] Duane C Wallace 1931-, US materials physicist.

[14] This is a requirement of the third law of thermodynamics.

[15] e.g. in a simulation of a shock impact

The combined first and second laws of thermodynamics for a solid was given in eq 2.17. For simplicity we consider a crystal of just one atomic species in which there is one atom associated with each lattice site[16]. The homogeneity of the crystal enables the combined first and second laws to be written as follows:

$$dE = TdS + \sigma_{ij}de_{ij}V_0, \tag{3.32}$$

where V_0 is the volume of the crystal in its current state of strain.

The Gibbs free energy is defined as $G = E - TS - \sigma_{ij}e_{ij}V_0$. Then we have:

$$dG = -SdT - e_{ij}d\sigma_{ij}V_0 \tag{3.33}$$

Since dG is an exact differential the following Maxwell relation holds:

$$\frac{1}{V_0}\left(\frac{\partial S}{\partial \sigma_{ij}}\right)_{T,\sigma'} = \left(\frac{\partial e_{ij}}{\partial T}\right)_{\sigma} = \alpha_{ij}, \tag{3.34}$$

where α_{ij} is the thermal strain tensor[17]. We will use this Maxwell relation below.

Writing the strain component as a function of all the stress tensor components and temperature, $e_{ij} = e_{ij}(\sigma, T)$, we have:

$$\begin{aligned} de_{ij} &= \left(\frac{\partial e_{ij}}{\partial \sigma_{kl}}\right)_T d\sigma_{kl} + \left(\frac{\partial e_{ij}}{\partial T}\right)_{\sigma} dT \\ &= s^T_{ijkl}d\sigma_{kl} + \alpha_{ij}dT, \end{aligned} \tag{3.35}$$

where s^T_{ijkl} is a component of the isothermal elastic compliance tensor. Since de_{ij} is an exact differential we have the following Maxwell relation:

$$\left(\frac{\partial s^T_{ijkl}}{\partial T}\right)_{\sigma} = \left(\frac{\partial \alpha_{ij}}{\partial \sigma_{kl}}\right)_{T,\sigma'}. \tag{3.36}$$

This Maxwell relation shows that the temperature dependence of the isothermal elastic compliance tensor at constant stress is determined by the stress dependence of the thermal strain tensor at constant temperature.

[16]If there is more than one atomic species present and/or more than one atom associated with each lattice site we have to include the relaxations of atoms not carried to their final positions by the strain tensor.

[17]In the partial derivative $(\partial S/\partial \sigma_{ij})_{T,\sigma'}$, all stress components except σ_{ij} and σ_{ji} are held constant as well as the temperature. In $(\partial e_{ij}/\partial T)_{\sigma}$ all stress components are held constant.

Writing the entropy as a function of the stress tensor and temperature, $S = S(\sigma, T)$, we have:

$$\begin{aligned} dS &= \left(\frac{\partial S}{\partial \sigma_{kl}}\right)_{T,\sigma'} d\sigma_{kl} + \left(\frac{\partial S}{\partial T}\right)_{\sigma} dT \\ &= V_0 \alpha_{kl} d\sigma_{kl} + \frac{C_\sigma}{T} dT, \end{aligned} \tag{3.37}$$

where we have used the Maxwell relation of eq 3.34 and C_σ is the specific heat at constant stress of the crystal. During an adiabatic change $dS = 0$. Using eq 3.37 we find that the corresponding change in temperature when the stress is changed adiabatically is as follows:

$$dT = -\frac{V_0 T}{C_\sigma} \alpha_{kl} d\sigma_{kl}. \tag{3.38}$$

Putting this expression for dT into eq 3.35 we obtain the following relationship between the adiabatic elastic compliance tensor s^S_{ijkl} and the isothermal elastic compliance tensor s^T_{ijkl}:

$$s^T_{ijkl} = s^S_{ijkl} + \frac{V_0 T}{C_\sigma} \alpha_{ij} \alpha_{kl}. \tag{3.39}$$

This equation shows that the isothermal and adiabatic elastic compliances of a crystal differ owing to the anharmonicity of atomic interactions, without which the thermal strain is zero. A similar conclusion was reached in section 2.8 where the anharmonicity of atomic interactions was shown to be responsible for the difference between isothermal and adiabatic stresses.

Exercise 3.7 By writing the stress as a function of strain and temperature, $\sigma_{ij} = \sigma_{ij}(e, T)$, show that:

$$d\sigma_{ij} = c^T_{ijkl} de_{kl} - \beta_{ij} dT, \tag{3.40}$$

where c^T_{ijkl} is a component of the isothermal elastic constant tensor and $\beta_{ij} = -(\partial \sigma^T_{ij}/\partial T)_e$ is the thermal stress tensor (see section 2.8). Setting $d\sigma_{ij} = 0$, show that:

$$\beta_{ij} = c^T_{ijkl} \alpha_{kl}, \tag{3.41}$$

$d\sigma_{ij}(e, T)$ is an exact differential. Show that the corresponding Maxwell relation is as follows:

$$\left(\frac{\partial c^T_{ijkl}}{\partial T}\right)_e = -\left(\frac{\partial \beta_{ij}}{\partial e_{kl}}\right)_{T,e'}. \tag{3.42}$$

The temperature dependence of the isothermal elastic constant tensor at constant strain is seen here to be determined by the strain dependence of the thermal stress tensor at constant temperature.

By considering the entropy S as a function of strain and temperature show that at constant entropy:

$$\mathrm{d}T = -\frac{V_0 T}{C_e}\beta_{kl}\mathrm{d}e_{kl}, \tag{3.43}$$

where $C_e = T(\partial S/\partial T)_e$ is the specific heat of the volume V_0 at constant strain. Inserting this expression for dT into eq 3.40 show that the isothermal and adiabatic elastic constants are related as follows:

$$c^T_{ijkl} = c^S_{ijkl} - \frac{V_0 T}{C_e}\beta_{ij}\beta_{kl}. \tag{3.44}$$

Substituting $\mathrm{d}\sigma_{kl} = c^T_{klmn}\mathrm{d}e_{mn} - \beta_{kl}\mathrm{d}T$ into eq 3.37, and using eq 3.41, show that:

$$C_\sigma - C_e = V_0 T \alpha_{ij} c^T_{ijkl}\alpha_{kl}. \tag{3.45}$$

Since $\alpha_{ij}c^T_{ijkl}\alpha_{kl} > 0$ for mechanical stability this equation demonstrates that C_σ is always larger than C_e.

For copper at 25 C the linear thermal expansion coefficient is 17.1×10^{-6} K^{-1}, the density is 8.96 g·cm^{-3} and the specific heat is 0.385 J·g^{-1}·K^{-1}. The adiabatic elastic constants at 25 C are $C^S_{11} = 1.684 \times 10^{11}$ Pa, $C^S_{12} = 1.214 \times 10^{11}$ Pa and $C^S_{44} = 0.754 \times 10^{11}$ Pa.

Using eqs 3.41 and 3.44 calculate the isothermal elastic constants C^T_{11}, C^T_{12}, C^T_{44} at 25 C.

Hence calculate $C_P - C_V$ for copper at 25 C, where C_P and C_V are the specific heats in units of J·g^{-1}·K^{-1} at constant pressure and constant volume respectively. ■

Solution Writing $\sigma_{ij} = \sigma_{ij}(e, T)$ then

$$\mathrm{d}\sigma_{ij} = \left(\frac{\partial \sigma_{ij}}{\partial e_{kl}}\right)_{T,e'} \mathrm{d}e_{kl} + \left(\frac{\partial \sigma_{ij}}{\partial T}\right)_e \mathrm{d}T = c^T_{ijkl}\mathrm{d}e_{kl} - \beta_{ij}\mathrm{d}T.$$

Setting $\mathrm{d}\sigma_{ij} = 0$ we obtain:

$$\beta_{ij} = c^T_{ijkl}\left(\frac{\partial e_{kl}}{\partial T}\right)_\sigma = c^T_{ijkl}\alpha_{kl},$$

where α_{kl} is the thermal strain tensor.

Returning to $\mathrm{d}\sigma_{ij} = c^T_{ijkl}\mathrm{d}e_{kl} - \beta_{ij}\mathrm{d}T$ the Maxwell relation is:

$$\left(\frac{\partial c^T_{ijkl}}{\partial T}\right)_e = -\left(\frac{\partial \beta_{ij}}{\partial e_{kl}}\right)_{e',T}.$$

Writing $S = S(e, T)$ we have:

$$\mathrm{d}S = \left(\frac{\partial S}{\partial e_{kl}}\right)_{T,e'} \mathrm{d}e_{kl} + \left(\frac{\partial S}{\partial T}\right)_e \mathrm{d}T = -V_0\left(\frac{\partial \sigma_{kl}}{\partial T}\right)_e \mathrm{d}e_{kl} + \frac{C_e}{T}\mathrm{d}T = V_0\beta_{kl}\mathrm{d}e_{kl} + \frac{C_e}{T}\mathrm{d}T.$$

where we have used eq 2.22:

$$-\left(\frac{\partial \sigma^T_{ij}}{\partial T}\right)_e = \frac{1}{V_0}\left(\frac{\partial S}{\partial e_{ij}}\right)_{T,e'} = \beta_{ij}.$$

When $\mathrm{d}S = 0$ we have

$$V_0\beta_{kl}\mathrm{d}e_{kl} = -\frac{C_e}{T}\mathrm{d}T$$

and therefore,

$$\mathrm{d}T = -\frac{TV_0}{C_e}\beta_{kl}\mathrm{d}e_{kl}$$

Substituting this into the expression for $\mathrm{d}\sigma_{ij}$ we obtain:

$$\mathrm{d}\sigma_{ij} = c^T_{ijkl}\mathrm{d}e_{kl} + \beta_{ij}\beta_{kl}\frac{TV_0}{C_e}\mathrm{d}e_{kl}$$

and hence

$$c^T_{ijkl} = c^S_{ijkl} - \frac{V_0 T}{C_e}\beta_{ij}\beta_{kl}.$$

Substituting $\mathrm{d}\sigma_{kl} = c^T_{klmn}\mathrm{d}e_{mn} - \beta_{kl}\mathrm{d}T$ into $\mathrm{d}S = V_0\alpha_{kl}\mathrm{d}\sigma_{kl} + C_\sigma \mathrm{d}T/T$, and using eq 3.41 we get:

$$\mathrm{d}S = V_0\alpha_{kl}\left(c^T_{klmn}\mathrm{d}e_{mn} - \beta_{kl}\mathrm{d}T\right) + \frac{C_\sigma}{T}\mathrm{d}T.$$

Therefore,

$$\left(\frac{\partial S}{\partial T}\right)_e = \frac{C_e}{T} = -V_0\alpha_{kl}\beta_{kl} + \frac{C_\sigma}{T},$$

but since $\beta_{kl} = c^T_{klmn}\alpha_{mn}$ we obtain finally:

$$C_\sigma - C_e = V_0 T\alpha_{kl}c^T_{klmn}\alpha_{mn} \geq 0.$$

Numerical part

Since copper has a cubic crystal structure

$$\beta_{11} = \beta_{22} = \beta_{33} = \left(c^T_{1111} + 2c^T_{1122}\right)\alpha$$

where the linear thermal expansion coefficients $\alpha_{11}, \alpha_{22}, \alpha_{33}$ have been equated to α ($= 17.1 \times 10^{-6}$ K^{-1}).

Therefore, eq 3.44 becomes:

$$c^T_{1111} = c^S_{1111} - \frac{V_0 T}{C_e}\alpha^2\left(c^T_{1111} + 2c^T_{1122}\right)^2$$

$$c^T_{1122} = c^S_{1122} - \frac{V_0 T}{C_e}\alpha^2\left(c^T_{1111} + 2c^T_{1122}\right)^2.$$

These are coupled quadratic equations in c^T_{1111} and c^T_{1122}. I calculate:

$c^T_{1111} = 1.643 \times 10^{11}$ Pa and $c^T_{1122} = 1.173 \times 10^{11}$ Pa.

Since copper is cubic the specific heats $C_\sigma \to C_P$ and $C_e \to C_V$. Therefore,

$$\begin{aligned}\frac{C_P - C_V}{V_0} &= 3T\left(c^T_{1111} + 2c^T_{1122}\right)\alpha^2 \\ &= 104 \text{ kJ K}^{-1}\text{m}^{-3} \\ C_P - C_V &= 0.012 \text{ J K}^{-1}\text{g}^{-1}.\end{aligned}$$

Comment

Notice that $c^T_{1212} = c^S_{1212}$, because $\alpha_{12} = \beta_{12} = 0$. It follows that the isothermal and adiabatic shear elastic constants C_{44} are equal. Similarly, $c^T_{1111} - c^T_{1122} = c^S_{1111} - c^S_{1122}$. Therefore, the isothermal and adiabatic shear elastic constants C' are also equal. A difference between adiabatic and isothermal elastic constants in a cubic crystal arises only when the corresponding deformation involves a change of volume.

3.9 Problems

Problem 3.1 In a face-centred cubic crystal with lattice constant a the 12 nearest neighbours of an atom are at $\pm a/2[110], \pm a/2[1\bar{1}0], \pm a/2[101], \pm a/2[10\bar{1}], \pm a/2[011], \pm a/2[01\bar{1}]$. Consider a model of the crystal in which the bonding is represented by linear springs, with spring constant k, between nearest neighbours only. By considering the elastic energy density of the crystal when an arbitrary small elastic strain is applied show that the elastic constants are $C_{11} = 2k/a$, $C_{12} = k/a$ and $C_{44} = k/a$. It is interesting to note that the anisotropy ratio for this simple model is 2, i.e. the crystal is elastically anisotropic.

Solution As shown in problem 1.2, when an arbitrary strain e_{ij} is applied to a bond **b** the bond length b changes by $\delta b = b_i e_{ij} b_j / b$. For example, the change in length of the $a/2[110]$ bond is as follows:

$$\delta b = \frac{(a/2)^2}{a/\sqrt{2}} \begin{pmatrix} 1 & 1 & 0 \end{pmatrix} \begin{pmatrix} e_{11} & e_{12} & e_{13} \\ e_{12} & e_{22} & e_{23} \\ e_{13} & e_{23} & e_{33} \end{pmatrix} \begin{pmatrix} 1 \\ 1 \\ 0 \end{pmatrix} = \frac{a}{2\sqrt{2}}(e_{11} + 2e_{12} + e_{22})$$

The changes in length of all 12 bonds are as follows:

For $\mathbf{b} = \pm(a/2)[110]$, $\delta b = (e_{11} + 2e_{12} + e_{22})a/(2\sqrt{2})$

For $\mathbf{b} = \pm(a/2)[1\bar{1}0]$, $\delta b = (e_{11} - 2e_{12} + e_{22})a/(2\sqrt{2})$

For $\mathbf{b} = \pm(a/2)[101]$, $\delta b = (e_{11} + 2e_{13} + e_{33})a/(2\sqrt{2})$

For $\mathbf{b} = \pm(a/2)[10\bar{1}]$, $\delta b = (e_{11} - 2e_{13} + e_{33})a/(2\sqrt{2})$

For $\mathbf{b} = \pm(a/2)[011]$, $\delta b = (e_{22} + 2e_{23} + e_{33})a/(2\sqrt{2})$

For $\mathbf{b} = \pm(a/2)[01\bar{1}]$, $\delta b = (e_{22} - 2e_{23} + e_{33})a/(2\sqrt{2})$

Therefore, the energy per atom caused by the strain is:

$$E = \frac{1}{2} \sum_{12 \text{ bonds}} \frac{1}{2} k(\delta b)^2,$$

where the factor of $1/2$ before the sum is because each bond is shared by two atoms. Writing out the sum:

$$\begin{aligned}
E \;=\; & \frac{ka^2}{16}\left((e_{11}+e_{22})^2 + \cancel{4e_{12}(e_{11}+e_{22})} + 4e_{12}^2\right) \\
+ \; & \frac{ka^2}{16}\left((e_{11}+e_{22})^2 - \cancel{4e_{12}(e_{11}+e_{22})} + 4e_{12}^2\right) \\
+ \; & \frac{ka^2}{16}\left((e_{11}+e_{33})^2 + \cancel{4e_{13}(e_{11}+e_{33})} + 4e_{13}^2\right) \\
+ \; & \frac{ka^2}{16}\left((e_{11}+e_{33})^2 - \cancel{4e_{13}(e_{11}+e_{33})} + 4e_{13}^2\right) \\
+ \; & \frac{ka^2}{16}\left((e_{22}+e_{33})^2 + \cancel{4e_{23}(e_{22}+e_{33})} + 4e_{23}^2\right) \\
+ \; & \frac{ka^2}{16}\left((e_{22}+e_{33})^2 - \cancel{4e_{23}(e_{22}+e_{33})} + 4e_{23}^2\right)
\end{aligned}$$

Multiplying out the brackets and collecting terms:

$$E = \frac{ka^2}{4}\left\{e_{11}^2 + e_{22}^2 + e_{33}^2 + e_{11}e_{22} + e_{22}e_{33} + e_{33}e_{11} + 2\left(e_{12}^2 + e_{23}^2 + e_{31}^2\right)\right\}.$$

This has to be equated to the elastic energy per atom in terms of the elastic constants

$$E_{el} = \frac{a^3}{4}\left\{\frac{C_{11}}{2}\left(e_{11}^2+e_{22}^2+e_{33}^2\right)+\frac{C_{12}}{2}\left(2e_{11}e_{22}+2e_{22}e_{33}+2e_{33}e_{11}\right)\right.$$
$$\left.+\frac{C_{44}}{2}\left(4e_{12}^2+4e_{23}^2+4e_{31}^2\right)\right\}$$

To convert the elastic energy density into an energy per atom we multiplied by $a^3/4$, which is the volume per atom. Equating E_{el} to E we obtain:

$$C_{11} = \frac{4}{a^3}\frac{\partial^2 E_{el}}{\partial e_{11}^2} = \frac{4}{a^3}\frac{\partial^2 E}{\partial e_{11}^2} = \frac{4}{a^3}\frac{ka^2}{2} = \frac{2k}{a}$$

$$C_{12} = \frac{4}{a^3}\frac{\partial^2 E_{el}}{\partial e_{11}\partial e_{22}} = \frac{4}{a^3}\frac{\partial^2 E}{\partial e_{11}\partial e_{22}} = \frac{4}{a^3}\frac{ka^2}{4} = \frac{k}{a}$$

$$C_{44} = \frac{1}{a^3}\frac{\partial^2 E_{el}}{\partial e_{23}^2} = \frac{1}{a^3}\frac{\partial^2 E}{\partial e_{23}^2} = \frac{1}{a^3}\frac{ka^2}{4}4 = \frac{k}{a}$$

Comment

The equality $C_{12} = C_{44}$ is an example of a result that applies to all models in which atoms interact in pairs: see problem 3.3.

Problem 3.2 This problem and the next are based on sections 5.3.1 and 5.3.2 of the book[18] by Finnis[19]. Let a homogeneous strain tensor be written as $e_{ij} = \gamma T_{ij}$, where γ is a scalar which scales the magnitude of the strain, and the matrix elements T_{ij} are constant. The component X_i of a vector $\mathbf{X}$ becomes $(\delta_{ij} + \gamma T_{ij})X_j$. If $X = |\mathbf{X}|$ show that:

$$\left(\frac{\mathrm{d}X}{\mathrm{d}\gamma}\right)_{\gamma=0} = \frac{X_i T_{ij} X_j}{X} \tag{3.46}$$

$$\left(\frac{\mathrm{d}^2X}{\mathrm{d}\gamma^2}\right)_{\gamma=0} = \frac{X_i T_{ij} T_{jk} X_k}{X} - \frac{\left(X_i T_{ij} X_j\right)^2}{X^3} \tag{3.47}$$

[18] Finnis, M W, *Interatomic forces in condensed matter*, (2003), Oxford University Press: Oxford. ISBN 978-0198509776.

[19] Michael William Finnis FRS 1950- , British materials physicist.

Solution Since $X_i \to (\delta_{ij} + \gamma T_{ij})X_j$ then

$$\begin{aligned}
X^2 &\to (\delta_{ij} + \gamma T_{ij})X_j(\delta_{ik} + \gamma T_{ik})X_k \\
&= X^2 + 2\gamma X_j T_{jk} X_k + \gamma^2 X_j T_{ji} T_{ik} X_k \\
&= X^2\left\{1 + 2\gamma \frac{X_j}{X} T_{jk} \frac{X_k}{X} + \gamma^2 \frac{X_j}{X} T_{ji} T_{ik} \frac{X_k}{X}\right\},
\end{aligned}$$

where we have used the symmetry $T_{ij} = T_{ji}$. Therefore $X = |\mathbf{X}|$ changes as follows:

$$\begin{aligned}
X &\to X\left\{1 + 2\gamma \frac{X_j}{X} T_{jk} \frac{X_k}{X} + \gamma^2 \frac{X_j}{X} T_{ji} T_{ik} \frac{X_k}{X}\right\}^{\frac{1}{2}} \\
&= X\left\{1 + \gamma \frac{X_j}{X} T_{jk} \frac{X_k}{X} + \frac{\gamma^2}{2} \frac{X_j}{X} T_{ji} T_{ik} \frac{X_k}{X} - \frac{\gamma^2}{2}\left(\frac{X_j}{X} T_{jk} \frac{X_k}{X}\right)^2 + O\left(\gamma^3\right)\right\}
\end{aligned}$$

Therefore, to second order in γ, the change in the length of $\mathbf{X}$ is:

$$\delta X = \gamma \frac{X_j T_{jk} X_k}{X} + \frac{\gamma^2}{2}\left(\frac{X_j T_{ji} T_{ik} X_k}{X} - \frac{(X_j T_{jk} X_k)^2}{X^3}\right).$$

Therefore,

$$\begin{aligned}
\left(\frac{\mathrm{d}X}{\mathrm{d}\gamma}\right)_{\gamma=0} &= \frac{X_j T_{jk} X_k}{X} \\
\left(\frac{\mathrm{d}^2X}{\mathrm{d}\gamma^2}\right)_{\gamma=0} &= \left(\frac{X_j T_{ji} T_{ik} X_k}{X} - \frac{(X_j T_{jk} X_k)^2}{X^3}\right).
\end{aligned}$$

Comment

Equation 3.46 is a restatement of the solution to problem 1.2.

Problem 3.3 In a pairwise interaction model of a crystal the interaction energy between any two atoms is a function of their separation only. The total energy is then the sum of all such pairwise interactions. Let $V(X)$ be the interaction energy between two atoms separated by X. For example, $V(X)$ might be a Lennard-Jones potential:

$$V(X) = \varepsilon\left[\left(\frac{X_0}{X}\right)^{12} - 2\left(\frac{X_0}{X}\right)^{6}\right],$$

which leads to repulsion between atoms separated by $X < X_0$, attraction when $X > X_0$, and $V(X)$ has a minimum at $X = X_0$ where $V(X) = -\varepsilon$. Consider a *cubic* crystal in which there is just one atom per lattice site. Let the origin of a cartesian coordinate system be located at an atomic site, and orient the axes along the $\langle 100 \rangle$ directions of the crystal. The energy of the atom at the origin is then:

$$E_c = \frac{1}{2}\sum_n V(X^{(n)})$$

where $X^{(n)}$ is the distance to atom n, and the factor of one half is because each pairwise interaction is shared by two atoms. In this question you will prove the following:

- The equilibrium volume of the crystal is determined by the equation:

$$\frac{1}{6\Omega}\sum_n X^{(n)} V'(X^{(n)}) = 0,$$

  where the prime denotes differentiation.
- The elastic constants C_{11}, C_{12} and C_{44} are as follows:

$$C_{11} = \frac{1}{6\Omega}\sum_n V'(X^{(n)})X^{(n)} + \left(V''(X^{(n)})\left(X^{(n)}\right)^2 - V'(X^{(n)})X^{(n)}\right) F(X_1^{(n)}, X_2^{(n)}, X_3^{(n)})$$

$$C_{12} = \frac{1}{6\Omega}\sum_n \left(V''(X^{(n)})\left(X^{(n)}\right)^2 - V'(X^{(n)})X^{(n)}\right) H(X_1^{(n)}, X_2^{(n)}, X_3^{(n)})$$

$$C_{44} = \frac{1}{12\Omega}\sum_n V'(X^{(n)})X^{(n)}$$

$$+ \frac{1}{6\Omega}\sum_n \left(V''(X^{(n)})\left(X^{(n)}\right)^2 - V'(X^{(n)})X^{(n)}\right) H(X_1^{(n)}, X_2^{(n)}, X_3^{(n)}),$$

  where

$$F(X_1^{(n)}, X_2^{(n)}, X_3^{(n)}) = \frac{\left(X_1^{(n)}\right)^4 + \left(X_2^{(n)}\right)^4 + \left(X_3^{(n)}\right)^4}{(X^{(n)})^4}$$

$$H(X_1^{(n)}, X_2^{(n)}, X_3^{(n)}) = \frac{\left(X_1^{(n)}\right)^2\left(X_2^{(n)}\right)^2 + \left(X_2^{(n)}\right)^2\left(X_3^{(n)}\right)^2 + \left(X_3^{(n)}\right)^2\left(X_1^{(n)}\right)^2}{(X^{(n)})^4}$$

$$= \frac{1}{2}\left(1 - F(X_1^{(n)}, X_2^{(n)}, X_3^{(n)})\right).$$

- At the equilibrium volume of the crystal it follows that $C_{12} = C_{44}$ for all pairwise interaction models of cubic crystals.

The equilibrium volume of the crystal is determined by the condition that the pressure arising from all the pairwise interactions is zero. Mathematically, this amounts to the condition that $\mathrm{d}E_c/\mathrm{d}\Omega = 0$ where Ω is the atomic volume. Setting the tensor T_{ij} of the previous question equal to δ_{ij} show that

$$\frac{\mathrm{d}E_c}{\mathrm{d}\Omega} = \frac{1}{3\Omega}\left(\frac{\mathrm{d}E_c}{\mathrm{d}\gamma}\right)_{\gamma=0}.$$

Hence show that

$$\frac{\mathrm{d}E_c}{\mathrm{d}\Omega} = \frac{1}{6\Omega}\sum_n X^{(n)} V'(X^{(n)}). \tag{3.48}$$

In a pairwise interaction model the volume per atom in a cubic crystal with just one per lattice site is determined by the condition that the above sum is zero.

For an arbitrary homogeneous strain $e_{ij} = \gamma T_{ij}$ the elastic energy per atom in the cubic crystal is

$$E_{el} = \frac{1}{2}\gamma^2 \Omega c_{ijkl} T_{ij} T_{kl},$$

so that

$$\frac{\mathrm{d}^2 E_{el}}{\mathrm{d}\gamma^2} = \Omega c_{ijkl} T_{ij} T_{kl}.$$

Show that

$$\frac{\mathrm{d}^2 E_c}{\mathrm{d}\gamma^2} = \frac{1}{2}\sum_n V''(X^{(n)})\left(\frac{\mathrm{d}X^{(n)}}{\mathrm{d}\gamma}\right)^2 + V'(X^{(n)})\frac{\mathrm{d}^2 X^{(n)}}{\mathrm{d}\gamma^2},$$

where the derivatives with respect to γ are given by eq 3.46 and eq 3.47.

Equating the change in E_c to the elastic energy we obtain:

$$c_{ijkl} T_{ij} T_{kl} = \frac{1}{2\Omega}\sum_n V''(X^{(n)})\left(\frac{\mathrm{d}X^{(n)}}{\mathrm{d}\gamma}\right)^2 + V'(X^{(n)})\frac{\mathrm{d}^2 X^{(n)}}{\mathrm{d}\gamma^2}. \tag{3.49}$$

By choosing $T_{ij} = \delta_{ij}$ show that

$$c_{ijkl} T_{ij} T_{kl} = 3(C_{11} + 2C_{12})$$

$$
\begin{aligned}
\frac{dX^{(n)}}{d\gamma} &= X^{(n)} \\
\frac{d^2X^{(n)}}{d\gamma^2} &= 0
\end{aligned}
$$

and hence

$$
C_{11}+2C_{12}=\frac{1}{6\Omega}\sum_n V''(X^{(n)})\left(X^{(n)}\right)^2, \tag{3.50}
$$

which is three times the bulk modulus.

To obtain an expression for $C'=\frac{1}{2}(C_{11}-C_{12})$ we may set $T_{ij}=\delta_{i1}\delta_{j1}-\delta_{i2}\delta_{j2}$. Then $c_{ijkl}T_{ij}T_{kl}=2(C_{11}-C_{12})=4C'$. Using cubic symmetry show that:

$$
\begin{aligned}
C' &= \frac{1}{24\Omega}\sum_n V''(X^{(n)})\left(X^{(n)}\right)^2\left(3F(X_1^{(n)},X_2^{(n)},X_3^{(n)})-1\right) \\
&+ 3V'(X^{(n)})X^{(n)}\left[1-F(X_1^{(n)},X_2^{(n)},X_3^{(n)})\right].
\end{aligned} \tag{3.51}
$$

Using eq 3.50 and eq 3.51 derive the formulae for C_{11} and C_{12} stated above.

To obtain C_{44} we may set $T_{ij}=\delta_{i1}\delta_{j2}+\delta_{i2}\delta_{j1}+\delta_{i2}\delta_{j3}+\delta_{i3}\delta_{j2}+\delta_{i3}\delta_{j1}+\delta_{i1}\delta_{j3}$. Show that:

$$
\begin{aligned}
c_{ijkl}T_{ij}T_{kl} &= 12C_{44} \\
\frac{dX^{(n)}}{d\gamma} &= \frac{2\left(X_1^{(n)}X_2^{(n)}+X_2^{(n)}X_3^{(n)}+X_3^{(n)}X_1^{(n)}\right)}{X^{(n)}} \\
\frac{d^2X^{(n)}}{d\gamma^2} &= \frac{2\left[\left(X_1^{(n)}\right)^2+\left(X_2^{(n)}\right)^2+\left(X_3^{(n)}\right)^2\right]}{X^{(n)}} \\
&+ \frac{2\left[\left(X_1^{(n)}\right)\left(X_2^{(n)}\right)+\left(X_2^{(n)}\right)\left(X_3^{(n)}\right)+\left(X_3^{(n)}\right)\left(X_1^{(n)}\right)\right]}{X^{(n)}} \\
&- \frac{4\left[\left(X_1^{(n)}\right)\left(X_2^{(n)}\right)+\left(X_2^{(n)}\right)\left(X_3^{(n)}\right)+\left(X_3^{(n)}\right)\left(X_1^{(n)}\right)\right]^2}{\left(X^{(n)}\right)^3}
\end{aligned}
$$

Hence, using cubic symmetry derive the formula stated above for C_{44}.

In problem 3.1 the harmonic springs between nearest neighbours are described by $V(X) = \frac{1}{2}k\left[X - (a/\sqrt{2})\right]^2$, where $a/\sqrt{2}$ is the equilibrium bond length. Show that the equations derived in this question for C_{11}, C_{12} and C_{44} are consistent with the elastic constants obtained in the first question.

Solution The atomic volume Ω becomes $(1+\gamma)^3\Omega$ when the strain $e_{ij} = \gamma\delta_{ij}$ is applied. Therefore,

$$\left(\frac{d\Omega}{d\gamma}\right)_{\gamma=0} = 3\Omega.$$

The derivative $(dE_c/d\gamma)_{\gamma=0}$ is evaluated as follows:

$$\begin{aligned}\left(\frac{dE_c}{d\gamma}\right)_{\gamma=0} &= \frac{1}{2}\sum_n \frac{dV(X^{(n)})}{dX^{(n)}}\left(\frac{dX^{(n)}}{d\gamma}\right)_{\gamma=0} \\ &= \frac{1}{2}\sum_n V'(X^{(n)})\frac{X_i^{(n)}\delta_{ij}X_j^{(n)}}{X^{(n)}} \\ &= \frac{1}{2}\sum_n V'(X^{(n)})X^{(n)}.\end{aligned}$$

It follows that:

$$\left(\frac{dE_c}{d\Omega}\right)_{\gamma=0} = \left(\frac{dE_c}{d\gamma}\right)_{\gamma=0} \Big/ \left(\frac{d\Omega}{d\gamma}\right)_{\gamma=0} = \frac{1}{6\Omega}\sum_n V'(X^{(n)})X^{(n)}.$$

Since

$$\left(\frac{dE_c}{d\gamma}\right)_{\gamma=0} = \frac{1}{2}\sum_n V'(X^{(n)})X^{(n)}$$

the second derivative of E_c with respect to γ is as follows:

$$\left(\frac{\mathrm{d}^2E_c}{\mathrm{d}\gamma^2}\right)_{\gamma=0}=\frac{1}{2}\sum_n\left\{V''(X^{(n)})\left(\frac{\mathrm{d}X^{(n)}}{\mathrm{d}\gamma}\right)^2_{\gamma=0}+V'(X^{(n)})\left(\frac{\mathrm{d}^2X^{(n)}}{\mathrm{d}\gamma^2}\right)_{\gamma=0}\right\}$$

When $T_{ij}=\delta_{ij}$ we have $c_{ijkl}T_{ij}T_{kl}=c_{iikk}$.

$$\begin{aligned}c_{iikk} &= c_{1111}+c_{1122}+c_{1133}\\ &+ c_{2211}+c_{2222}+c_{2233}\\ &+ c_{3311}+c_{3322}+c_{3333}\\ &= 3(c_{1111}+2c_{1122})=3(C_{11}+2C_{12})\end{aligned}$$

Using the expressions we derived in the previous problem for $(\mathrm{d}X/\mathrm{d}\gamma)_{\gamma=0}$ and $(\mathrm{d}^2X/\mathrm{d}\gamma^2)_{\gamma=0}$, and $T_{ij}=\delta_{ij}$, we obtain:

$$\begin{aligned}\left(\frac{\mathrm{d}X^{(n)}}{\mathrm{d}\gamma}\right)_{\gamma=0} &= \frac{X_j^{(n)}\delta_{jk}X_k^{(n)}}{X^{(n)}}=X^{(n)}\\ \left(\frac{\mathrm{d}^2X^{(n)}}{\mathrm{d}\gamma^2}\right)_{\gamma=0} &= \frac{X_j^{(n)}\delta_{jk}X_k^{(n)}}{X^{(n)}}-\frac{\left(X_j^{(n)}\delta_{jk}X_k^{(n)}\right)^2}{\left(X^{(n)}\right)^3}\\ &= X^{(n)}-\frac{\left(X^{(n)}\right)^4}{\left(X^{(n)}\right)^3}=0.\end{aligned}$$

Therefore,

$$C_{11}+2C_{12}=\frac{1}{6\Omega}\sum_n V''(X^{(n)})\left(X^{(n)}\right)^2.$$

To obtain an expression for $C'=\frac{1}{2}(C_{11}-C_{12})$ we choose $T_{ij}=\delta_{i1}\delta_{j1}-\delta_{i2}\delta_{j2}$:

$$\begin{aligned}c_{ijkl}T_{ij}T_{kl} &= c_{ijkl}(\delta_{i1}\delta_{j1}-\delta_{i2}\delta_{j2})(\delta_{k1}\delta_{l1}-\delta_{k2}\delta_{l2})\\ &= c_{ijkl}(\delta_{i1}\delta_{j1}\delta_{k1}\delta_{l1}+\delta_{i2}\delta_{j2}\delta_{k2}\delta_{l2}-\delta_{i1}\delta_{j1}\delta_{k2}\delta_{l2}-\delta_{i2}\delta_{j2}\delta_{k1}\delta_{l1})\end{aligned}$$

$$= \quad c_{1111} + c_{2222} - c_{1122} - c_{2211} = 2(C_{11} - C_{12}) = 4C'.$$

The derivatives of $X^{(n)}$ are as follows:

$$\left(\frac{\mathrm{d}X^{(n)}}{\mathrm{d}\gamma}\right)_{\gamma=0} = \frac{X_j^{(n)}(\delta_{j1}\delta_{k1} - \delta_{j2}\delta_{k2})X_k^{(n)}}{X^{(n)}} = \frac{\left(X_1^{(n)}\right)^2 - \left(X_2^{(n)}\right)^2}{X^{(n)}}$$

$$\left(\frac{\mathrm{d}^2X^{(n)}}{\mathrm{d}\gamma^2}\right)_{\gamma=0} = \frac{X_i^{(n)}(\delta_{i1}\delta_{j1} - \delta_{i2}\delta_{j2})(\delta_{j1}\delta_{k1} - \delta_{j2}\delta_{k2})X_k^{(n)}}{X^{(n)}}$$

$$- \frac{\left(X_j^{(n)}(\delta_{j1}\delta_{k1} - \delta_{j2}\delta_{k2})X_k^{(n)}\right)^2}{\left(X^{(n)}\right)^3}$$

$$= \frac{\left(X_1^{(n)}\right)^2 + \left(X_2^{(n)}\right)^2}{X^{(n)}} - \frac{\left(\left(X_1^{(n)}\right)^2 - \left(X_2^{(n)}\right)^2\right)^2}{\left(X^{(n)}\right)^3}$$

Therefore,

$$4C' = \frac{1}{2\Omega}\sum_n V''(X^{(n)})\left[\frac{\left(X_1^{(n)}\right)^2 - \left(X_2^{(n)}\right)^2}{X^{(n)}}\right]^2$$

$$+\frac{1}{2\Omega}\sum_n V'(X^{(n)})\left[\frac{\left(X_1^{(n)}\right)^2 + \left(X_2^{(n)}\right)^2}{X^{(n)}} - \frac{\left(\left(X_1^{(n)}\right)^2 - \left(X_2^{(n)}\right)^2\right)^2}{\left(X^{(n)}\right)^3}\right]$$

Since the crystal has cubic symmetry, and there is just one atom per lattice site, we may express C' in a more symmetrical way as follows:

$$24\Omega C' = \sum_n \frac{V''(X^{(n)})}{\left(X^{(n)}\right)^2}\left[\left(\left(X_1^{(n)}\right)^2 - \left(X_2^{(n)}\right)^2\right)^2 + \left(\left(X_2^{(n)}\right)^2 - \left(X_3^{(n)}\right)^2\right)^2\right.$$

$$+\left(\left(X_3^{(n)}\right)^2-\left(X_1^{(n)}\right)^2\right)^2\Bigg]$$

$$\sum_n \frac{V'(X^{(n)})}{X^{(n)}}\left[\left(X_1^{(n)}\right)^2+\left(X_2^{(n)}\right)^2+\left(X_2^{(n)}\right)^2+\left(X_3^{(n)}\right)^2\right.$$

$$\left.+\left(X_3^{(n)}\right)^2+\left(X_1^{(n)}\right)^2\right]$$

$$-\sum_n \frac{V'(X^{(n)})}{(X^{(n)})^3}\left[\left(\left(X_1^{(n)}\right)^2-\left(X_2^{(n)}\right)^2\right)^2+\left(\left(X_2^{(n)}\right)^2-\left(X_3^{(n)}\right)^2\right)^2\right.$$

$$\left.+\left(\left(X_3^{(n)}\right)^2-\left(X_1^{(n)}\right)^2\right)^2\right]$$

Simplifying we get:

$$24\Omega C' = 2\sum_n \left(X^{(n)}\right)^2 V''(X^{(n)})\left[\frac{\left(X_1^{(n)}\right)^4+\left(X_2^{(n)}\right)^4+\left(X_3^{(n)}\right)^4}{(X^{(n)})^4}\right]$$

$$-\ 2\sum_n \left(X^{(n)}\right)^2 V''(X^{(n)})$$

$$\times\left[\frac{\left\{\left(X_1^{(n)}\right)^2\left(X_2^{(n)}\right)^2+\left(X_2^{(n)}\right)^2\left(X_3^{(n)}\right)^2+\left(X_3^{(n)}\right)^2\left(X_2^{(n)}\right)^2\right\}}{(X^{(n)})^4}\right]$$

$$+2\sum_n X^{(n)}V'(X^{(n)})$$

$$-2\sum_n X^{(n)}V'(X^{(n)})\left[\frac{\left(X_1^{(n)}\right)^4+\left(X_2^{(n)}\right)^4+\left(X_3^{(n)}\right)^4}{(X^{(n)})^4}\right]$$

$$+2\sum_n X^{(n)}V'(X^{(n)})$$

$$
\begin{aligned}
&\times\left[\frac{\left\{\left(X_1^{(n)}\right)^2\left(X_2^{(n)}\right)^2+\left(X_2^{(n)}\right)^2\left(X_3^{(n)}\right)^2+\left(X_3^{(n)}\right)^2\left(X_2^{(n)}\right)^2\right\}}{(X^{(n)})^4}\right] \\
&= 2\sum_n\left[\left(X^{(n)}\right)^2 V''(X^{(n)})-X^{(n)}V'(X^{(n)})\right] \\
&\quad\times\left(F(X_1^{(n)},X_2^{(n)},X_3^{(n)})-H(X_1^{(n)},X_2^{(n)},X_3^{(n)})\right) \\
&\quad+2\sum_n X^{(n)}V'(X^{(n)}) \\
&= \sum_n\left[\left(X^{(n)}\right)^2 V''(X^{(n)})-X^{(n)}V'(X^{(n)})\right]\left(3F(X_1^{(n)},X_2^{(n)},X_3^{(n)})-1\right) \\
&\quad+2\sum_n X^{(n)}V'(X^{(n)}) \\
&= \sum_n\left(X^{(n)}\right)^2 V''(X^{(n)})\left[3F(X_1^{(n)},X_2^{(n)},X_3^{(n)})-1\right] \\
&\quad+\sum_n 3V'(X^{(n)})X^{(n)}\left[1-F(X_1^{(n)},X_2^{(n)},X_3^{(n)})\right]
\end{aligned}
$$

Using this equation and the equation we have derived for $C_{11}+2C_{12}$ we obtain the following expressions for C_{11} and C_{12}:

$$
\begin{aligned}
C_{11} &= \frac{1}{6\Omega}\sum_n\left\{V'(X^{(n)})X^{(n)}\right. \\
&\quad\left.+\left(V''(X^{(n)})\left(X^{(n)}\right)^2-V'(X^{(n)})X^{(n)}\right)F(X_1^{(n)},X_2^{(n)},X_3^{(n)})\right\} \\
C_{12} &= \frac{1}{6\Omega}\sum_n\left(V''(X^{(n)})\left(X^{(n)}\right)^2-V'(X^{(n)})X^{(n)}\right)H(X_1^{(n)},X_2^{(n)},X_3^{(n)})
\end{aligned}
$$

To evaluate C_{44} we use the following matrices **T** and $\mathbf{T}^2$:

$$\mathbf{T} = \begin{pmatrix} 0 & 1 & 1 \\ 1 & 0 & 1 \\ 1 & 1 & 0 \end{pmatrix}$$

$$\mathbf{T}^2 = \begin{pmatrix} 2 & 1 & 1 \\ 1 & 2 & 1 \\ 1 & 1 & 2 \end{pmatrix}$$

Then we get:

$$\left(\frac{dX^{(n)}}{d\gamma}\right)_{\gamma=0} = \frac{1}{X^{(n)}}\begin{pmatrix} X_1^{(n)} & X_2^{(n)} & X_3^{(n)} \end{pmatrix}\begin{pmatrix} 0 & 1 & 1 \\ 1 & 0 & 1 \\ 1 & 1 & 0 \end{pmatrix}\begin{pmatrix} X_1^{(n)} \\ X_2^{(n)} \\ X_3^{(n)} \end{pmatrix}$$

$$= \frac{2\left(X_1^{(n)}X_2^{(n)} + X_2^{(n)}X_3^{(n)} + X_3^{(n)}X_1^{(n)}\right)}{X^{(n)}}$$

and in a cubic environment this leads to:

$$\left(\frac{dX^n}{d\gamma}\right)^2_{\gamma=0} = \frac{4\left(\left(X_1^{(n)}\right)^2\left(X_2^{(n)}\right)^2 + \left(X_2^{(n)}\right)^2\left(X_3^{(n)}\right)^2 + \left(X_3^{(n)}\right)^2\left(X_1^{(n)}\right)^2\right)}{\left(X^{(n)}\right)^2}$$

and $(d^2X^{(n)}/d\gamma^2)_{\gamma=0}$ is evaluated as follows:

$$\left(\frac{d^2X^{(n)}}{d\gamma^2}\right)_{\gamma=0} = \frac{1}{X^{(n)}}\begin{pmatrix} X_1^{(n)} & X_2^{(n)} & X_3^{(n)} \end{pmatrix}\begin{pmatrix} 2 & 1 & 1 \\ 1 & 2 & 1 \\ 1 & 1 & 2 \end{pmatrix}\begin{pmatrix} X_1^{(n)} \\ X_2^{(n)} \\ X_3^{(n)} \end{pmatrix}$$

$$- \frac{1}{\left(X^{(n)}\right)^3}\left[\begin{pmatrix} X_1^{(n)} & X_2^{(n)} & X_3^{(n)} \end{pmatrix}\begin{pmatrix} 2 & 1 & 1 \\ 1 & 2 & 1 \\ 1 & 1 & 2 \end{pmatrix}\begin{pmatrix} X_1^{(n)} \\ X_2^{(n)} \\ X_3^{(n)} \end{pmatrix}\right]^2 .$$

In a cubic environment this is

$$\left(\frac{d^2X^{(n)}}{d\gamma^2}\right)_{\gamma=0} = 2X^{(n)} - \frac{4\left(\left(X_1^{(n)}\right)^2\left(X_2^{(n)}\right)^2+\left(X_2^{(n)}\right)^2\left(X_3^{(n)}\right)^2+\left(X_3^{(n)}\right)^2\left(X_1^{(n)}\right)^2\right)}{\left(X^{(n)}\right)^3}$$

$c_{ijkl}T_{ij}T_{kl}$ is evaluated as follows:

$$\begin{aligned} c_{ijkl}T_{ij}T_{kl} &= \left(\delta_{i1}\delta_{j2}+\delta_{i2}\delta_{j1}+\delta_{i2}\delta_{j3}+\delta_{i3}\delta_{j2}+\delta_{i3}\delta_{j1}+\delta_{i1}\delta_{j3}\right) \\ &\quad \times c_{ijkl} \\ &\quad \times \left(\delta_{k1}\delta_{l2}+\delta_{k2}\delta_{l1}+\delta_{k2}\delta_{l3}+\delta_{k3}\delta_{l2}+\delta_{k3}\delta_{l1}+\delta_{k1}\delta_{l3}\right) \\ &= 4\left(c_{1212}+c_{2323}+c_{3131}\right) \\ &= 12C_{44} \end{aligned}$$

where we have again used the symmetry of the cubic environment to equate some terms to zero.

Therefore,

$$\begin{aligned} 12C_{44} &= \frac{1}{2\Omega}\sum_n V''(X^{(n)})\left(\frac{dX^{(n)}}{d\gamma}\right)^2 + V'(X^{(n)})\frac{d^2X^{(n)}}{d\gamma^2} \\ &= \frac{1}{2\Omega}\sum_n 4V''(X^{(n)})\left(X^{(n)}\right)^2 H(X_1^{(n)},X_2^{(n)},X_3^{(n)}) \\ &\quad + 2X^{(n)}V'(X^{(n)}) - 4X^{(n)}V'(X^{(n)})H(X_1^{(n)},X_2^{(n)},X_3^{(n)}). \end{aligned}$$

Simplifying we obtain the final expression for C_{44}:

$$C_{44} = \frac{1}{6\Omega}\sum_n\left(V''(X^{(n)})\left(X^{(n)}\right)^2 - X^{(n)}V'(X^{(n)})\right)H(X_1^{(n)},X_2^{(n)},X_3^{(n)}) + \frac{1}{2}X^{(n)}V'(X^{(n)}).$$

For $V(X) = \frac{1}{2}k\left[X - (a/\sqrt{2})\right]^2$ we have $V'(X) = 0$ at $X = a/\sqrt{2}$ and $V''(X) = k$ at all X. In the fcc crystal $F = 1/2$ and $H = 1/4$ for each of the 12 nearest neighbours. The volume per atom is $\Omega = a^3/4$. Using the formulae we have derived we obtain $C_{11} = 2k/a$ and $C_{12} = C_{44} = k/a$, reproducing the answers to problem 3.1.

Comments

(i) The difference $C_{12} - C_{44} = -(1/12\Omega)\sum_n V'(X^{(n)})X^{(n)}$ is called the *Cauchy pressure.* In a model where all the cohesion is provided by only pairwise interactions the Cauchy pressure is zero when the crystal is at its equilibrium volume. Therefore, $C_{12} = C_{44}$ for all pairwise interaction models of cubic crystals at their equilibrium volumes. The experimental fact that C_{12} and C_{44} are not equal to each other in any cubic crystal at equilibrium indicates that the pairwise interaction model fails as a description of atomic interactions. Not knowing any better in the 19th century Cauchy assumed a pairwise interaction model for all atomic interactions and showed there are six such relations between the elastic constants. They are: $C_{23} = C_{44}$; $C_{14} = C_{56}$; $C_{31} = C_{55}$; $C_{25} = C_{46}$; $C_{12} = C_{66}$; $C_{45} = C_{36}$. In a cubic crystal $C_{23} = C_{31} = C_{12}$ and $C_{66} = C_{55} = C_{44}$ so we expect C_{12} to equal C_{44}. These six equalities are known as the Cauchy relations and Cauchy used them to argue (incorrectly) that the maximum number of independent elastic constants is not 21 but 15. The Cauchy relations are now only of historical significance. But the extent to which they are violated is an indication of how well atomic interactions in a crystal may be described by pair potentials only.

(ii) In metals with nearly free electrons, such as aluminium and the alkali metals, the cohesive energy of the crystal is described quite accurately by a term that depends only on the average electron density in the metal together with a sum of pairwise interaction energies which are themselves dependent on the local electron density[a]. The presence of the density dependent energy and the density dependence of the pairwise interactions ensures the Cauchy pressure is not zero, and that $C_{12} \neq C_{44}$. Similarly, in transition metals the Finnis-Sinclair model[b] also ensures $C_{12} \neq C_{44}$ through the addition to a sum of pairwise interaction energies a new term that is the square root of a sum of pairwise interactions. The square root ensures the Cauchy pressure is not zero.

[a] see Chapter 6 of the book by Finnis.
[b] see section 7.10 of the book by Finnis.

Problem 3.4 With respect to arbitrary rotations of the coordinate system prove that c_{ijij} and c_{iijj} are invariant in all crystals. In a cubic crystal show that the invariant

$c_{ijij} - c_{iijj}$ is equal to $6\,(C_{12} - C_{44})$.

Solution Under an arbitrary rotation of the coordinate system c_{ijij} becomes:

$$\begin{aligned} c'_{ijij} &= R_{im}R_{jn}R_{ip}R_{jq}c_{mnpq} \\ &= R_{im}R_{ip}R_{jn}R_{jq}c_{mnpq} \\ &= R^T_{mi}R_{ip}R^T_{nj}R_{jq}c_{mnpq} \\ &= \delta_{mp}\delta_{nq}c_{mnpq} \\ &= c_{mnmn} \end{aligned}$$

where we have used the orthogonality of the rotation matrix twice. We see that c_{ijij} is an invariant of any fourth rank tensor.

Similarly, c_{iijj} becomes:

$$\begin{aligned} c'_{iijj} &= R_{im}R_{in}R_{jp}R_{jq}c_{mnpq} \\ &= R^T_{mi}R_{in}R^T_{pj}R_{jq}c_{mnpq} \\ &= \delta_{mn}\delta_{pq}c_{mnpq} \\ &= c_{mmnn}, \end{aligned}$$

which proves that c_{iijj} is an invariant of all fourth rank tensors.
In any crystal symmetry:

$$\begin{aligned} c_{ijij} &= c_{1111} + c_{1212} + c_{1313} + \\ & \quad c_{2121} + c_{2222} + c_{2323} + \\ & \quad c_{3131} + c_{3232} + c_{3333} \\ &= C_{11} + C_{22} + C_{33} + 2(C_{44} + C_{55} + C_{66}). \end{aligned}$$

Hence, in a cubic crystal $c_{ijij} = 3(C_{11} + 2C_{44})$.

In any crystal symmetry:

$$c_{iijj} = c_{1111} + c_{1122} + c_{1133} +$$

$$c_{2211} + c_{2222} + c_{2233} +$$

$$c_{3311} + c_{3322} + c_{3333}$$

$$= \quad C_{11} + C_{22} + C_{33} + 2(C_{12} + C_{23} + C_{31}).$$

Hence, in a cubic crystal $c_{iijj} = 3(C_{11} + 2C_{12})$.

Therefore, in a cubic crystal $c_{mnmn} - c_{mmpp} = 6(C_{44} - C_{12})$.

Comment

In a cubic crystal $\frac{1}{2}(C_{12} - C_{44})$ is called the Cauchy pressure. In a model where atoms interact through pair potentials we have seen in problem 3.3 that the contribution they make to $C_{12} - C_{44}$ is $-\frac{1}{12\Omega}\sum_n X^{(n)}V'(X^{(n)})$. If there is no other contribution to the cohesion of the crystal then $C_{12} = C_{44}$ because $P_{pair} = -(\mathrm{d}E_c/\mathrm{d}\Omega)_{\gamma=0} = -\frac{1}{6\Omega}\sum_n X^{(n)}V'(X^{(n)}) = 0$ is the pressure arising from the pair potentials, which must be zero when the crystal is at its equilibrium volume. However, if there is an additional contribution to the cohesive energy of the crystal, such as a term that is purely density dependent, then P_{pair} can be non-zero at the equilibrium volume as long as it is cancelled by the contribution to the pressure arising from the additional term. In that case the Cauchy pressure is non-zero. A difference between experimentally measured values of C_{12} and C_{44} in a cubic crystal indicates that the cohesion of the crystal cannot be attributed to pair potentials alone.

Problem 3.5 Show that $c_{ijij} = C_{11} + C_{22} + C_{33} + 2(C_{44} + C_{55} + C_{66})$ and $c_{iijj} = C_{11} + C_{22} + C_{33} + 2(C_{12} + C_{13} + C_{23})$. Show that c_{iijj} is directly related to the bulk modulus in any crystal structure.

Solution The first part of the problem was answered in the previous solution. The bulk modulus B relates the hydrostatic pressure, p to the dilation, $e = e_{kk}$:

$$p = -Be.$$

Starting from Hooke's law, $\sigma_{ij} = c_{ijkl}e_{kl}$, the stress generated by a dilation $e_{kk} = e_{kl}\delta_{lk}$, is:

$$\sigma_{ij} = c_{ijkl}e_{kl}\delta_{lk} = c_{ijkk}e_{kk}.$$

The hydrostatic stress $\sigma_{ii}/3 = \sigma_{ij}\delta_{ji}/3$. Therefore,

$$p = -\frac{\sigma_{ii}}{3} = -\frac{1}{3}\sigma_{ij}\delta_{ij} = -\frac{1}{3}c_{ijkk}\delta_{ij}e_{kk} = -\frac{1}{3}c_{iikk}e_{kk} = -Be_{kk}.$$

Therefore $B = c_{iikk}/3$.

Comments

It is no surprise the bulk modulus is an invariant of the elastic constant tensor because it relates an invariant of the stress tensor to an invariant of the strain tensor. In other words, the bulk modulus relates two scalar quantities, pressure and dilation, and therefore it must itself be a scalar and therefore invariant.

The Cauchy relations $C_{44} = C_{23}$, $C_{55} = C_{31}$, $C_{66} = C_{12}$ apply in any crystal structure where cohesion is provided by only pair potentials. In that case the invariance of c_{ijij} follows from the invariance of c_{iijj} and the invariance of the difference $c_{mmpp} - c_{mnmn} = 2(C_{12} + C_{23} + C_{31}) - 2(C_{44} + C_{55} + C_{66})$, which is always zero.

Problem 3.6 Consider a polycrystal in which all crystals are elastically anisotropic and identical except for the orientations of their crystal axes. In general, when an arbitrary homogeneous stress is applied to the polycrystal the strain generated within each crystal is different owing to the elastic anisotropy. Consequently to maintain continuity of displacements and tractions at the grain boundaries additional stress fields are generated. These additional stresses are called *compatibility stresses*, and they are often of the same order of magnitude as the applied stress. However, there is a notable exception to this general rule. Show that if a purely hydrostatic stress is applied to a polycrystal, in which all crystals have the same *cubic* structure, there are no compatibility stresses required at the grain boundaries because the strain in all crystals is the same pure dilation.

Solution The absence of compatibility stresses is a consequence of the invariance of $s_{iijj} = 3(s_{11} + s_{12})$, where s_{ijkl} is the elastic compliance tensor.

Consider a grain in the polycrystal where the elastic compliance tensor is s_{ijkl} expressed in a coordinate frame aligned with the crystal axes. The applied hydrostatic stress is described by $\sigma_{ij} = h\delta_{ij}$. The strain generated in the grain by this hydrostatic stress is:

$$e_{ij} = s_{ijkl}\sigma_{kl} = s_{ijkl}h\delta_{kl}$$

$$= \left\{s_{12}\delta_{ij}\delta_{kl} + s_{44}(\delta_{ik}\delta_{jl} + \delta_{il}\delta_{jk}) + (s_{11} - s_{12} - 2s_{44})\delta_{ij}\delta_{jk}\delta_{kl}\right\} h\delta_{kl}$$

$$= \left\{3s_{12}\delta_{ij} + 2s_{44}\delta_{ij} + (s_{11} - s_{12} - 2s_{44})\delta_{ij}\right\} h$$

$$= (s_{11} + 2s_{12})\delta_{ij}h$$

Therefore the hydrostatic stress produces a pure dilation of $3(s_{11} + 2s_{12})h$ in this grain. Since $s_{11} + 2s_{12}$ is invariant in cubic crystals the same dilation is produced in all crystals, irrespective of the orientation of their crystal axes. Hence there are no incompatibilities at the grain boundaries.

Problem 3.7 In a monoclinic crystal the only rotational symmetry is a two-fold rotation axis along x_3. Prove that the following elastic 8 constants are zero:
$C_{14}, C_{15}, C_{24}, C_{25}, C_{34}, C_{35}, C_{46}, C_{56}$.

Solution Under a two-fold rotation about x_3: $x_1 \rightarrow -x_1, x_2 \rightarrow -x2, x_3 \rightarrow x_3$. Therefore, any c_{ijkl} is zero in which the total number of indices equal to 1 or 2 is odd:

$$\begin{aligned}
C_{14} &= c_{1123} \rightarrow -c_{1123} = 0 \\
C_{15} &= c_{1113} \rightarrow -c_{1113} = 0 \\
C_{24} &= c_{2223} \rightarrow -c_{2223} = 0 \\
C_{25} &= c_{2213} \rightarrow -c_{2213} = 0 \\
C_{34} &= c_{3323} \rightarrow -c_{3323} = 0 \\
C_{35} &= c_{3313} \rightarrow -c_{3313} = 0 \\
C_{46} &= c_{2312} \rightarrow -c_{2312} = 0 \\
C_{56} &= c_{1312} \rightarrow -c_{1312} = 0
\end{aligned}$$

The elastic constant tensor has the following form:

$$\mathbf{C} = \begin{pmatrix} C_{11} & C_{12} & C_{13} & 0 & 0 & C_{16} \\ C_{12} & C_{22} & C_{23} & 0 & 0 & C_{26} \\ C_{13} & C_{23} & C_{33} & 0 & 0 & C_{36} \\ 0 & 0 & 0 & C_{44} & C_{45} & 0 \\ 0 & 0 & 0 & C_{45} & C_{55} & 0 \\ C_{16} & C_{26} & C_{36} & 0 & 0 & C_{66} \end{pmatrix}.$$

There are 13 independent elastic constants in a monoclinic crystal.

Problem 3.8 In an orthorhombic crystal there are two-fold rotation axes along each of the cartesian axes x_1,x_2,x_3. Starting from the reduced elastic constant matrix of the monoclinic crystal prove that the effect of the two-fold rotation axes along x_1 or x_2 is to make the following 4 additional elastic constants zero: $C_{16},C_{26},C_{36},C_{45}$. Hence show there are 9 independent elastic constants in an orthorhombic crystal.

Solution Consider a two-fold rotation axis along x_2: $x_1 \to -x_1, x_2 \to x_2, x_3 \to -x_3$. Therefore, any c_{ijkl} is zero in which the total number of indices equal to 1 or 3 is odd. This leads to the following additional elastic constants equalling zero:

$$\begin{aligned}
C_{16} &= c_{1112} \to -c_{1112} = 0 \\
C_{26} &= c_{2212} \to -c_{2212} = 0 \\
C_{36} &= c_{3312} \to -c_{3312} = 0 \\
C_{45} &= c_{2313} \to -c_{2313} = 0
\end{aligned}$$

No further elastic constants are eliminated by considering the effect of the two-fold rotation axis along x_1. The elastic constant tensor in an orthorhombic crystal has the following form:

$$\mathbf{C} = \begin{pmatrix}
C_{11} & C_{12} & C_{13} & 0 & 0 & 0 \\
C_{12} & C_{22} & C_{23} & 0 & 0 & 0 \\
C_{13} & C_{23} & C_{33} & 0 & 0 & 0 \\
0 & 0 & 0 & C_{44} & 0 & 0 \\
0 & 0 & 0 & 0 & C_{55} & 0 \\
0 & 0 & 0 & 0 & 0 & C_{66}
\end{pmatrix}$$

There are 9 independent elastic constants in an orthorhombic crystal.

Problem 3.9 In a tetragonal crystal the x_3-axis is a four-fold rotational symmetry axis. Starting from the reduced elastic constant matrix of the orthorhombic crystal prove that the effect of the four-fold rotation axis is to make: $C_{11} = C_{22}$; $C_{13} = C_{23}$; $C_{44} = C_{55}$. Hence show that there are 6 independent elastic constants in a tetragonal crystal with point group 422 (D_4): $C_{11},C_{12},C_{13},C_{33},C_{44},C_{66}$. The only difference between this elastic constant matrix and the elastic constant matrix for hexagonal crystals is that C_{66} is no longer independent in a hexagonal crystal: $C_{66} = \frac{1}{2}(C_{11} - C_{12})$.

Solution Putting the four-fold rotation axis of the tetragonal crystal along x_3, a clockwise rotation of $\pi/4$ about the positive x_3 axis sends: $x_1 \rightarrow x_2, x_2 \rightarrow -x_1, x_3 \rightarrow x_3$:

$$\begin{aligned} C_{11} &= c_{1111} \rightarrow c_{2222} = C_{22} \\ C_{13} &= c_{1133} \rightarrow c_{2233} = C_{23} \\ C_{44} &= c_{2323} \rightarrow c_{1313} = C_{55} \end{aligned}$$

The elastic constant tensor in a tetragonal crystal has the following form:

$$\mathbf{C} = \begin{pmatrix} C_{11} & C_{12} & C_{13} & 0 & 0 & 0 \\ C_{12} & C_{11} & C_{13} & 0 & 0 & 0 \\ C_{13} & C_{13} & C_{33} & 0 & 0 & 0 \\ 0 & 0 & 0 & C_{44} & 0 & 0 \\ 0 & 0 & 0 & 0 & C_{44} & 0 \\ 0 & 0 & 0 & 0 & 0 & C_{66} \end{pmatrix}$$

There are 6 independent elastic constants in a tetragonal crystal.

Problem 3.10 Show that in hexagonal crystals the elastic constant matrix is invariant with respect to rotations about the x_3 axis. The elastic constant matrix in a hexagonal crystal is as follows:

$$\mathbf{C} = \begin{pmatrix} C_{11} & C_{12} & C_{13} & 0 & 0 & 0 \\ C_{12} & C_{11} & C_{13} & 0 & 0 & 0 \\ C_{13} & C_{13} & C_{33} & 0 & 0 & 0 \\ 0 & 0 & 0 & C_{44} & 0 & 0 \\ 0 & 0 & 0 & 0 & C_{44} & 0 \\ 0 & 0 & 0 & 0 & 0 & \frac{1}{2}(C_{11} - C_{12}) \end{pmatrix} \tag{3.52}$$

This is why hexagonal crystals are described as *transversely isotropic.*

Solution A rotation by θ about the x_3 axis is represented by the following rotation matrix:

$$\mathbf{R} = \begin{pmatrix} \cos\theta & -\sin\theta & 0 \\ \sin\theta & \cos\theta & 0 \\ 0 & 0 & 1 \end{pmatrix}$$

Using the transformation law:

$$c'_{mnpq} = R_{mi}R_{nj}R_{pk}R_{ql}c_{ijkl}$$

we obtain:

$$\begin{aligned}
C'_{11} = c'_{1111} &= R_{11}R_{11}R_{11}R_{11}c_{1111} + R_{11}R_{11}R_{12}R_{12}c_{1122} \\
&+ R_{12}R_{12}R_{11}R_{11}c_{2211} + R_{11}R_{12}R_{11}R_{12}c_{1212} \\
&+ R_{12}R_{11}R_{11}R_{12}c_{2112} + R_{11}R_{12}R_{12}R_{11}c_{1221} \\
&+ R_{12}R_{11}R_{12}R_{11}c_{2121} + R_{12}R_{12}R_{12}R_{12}c_{2222} \\
&= \cos^4\theta C_{11} + \sin^4\theta C_{11} + 2\sin^2\theta\cos^2\theta C_{12} \\
&\quad +4\sin^2\theta\cos^2\theta\frac{1}{2}(C_{11} - C_{12}) \\
&= (\cos^2\theta + \sin^2\theta)^2 C_{11} \\
&= C_{11}
\end{aligned}$$

$$\begin{aligned}
C'_{12} = c'_{1122} &= R_{11}R_{11}R_{21}R_{21}c_{1111} + R_{11}R_{11}R_{22}R_{22}c_{1122} \\
&+ R_{12}R_{12}R_{21}R_{21}c_{2211} + R_{11}R_{12}R_{21}R_{22}c_{1212} \\
&+ R_{12}R_{11}R_{21}R_{22}c_{2112} + R_{11}R_{12}R_{22}R_{21}c_{1221} \\
&+ R_{12}R_{11}R_{22}R_{21}c_{2121} + R_{12}R_{12}R_{22}R_{22}c_{2222} \\
&= 2\sin^2\theta\cos^2\theta C_{11} + (\cos^4\theta + \sin^4\theta)C_{12} \\
&\quad -4\sin^2\theta\cos^2\theta\frac{1}{2}(C_{11} - C_{12}) \\
&= (\cos^2\theta + \sin^2\theta)^2 C_{12} \\
&= C_{12}
\end{aligned}$$

$$
\begin{aligned}
C'_{13} = c'_{1133} &= R_{11}R_{11}R_{33}R_{33}c_{1133} + R_{12}R_{12}R_{33}R_{33}c_{2233} \\
&= (\cos^2\theta + \sin^2\theta)C_{13} \\
&= C_{13}
\end{aligned}
$$

$$
\begin{aligned}
C'_{33} = c'_{3333} &= R_{33}R_{33}R_{33}R_{33}c_{3333} \\
&= C_{33}
\end{aligned}
$$

$$
\begin{aligned}
C'_{44} = c'2323 &= R_{22}R_{33}R_{22}R_{33}c_{2323} + R_{21}R_{33}R_{21}R_{33}c_{1133} \\
&= (\cos^2\theta + \sin^2\theta)C_{44} \\
&= C_{44}
\end{aligned}
$$

$C'_{66} = C_{66}$ because we have already shown $C'_{11} = C_{11}$ and $C'_{12} = C_{12}$.
Therefore, the elastic constant tensor in a hexagonal crystal is invariant with respect to arbitrary rotations about the x_3 axis.

Problem 3.11 Show that the following restrictions must apply to the elastic constants in a hexagonal crystal: $C_{11} > 0$; $C_{11} > C_{12}$; $C_{44} > 0$; $C_{33}(C_{11} + C_{12}) > 2C_{13}^2$. Hence deduce that $C_{33} > 0$; $C_{11} + C_{12} > 0$.

Solution The subdeterminants of the matrix of elastic constants all have to be positive. Therefore, $C_{11} > 0$, $C_{11} > C_{12}$, $C_{44} > 0$ follow immediately.

This leaves the following 3×3 determinant which must be positive:

$$
\begin{vmatrix} C_{11} & C_{12} & C_{13} \\ C_{12} & C_{11} & C_{13} \\ C_{13} & C_{13} & C_{33} \end{vmatrix} = (C_{11} - C_{12})(C_{11}C_{33} + C_{12}C_{33} - 2C_{13}^2) > 0
$$

Therefore, since $C_{11} > C_{12}$ we must also have $C_{33}(C_{11} + C_{12}) > 2C_{13}^2$.
Since $C_{33}(C_{11}^2 - C_{12}^2) > 2C_{13}^2(C_{11} - C_{12}) > 0$ and $C_{11} > C_{12}$ we must have $C_{33} > 0$ and $C_{11} + C_{12} > 0$.

To summarise, the following inequalities must be satisfied for the elastic energy density to be positive definite: $C_{11} > 0$, $C_{11}^2 > C_{12}^2$, $C_{33} > 0$, $C_{44} > 0$.

Problem 3.12 If the elastic constants of a crystal are averaged over all possible orientations with respect to a fixed coordinate system the elastic constants obtained are those of an isotropic medium. To obtain the elastic constants λ and μ of this isotropic medium we can use the two invariants of the elastic constant tensor c_{ijij} and c_{iijj} mentioned in problem 3.4.

Show that in an isotropic medium $c_{ijij} = 3\lambda + 12\mu$ and $c_{iijj} = 9\lambda + 6\mu$.

In a cubic crystal show that $c_{ijij} = 3C_{11} + 6C_{44}$ and $c_{iijj} = 3C_{11} + 6C_{12}$.

Hence show that:

$$\begin{aligned}\lambda &= C_{12} + \frac{1}{5}(C_{11} - C_{12} - 2C_{44}) \\ \mu &= C_{44} + \frac{1}{5}(C_{11} - C_{12} - 2C_{44})\end{aligned}$$

Hence show that the average values of C_{11}, C_{12} and C_{44} over all orientations of a cubic crystal are given by:

$$\begin{aligned}5\langle C_{11}\rangle &= 3C_{11} + 2C_{12} + 4C_{44} \\ 5\langle C_{12}\rangle &= C_{11} + 4C_{12} - 2C_{44} \\ 5\langle C_{44}\rangle &= C_{11} - C_{12} + 3C_{44} \qquad (3.53)\end{aligned}$$

where $\langle C_{11}\rangle = \langle C_{12}\rangle + 2\langle C_{44}\rangle$. These averages may be obtained directly from eqs 3.24, 3.27 and 3.26 respectively.

Solution In an isotropic medium:

$$\begin{aligned}c_{ijij} &= c_{ijkl}\delta_{ik}\delta_{jl} = \left(\lambda\delta_{ij}\delta_{kl} + \mu(\delta_{ik}\delta_{jl} + \delta_{il}\delta_{jk})\right)\delta_{ik}\delta_{jl} \\ &= 3(\lambda + 4\mu) \\ c_{iijj} &= c_{ijkl}\delta_{ij}\delta_{kl} = \left(\lambda\delta_{ij}\delta_{kl} + \mu(\delta_{ik}\delta_{jl} + \delta_{il}\delta_{jk})\right)\delta_{ij}\delta_{kl} \\ &= 3(3\lambda + 2\mu)\end{aligned}$$

In a cubic crystal $c_{ijij} = 3(C_{11} + 2C_{44})$ and $c_{iijj} = 3(C_{11} + 2C_{12})$. Equating $\lambda + 4\mu = C_{11} + 2C_{44}$ and $3\lambda + 2\mu = C_{11} + 2C_{12}$, we find:

$$\begin{aligned}\lambda &= C_{12} + \frac{1}{5}(C_{11} - C_{12} - 2C_{44}) \\ \mu &= C_{44} + \frac{1}{5}(C_{11} - C_{12} - 2C_{44}).\end{aligned}$$

When the elastic constants are averaged over all orientations the averages are the elastic constants of an isotropic medium. That is, $\langle C_{11}\rangle = \langle C_{12}\rangle + 2\langle C_{44}\rangle$. Then

$$\begin{aligned}\langle C_{12}\rangle &= \lambda = C_{12} + \frac{1}{5}(C_{11} - C_{12} - 2C_{44}) \\ \langle C_{44}\rangle &= \mu = C_{44} + \frac{1}{5}(C_{11} - C_{12} - 2C_{44}) \\ \langle C_{11}\rangle &= \lambda + 2\mu = C_{11} - \frac{2}{5}(C_{11} - C_{12} - 2C_{44}).\end{aligned}$$

4. The Green's function

4.1 Differential equation for the displacement field

Imagine you are applying pressure with your finger against a piece of supported rubber. You see the rubber distorts until you can't push any further. Equilibrium has been achieved between the force you are applying through your finger and the stresses created within the rubber by the distortion. This is the physics of eq 2.15: $\sigma_{ij,j}(\mathbf{x}) + f_i(\mathbf{x}) = 0$. The force f_i acting at $\mathbf{x}$ creates a stress field with a divergence that exactly balances the force and keeps the body in equilibrium. We can make this more explicit by using the divergence theorem. Let the body force be a point force located at $\mathbf{x}_0$, such that $f_i(\mathbf{x}) = \mathcal{F}_i\delta(\mathbf{x}-\mathbf{x}_0)$. Then we can write:

$$\int_{\mathcal{R}} f_i \,\mathrm{d}V = \int_{\mathcal{R}} \mathcal{F}_i\delta(\mathbf{x}-\mathbf{x}_0)\,\mathrm{d}V = \mathcal{F}_i = -\int_{\mathcal{R}} \sigma_{ij,j}\,\mathrm{d}V = -\int_{\mathcal{S}} \sigma_{ij}n_j\,\mathrm{d}S. \tag{4.1}$$

Here $\mathcal{R}$ is any region in the continuum containing the body force at $\mathbf{x}_0$. The surface of $\mathcal{R}$ is $\mathcal{S}$. The stresses created by the distortion of the medium in response to the body force at $\mathbf{x}_0$ give rise to tractions, $\sigma_{ij}n_j$, on the surface $\mathcal{S}$, which when integrated over the whole surface surrounding $\mathcal{R}$ exactly cancel the force $\mathcal{F}_i$.

If we substitute Hooke's law into the equilibrium condition $\sigma_{ij,j}(\mathbf{x}) + f_i(\mathbf{x}) = 0$ we obtain a differential equation for the strain field created by the body force: $c_{ijkl}\,e_{kl,j}(\mathbf{x}) + f_i(\mathbf{x}) = 0$. If we then use the relationship between the strain and the displacement field, $e_{kl}(\mathbf{x}) = \frac{1}{2}(u_{k,l}(\mathbf{x}) + u_{l,k}(\mathbf{x}))$, we obtain a second order partial differential equation for the displacement field:

$$c_{ijkl}\,u_{k,lj}(\mathbf{x}) + f_i(\mathbf{x}) = 0, \tag{4.2}$$

where we have used the symmetry of the elastic constant tensor to write $c_{ijkl}\,u_{l,k} = c_{ijlk}\,u_{l,k}$. The assertion that the strain field is the symmetrized gradient of the displacement field, $e_{kl}(\mathbf{x}) = \frac{1}{2}(u_{k,l}(\mathbf{x}) + u_{l,k}(\mathbf{x}))$, ensures that the material surrounding the region of the body force fits together compatibly, that is with no holes or overlapping material[1]. This assertion cannot be made at points where body forces act.

Equation 4.2 plays the same role in elasticity as Poisson's equation in electrostatics. The equivalent in elasticity of Laplace's equation in electrostatics is $c_{ijkl}u_{k,lj} = 0$. Whereas electrostatics relies on invisible electric fields to transmit forces between charges, in elasticity the forces acting between defects are conveyed by the elastic displacement field which can sometimes be made visible in an electron microscope using modern imaging techniques.

The apparent simplicity of eq 4.2 is deceptive. It comprises three equations, one for each component of the body force. Each of these equations involves second derivatives of all three components of the displacement field. Below we consider the simplest case, which arises when the isotropic elastic approximation is made.

Boundary value problems in elasticity amount to finding solutions of eq 4.2 subject to boundary conditions of three principal types. The first is where displacements are prescribed on the surface of the body. The second is where tractions on the surface of the body are prescribed. The third is a mixture of the first two, where displacements are prescribed on parts of the surface of the body and tractions on the surface of the body where displacements are not prescribed. There are uniqueness theorems for the solutions of eq 4.2 for both simply connected and multiply connected bodies[2].

4.1.1 Navier's equation

The elastic constant tensor in isotropic elasticity is conveniently expressed in eq 3.11. When this is inserted into eq 4.2 we obtain the following differential equation:

$$\mu u_{i,jj} + (\lambda + \mu) u_{k,ki} + f_i = 0, \tag{4.3}$$

which can be expressed in vector form as follows:

$$\mu \nabla^2 \mathbf{u} + (\lambda + \mu)\nabla(\nabla \cdot \mathbf{u}) + \mathbf{f} = 0. \tag{4.4}$$

This is known as Navier's equation.

Exercise 4.1 Derive eq 4.3 from eq 3.11 and eq 4.2. ■

---

[1]For an illuminating discussion of compatibility and incompatibility see section 10 p.65-73 of Mura, T, *Micromechanics of defects in solids* 2nd ed. (1991) Kluwer: Dordrecht. ISBN 90-247-3256-5. Toshio Mura 1925-2009. Japanese born US scientist and engineer.

[2]see Teodosiu, C, *Elastic models of crystal defects*, section 6.2, Springer Verlag: Berlin (1982). ISBN 0-387-11226-X. Cristian Victor Teodosiu 1937- , Romanian mathematician, physicist and engineer.

Solution Substituting the expression for the isotropic elastic constant tensor into the equation of motion we obtain:

$$\begin{aligned} 0 &= \left(\lambda\delta_{ij}\delta_{kl} + \mu(\delta_{ik}\delta_{jl} + \delta_{il}\delta_{jk}\right) u_{k,lj} + f_i \\ &= \lambda u_{k,ki} + \mu(u_{i,jj} + u_{j,ji}) + f_i \\ &= (\lambda + \mu) u_{j,ji} + \mu u_{i,jj} + f_i. \end{aligned}$$

4.2 The Green's function in elasticity

4.2.1 The physical meaning of the Green's function

Consider an infinite homogeneous elastic medium in which there is a point force **f** acting at $\mathbf{x} = \mathbf{x}_0$. In linear elasticity the Green's function gives the elastic displacement field created by this point force:

$$u_i(\mathbf{x}) = G_{ij}(\mathbf{x} - \mathbf{x}_0) f_j(\mathbf{x}_0). \tag{4.5}$$

This may be taken as the definition of the Green's function in linear elasticity, although it does not by itself enable the Green's function to be evaluated. Since eq 4.2 is linear the displacement due to a distribution of body forces $\mathbf{f}(\mathbf{x})$ is found by linear superposition:

$$u_i(\mathbf{x}) = \int G_{ij}(\mathbf{x} - \mathbf{x}') f_j(\mathbf{x}') \mathrm{d}^3 x'. \tag{4.6}$$

The elastic fields of structural defects in crystals arise from forces between atoms in the centre of the defect, otherwise known as the core of the defect. Atoms move from their perfect crystal positions until the forces on them are counteracted by forces from their neighbours, including those further from the defect. The neighbours of the neighbours move from their ideal crystal positions until the net forces on them return to zero, and so the displacement field spreads from the defect. The displacements each neighbour shell undergoes decay with distance from the defect because the forces are distributed among more atoms. For example, consider a missing atom in a crystal, which is called a vacancy. The atoms neighbouring the vacancy experience a net force as a result of the missing atom. They move in response to the net forces acting on them, which sets up forces on their neighbours, and so their neighbours also move but by generally smaller amounts. When equilibrium is re-established the net force acting on any atom is zero and an elastic displacement field is set up in the crystal which decays, in an infinite crystal, as the inverse square of the distance from the vacancy.

The Green's function is unlikely to estimate accurately the displacements of atoms in the core of the defect where the forces may be very large. In this region it is best to use an atomistic model with an accurate model of atomic interactions. But it is generally found that at distances of no more than a nanometre or so from the defect core the description of the relaxation displacements is quite accurately described by linear anisotropic elasticity. Conversely, atomistic models are inappropriate for long-range interactions between defects because elasticity theory is likely to be more accurate for long-range interactions and provide more physical insight.

Before we derive the equation that determines the Green's function in linear elasticity we note some of its properties. First, since the medium is assumed to be homogeneous it follows that G_{ij} depends only on the relative position of the field point $\mathbf{x}$ and the location $\mathbf{x}_0$ of the point force. If the material were inhomogeneous then G_{ij} would become a function of both $\mathbf{x}$ and $\mathbf{x}_0$. Secondly, $G_{ij}(\mathbf{x}) = G_{ij}(-\mathbf{x})$ because the infinite, homogeneous elastic continuum is everywhere centrosymmetric.

Thirdly, $G_{ij}(\mathbf{x}) = G_{ji}(\mathbf{x})$. This follows from a result in mechanics, known as Maxwell's reciprocity theorem[3], which may be stated as follows: the work done by a force $\mathbf{F}^{(1)}$ when its point of application is displaced by $\mathbf{u}^{(2)}$ due to another force $\mathbf{F}^{(2)}$ is equal to the work done by the force $\mathbf{F}^{(2)}$ when its point of application is displaced by $\mathbf{u}^{(1)}$ due to the force $\mathbf{F}^{(1)}$.

To prove this theorem we apply a point force $\mathbf{F}^{(1)}$ gradually at $\mathbf{x}^{(1)}$. It produces a displacement $\mathbf{u}^{(1)}$ at $\mathbf{x}^{(1)}$, and the work done is $W^{(1)} = \frac{1}{2}\mathbf{F}^{(1)} \cdot \mathbf{u}^{(1)}(\mathbf{x}^{(1)})$. The factor of a half is because the force is gradually built up from zero to $\mathbf{F}^{(1)}$, during which the displacement of $\mathbf{x}^{(1)}$ increases linearly from zero to its final value $\mathbf{u}^{(1)}$.

Now we introduce the force $\mathbf{F}^{(2)}$ gradually at $\mathbf{x}^{(2)}$. The additional work done is $W^{(2)} + W^{(12)}$ where $W^{(2)} = \frac{1}{2}\mathbf{F}^{(2)} \cdot \mathbf{u}^{(2)}(\mathbf{x}^{(2)})$ and $W^{(12)} = \mathbf{F}^{(1)} \cdot \mathbf{u}^{(2)}(\mathbf{x}^{(1)})$. Note the absence of the factor of a half in $W^{(12)}$ because the force $\mathbf{F}^{(1)}$ at $\mathbf{x}^{(1)}$ already exists in full. Thus the total work done is $\frac{1}{2}\mathbf{F}^{(1)} \cdot \mathbf{u}^{(1)}(\mathbf{x}^{(1)}) + \frac{1}{2}\mathbf{F}^{(2)} \cdot \mathbf{u}^{(2)}(\mathbf{x}^{(2)}) + \mathbf{F}^{(1)} \cdot \mathbf{u}^{(2)}(\mathbf{x}^{(1)})$.

Repeat the process, but introduce first the force $\mathbf{F}^{(2)}$ gradually at $\mathbf{x}^{(2)}$ and then introduce the force $\mathbf{F}^{(1)}$ gradually at $\mathbf{x}^{(1)}$. The total work done must be the same, but now it is $\frac{1}{2}\mathbf{F}^{(1)} \cdot \mathbf{u}^{(1)}(\mathbf{x}^{(1)}) + \frac{1}{2}\mathbf{F}^{(2)} \cdot \mathbf{u}^{(2)}(\mathbf{x}^{(2)}) + \mathbf{F}^{(2)} \cdot \mathbf{u}^{(1)}(\mathbf{x}^{(2)})$. Thus, we arrive at Maxwell's reciprocity relation:

$$\mathbf{F}^{(1)} \cdot \mathbf{u}^{(2)}(\mathbf{x}^{(1)}) = \mathbf{F}^{(2)} \cdot \mathbf{u}^{(1)}(\mathbf{x}^{(2)}) \tag{4.7}$$

Since $u_i^{(1)}(\mathbf{x}^{(2)}) = G_{ij}(\mathbf{x}^{(2)} - \mathbf{x}^{(1)})F_j^{(1)}$ and $u_i^{(2)}(\mathbf{x}^{(1)}) = G_{ij}(\mathbf{x}^{(1)} - \mathbf{x}^{(2)})F_j^{(2)}$ then the left hand side of eq 4.7 becomes:

$$F_i^{(1)}u_i^{(2)}(\mathbf{x}^{(1)}) = F_i^{(1)}G_{ij}(\mathbf{x}^{(1)} - \mathbf{x}^{(2)})F_j^{(2)}$$

[3]James Clerk Maxwell FRS 1831-1879. The theorem is in this paper: Maxwell, J C, *On the calculation of the equilibrium and stiffness of frames*, Philosophical Magazine, **27**, 294-299 (1864).

and the right-hand side becomes

$$F_i^{(2)} u_i^{(1)}(\mathbf{x}^{(2)}) = F_i^{(2)} G_{ij}(\mathbf{x}^{(2)} - \mathbf{x}^{(1)}) F_j^{(1)}.$$

Swopping the dummy indices i and j in the last line and using $G_{ij}(\mathbf{x}) = G_{ij}(-\mathbf{x})$ the right hand side becomes $F_j^{(2)} G_{ji}(\mathbf{x}^{(1)} - \mathbf{x}^{(2)}) F_i^{(1)} = F_i^{(1)} G_{ji}(\mathbf{x}^{(1)} - \mathbf{x}^{(2)}) F_j^{(2)}$. Equating this to the left hand-side, $F_i^{(1)} G_{ij}(\mathbf{x}^{(1)} - \mathbf{x}^{(2)}) F_j^{(2)}$, we obtain the final result, $G_{ij}(\mathbf{x}) = G_{ji}(\mathbf{x})$.

Green's functions can be defined for other *linear* differential equations. For example for Poisson's equation in a vacuum, $\nabla^2 V(\mathbf{x}) = -\rho(\mathbf{x})/\varepsilon_0$, the Green's function is the familiar potential at $\mathbf{x}$ of a unit point charge at $\mathbf{x}'$: $G(\mathbf{x} - \mathbf{x}') = 1/(4\pi\varepsilon_0 |\mathbf{x} - \mathbf{x}'|)$. The linearity of Poisson's equation enables the potential at $\mathbf{x}$ of a charge density $\rho(\mathbf{x}')$ to be calculated as a linear superposition of the potentials from the charges $\rho(\mathbf{x}')\mathrm{d}^3x'$ in the distribution:

$$\begin{aligned} V(\mathbf{x}) &= \int G(\mathbf{x} - \mathbf{x}')\rho(\mathbf{x}')\mathrm{d}^3x' \\ &= \int \frac{\rho(\mathbf{x}')}{4\pi\varepsilon_0 |\mathbf{x} - \mathbf{x}'|} \mathrm{d}^3x'. \end{aligned}$$

4.2.2 The equation for the Green's function in an infinite medium

Consider a point force $\mathbf{F}$ applied at $\mathbf{x}_0$ in an infinite, homogeneous elastic continuum. From the definition of the Green's function, eq 4.5, this force sets up an elastic displacement field given by $u_i(\mathbf{x}) = G_{ij}(\mathbf{x} - \mathbf{x}_0) F_j(\mathbf{x}_0)$. Differentiating this displacement field and using Hooke's law we find the stress field associated with this displacement field is as follows:

$$\sigma_{kp}(\mathbf{x}) = c_{kpim} G_{ij,m}(\mathbf{x} - \mathbf{x}_0) F_j(\mathbf{x}_0),$$

where the derivative of the Green's function is with respect to x_m. Consider any region $\mathcal{R}$, with surface $\mathcal{S}$, containing $\mathbf{x}_0$. For mechanical equilibrium we must have:

$$F_k(\mathbf{x}_0) + \int_{\mathcal{S}} \sigma_{kp}(\mathbf{x}) n_p \, \mathrm{d}S(\mathbf{x}) = 0,$$

where n_p is the outward normal at $\mathbf{x}$ to the surface $\mathcal{S}$. Therefore,

$$F_k(\mathbf{x}_0) + \int_{\mathcal{S}} c_{kpim} G_{ij,m}(\mathbf{x} - \mathbf{x}_0) F_j(\mathbf{x}_0) n_p \, \mathrm{d}S(\mathbf{x}) = 0.$$

Applying the divergence theorem to the surface integral we transform it into a volume integral over $\mathcal{R}$:

$$F_k(\mathbf{x}_0) + \int_{\mathcal{R}} c_{kpim} G_{ij,mp}(\mathbf{x}-\mathbf{x}_0) F_j(\mathbf{x}_0)\, \mathrm{d}V = 0.$$

We bring $F_k(\mathbf{x}_0)$ inside the volume integral by writing it as $\int_{\mathcal{R}} \delta_{jk}\, \delta(\mathbf{x}-\mathbf{x}_0) F_j(\mathbf{x}_0) \mathrm{d}V$, where the integration is again over $\mathbf{x}$ inside the region $\mathcal{R}$. We then obtain:

$$\int_{\mathcal{R}} \left[c_{kpim} G_{ij,mp}(\mathbf{x}-\mathbf{x}_0) + \delta_{jk}\, \delta(\mathbf{x}-\mathbf{x}_0) \right] F_j(\mathbf{x}_0)\, \mathrm{d}V = 0.$$

Since this must hold for all point forces at $\mathbf{x}_0$, and for all regions $\mathcal{R}$ containing $\mathbf{x}_0$, the expression in square brackets must be zero:

$$c_{kpim} G_{ij,mp}(\mathbf{x}-\mathbf{x}_0) + \delta_{jk}\, \delta(\mathbf{x}-\mathbf{x}_0) = 0. \tag{4.8}$$

This is the partial differential equation that enables us to calculate the Green's function.

Exercise 4.2 Show that $\delta(h\mathbf{x}) = (1/|h|^3)\delta(\mathbf{x})$, where h is any scaling factor. Therefore the delta function in eq 4.8 behaves as a homogeneous function of degree -3. Hence show that

$$G_{ij}(\mathbf{x}-\mathbf{x}_0) = \frac{1}{|\mathbf{x}-\mathbf{x}_0|} g_{ij}, \tag{4.9}$$

where g_{ij} depends only on the orientation of $\mathbf{x}-\mathbf{x}_0$. This separation of the Green's function into radial and orientational dependencies applies in all cases regardless of the degree of anisotropy. ■

Solution Consider the integral

$$f(x) = \int_{-\infty}^{\infty} f(u)\delta(u-x)\mathrm{d}u$$

where $f(x)$ is an arbitrary function. Let $x = h\alpha$ and $u = h\beta$ where h is a scaling factor. We assume first that h is positive. Then we have

$$f(h\alpha) = \int_{-\infty}^{\infty} f(h\beta)\, h\, \delta(h(\beta-\alpha))\, \mathrm{d}\beta.$$

If h is negative let us write $h = -H$ where $H > 0$.

$$\begin{aligned} f(-H\alpha) &= \int_{\infty}^{-\infty} f(-H\beta)\,(-H)\,\delta(H(\alpha-\beta))\,\mathrm{d}\beta \\ &= \int_{-\infty}^{\infty} f(-H\beta)\, H\, \delta(H(\alpha-\beta))\,\mathrm{d}\beta \end{aligned}$$

Therefore, regardless of the sign of h we have:

$$f(h\alpha) = \int_{-\infty}^{\infty} f(h\beta)\, |h|\, \delta(h(\beta-\alpha))\, \mathrm{d}\beta.$$

We may also write

$$f(h\alpha) = \int_{\infty}^{-\infty} f(h\beta)\delta(\beta-\alpha)\mathrm{d}\beta$$

Comparing the last two integrals we deduce that

$$\delta(h(\beta-\alpha)) = \frac{1}{|h|}\delta(\beta-\alpha).$$

Thus, the delta function $\delta(\beta-\alpha)$ is a homogeneous function of degree -1. Since,

$$\delta(\mathbf{u}-\mathbf{x}) = \delta(u_1-x_1)\delta(u_2-x_2)\delta(u_3-x_3)$$

then

$$\begin{aligned} \delta\,(h(\mathbf{u}-\mathbf{x})) &= \delta\,(h(u_1-x_1))\,\delta\,(h(u_2-x_2))\,\delta\,(h(u_3-x_3)) \\ &= \frac{\delta(u_1-x_1)\delta(u_2-x_2)\delta(u_3-x_3)}{|h|^3} \end{aligned}$$

It follows that $\delta(\mathbf{x}-\mathbf{x}_0)$ is a homogeneous function of $|\mathbf{x}-\mathbf{x}_0|$ of degree -3. Therefore, the distance dependence of the second derivative, $G_{ij,mp}(\mathbf{x}-\mathbf{x}_0)$, must vary as $|\mathbf{x}-\mathbf{x}_0|^{-3}$, and hence the distance dependence of $G_{ij}(\mathbf{x}-\mathbf{x}_0)$ must be $|\mathbf{x}-\mathbf{x}_0|^{-1}$. It follows that,

$$G_{ij}(\mathbf{x}-\mathbf{x}_0) = \frac{1}{|\mathbf{x}-\mathbf{x}_0|} g_{ij},$$

where g_{ij} depends only on the orientation of $\mathbf{x}-\mathbf{x}_0$.

4.2.3 Solving elastic boundary value problems with the Green's function

This section illustrates the usefulness of the Green's function for solving boundary value problems. To do this we do not need explicit functional forms for the Green's function, only its defining differential equation, eq 4.8. We will see how and why we can use the Green's function for an infinite medium even when we are dealing with a finite medium, which of course has a surface. We shall use two spatial variables $\mathbf{x}$ and $\mathbf{x}'$. Differentiation with respect to the primed variable will be indicated by a prime on the subscript, thus $\partial f/\partial x'_m \equiv f_{,m'}$. If there is no prime on a subscript after a comma it signifies differentiation with respect to the corresponding component of $\mathbf{x}$.

All linear elastic fields must satisfy the equation of mechanical equilibrium, eq 4.2:

$$c_{kpim}\, u_{i,m'p'}(\mathbf{x}') + f_k(\mathbf{x}') = 0. \tag{4.10}$$

Replacing $\mathbf{x}_0$ in eq 4.8 with $\mathbf{x}'$, the Green's function $G_{ij}(\mathbf{x}-\mathbf{x}')$ satisfies:

$$c_{kpim} G_{ij,mp}(\mathbf{x}-\mathbf{x}') + \delta_{jk}\,\delta(\mathbf{x}-\mathbf{x}') = 0. \tag{4.11}$$

Multiplying eq 4.11 by $u_k(\mathbf{x}')$ and eq 4.10 by $G_{kj}(\mathbf{x}-\mathbf{x}')$ and subtracting we obtain:

$$c_{kpim} G_{ij,mp}(\mathbf{x}-\mathbf{x}') u_k(\mathbf{x}') + \delta_{jk}\,\delta(\mathbf{x}-\mathbf{x}') u_k(\mathbf{x}') - c_{kpim}\, u_{i,m'p'}(\mathbf{x}') G_{kj}(\mathbf{x}-\mathbf{x}')$$
$$- f_k(\mathbf{x}') G_{kj}(\mathbf{x}-\mathbf{x}') = 0.$$

Integrating this equation with respect to $\mathbf{x}'$ over a region $\mathcal{R}$ containing $\mathbf{x}$ we get:

$$\begin{aligned} u_j(\mathbf{x}) \;=\; & \int_{\mathcal{R}} G_{jk}(\mathbf{x}-\mathbf{x}') f_k(\mathbf{x}') \mathrm{d}^3x' \\ +\; & \int_{\mathcal{R}} c_{kpim}\, u_{i,m'p'}(\mathbf{x}') G_{kj}(\mathbf{x}-\mathbf{x}') \mathrm{d}^3x' \end{aligned}$$

$$- \int_{\mathcal{R}} c_{kpim} G_{ij,m'p'}(\mathbf{x}-\mathbf{x}') u_k(\mathbf{x}') \mathrm{d}^3x',$$

where we have used $G_{ij,mp}(\mathbf{x}-\mathbf{x}') = G_{ij,m'p'}(\mathbf{x}-\mathbf{x}')$. We can combine the last two volume integrals as follows:

$$\int_{\mathcal{R}} c_{kpim} u_{i,m'p'}(\mathbf{x}') G_{kj}(\mathbf{x}-\mathbf{x}') \mathrm{d}^3x' - \int_{\mathcal{R}} c_{kpim} G_{ij,m'p'}(\mathbf{x}-\mathbf{x}') u_k(\mathbf{x}') \mathrm{d}^3x' =$$
$$\int_{\mathcal{R}} c_{kpim} \left[u_{i,m'}(\mathbf{x}') G_{kj}(\mathbf{x}-\mathbf{x}') - u_k(\mathbf{x}') G_{ij,m'}(\mathbf{x}-\mathbf{x}')\right]_{,p'} \mathrm{d}^3x',$$

where we have used the symmetry of the elastic constant tensor $c_{kpim} = c_{imkp}$ to achieve a cancellation of two additional terms that arise in the differentiation of the expression in square brackets. Applying the divergence theorem to the resulting volume integral we obtain the following surface integrals over the surface $\mathcal{S}$ of the region $\mathcal{R}$:

$$\int_{\mathcal{R}} c_{kpim} \left[u_{i,m'}(\mathbf{x}') G_{kj}(\mathbf{x}-\mathbf{x}') - u_k(\mathbf{x}') G_{ij,m'}(\mathbf{x}-\mathbf{x}')\right]_{,p'} \mathrm{d}^3x' =$$
$$\int_{\mathcal{S}} G_{jk}(\mathbf{x}-\mathbf{x}') c_{kpmi} u_{i,m'}(\mathbf{x}') n_{p'} \mathrm{d}^2x' - \int_{\mathcal{S}} G_{ji,m'}(\mathbf{x}-\mathbf{x}') c_{mikp} u_k(\mathbf{x}') n_{p'} \mathrm{d}^2x',$$

where we observe that $c_{kpmi} u_{i,m'} n_{p'} = \sigma_{kp}(\mathbf{x}') n_{p'} = t_k(\mathbf{x}')$ is the traction acting on $\mathcal{S}$ at $\mathbf{x}'$. Replacing the last two volume integrals in the above equation for $u_j(\mathbf{x})$ with these two surface integrals we obtain finally:

$$\begin{aligned} u_j(\mathbf{x}) &= \int_{\mathcal{R}} G_{jk}(\mathbf{x}-\mathbf{x}') f_k(\mathbf{x}') \mathrm{d}^3x' \\ &+ \int_{\mathcal{S}} G_{jk}(\mathbf{x}-\mathbf{x}') t_k(\mathbf{x}') \mathrm{d}^2x' \\ &- \int_{\mathcal{S}} G_{ji,m'}(\mathbf{x}-\mathbf{x}') c_{mikp} u_k(\mathbf{x}') n_{p'} \mathrm{d}^2x'. \end{aligned} \tag{4.12}$$

Here are some observations to illustrate the usefulness of this result. First, consider a distribution of body force, such as that created by a structural defect, inside a finite body with free surfaces. If we use only the first line of eq 4.12 to evaluate the displacement field in the finite body the answer will be wrong because the Green's function is constructed for an infinite body. That is because the displacement field will predict tractions at the surface of the finite body, when the surface should be free of such tractions. To correct this we may calculate the displacement field caused by an equal and opposite distribution of surface tractions using the second line of eq 4.12. We may then use the superposition principle and add this to the displacements caused by the distribution of body forces in the first line to obtain the correct solution with surfaces free of tractions. In this way we are able to use the Green's function for an infinite body to solve a problem in a finite body, which is very convenient because all bodies are finite in practice. Secondly, if there are additional tractions applied to the surface of the body they may be included in the surface integral on the second line. The surface integral on the third line is used when displacements are specified on the surface $\mathcal{S}$. Here the surface may include a cut made from the external surface of the body into some point inside the body, where there is a discontinuity in the displacement across the cut. This will be very useful when we discuss dislocations in Chapter 6.

It may be puzzling that the derivative of the Green's function appears in the third line of eq 4.12 to satisfy the boundary condition where a displacement is specified. To get some insight into this we offer a slightly modified version of an argument from the book[4] by Landau[5] and Lifshitz[6]. Consider an infinite planar fault with unit normal $\hat{\mathbf{n}}$ in an infinite continuum. The half-space into which $\hat{\mathbf{n}}$ points is designated positive, the other half-space negative. The fault introduces a rigid translation by $\boldsymbol{\tau}$ of the entire negative half-space relative to the entire positive half-space. The relative translation is established by displacing the negative half-space by $\mathbf{u} = \boldsymbol{\tau}/2$, and displacing the positive half space by $\mathbf{u} = -\boldsymbol{\tau}/2$. To deal with the singularity in the displacement gradient in the plane of the fault we assign a finite thickness Δ to the fault, and subsequently take the limit $\Delta \longrightarrow 0$. The displacement gradient is then $u_{i,k} = -\tau_i n_k/\Delta$ inside the fault and zero outside. The strain tensor is the constant value $e_{ik} = -\frac{1}{2}(\tau_i n_k + n_i \tau_k)/\Delta$ inside the fault and zero outside. Thus the stress changes discontinuously at the planes where the displacement gradient begins and ends. Discontinuous changes in stress are unphysical because they correspond to infinite forces. Therefore we use the equilibrium condition $\sigma_{im,m} + f_i = 0$ to introduce sheets of body force to cancel these discontinuities. The displacement field is then given by:

$$u_i(\mathbf{x}) = \int_{\text{all space}} G_{ij}(\mathbf{x}-\mathbf{x}')f_j(\mathbf{x}')\mathrm{d}^3x'$$

[4]Landau, L D, and Lifshitz, E M, *Theory of Elasticity*, Pergamon Press: Oxford, 3rd edition (1986), section 27, p.111. ISBN 0-08-033916-6.

[5]Lev Davidovich Landau 1908-1968. Nobel prize winning Soviet physicist

[6]Evgeny Mikhailovich Lifshitz ForMemRS 1915-1985. Soviet physicist

$$
\begin{aligned}
&= -\int_{\text{all space}} G_{ij}(\mathbf{x}-\mathbf{x}')\sigma_{jm,m'}(\mathbf{x}')\mathrm{d}^3x' \\
&= -\int_{\text{all space}} G_{ij}(\mathbf{x}-\mathbf{x}')c_{jmkl}e_{kl,m'}(\mathbf{x}')\mathrm{d}^3x' \\
&= -\int_{\text{all space}} \left\{\left(G_{ij}(\mathbf{x}-\mathbf{x}')c_{jmkl}e_{kl}(\mathbf{x}')\right)_{,m'} - G_{ij,m'}(\mathbf{x}-\mathbf{x}')c_{jmkl}e_{kl}(\mathbf{x}')\right\}\mathrm{d}^3x' \\
&= -\int_{\text{fault surfaces}} \left(G_{ij}(\mathbf{x}-\mathbf{x}')c_{jmkl}e_{kl}(\mathbf{x}')\right)n_{m'}\mathrm{d}^2x' \\
&\quad + \int_{\text{all space}} G_{ij,m'}(\mathbf{x}-\mathbf{x}')c_{jmkl}e_{kl}(\mathbf{x}')\mathrm{d}^3x'.
\end{aligned}
$$

The surface integral is over the surfaces of the fault region where the displacement gradient begins and ends. In the limit $\Delta \longrightarrow 0$ these surface integrals cancel because the surface normal changes sign on either side of the fault. The region of integration in the volume integral in the last line reduces to the volume occupied by the fault, which is Δ per unit area; this cancels the Δ in the denominator of e_{kl}. As $\Delta \longrightarrow 0$ we obtain:

$$
\begin{aligned}
u_i(\mathbf{x}) &= -\int_{\text{fault plane}} G_{ij,m'}(\mathbf{x}-\mathbf{x}')c_{jmkl}\frac{1}{2}(\tau_k n_l + n_k \tau_l)\mathrm{d}^2x' \\
&= -\int_{\text{fault plane}} G_{ij,m'}(\mathbf{x}-\mathbf{x}')c_{jmkl}\tau_k n_l \mathrm{d}^2x', \qquad (4.13)
\end{aligned}
$$

where we have used the symmetry of the elastic constant tensor $c_{jmkl} = c_{jmlk}$ in the final integral.

In summary, to create a fault with normal $\hat{\mathbf{n}}$ and a constant relative translation of $\boldsymbol{\tau}$ we introduce dipolar sheets of forces on either side of the fault. It is because the constant displacement is created by force dipoles that the derivative of the Green's function is involved. There is an analogy here with a jump in the electrostatic potential on either side of an interface at which there are dipolar sheets of charges.

4.3 A general formula for the Green's function

We turn now to deriving a general formula for the Green's function in an anisotropic elastic continuum. We shall follow the treatment given in the review[7] by Bacon, Barnett and Scattergood. There are only two cases where the Green's function can be obtained exactly and analytically for any direction within a crystal, and they are the cases of elastic isotropy and hexagonal symmetry. However, in problem 4.3 perturbation theory is used to derive an approximate analytic form of the elastic Green's function in a cubic crystal when the anisotropy ratio differs only slightly from unity.

We start from eq 4.8:

$$c_{jlms} G_{mp,ls}(\mathbf{x}) + \delta_{jp}\,\delta(\mathbf{x}) = 0,$$

where we have used the translational invariance of the infinite, homogeneous continuum to replace $\mathbf{x} - \mathbf{x}_0$ with just $\mathbf{x}$. Taking the Fourier transform of this equation we obtain:

$$c_{jlms} k_l k_s \tilde{G}_{mp}(\mathbf{k}) = \delta_{jp},$$

where

$$\begin{aligned}\tilde{G}_{mp}(\mathbf{k}) &= \int G_{mp}(\mathbf{x}) e^{i\mathbf{k}\cdot\mathbf{x}} \mathrm{d}^3x \\ G_{mp}(\mathbf{x}) &= \frac{1}{(2\pi)^3}\int \tilde{G}_{mp}(\mathbf{k}) e^{-i\mathbf{k}\cdot\mathbf{x}} \mathrm{d}^3k.\end{aligned}$$

We now introduce a matrix (kk) that plays a central role in many aspects of anisotropic elasticity. It is where all the information about the elastic constants is stored:

$$(kk)_{jm} = c_{jlms} k_l k_s, \tag{4.14}$$

Note that (kk) is a symmetric matrix. The 3×3 matrix of Fourier transforms is just the inverse of the matrix (kk):

$$\tilde{G}_{mp}(\mathbf{k}) = \left[(kk)^{-1}\right]_{mp},$$

where we put square brackets around $(kk)^{-1}$ to make clear that $\tilde{G}_{mp}(\mathbf{k})$ is the mp-element of the inverse matrix of (kk) not the inverse of the matrix element $(kk)_{mp}$. Taking the inverse transform we obtain:

$$G_{mp}(\mathbf{x}) = \frac{1}{(2\pi)^3}\int \left[(kk)^{-1}\right]_{mp} e^{-i\mathbf{k}\cdot\mathbf{x}} \mathrm{d}^3k \tag{4.15}$$

[7] Bacon, D J, Barnett, D M and Scattergood R O, *Anisotropic continuum theory of lattice defects*, Progress in Materials Science, **23**, 51-262 (1979). David J Bacon, British materials physicist and engineer. David M Barnett, US materials physicist and engineer. Ronald O Scattergood, US materials engineer.

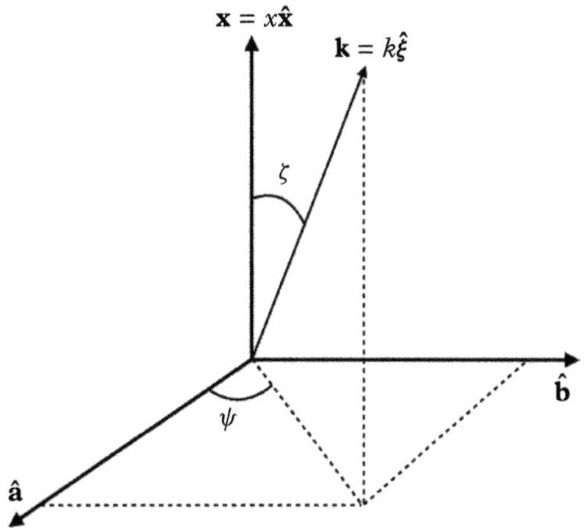

Figure 4.1: To illustrate the vectors and angles used to define **x** and **k**.

We introduce some additional vectors to help us evaluate this triple integral. Let the unit vectors $\hat{\mathbf{e}}_1, \hat{\mathbf{e}}_2, \hat{\mathbf{e}}_3$ be aligned with the cartesian axes used to define the elastic constant tensor. We may express $\mathbf{x}$ in spherical polar coordinates defined with respect to $\hat{\mathbf{e}}_1, \hat{\mathbf{e}}_2, \hat{\mathbf{e}}_3$ as follows:

$$\mathbf{x} = x\,\hat{\boldsymbol{\rho}} = x\,[\sin\theta\cos\phi\,\hat{\mathbf{e}}_1 + \sin\theta\sin\phi\,\hat{\mathbf{e}}_2 + \cos\theta\,\hat{\mathbf{e}}_3],$$

where we have also introduced x as the length of $\mathbf{x}$ and $\hat{\boldsymbol{\rho}}$ as the direction of $\mathbf{x}$. It is useful to introduce two more orthonormal vectors $\hat{\mathbf{a}}$ and $\hat{\mathbf{b}}$ which are perpendicular to $\mathbf{x}$, such that $\hat{\mathbf{a}}, \hat{\mathbf{b}}$ and $\hat{\boldsymbol{\rho}}$ form a right-handed set, $\hat{\mathbf{a}} \times \hat{\mathbf{b}} = \hat{\boldsymbol{\rho}}$:

$$\begin{aligned}
\hat{\mathbf{a}} &= \sin\phi\hat{\mathbf{e}}_1 - \cos\phi\hat{\mathbf{e}}_2 \\
\hat{\mathbf{b}} &= \cos\theta\cos\phi\hat{\mathbf{e}}_1 + \cos\theta\sin\phi\hat{\mathbf{e}}_2 - \sin\theta\hat{\mathbf{e}}_3 \\
\hat{\boldsymbol{\rho}} &= \sin\theta\cos\phi\hat{\mathbf{e}}_1 + \sin\theta\sin\phi\hat{\mathbf{e}}_2 + \cos\theta\hat{\mathbf{e}}_3.
\end{aligned}$$

Let k be the length of the vector $\mathbf{k}$ and let $\hat{\boldsymbol{\xi}}$ be the direction of $\mathbf{k}$. Then $\mathbf{k}$ may be expressed in the coordinate system $\hat{\mathbf{a}}, \hat{\mathbf{b}}, \hat{\boldsymbol{\rho}}$ as follows:

$$\mathbf{k} = k\hat{\boldsymbol{\xi}} = k\left[\sin\zeta\cos\psi\hat{\mathbf{a}} + \sin\zeta\sin\psi\hat{\mathbf{b}} + \cos\zeta\hat{\boldsymbol{\rho}}\right],$$

where ζ is the angle between $\mathbf{x}$ and $\mathbf{k}$, see Fig 4.1.

At this point we recover the separation of variables in eq 4.9. Since $(kk) = k^2(\xi\xi)$ then the Fourier transform $\tilde{G}_{mp}(\mathbf{k})$ is given by:

$$\tilde{G}_{mp}(\mathbf{k}) = [(kk)^{-1}]_{mp} = \frac{1}{k^2}\left[(\xi\xi)^{-1}\right]_{mp},$$

where we recognise $1/k^2$ as being proportional to the Fourier transform of $1/x$ and the directional dependence of g_{mp} is contained in the inverse transform of $\left[(\xi\xi)^{-1}\right]_{mp}$. We note that since (kk) is symmetric, so is its inverse, and therefore $G_{mp}(\mathbf{x})$ is also symmetric, as required by the Maxwell reciprocity theorem. Since $\mathrm{d}^3k = k^2 \sin\zeta\, \mathrm{d}k\, \mathrm{d}\zeta\, \mathrm{d}\psi$ and $\mathbf{k}\cdot\mathbf{x} = kx\cos\zeta$, eq 4.15 becomes:

$$G_{mp}(\mathbf{x}) = \frac{1}{(2\pi)^3}\int_0^{2\pi}\mathrm{d}\psi\int_0^{\pi}\mathrm{d}\zeta\,\sin\zeta\left[(\xi\xi)^{-1}\right]_{mp}\int_0^{\infty}\mathrm{d}k\frac{1}{\cancel{k^2}}\cancel{k^2}\cos(kx\cos\zeta).$$

The integration over k may be evaluated as follows:

$$\int_0^{\infty}\mathrm{d}k\cos(kx\cos\zeta) = \frac{1}{2}\int_{-\infty}^{\infty}\mathrm{d}k\,e^{ikx\cos\zeta} = \pi\delta(x\cos\zeta) = \frac{\pi}{x\sin\zeta}\delta(\zeta-\pi/2).$$

This simplifies the evaluation of $G_{mp}(\mathbf{x})$ considerably:

$$\begin{aligned} G_{mp}(\mathbf{x}) &= \frac{1}{(2\pi)^3}\int_0^{2\pi}\mathrm{d}\psi\int_0^{\pi}\mathrm{d}\zeta\,\cancel{\sin\zeta}\left[(\xi\xi)^{-1}\right]_{mp}\frac{\pi}{x\cancel{\sin\zeta}}\delta(\zeta-\pi/2) \\ &= \frac{1}{8\pi^2 x}\int_0^{2\pi}\left[(\xi\xi)^{-1}\right]_{mp}\mathrm{d}\psi. \end{aligned} \tag{4.16}$$

The anticipated separation into radial $(1/x)$ and orientational dependencies is now explicit. The delta function $\delta(\zeta-\pi/2)$ requires that the integration with respect to ψ is around the unit circle perpendicular to the vector $\mathbf{x}$, with its centre at the origin. On this circle $\hat{\boldsymbol{\xi}} = \hat{\mathbf{a}}\cos\psi + \hat{\mathbf{b}}\sin\psi$. Thus, the orientational dependence of the Green's function is contained in this line integral. In most cases of elastic anisotropy the line integral may be evaluated only numerically. But it is only one integral to be evaluated for each direction $\hat{\rho}$, in contrast to the three integrals in the inverse Fourier transform of eq 4.15, and it is numerically stable. Finally, it is stressed that eq 4.16 is valid in *all* crystal symmetries. This remarkable result was first derived by Synge[8].

4.4 The Green's function in an isotropic elastic medium

In an isotropic medium the Green's function[9] can be determined directly by taking the inverse Fourier transform in eq 4.15. Using eq 3.11 for the elastic constants in an

[8]John Lighton Synge FRS 1897-1995, Irish mathematician and physicist. The derivation of eq 4.16 appears in his book *The hypercircle in mathematical physics* (1957) CUP: Cambridge, p.411-413.

[9]The solution in an isotropic medium for the displacement field of a point force was first derived by Lord Kelvin FRS in 1848, in a short paper entitled *Note on the integration of the equations of equilibrium*

isotropic elastic medium the matrix element $(kk)_{jm}$ is as follows:

$$(kk)_{jm} = \mu k^2 \delta_{jm} + (\lambda + \mu) k_j k_m \tag{4.17}$$

As we have seen in the previous section, the inverse of this matrix is the matrix of Fourier transforms $(kk)_{jm}\tilde{G}_{mp}(\mathbf{k}) = \delta_{jp}$. Rather than taking the brute force approach and evaluating the inverse of (kk) directly, we may multiply both sides of $(kk)_{jm}\tilde{G}_{mp}(\mathbf{k}) = \delta_{jp}$ by k_j and obtain an equation for $k_m\tilde{G}_{mp}(\mathbf{k})$:

$$k_m \tilde{G}_{mp}(\mathbf{k}) = \frac{k_p}{(\lambda + 2\mu)k^2}$$

When this is substituted into $(\mu k^2 \delta_{jm} + (\lambda + \mu) k_j k_m)\tilde{G}_{mp}(\mathbf{k}) = \delta_{jp}$ we find:

$$\tilde{G}_{jp}(\mathbf{k}) = \frac{\delta_{jp}}{\mu k^2} - \frac{(\lambda + \mu) k_j k_p}{\mu(\lambda + 2\mu) k^4} \tag{4.18}$$

It may be shown by contour integration that:

$$\begin{aligned} \int \frac{1}{k^2} e^{-i\mathbf{k}\cdot\mathbf{x}} \mathrm{d}^3 k &= \frac{2\pi^2}{x} \\ \int \frac{k_j k_p}{k^4} e^{-i\mathbf{k}\cdot\mathbf{x}} \mathrm{d}^3 k &= \pi^2 \left(\frac{\delta_{jp}}{x} - \frac{x_j x_p}{x^3} \right), \end{aligned}$$

where $x = |\mathbf{x}|$. Using these integrals to evaluate the inverse Fourier transform we obtain:

$$\begin{aligned} G_{jp}(\mathbf{x}) &= \frac{1}{8\pi\mu(\lambda + 2\mu)x} \left((\lambda + 3\mu)\delta_{jp} + (\lambda + \mu)\frac{x_j x_p}{x^2} \right) \\ &= \frac{1}{16\pi\mu(1-\nu)x} \left((3 - 4\nu)\delta_{jp} + \frac{x_j x_p}{x^2} \right). \end{aligned} \tag{4.19}$$

We shall use this result extensively.

4.5 The multipole expansion

Consider a point defect in a crystal such as a foreign atom occupying a site that would normally be occupied by a host atom, i.e. a substitutional point defect. Let $\mathbf{f}^{(n)}$ be the excess force exerted on the n'th host atom when the site is occupied by the foreign

of an elastic solid, which can be found in Article 37 of Volume 1 of his *Mathematical and Physical Papers* p.97.

atom as compared to when the site is occupied by a host atom. At equilibrium the net force on each atom is zero. But as we saw in section 2.4 the excess forces exerted by the defect on the host atoms in this equilibrium state are the source of the stress field it creates in the host. We call the excess force that the defect exerts on a host atom a *defect force*. In this equilibrium state the defect may be viewed as a collection of defect forces that sum to zero because their sum is the negative of the net force on the defect atom, which is zero at equilibrium. The elastic displacement field generated by this collection of defect forces may be computed using the superposition principle and the Green's function. This does involve approximations because the Green's function and the superposition principle assume linear elasticity is valid. But beyond some distance from the defect the predicted displacement, stress and strain fields are expected to be accurate. Indeed, when we wish to calculate interaction energies between defects over distances much larger than the spatial extent of their defect forces, elasticity theory is very accurate. It provides also unrivalled physical insight into such interactions.

Let the foreign atom be located at $\mathbf{d}$. Let the position of the n'th neighbour relative to the foreign atom be $\mathbf{X}^{(n)}$, and let the defect force it experiences be $\mathbf{f}^{(n)}$. Therefore the displacement at $\mathbf{x}$ caused by the defect is:

$$u_i(\mathbf{x}) = \sum_n G_{ij}(\mathbf{x}-\mathbf{d}-\mathbf{X}^{(n)})f_j^{(n)} \tag{4.20}$$

Within linear elasticity this is exact, but the approximations of linear elasticity are such that this expression becomes accurate only when $|\mathbf{x}-\mathbf{d}| \gg |\mathbf{X}^{(n)}|$. When this condition is satisfied we can expand the Green's function as a Taylor series:

$$G_{ij}(\mathbf{x}-\mathbf{d}-\mathbf{X}^{(n)}) = G_{ij}(\mathbf{x}-\mathbf{d}) - G_{ij,k}(\mathbf{x}-\mathbf{d})X_k^{(n)} + \frac{1}{2}G_{ij,kl}(\mathbf{x}-\mathbf{d})X_k^{(n)}X_l^{(n)} - \dots$$

Inserting this expansion into eq 4.20 we obtain the following series:

$$\begin{aligned} u_i(\mathbf{x}) &= G_{ij}(\mathbf{x}-\mathbf{d})\sum_n f_j^{(n)} \\ &\quad -G_{ij,k}(\mathbf{x}-\mathbf{d})\sum_n X_k^{(n)} f_j^{(n)} \\ &\quad +\frac{1}{2}G_{ij,kl}(\mathbf{x}-\mathbf{d})\sum_n X_k^{(n)}X_l^{(n)} f_j^{(n)} \\ &\quad -\frac{1}{6}G_{ij,klm}(\mathbf{x}-\mathbf{d})\sum_n X_k^{(n)}X_l^{(n)}X_m^{(n)} f_j^{(n)} \dots \end{aligned} \tag{4.21}$$

This equation is called the multipole expansion of the displacement field of a point defect. At equilibrium the first term on the right hand side of eq 4.21 is zero because the defect forces sum to zero. The next term involves the first moment of the defect forces, which is called the elastic dipole tensor: $\rho_{kj} = \sum_n X_k^{(n)} f_j^{(n)}$. The next term involves the second moment of the defect forces, $q_{klj} = \sum_n X_k^{(n)} X_l^{(n)} f_j^{(n)}$ and is called the elastic quadrupole tensor. The next term involves the third moment of the defect forces $o_{klmj} = \sum_n X_k^{(n)} X_l^{(n)} X_m^{(n)} f_j^{(n)}$, and is called the elastic octupole tensor, and so on.

Exercise 4.3 Equilibrium requires not only that the net force on the point defect is zero but also that the net torque exerted on the defect atom is zero. Show that this condition is satisfied provided the elastic dipole tensor is symmetric: $\rho_{ij} = \rho_{ji}$. ▪

Solution The torque on the point defect is $\mathbf{T} = \sum_n \mathbf{X}^{(n)} \times \mathbf{f}^{(n)}$. In component form:

$$
\begin{aligned}
T_i &= \sum_n \varepsilon_{ijk} X_j^{(n)} f_k^{(n)} \\
&= \frac{1}{2} \sum_n \varepsilon_{ijk} X_j^{(n)} f_k^{(n)} + \varepsilon_{ikj} X_k^{(n)} f_j^{(n)} \\
&= \frac{1}{2} \sum_n \varepsilon_{ijk} \left(X_j^{(n)} f_k^{(n)} - X_k^{(n)} f_j^{(n)} \right) = \frac{1}{2} \varepsilon_{ijk} \sum_n \left(X_j^{(n)} f_k^{(n)} - X_k^{(n)} f_j^{(n)} \right) \\
&= 0 \text{ if } \sum_n X_j^{(n)} f_k^{(n)} - X_k^{(n)} f_j^{(n)} = 0, \text{ i.e. } \rho_{jk} - \rho_{kj} = 0.
\end{aligned}
$$

The multipole expansion reveals how the elastic displacement field of a point defect decays with distance x. The dipole tensor gives rise to a displacement field that decays as $1/x^2$, the quadrupole as $1/x^3$, the octupole as $1/x^4$ and so on. The displacement fields from higher order poles become increasingly significant as the defect is approached. The orientational dependence of each of these contributions to the displacement field is contained in the derivatives of the Green's functions.

The multipole expansion reflects the symmetry of the atomic environment surrounding the point defect. This is very significant for those point defects that have particular symmetries, such as crowdions which have a distinct axial symmetry.

The multipole expansion shows that the long range elastic field caused by a distribution of defect forces, with zero resultant, is determined by the lowest order non-zero moment of the distribution. By 'long range' we mean at separations from the distribution of defect forces larger than its characteristic length scale. Higher order moments of the distribution of defect forces make significant contributions only at smaller separations. Furthermore to calculate the long-range field we can replace the actual force distribution, which may be complicated, by a simple force distribution with the same lowest order non-zero moment. This very useful insight is called *St. Venant's principle*. For example, it provides a simple but accurate approximation to the elastic field of a cluster of point defects at distances larger than the cluster itself. If the cluster has a non-zero dipole moment its displacement field at distances beyond the cluster size may be approximated very well using only the second line of eq 4.21. Point defect clusters occur often in irradiated materials.

Equation 4.21 is an exemplar of *multi-scale modelling*. It takes information from the atomic scale, namely the defect forces, and applies the theory of elasticity to determine the elastic field of the point defect at long range. It is a bridge between the atomic and continuum length scales, and perhaps this is why it was described by Leibfried and Breuer as the most important new concept in defect physics[10]. Kröner[11] was the first to introduce a systematic description of point defects in terms of force multipoles[12].

4.6 Green's functions for an elastic continuum and a crystal lattice

Consider a crystal containing N lattice sites, with one atom at each lattice site. In harmonic lattice theory the potential energy of the distorted crystal is expanded to second order in the displacements of atoms from their perfect crystal positions. Using a slightly different notation from eq 3.28 the potential energy is as follows:

$$E = -\sum_n f_i(\mathbf{X}^{(n)})u_i(\mathbf{X}^{(n)}) + \frac{1}{2}\sum_n\sum_p S_{ij}(\mathbf{X}^{(n)} - \mathbf{X}^{(p)})u_i(\mathbf{X}^{(n)})u_j(\mathbf{X}^{(p)}). \qquad (4.22)$$

The force $f_i(\mathbf{X}^{(n)})$ is not zero this time because now there is a crystal defect that displaces the atom from its perfect crystal position at $\mathbf{X}^{(n)}$. It is equivalent to the body forces we have discussed in elasticity. This is different from eq 3.28, where the forces were all set to zero, because the perfect crystal configuration was being considered with no defects present.

By minimising the energy in eq 4.22 with respect to the displacements we obtain the following equation:

$$f_m(\mathbf{X}^{(n)}) = \sum_p S_{mj}(\mathbf{X}^{(n)} - \mathbf{X}^{(p)})u_j(\mathbf{X}^{(p)}). \qquad (4.23)$$

[10]Leibfried, G and Breuer, N, *Point defect in Metals I*, Springer-Verlag: Berlin, p.146 (1978). ISBN 978-3662154489. Günther Leibfried 1915-1977, German physicist

[11]Ekkehart Kröner 1919-2000. German physicist.

[12]Kröner, E, *Kontinuumstheorie der Versetzungen und Eigenspannungen*, Springer-Verlag: Berlin (1958). ISBN 978-3540022619.

We want the inverse of this relationship, that is we want an expression for the displacements in terms of the forces:

$$u_l(\mathbf{X}^{(n)}) = \sum_p G_{lh}(\mathbf{X}^{(n)} - \mathbf{X}^{(p)}) f_h(\mathbf{X}^{(p)}) \tag{4.24}$$

This equation is equivalent in the discrete lattice to eq 4.6 in the continuum, and it defines the crystal lattice Green's function. In this section we investigate the relationship between this crystal lattice Green's function and the Green's function in linear elasticity. To invert eq 4.23 we use discrete Fourier transforms defined by:

$$\tilde{W}(\mathbf{k}) = \sum_{n=1}^{N} W(\mathbf{X}^{(n)}) e^{i\mathbf{k}\cdot\mathbf{X}^{(n)}} \tag{4.25}$$

$$W(\mathbf{X}^{(p)}) = \frac{1}{N} \sum_{\mathbf{k}} \tilde{W}(\mathbf{k}) e^{-i\mathbf{k}\cdot\mathbf{X}^{(p)}}. \tag{4.26}$$

Born von Karman periodic boundary conditions are applied to the crystal faces, and there are N wave vectors $\mathbf{k}$ in the Brillouin zone. Taking the discrete Fourier transform of eq 4.23 we obtain:

$$\tilde{f}_m(\mathbf{k}) = \tilde{S}_{mj}(\mathbf{k}) \tilde{u}_j(\mathbf{k}), \tag{4.27}$$

where $\tilde{S}_{mj}(\mathbf{k})$ is a 3×3 matrix for each wave vector $\mathbf{k}$ in the Brillouin zone. Inverting this equation we obtain:

$$\tilde{u}_l(\mathbf{k}) = \left[\tilde{S}^{-1}(\mathbf{k})\right]_{lh} \tilde{f}_h(\mathbf{k}). \tag{4.28}$$

Taking the discrete inverse Fourier transform we find:

$$\begin{aligned}
u_l(\mathbf{X}^{(n)}) &= \frac{1}{N} \sum_{\mathbf{k}} \left[\tilde{S}^{-1}(\mathbf{k})\right]_{lh} \tilde{f}_h(\mathbf{k}) e^{-i\mathbf{k}\cdot\mathbf{X}^{(n)}} \\
&= \frac{1}{N} \sum_{\mathbf{k}} \left[\tilde{S}^{-1}(\mathbf{k})\right]_{lh} \left(\sum_p f_h(\mathbf{X}^{(p)}) e^{i\mathbf{k}\cdot\mathbf{X}^{(p)}} \right) e^{-i\mathbf{k}\cdot\mathbf{X}^{(n)}} \\
&= \sum_p \left(\frac{1}{N} \sum_{\mathbf{k}} \left[\tilde{S}^{-1}(\mathbf{k})\right]_{lh} e^{-i\mathbf{k}\cdot(\mathbf{X}^{(n)} - \mathbf{X}^{(p)})} \right) f_h(\mathbf{X}^{(p)})
\end{aligned}$$

Comparing this with eq 4.24 we obtain the desired expression for the crystal lattice Green's function[13].

$$G_{lh}(\mathbf{X}^{(n)} - \mathbf{X}^{(p)}) = \frac{1}{N} \sum_{\mathbf{k}} \left[\tilde{S}^{-1}(\mathbf{k})\right]_{lh} e^{-i\mathbf{k}\cdot(\mathbf{X}^{(n)} - \mathbf{X}^{(p)})}. \tag{4.29}$$

[13]This Green's function has been used to calculate the displacement fields of defects in discrete lattices where atoms interact by harmonic potentials. See Tewary, V K, Advances in Physics **22**, 757-810 (1973).

To aid comparison we quote here the Green's function from eq 4.15:

$$G_{mp}(\mathbf{x}) = \frac{1}{(2\pi)^3}\int \left[(kk)^{-1}\right]_{mp} e^{-i\mathbf{k}\cdot\mathbf{x}}\,\mathrm{d}^3k$$

Since $S_{mj}(\mathbf{X}^{(n)}) = S_{mj}(-\mathbf{X}^{(n)})$, and $\sum_n S_{mj}(\mathbf{X}^{(n)}) = 0$, we may write[14]:

$$\begin{aligned}
\tilde{S}_{mj}(\mathbf{k}) &= \sum_n S_{mj}(\mathbf{X}^{(n)}) e^{i\mathbf{k}\cdot\mathbf{X}^{(n)}} \\
&= \frac{1}{2}\sum_n S_{mj}(\mathbf{X}^{(n)}) \left(e^{i\mathbf{k}\cdot\mathbf{X}^{(n)}} + e^{-i\mathbf{k}\cdot\mathbf{X}^{(n)}} - 2\right) \\
&= \frac{1}{2}\sum_n S_{mj}(\mathbf{X}^{(n)}) \left(2\cos\left(\mathbf{k}\cdot\mathbf{X}^{(n)}\right) - 2\right) \\
&= -2\sum_n S_{mj}(\mathbf{X}^{(n)}) \sin^2\left(\frac{\mathbf{k}\cdot\mathbf{X}^{(n)}}{2}\right)
\end{aligned}$$

In the limit of long wavelengths the magnitude of the wave vector **k** becomes very small. In that limit we may write:

$$\begin{aligned}
\tilde{S}_{mj}(\mathbf{k}) &\to -\frac{1}{2}\sum_n S_{mj}(\mathbf{X}^{(n)}) \left(\mathbf{k}\cdot\mathbf{X}^{(n)}\right)^2 \\
&= \left[-\frac{1}{2}\sum_n S_{mj}(\mathbf{X}^{(n)}) X_h^{(n)} X_l^{(n)}\right] k_h k_l .
\end{aligned}$$

Comparing with the expression for the elastic constant tensor c_{ikjl} in eq 3.31 we recognise the term in square brackets as being Ωc_{mhjl}, and therefore in the limit of long wavelengths $\tilde{S}_{mj}(\mathbf{k}) \to \Omega(kk)_{mj}$.

Each **k**-point in the Brillouin zone is associated with a volume in **k**-space of $(2\pi)^3/(N\Omega)$. Replacing the sum over **k**-points in eq 4.29 with a continuous integral over all **k**-space we find the crystal lattice Green's function, in the limit of long wavelengths, becomes:

$$G_{mp}(\mathbf{X}^{(n)}) \to \frac{1}{N\Omega}\frac{N\Omega}{(2\pi)^3}\int \mathrm{d}^3k\,\left[(kk)^{-1}\right]_{mp} e^{-i\mathbf{k}\cdot\mathbf{X}^{(n)}} = \frac{1}{(2\pi)^3}\int \mathrm{d}^3k\,\left[(kk)^{-1}\right]_{mp} e^{-i\mathbf{k}\cdot\mathbf{X}^{(n)}}, \tag{4.30}$$

which is identical to the Green's function in linear elasticity. We conclude that the elastic and crystal lattice Green's functions are equivalent only in the limit of long

[14]see Ashcroft, N W and Mermin, N D *Solid state physics*, p437-440 (1976). ISBN 978-0030839931.

wavelengths compared to the distance from the force creating the displacement, so that $\mathbf{k}\cdot\mathbf{x} \ll 1$. At smaller wavelengths the lattice Green's function may be expected to deviate significantly from the Green's function in linear elasticity. The implication is that when we are studying the elastic displacement fields of defects with the Green's function in linear elasticity the fields nearer to the defect are less accurate than those further away. At larger distances from the defect the smaller wavelength contributions to the crystal lattice Green's function tend to cancel out, leaving only the longer wavelength components.

4.7 The equation of motion and elastic waves

We shall use Hamilton's principle of classical mechanics[15] to derive the equation of motion of the continuum. Hamilton's principle states that the following integral is stationary with respect to variations δu_i of the displacement field which vanish at times $t = t_1$ and $t = t_2$ and on the surface of the volume V:

$$\delta \int_{t_1}^{t_2} \mathrm{d}t \int_V \mathrm{d}V \mathcal{L}(u_i, \dot{u}_i, u_{i,j}) = 0. \tag{4.31}$$

$\mathcal{L}$ is the Lagrangian density, which is the difference between the kinetic energy and the potential energy densities:

$$\mathcal{L} = \frac{1}{2}\rho\dot{u}_i^2 - \left(-f_i u_i + \frac{1}{2}c_{ijkl}u_{i,j}u_{k,l}\right).$$

In this expression ρ is the mass density, $\dot{u}_i$ is the velocity of the medium at $\mathbf{x}$, and we have included a contribution from body forces f_i in the potential energy density. Substituting this Lagrangian density into eq 4.31 and performing the variation we obtain:

$$\delta \int_{t_1}^{t_2} \mathrm{d}t \int_V \mathrm{d}V \mathcal{L}(u_i, \dot{u}_i, u_{i,j}) = \int_{t_1}^{t_2} \mathrm{d}t \int_V \mathrm{d}V \, \rho\dot{u}_i\delta\dot{u}_i + f_i\delta u_i - c_{ijkl}u_{i,j}\delta u_{k,l}$$

$$= \int_V \mathrm{d}V [\rho\dot{u}_i\delta u_i]_{t=t_1}^{t_2} - \int_{t_1}^{t_2} \mathrm{d}t \int_V \mathrm{d}V \rho\ddot{u}_i\delta u_i - f_i\delta u_i + \left(c_{ijkl}u_{i,j}\delta u_k\right)_{,l} - c_{ijkl}u_{i,jl}\delta u_k$$

[15] for example, see Goldstein, H, *Classical mechanics*, Addison-Wesley: Reading Mass (1980). ISBN 0-201-02969-3.

$$= -\int_{t_1}^{t_2}\int_{S} \cancel{c_{ijkl} u_{i,j} \delta u_k n_l \mathrm{d}S} - \int_{t_1}^{t_2} \mathrm{d}t \int_{V} \mathrm{d}V \left(\rho \ddot{u}_i - f_i - \sigma_{ij,j}\right) \delta u_i \tag{4.32}$$

We have used $\delta u_i = 0$ at $t = t_1$ and $t = t_2$, and on the surface S of the body at all times. We have also used the divergence theorem to convert a volume integral into a surface integral. In order for the variation to be zero the integrand must be zero:

$$\rho \ddot{u}_i = f_i + \sigma_{ij,j}, \tag{4.33}$$

which is the equation of motion of the continuum.

4.7.1 Elastic waves

In the absence of forces in the equation of motion, eq 4.33, we try a wave solution of the form $u_i(\mathbf{x}, t) = A_i(\mathbf{k}, \omega) e^{i(\mathbf{k}\cdot\mathbf{x} - \omega t)}$, which leads to the equation:

$$\left(\rho \omega^2 \delta_{ij} - (kk)_{ij}\right) A_j(\mathbf{k}, \omega) = 0. \tag{4.34}$$

This is a set of three equations in the wave amplitudes $A_i(\mathbf{k}, \omega)$, and the condition for non-trivial solutions is that the following 3×3 determinant is zero:

$$\begin{vmatrix} (kk)_{11} - \rho\omega^2 & (kk)_{12} & (kk)_{13} \\ (kk)_{12} & (kk)_{22} - \rho\omega^2 & (kk)_{23} \\ (kk)_{13} & (kk)_{23} & (kk)_{33} - \rho\omega^2 \end{vmatrix} = 0. \tag{4.35}$$

In general this provides a set of three solutions for $\omega^2 = \omega^2(\mathbf{k})$, which are known as the three branches of the dispersion relations. By writing $(kk) = k^2(\xi\xi)$ and by making ω/k the quantity to be determined the determinant can be rewritten as follows:

$$\begin{vmatrix} (\xi\xi)_{11} - \rho\omega^2/k^2 & (\xi\xi)_{12} & (\xi\xi)_{13} \\ (\xi\xi)_{12} & (\xi\xi)_{22} - \rho\omega^2/k^2 & (\xi\xi)_{23} \\ (\xi\xi)_{13} & (\xi\xi)_{23} & (\xi\xi)_{33} - \rho\omega^2/k^2 \end{vmatrix} = 0. \tag{4.36}$$

This makes it clear that ω is a linear function of k and that ω varies in general with the direction of $\mathbf{k}$. Thus, in general, the group velocity of the wave, $\mathbf{v}_g = \nabla_{\mathbf{k}}\omega(\mathbf{k})$, depends on the direction of the wave but not on its frequency. In particular, we note that in general the wave does not travel in the direction of its wave vector because in general $\mathbf{v}_g$ is not parallel to $\mathbf{k}$. Since the $(\xi\xi)$ matrix is symmetric its eigenvalues are always real and its eigenvectors are always perpendicular to each other. The eigenvectors are the amplitudes $\mathbf{A}(\mathbf{k},\omega)$ in eq 4.34 and their directions relative to the wave vector $\mathbf{k}$ determine whether the waves are longitudinal or transverse or a mixture of the two.

In the isotropic elastic approximation (kk) is given by eq 4.17. One solution of the dispersion equations, eq 4.35, is $c_l^2 = \omega^2/k^2 = (\lambda+2\mu)/\rho$, and is independent of the direction of $\mathbf{k}$, as expected in an isotropic medium. When this solution is substituted back into eq 4.34 it is found that the wave amplitude is parallel to the wave vector and it is called the longitudinal wave or *P-wave*. This wave produces alternating compression and dilation of the medium along the direction of propagation of the wave. The other two solutions are degenerate with $c_t^2 = \omega^2/k^2 = \mu/\rho$, and they are also independent of the direction of the wave vector. In these cases the wave amplitudes are perpendicular to each other and to the wave vector, and they are called transverse waves, or *S-waves*. These waves produce shears of the medium and are not associated with local volume changes. For all three waves the group velocity is always parallel to the wave vector, and there is always one purely longitudinal wave and two purely transverse waves for all wave vectors. This combination of features of the propagation of elastic waves in isotropic materials is unique, and may be found in anisotropic materials only when the wave vector coincides with a direction of high rotational symmetry.

In section 4.6 we saw that the crystal lattice Green's function converged to the elastic Green's function in the limit of long wavelengths. For the same reason the dispersion relations for the crystal lattice converge to those of the elastic continuum in the limit of long wavelengths, i.e. as $|\mathbf{k}| \to 0$.

4.8 The elastodynamic Green's function

Having analysed the properties of elastic waves we are in a position to tackle the time-dependent Green's function of elastodynamics. The time-dependent generalisation of eq 4.8 for the elastostatic Green's function is to consider a point force that is applied as an infinitely abrupt pulse at $t = 0$ that generates a displacement field that propagates away. In an isotropic medium we can expect a spherical longitudinal wave to travel outwards from the point pulse followed by two degenerate spherical shear waves. Thus in the equation of motion, eq 4.33, we set $f_i = \delta_{im}\delta(\mathbf{x})\delta(t)$ to describe a pulse of force of unit magnitude along the x_m-axis at the origin at time $t = 0$. The elastodynamic Green's function in an infinite, homogeneous elastic medium is then defined by the following partial differential equation:

$$c_{ijkl}G_{km,lj}(\mathbf{x},t) + \delta_{im}\delta(\mathbf{x})\delta(t) = \rho\ddot{G}_{im}(\mathbf{x},t) \tag{4.37}$$

where as usual the two dots over the Green's function signify the second derivative with respect to time.

To solve this equation we exploit the translational symmetry of the medium in space and time through four-dimensional Fourier transforms:

$$\tilde{f}(\mathbf{k},\omega) = \iint f(\mathbf{x},t)\, e^{i(\mathbf{k}\cdot\mathbf{x}-\omega t)}\, \mathrm{d}^3x\, \mathrm{d}t \tag{4.38}$$

$$f(\mathbf{x},t) = \frac{1}{(2\pi)^4} \iint \tilde{f}(\mathbf{k},\omega)\, e^{-i(\mathbf{k}\cdot\mathbf{x}-\omega t)}\, \mathrm{d}^3k\, \mathrm{d}\omega \tag{4.39}$$

Applying the four-dimensional Fourier transform to eq 4.37 we obtain:

$$\left((kk)_{ik} - \rho\omega^2\delta_{ik}\right)\tilde{G}_{km}(\mathbf{k},\omega) = \delta_{im}, \tag{4.40}$$

from which it follows that

$$G_{km}(\mathbf{x},t) = \iint \left[((kk) - \rho\omega^2 I)^{-1}\right]_{km} e^{-i(\mathbf{k}\cdot\mathbf{x}-\omega t)}\, \mathrm{d}^3k\, \mathrm{d}\omega$$

where I is the 3×3 identity matrix.

As with the elastostatic Green's function we resort to the isotropic elastic approximation to make further progress[16] whereupon we have:

$$((kk) - \rho\omega^2 I)_{ik} = (\mu k^2 - \rho\omega^2)\delta_{ik} + (\lambda+\mu)k_i k_k$$

Exercise 4.4 Using the same procedure to invert (kk) following eq 4.17 show that

$$\tilde{G}_{im}(\mathbf{k},\omega) = \left[((kk) - \rho\omega^2 I)^{-1}\right]_{im} = \frac{1}{\rho}\left[\frac{\delta_{im} - k_i k_m/k^2}{c_t^2 k^2 - \omega^2} + \frac{k_i k_m/k^2}{c_l^2 k^2 - \omega^2}\right]$$

■

Solution In $\mathbf{k}$-space the isotropic elastodynamic Green's function is defined by:

$$\left\{(\mu k^2 - \rho\omega^2)\delta_{ik} + (\lambda+\mu)k_i k_k\right\}\tilde{G}_{kl}(\mathbf{k},\omega) = \delta_{il}.$$

Multiplying this equation by k_i and summing on i we deduce:

$$k_k \tilde{G}_{kl}(\mathbf{k},\omega) = \frac{k_l}{(\lambda+2\mu)k^2 - \rho\omega^2}.$$

[16] This was first done by Stokes, G G, Trans. Camb. Phil. Soc. **9** (1849), 1-62. Sir George Gabriel Stokes PRS 1819-1903, Irish mathematician and physicist.

Therefore,

$$\tilde{G}_{il}(\mathbf{k},\omega) = \frac{1}{\mu k^2 - \rho\omega^2}\left(\delta_{il} - \frac{(\lambda+\mu)k_i k_l}{(\lambda+2\mu)k^2 - \rho\omega^2}\right).$$

With $c_t^2 = \mu/\rho$ and $c_l^2 = (\lambda+2\mu)/\rho$ this becomes:

$$\tilde{G}_{il}(\mathbf{k},\omega) = \frac{1}{\rho}\left(\frac{\delta_{il}}{(c_t^2k^2-\omega^2)} - \frac{(c_l^2-c_t^2)k_ik_l}{(c_t^2k^2-\omega^2)(c_l^2k^2-\omega^2)}\right).$$

Splitting the last term into partial fractions we obtain:

$$\begin{aligned}\tilde{G}_{il}(\mathbf{k},\omega) &= \frac{1}{\rho}\left(\frac{\delta_{il}}{(c_t^2k^2-\omega^2)} - \frac{k_ik_l/k^2}{(c_t^2k^2-\omega^2)} + \frac{k_ik_l/k^2}{(c_l^2k^2-\omega^2)}\right)\\ &= \frac{1}{\rho}\left(\frac{\delta_{il}-k_ik_l/k^2}{(c_t^2k^2-\omega^2)} - \frac{k_ik_l/k^2}{(c_l^2k^2-\omega^2)}\right).\end{aligned}$$

Taking the inverse Fourier transform we obtain the following Green's function for $t > 0$ for the point force pulse at the origin:

$$\begin{aligned}4\pi\rho x G_{ij}(\mathbf{x},t) &= \frac{\delta(x-c_lt)}{c_l}\frac{x_ix_j}{x^2} + \frac{\delta(x-c_tt)}{c_t}\left(\delta_{ij} - \frac{x_ix_j}{x^2}\right)\\ &+ [H(x-c_lt) - H(x-c_tt)]\,\frac{t}{x^2}\left(\delta_{ij} - \frac{3x_ix_j}{x^2}\right),\end{aligned} \tag{4.41}$$

where $H(x)$ is the Heaviside step function: $H(x) = 1$ if $x > 0$, $H(x) = 0$ if $x < 0$. We see in this solution two spherical waves emanating from the point force pulse at the origin, with the longitudinal wave expanding faster than the shear wave. The material beyond the longitudinal wave front is undeformed.

4.9 Problems

Problem 4.1 Derive the Green's function in linear isotropic elasticity using the line integral in eq 4.16.

Solution We start with

$$\tilde{G}_{mp}(\mathbf{k}) = \left[(kk)^{-1}\right]_{mp} = \frac{1}{k^2}\left[(\xi\xi)^{-1}\right]_{mp} = \frac{1}{k^2}\frac{1}{\mu}\left[\delta_{mp} - \frac{\lambda+\mu}{\lambda+2\mu}\xi_i\xi_m\right].$$

Therefore,

$$\left[(\xi\xi)^{-1}\right]_{mp} = \frac{1}{\mu}\left[\delta_{mp} - \frac{\lambda+\mu}{\lambda+2\mu}\xi_i\xi_m\right].$$

The unit vector $\hat{\xi}$ is $\hat{\mathbf{a}}\cos\psi + \hat{\mathbf{b}}\sin\psi$ where $\hat{\mathbf{a}}$ and $\hat{\mathbf{b}}$ are perpendicular to $\mathbf{x}$. The three vectors $\hat{\rho}$, $\hat{\mathbf{a}}$ and $\hat{\mathbf{b}}$ form a right-handed orthonormal set. $\hat{\mathbf{a}}$ and $\hat{\mathbf{b}}$ are otherwise arbitrary. The matrix element $(\xi\xi)_{im}$ is as follows:

$$\begin{aligned}(\xi\xi)_{im} &= \xi_i\xi_m = (a_i\cos\psi + b_i\sin\psi)(a_m\cos\psi + b_m\sin\psi)\\ &= a_ia_m\cos^2\psi + b_ib_m\sin^2\psi + (a_ib_m + a_mb_i)\sin\psi\cos\psi.\end{aligned}$$

Therefore,

$$\begin{aligned}G_{mp}(\mathbf{x}) &= \frac{1}{8\pi^2 x}\int_0^{2\pi}[(\xi\xi)^{-1}]_{mp}\mathrm{d}\psi\\ &= \frac{1}{8\pi^2 x}\int_0^{2\pi}\frac{\mathrm{d}\psi}{\mu}\Bigg[\delta_{mp}\\ &\quad -\frac{\lambda+\mu}{\lambda+2\mu}\left(a_ma_p\cos^2\psi + b_mb_p\sin^2\psi + (a_mb_p + a_pb_m)\sin\psi\cos\psi\right)\Bigg]\\ &= \frac{1}{8\pi^2\mu x}\left[2\pi\delta_{mp} - \pi\frac{\lambda+\mu}{\lambda+2\mu}(a_ma_p + b_mb_p)\right]\end{aligned}$$

Since $\hat{\rho}$, $\hat{\mathbf{a}}$ and $\hat{\mathbf{b}}$ form a complete set of orthonormal unit vectors then[a] there is the following relation between their components:

$$a_ma_p + b_mb_p + \rho_m\rho_p = \delta_{mp}.$$

Using this relation and $(\lambda+\mu)/(\lambda+2\mu) = 1/[2(1-\nu)]$ we obtain:

$$G_{mp}(\mathbf{x}) = \frac{1}{8\pi^2\mu x}\pi\left(2\delta_{mp} - \frac{1}{2(1-\nu)}(\delta_{mp} - \rho_m\rho_p)\right)$$
$$= \frac{1}{16\pi\mu(1-\nu)x}\left((3-4\nu)\delta_{mp} + \rho_m\rho_p\right).$$

[a]see the first comment to the solution of problem 1.3

Problem 4.2 A *dumb-bell* interstitial defect[17] has axial symmetry and is located at the origin of a cartesian coordinate system in an elastically isotropic medium. With its axis along the unit vector $\hat{\mathbf{e}}$ it exerts a defect force $f\hat{\mathbf{e}}$ at $h\hat{\mathbf{e}}$ and another defect force $-f\hat{\mathbf{e}}$ at $-h\hat{\mathbf{e}}$.

Write down the dipole tensor for the defect.

Show that in an isotropic medium

$$G_{ij,k}(\mathbf{x}) = \frac{-1}{16\pi\mu(1-\nu)x^2}\left((3-4\nu)\rho_k\delta_{ij} - (\rho_j\delta_{ik} + \rho_i\delta_{jk}) + 3\rho_i\rho_j\rho_k\right), \quad (4.42)$$

where $\rho_i = x_i/x$.

Hence show that for $x \gg h$ the displacement field at $\mathbf{x}$ of the dumb-bell defect is given by

$$u_i(\mathbf{x}) = \frac{2fh}{16\pi\mu(1-\nu)x^2}\left[(3\cos^2\alpha - 1)\rho_i + 2(1-2\nu)(\cos\alpha)e_i\right]$$

where $\cos\alpha = e_j\rho_j$.

Hence show that at a given distance x from the point defect the elastic displacement along the axis of the defect is $4(1-\nu)$ larger than the displacement along any direction perpendicular to the defect axis, and of opposite sign.

[17]An interstitial defect occupies a site between those of the host crystal. The defect force for an interstitial defect is therefore the entire force it exerts on a neighbour in the relaxed configuration.

Solution There are only two sites on which the defect forces act. Therefore, the dipole tensor is as follows:

$$d_{ij} = \sum_n x_i^{(n)} f_j^{(n)} = (he_i)(fe_j) + (-he_i)(-fe_j) = 2hfe_ie_j.$$

The derivative of the isotropic elastic Green's function is evaluated as follows:

$$\begin{aligned} G_{ij}(\mathbf{x}) &= \frac{1}{16\pi\mu(1-\nu)}\left((3-4\nu)\delta_{ij}\frac{1}{x} + \frac{x_ix_j}{x^3}\right) \\ G_{ij,k}(\mathbf{x}) &= \frac{1}{16\pi\mu(1-\nu)}\left((3-4\nu)\delta_{ij}\frac{-x_k}{x^3} + \frac{\delta_{ik}x_j + x_i\delta_{jk}}{x^3} - 3\frac{x_ix_jx_k}{x^5}\right) \\ &= \frac{-1}{16\pi\mu(1-\nu)x^2}\left((3-4\nu)\rho_k\delta_{ij} - (\rho_j\delta_{ik} + \rho_i\delta_{jk}) + 3\rho_i\rho_j\rho_k\right) \end{aligned}$$

where $\rho_i = x_i/x$.

The displacement field of the dumb-bell interstitial defect is as follows:

$$\begin{aligned} u_i(\mathbf{x}) &= -G_{ij,k}(\mathbf{x})d_{kj} \\ &= \frac{-2fh}{16\pi\mu(1-\nu)x^2}\left((3-4\nu)\rho_k\delta_{ij} - (\rho_j\delta_{ik} + \rho_i\delta_{jk}) + 3\rho_i\rho_j\rho_k\right)e_ke_j \\ &= \frac{-2fh}{16\pi\mu(1-\nu)x^2}\left((3-4\nu)\cos\alpha\, e_i - \cos\alpha\, e_i - \rho_i + 3\cos^2\alpha\, \rho_i\right) \\ &= \frac{2fh}{16\pi\mu(1-\nu)x^2}\left((3\cos^2\alpha - 1)\rho_i + 2(1-2\nu)\cos\alpha\, e_i\right) \end{aligned}$$

When $\mathbf{x}$ is along the axis of the defect $\alpha = 0$ and $\rho_i = e_i$. Then

$$u_i(\mathbf{x}) = \frac{2fh}{16\pi\mu(1-\nu)x^2}4(1-\nu)\rho_i.$$

When $\mathbf{x}$ is perpendicular to the axis of the defect $\alpha = \pi/2$ and $\cos\alpha = 0$. In that case

$$u_i(\mathbf{x}) = -\frac{2fh}{16\pi\mu(1-\nu)x^2}\rho_i.$$

Thus, we have shown that a given distance x from the point defect the elastic displacement along the axis of the defect is $4(1-\nu)$ larger than the displacement along any direction perpendicular to the defect axis, and of opposite sign.

Comment

This problem illustrates just how non-spherical the elastic displacement fields of point defects can be even in an elastically isotropic material. A crowdion is another point defect with pronounced axial symmetry.

Problem 4.3 *An approximate analytic Green's function for a cubic crystal with small elastic anisotropy.*

Consider a cubic crystal where the elastic anisotropy ratio is close to 1. Then $D = C_{11} - C_{12} - 2C_{44}$ is small compared to C' and C_{44}. Equation 3.22 for the elastic constant tensor in a cubic crystal may always be expressed as

$$c_{ijkl} = c^0_{ijkl} + \delta_{ij}\delta_{jk}\delta_{kl}D,$$

where c^0_{ijkl} is the elastic constant tensor for an isotropic crystal where we choose $C^0_{12} = C_{12} = \lambda$, $C^0_{44} = C_{44} = \mu$. Let $(KK)_{jm} = c_{jlms}k_l k_s$ and $(kk)_{jm} = c^0_{jlms}k_l k_s$. Show that:

$$(KK)^{-1} = (kk)^{-1} - (kk)^{-1}V(KK)^{-1},$$

where $V_{jm} = D(k_1^2\delta_{j1}\delta_{m1} + k_2^2\delta_{j2}\delta_{m2} + k_3^2\delta_{j3}\delta_{m3})$. This equation[18] is exact whatever the magnitude of D. Although it can be solved for $(KK)^{-1}$ the inverse Fourier transform requires the solution of a sextic equation to locate the poles, which can be done only numerically. However, since D/μ is small it suggests a perturbation expansion:

$$(KK)^{-1} = (kk)^{-1} - (kk)^{-1}V(kk)^{-1} + (kk)^{-1}V(kk)^{-1}V(kk)^{-1} - \ldots$$

Using eq 4.18 for $\left[(kk)^{-1}\right]_{jp}$ show that to first order in D/C_{44} the Green's function tensor in the cubic crystal in k-space becomes:

$$\begin{aligned}\tilde{G}_{ij}(\mathbf{k}) &= \left[(KK)^{-1}\right]_{ij} \\ &= \frac{1}{\mu}\frac{\delta_{ij}}{k^2} - \frac{1}{\mu}\left(\frac{\lambda+\mu}{\lambda+2\mu}\right)\frac{k_i k_j}{k^4} \\ &\quad + \frac{D}{\mu}\frac{1}{\mu}\frac{\left(\delta_{i1}k_1^2\delta_{j1} + \delta_{i2}k_2^2\delta_{j2} + \delta_{i3}k_3^2\delta_{j3}\right)}{k^4}\end{aligned}$$

[18] This is a Dyson equation - see Economou, E N, *Green's functions in quantum physics*, Springer-Verlag: Berlin (1983). ISBN 978-3642066917.

$$
-\frac{D}{\mu}\frac{1}{\mu}\left(\frac{\lambda+\mu}{\lambda+2\mu}\right)\frac{\left(k_1^3(\delta_{i1}k_j+k_i\delta_{j1})+k_2^3(\delta_{i2}k_j+k_i\delta_{j2})+k_3^3(\delta_{i3}k_j+k_i\delta_{j3})\right)}{k^6}
$$

$$
+\frac{D}{\mu}\frac{1}{\mu}\left(\frac{\lambda+\mu}{\lambda+2\mu}\right)^2 k_i k_j\frac{(k_1^4+k_2^4+k_3^4)}{k^8} \qquad (4.43)
$$

Hence show that the first order corrections to the Green's functions in isotropic elasticity are:

$$
\Delta\tilde{G}_{11}(\mathbf{k}) = \frac{D}{\mu}\frac{1}{\mu}\left(\frac{k_1^2}{k^4}-\chi\frac{2k_1^4}{k^6}+\chi^2\frac{k_1^2(k_1^4+k_2^4+k_3^4)}{k^8}\right)
$$

$$
\Delta\tilde{G}_{12}(\mathbf{k}) = \frac{D}{\mu}\frac{1}{\mu}\left(-\chi\frac{k_1^3k_2+k_1k_2^3}{k^6}+\chi^2\frac{k_1k_2(k_1^4+k_2^4+k_3^4)}{k^8}\right)
$$

where $\chi=(\lambda+\mu)/(\lambda+2\mu)$. To invert these Fourier transforms the following integral is useful (it may be derived by contour integration):

$$
\frac{1}{(2\pi)^3}\int\frac{1}{k^{2n}}e^{-i\mathbf{k}\cdot\mathbf{x}}\mathrm{d}^3k = -\frac{1}{4\pi}\frac{(-1)^n x^{2n-3}}{(2n-2)!},
$$

where $n=1,2,3,\ldots$. Other integrals needed to evaluate the inverse transforms may by obtained by differentiating this integral with respect to components of $\mathbf{x}$. Thus, we obtain the following inverse transforms:

$$
\frac{1}{(2\pi)^3}\int\frac{k_1^2}{k^4}e^{-i\mathbf{k}\cdot\mathbf{x}}\mathrm{d}^3k = \frac{1}{8\pi x}\left(1-\frac{x_1^2}{x^2}\right)
$$

$$
\frac{1}{(2\pi)^3}\int\frac{k_1^3k_j}{k^6}e^{-i\mathbf{k}\cdot\mathbf{x}}\mathrm{d}^3k = \frac{3}{32\pi x}\left(\delta_{1j}-\frac{x_1x_j}{x^2}\right)\left(1-\frac{x_1^2}{x^2}\right)
$$

$$
\frac{1}{(2\pi)^3}\int\frac{k_1^5k_j}{k^8}e^{-i\mathbf{k}\cdot\mathbf{x}}\mathrm{d}^3k = \frac{5}{64\pi x}\left(\delta_{1j}-\frac{x_1x_j}{x^2}\right)\left(1-\frac{x_1^2}{x^2}\right)^2
$$

$$
\frac{1}{(2\pi)^3}\int\frac{k_1^2k_2^4}{k^8}e^{-i\mathbf{k}\cdot\mathbf{x}}\mathrm{d}^3k = \frac{1}{64\pi x}\left(1-\frac{x_2^2}{x^2}\right)\left(\frac{x_3^2}{x^2}+5\frac{x_1^2x_2^2}{x^4}\right)
$$

$$
\frac{1}{(2\pi)^3}\int\frac{k_1k_2k_3^4}{k^8}e^{-i\mathbf{k}\cdot\mathbf{x}}\mathrm{d}^3k = -\frac{1}{64\pi x}\frac{x_1x_2}{x^2}\left(1-\frac{x_3^2}{x^2}\right)\left(1-5\frac{x_3^2}{x^2}\right)
$$

Armed with these integrals we may derive the following first order corrections to the isotropic Green's function in a weakly anisotropic cubic crystal:

$$\begin{aligned}\Delta G_{11}(\mathbf{x}) &= \frac{1}{\mu}\frac{D}{\mu}\frac{1}{8\pi x}\left[\left(1-\frac{x_1^2}{x^2}\right)-\frac{3\chi}{2}\left(1-\frac{x_1^2}{x^2}\right)^2\right.\\ &\left.+\frac{\chi^2}{8}\left(5\left(1-\frac{x_1^2}{x^2}\right)^3+\left(1-\frac{x_2^2}{x^2}\right)\left(\frac{x_3^2}{x^2}+5\frac{x_1^2x_2^2}{x^4}\right)+\left(1-\frac{x_3^2}{x^2}\right)\left(\frac{x_2^2}{x^2}+5\frac{x_1^2x_3^2}{x^4}\right)\right)\right]\\ \Delta G_{12}(\mathbf{x}) &= \frac{1}{\mu}\frac{D}{\mu}\frac{1}{8\pi x}\frac{x_1x_2}{x^2}\left[\frac{3\chi}{4}\left(1+\frac{x_3^2}{x^2}\right)\right.\\ &\left.-\frac{\chi^2}{8}\left(5\left(\left(1-\frac{x_1^2}{x^2}\right)^2+\left(1-\frac{x_2^2}{x^2}\right)^2\right)+\left(1-\frac{x_3^2}{x^2}\right)\left(1-5\frac{x_3^2}{x^2}\right)\right)\right]\end{aligned}$$

The other elements of the Green's function are found by permuting the subscripts.

Solution With $(KK)_{jm}=c_{jlms}k_lk_s$, $(kk)_{jm}=c^0_{jlms}k_lk_s$ we have:

$$V_{jm}=D\delta_{jl}\delta_{lm}\delta_{ms}k_lk_s=D(k_1^2\delta_{j1}\delta_{m1}+k_2^2\delta_{j2}\delta_{m2}+k_3^2\delta_{j3}\delta_{m3}).$$

and

$$(KK)_{jm}=(kk)_{jm}+V_{jm}.$$

In terms of the 3×3 matrices (KK), (kk) and V we have:

$$\begin{aligned}(KK) &= (kk)+V\\ &= (kk)\left(I+(kk)^{-1}V\right)\\ (KK)^{-1} &= \left(I+(kk)^{-1}V\right)^{-1}(kk)^{-1}\\ (kk)^{-1} &= \left(I+(kk)^{-1}V\right)(KK)^{-1}\\ (KK)^{-1} &= (kk)^{-1}-(kk)^{-1}V(KK)^{-1}.\end{aligned}$$

Expanding this Dyson equation we obtain:

$$(KK)^{-1}=(kk)^{-1}-(kk)^{-1}V(kk)^{-1}+(kk)^{-1}V(kk)^{-1}V(kk)^{-1}-\ldots$$

Provided D is small we may obtain an approximation to $(KK)^{-1}$ by retaining only the first order term in V. In that case we have to evaluate $[(kk)^{-1}V(kk)^{-1}]_{ij}$:

$$\begin{aligned}
\left[(kk)^{-1}V(kk)^{-1}\right]_{ij} &= \left[(kk)^{-1}\right]_{ip} V_{pq} \left[(kk)^{-1}\right]_{qj} \\
&= D\left(\frac{\delta_{ip}}{\mu k^2} - \frac{\chi}{\mu}\frac{k_i k_p}{k^4}\right)\left(k_1^2\delta_{p1}\delta_{q1} + k_2^2\delta_{p2}\delta_{q2} + k_3^2\delta_{p3}\delta_{q3}\right)\left(\frac{\delta_{qj}}{\mu k^2} - \frac{\chi}{\mu}\frac{k_q k_j}{k^4}\right) \\
&= \frac{D}{\mu^2 k^4}\left(\delta_{i1}\delta_{j1}k_1^2 + \delta_{i2}\delta_{j2}k_2^2 + \delta_{i3}\delta_{j3}k_3^2\right) \\
&\quad - \frac{D\chi}{\mu^2 k^6}\left(k_1^3(\delta_{i1}k_j + \delta_{j1}k_i) + k_2^3(\delta_{i2}k_j + \delta_{j2}k_i) + k_3^3(\delta_{i3}k_j + \delta_{j3}k_i)\right) \\
&\quad + \frac{D\chi^2}{\mu^2}\frac{k_i k_j}{k^8}\left(k_1^4 + k_2^4 + k_3^4\right),
\end{aligned}$$

where $\chi = (\lambda+\mu)/(\lambda+2\mu) = 1/[2(1-\nu)]$.

The first order corrections to the isotropic elastic Green's functions are therefore as follows:

$$\begin{aligned}
\Delta\tilde{G}_{11}(\mathbf{k}) &= \frac{D}{\mu^2 k^2}\left(\frac{k_1^2}{k^2} - 2\chi\frac{k_1^4}{k^4} + \chi^2\frac{k_1^2(k_1^4 + k_2^4 + k_3^4)}{k^6}\right) \\
\Delta\tilde{G}_{12}(\mathbf{k}) &= \frac{D\chi}{\mu^2 k^2}\left(-\frac{k_1 k_2}{k^2}\frac{(k_1^2 + k_2^2)}{k^2} + \chi\frac{k_1 k_2}{k^2}\frac{(k_1^4 + k_2^4 + k_3^4)}{k^4}\right).
\end{aligned}$$

Other elements of $\Delta\tilde{G}_{ij}(\mathbf{k})$ are found by cyclic permutation of the subscripts. Using the inverse transforms provided in the problem yields the stated expressions for $\Delta G_{11}(\mathbf{x})$ and $\Delta G_{12}(\mathbf{x})$.

Comment

The correction terms are a factor of D/μ smaller than the terms in the isotropic elastic Green's function, which are proportional to $1/\mu$.

Problem 4.4 In a hexagonal crystal show that the (kk) matrix is as follows:

$$(kk) = \begin{pmatrix} C_{11}k_1^2 + C_{66}k_2^2 + C_{44}k_3^2 & \frac{1}{2}(C_{11}+C_{12})k_1k_2 & (C_{44}+C_{13})k_1k_3 \\ \frac{1}{2}(C_{11}+C_{12})k_1k_2 & C_{66}k_1^2 + C_{11}k_2^2 + C_{44}k_3^2 & (C_{44}+C_{13})k_2k_3 \\ (C_{44}+C_{13})k_1k_3 & (C_{44}+C_{13})k_2k_3 & C_{44}(k_1^2+k_2^2)+C_{33}k_3^2 \end{pmatrix}$$

where $C_{66} = \frac{1}{2}(C_{11} - C_{12})$.

Since the elastic constant tensor is invariant with respect to rotations about the hexagonal axis (which is the x_3-axis), we may choose $\mathbf{k} = (0, k_2, k_3)$. With this simplification the determinant in eq 4.35 can be factorised. Determine the three solutions for $\omega^2 = \omega^2(0, k_2, k_3)$.

Determine the frequencies and polarisations of the three waves with $\mathbf{k}$ in the basal plane and normal to the basal plane.

Solution The following elastic constants are non-zero in a hexagonal crystal:

$$\begin{aligned} C_{11} &= c_{1111} = c_{2222} \\ C_{12} &= c_{1122} = c_{2211} \\ C_{13} &= c_{1133} = c_{3311} = c_{2233} = c_{3322} \\ C_{33} &= c_{3333} \\ C_{44} &= c_{2323} = c_{2332} = c_{3223} = c_{3232} = c_{1313} = c_{1331} = c_{3113} = c_{3131} \\ C_{66} &= c_{1212} = c_{1221} = c_{2112} = c_{2121} \end{aligned}$$

Therefore, elements of the symmetric matrix $(kk) = c_{jlms}k_lk_s = c_{mljs}k_lk_s$ are as follows:

$$\begin{aligned} (kk)_{11} &= c_{1l1s}k_lk_s = c_{1111}k_1^2 + c_{1212}k_2^2 + c_{1313}k_3^2 = C_{11}k_1^2 + C_{66}k_2^2 + C_{44}k_3^2 \\ (kk)_{12} &= c_{1l2s}k_lk_s = c_{1122}k_1k_2 + c_{1221}k_1k_2 = \frac{1}{2}(C_{11}+C_{12})k_1k_2 \\ (kk)_{13} &= c_{1l3s}k_lk_s = c_{1133}k_1k_3 + c_{1331}k_1k_3 = (C_{13}+C_{44})k_1k_3 \\ (kk)_{22} &= c_{2l2s}k_lk_s = c_{2222}k_2^2 + c_{2323}k_3^2 + c_{2121}k_1^2 = C_{11}k_2^2 + C_{44}k_3^2 + C_{66}k_1^2 \\ (kk)_{23} &= c_{2l3s}k_lk_s = c_{2233}k_2k_3 + c_{2332}k_3k_2 = (C_{13}+C_{44})k_2k_3 \\ (kk)_{33} &= c_{3l3s}k_lk_s = c_{3131}k_1^2 + c_{3232}k_2^2 + c_{3333}k_3^2 = C_{44}(k_1^2+k_2^2) + C_{33}k_3^2 \end{aligned}$$

With $\mathbf{k} = (0, k_2, k_3)$ the equation $\det((kk) - \rho\omega^2 I) = 0$ becomes:

$$\begin{vmatrix} C_{66}k_2^2 + C_{44}k_3^2 - \rho\omega^2 & 0 & 0 \\ 0 & C_{11}k_2^2 + C_{44}k_3^2 - \rho\omega^2 & (C_{13} + C_{44})k_2k_3 \\ 0 & (C_{13} + C_{44})k_2k_3 & C_{44}k_2^2 + C_{33}k_3^2 - \rho\omega^2 \end{vmatrix} = 0.$$

Therefore, either

$$\rho\omega^2 = C_{66}k_2^2 + C_{44}k_3^2$$

or

$$(C_{11}k_2^2 + C_{44}k_3^2 - \rho\omega^2)(C_{44}k_2^2 + C_{33}k_3^2 - \rho\omega^2) = (C_{13} + C_{44})^2 k_2^2 k_3^2.$$

The solutions of this quadratic equation are:

$$\begin{aligned} \rho\omega^2 &= (C_{11} + C_{44})k_2^2 + (C_{33} + C_{44})k_3^2 \\ &\pm \frac{1}{2}\Big[\left((C_{11} + C_{44})k_2^2 + (C_{33} + C_{44})k_3^2\right)^2 \\ &- 4(C_{11}k_2^2 + C_{44}k_3^2)(C_{44}k_2^2 + C_{33}k_3^2) + 4(C_{13} + C_{44})^2 k_2^2 k_3^2\Big]^{\frac{1}{2}}. \end{aligned}$$

In the basal plane $\mathbf{k} = (0, k_2, 0)$. Then either

$$\rho\omega^2 = C_{66}k_2^2, \text{ which is a transverse wave}$$

or

$$\rho\omega^2 = C_{11}k_2^2, \text{ which is a longitudinal wave}$$

or

$$\rho\omega^2 = C_{44}k_2^2, \text{ which is a transverse wave}$$

Normal to the basal plane $\mathbf{k} = (0, 0, k_3)$ either $\rho\omega^2 = C_{33}k_3^2$, which is a longitudinal wave, or $\rho\omega^2 = C_{44}k_3^2$, which corresponds to two degenerate transverse waves.

Problem 4.5 Consider the Green's function, $\gamma_{km}(\mathbf{x}, t)$ in isotropic elasticity, for an oscillatory unit body force located at the origin described by $f_i(t) = \delta_{im}\delta(\mathbf{x})\cos(\omega t)$.

The equation of motion for $\gamma_{km}(\mathbf{x},t)$ is as follows:

$$c_{ijkl}\gamma_{km,lj}(\mathbf{x},t)+\delta_{im}\delta(\mathbf{x})\cos(\omega t)=\rho\ddot{\gamma}_{im}(\mathbf{x},t).$$

To satisfy the time dependence in this equation we must have[19] $\gamma_{im}(\mathbf{x},t)=\chi_{im}(\mathbf{x})\cos\omega t$, where $\chi(\mathbf{x})$ satisfies following equation:

$$c_{ijkl}\chi_{km,lj}(\mathbf{x})+\delta_{im}\delta(\mathbf{x})=-\rho\omega^2\chi_{im}(\mathbf{x})$$

Show that the Fourier transform $\tilde{\chi}(\mathbf{k})$satisfies:

$$\left[(kk)_{ik}-\rho\omega^2\delta_{ik}\right]\tilde{\chi}_{km}=\delta_{im}$$

and therefore, in an isotropic medium,

$$\tilde{\chi}_{im}=\frac{1}{\rho}\left[\frac{\delta_{im}-k_ik_m/k^2}{c_t^2k^2-\omega^2}+\frac{k_ik_m/k^2}{c_l^2k^2-\omega^2}\right].$$

Hence show that

$$\chi_{im}(\mathbf{x})=\frac{\delta_{im}}{4\pi\rho c_t^2}\frac{\cos(\omega x/c_t)}{x}+\frac{1}{4\pi\rho\omega^2 x}\frac{\partial}{\partial x_i}\frac{\partial}{\partial x_m}\left(\cos(\omega x/c_t)-\cos(\omega x/c_l)\right).$$

Solution Taking the Fourier transform of the equation $c_{ijkl}\chi_{km,lj}(\mathbf{x})+\delta_{im}\delta(\mathbf{x})=-\rho\omega^2\chi_{im}(\mathbf{x})$ we obtain:

$$\begin{aligned}-c_{ijkl}\tilde{\chi}_{km}k_lk_j+\delta_{im}&=-\rho\omega^2\tilde{\chi}_{im}\\ \left((kk)_{ik}-\rho\omega^2\right)\tilde{\chi}_{km}&=\delta_{im}.\end{aligned}$$

Since $(kk)_{ik}=\mu k^2\delta_{ik}+(\lambda+\mu)k_ik_k$ in the isotropic elastic approximation, we have:

$$\left[(\mu k^2-\rho\omega^2)\delta_{ik}+(\lambda+\mu)k_ik_k\right]\tilde{\chi}_{km}=\delta_{im}.$$

Multiplying both sides of this equation by k_i and summing on i we obtain:

$$k_k\tilde{\chi}_{km}=\frac{k_m}{(\lambda+2\mu)k^2-\rho\omega^2}.$$

[19] if damping were included in the equation of motion there would be a phase difference between the vibrations of the medium and those of the oscillator.

Therefore,

$$\begin{aligned}\tilde{\chi}(\mathbf{k}) &= \frac{\delta_{im}}{(\mu k^2-\rho\omega^2)} - \frac{(\lambda+\mu)k_ik_m}{(\mu k^2-\rho\omega^2)((\lambda+2\mu)k^2-\rho\omega^2)}\\ &= \frac{\delta_{im}/\rho}{c_t^2k^2-\omega^2} - \frac{(c_l^2-c_t^2)k_ik_m/\rho}{(c_t^2k^2-\omega^2)(c_l^2k^2-\omega^2)}\\ &= \frac{1}{\rho}\left\{\frac{\delta_{im}-k_ik_m/k^2}{c_t^2k^2-\omega^2}+\frac{k_ik_m/k^2}{c_l^2k^2-\omega^2}\right\},\end{aligned}$$

where $c_t^2 = \mu/\rho$ and $c_l^2 = (\lambda+2\mu)/\rho$.

Taking the inverse transform we have:

$$\begin{aligned}\chi_{im}(\mathbf{x}) &= \frac{1}{(2\pi)^3}\frac{1}{\rho}\int \mathrm{d}^3k\left\{\frac{\delta_{im}-k_ik_m/k^2}{c_t^2k^2-\omega^2}+\frac{k_ik_m/k^2}{c_l^2k^2-\omega^2}\right\}e^{-i\mathbf{k}\cdot\mathbf{x}}\\ &= \frac{1}{(2\pi)^2}\frac{1}{\rho}\int_0^\infty \mathrm{d}k k^2\left\{\frac{\delta_{im}-k_ik_m/k^2}{c_t^2k^2-\omega^2}+\frac{k_ik_m/k^2}{c_l^2k^2-\omega^2}\right\}\int_{-1}^{1}\mathrm{d}(\cos\theta)e^{-ikx\cos\theta}\\ &= \frac{2}{(2\pi)^2}\frac{1}{\rho}\int_0^\infty \mathrm{d}k\left\{\frac{k^2\delta_{im}-k_ik_m}{c_t^2k^2-\omega^2}+\frac{k_ik_m}{c_l^2k^2-\omega^2}\right\}\frac{\sin(kx)}{kx}\\ &= \frac{1}{(2\pi)^2}\frac{1}{\rho}\,\mathrm{Im}\left[\int_{-\infty}^{\infty}\mathrm{d}k\left\{\frac{k^2\delta_{im}-k_ik_m}{c_t^2k^2-\omega^2}+\frac{k_ik_m}{c_l^2k^2-\omega^2}\right\}\frac{e^{ikx}}{kx}\right],\end{aligned}$$

where Im[z] means the imaginary part of z. Consider first the integral, I_1, involving the Kronecker delta:

$$I_1 = \frac{\delta_{im}}{4\pi^2\rho c_t^2 x}\,\mathrm{Im}\left[P\int_{-\infty}^{\infty}\mathrm{d}k\frac{ke^{ikx}}{k^2-\omega^2/c_t^2}\right]$$

The P in front of the integral sign signifies it is to be understood as a principal value integral. It may be evaluated as a contour integral in the complex plane along the real axis from $= -\infty$ to $+\infty$ with simple poles on the real axis at $k = \pm\omega/c_t$, and closed by an infinite semi-circle in the upper half plane. Jordan's lemma indicates the contribution from the semi-circle is zero. The contour is shown in Fig 4.2.

Since there are no singularities inside the contour Cauchy's theorem tells us the contour integral is zero:

$$P\int_{-\infty}^{\infty} \mathrm{d}k \frac{k e^{ikx}}{\left(k+\frac{\omega}{c_t}\right)\left(k-\frac{\omega}{c_t}\right)} - \pi i \underbrace{\left[\frac{\cancel{(\omega/c_t)}e^{i\omega x/c_t}}{2\cancel{(\omega/c_t)}} + \frac{\cancel{(-\omega/c_t)}e^{-i\omega x/c_t}}{2\cancel{(-\omega/c_t)}}\right]}_{=\cos(\omega x/c_t)} = 0$$

It follows that,

$$I_1 = \frac{\delta_{im}}{4\pi\rho c_t^2} P\int_{-\infty}^{\infty} \mathrm{d}k \frac{k e^{ikx}}{\left(k+\frac{\omega}{c_t}\right)\left(k-\frac{\omega}{c_t}\right)} = \frac{\delta_{im}}{4\pi\rho c_t^2}\frac{\cos(\omega x/c_t)}{x}.$$

The second integral to be evaluated is:

$$\begin{aligned} I_2 &= \frac{1}{(2\pi)^3}\frac{1}{\rho}\int \mathrm{d}^3k \frac{k_i k_m/k^2}{c_l^2 k^2 - \omega^2} e^{-i\mathbf{k}\cdot\mathbf{x}} \\ &= -\frac{1}{(2\pi)^3}\frac{1}{\rho}\frac{\partial}{\partial x_i}\frac{\partial}{\partial x_m}\int \mathrm{d}^3k \frac{1/k^2}{c_l^2 k^2 - \omega^2} e^{-i\mathbf{k}\cdot\mathbf{x}} \\ &= -\frac{1}{4\pi^2\rho c_l^2 x}\frac{\partial}{\partial x_i}\frac{\partial}{\partial x_m}\mathrm{Im}\left[P\int_{-\infty}^{\infty}\mathrm{d}k \frac{1}{k^2-\omega^2/c_l^2}\frac{e^{ikx}}{k}\right]. \end{aligned}$$

This integral may also be evaluated as a contour integral in the complex plane, with simple poles on the real axis at $k = 0$ and $k = \pm\omega/c_l$. The contour is shown in Fig 4.3. Using Cauchy's theorem, we obtain:

$$P\int_{-\infty}^{\infty}\mathrm{d}k \frac{1}{k^2-\omega^2/c_l^2}\frac{e^{ikx}}{k} - \pi i\left(-\frac{c_l^2}{\omega^2} + \frac{c_l^2}{\omega^2}\cos(\omega x/c_l)\right) = 0.$$

It follows that:

$$I_2 = -\frac{1}{4\pi\rho\omega^2}\frac{1}{x}\frac{\partial}{\partial x_i}\frac{\partial}{\partial x_m}\cos(\omega x/c_l).$$

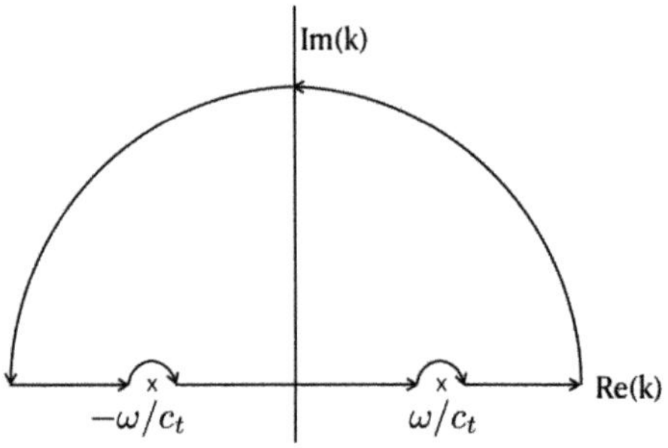

Figure 4.2: Contour in the complex plane used to evaluate the integral in I_1 of problem 4.5.

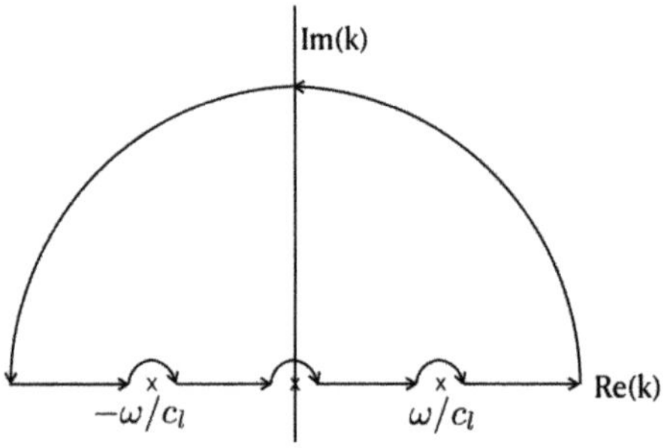

Figure 4.3: Contour in the complex plane used to evaluate the integral in I_2 of problem 4.5.

The final integral to be evaluated is:

$$I_3 = -\frac{1}{(2\pi)^3}\frac{1}{\rho}\int \mathrm{d}^3k \frac{k_i k_m/k^2}{c_t^2 k^2 - \omega^2} e^{-i\mathbf{k}\cdot\mathbf{x}}$$

I_3 is the same as I_2 except c_l is replaced by c_t:

$$I_3 = -\frac{1}{4\pi\rho\omega^2}\frac{1}{x}\frac{\partial}{\partial x_i}\frac{\partial}{\partial x_m}\cos(\omega x/c_t).$$

Putting it all together we have

$$\chi_{im}(\mathbf{x}) = \frac{1}{4\pi\rho x}\left(\delta_{im}\frac{\cos(\omega x/c_t)}{c_t^2} + \frac{1}{\omega^2}\frac{\partial}{\partial x_i}\frac{\partial}{\partial x_m}(\cos(\omega x/c_t) - \cos(\omega x/c_l))\right)$$

5. Point defects

The theory perhaps suffers from the disadvantage that its limitations are more immediately obvious than are those of other approximate methods which have to be used in dealing with the solid state... A simple treatment at cottage-industry level may sometimes provide a certain degree of insight into some phenomenon or other when an exact calculation, based on a theoretical model which apes reality precisely, will give accurate answers but may be too unsurveyable, or too numerical, to give much insight.

J D Eshelby[1] in *Point defect behaviour and diffusional processes*, eds. R E Smallman and E Harris, pp.3-10, The Metals Society, London (1977).

5.1 Introduction

Point defects are ubiquitous in crystalline materials. They are classified as either intrinsic or extrinsic. Intrinsic point defects are vacancies and self-interstitials. A vacancy is a vacant atomic site that would normally be occupied by a host atom. A self-interstitial is created when one of the atoms of the crystal is occupying a site in the space, called the *interstice*, between usual crystal sites. Extrinsic point defects are foreign atoms that have been incorporated into the crystal structure either on substitutional or interstitial sites.

Unlike linear and planar defects there may be populations of point defects in thermodynamic equilibrium in a crystal. For example the equilibrium concentration of vacancies in a crystal at a given temperature and pressure is approximately $\exp(-G^f/k_BT)$ where G^f is the Gibbs free energy of formation of the point defect. The thermodynamic equilibrium concentration of foreign atoms in a crystal depends

[1] John Douglas Eshelby, FRS 1916-1981. British physicist.

on their chemical potentials, which in turn depend on the partial pressures of these impurities in the vapour phase in equilibrium with the crystal.

Thermodynamic equilibrium is seldom achieved in crystals. But its existence is important because it indicates the state of the system towards which it will naturally evolve under prescribed environmental conditions, provided it is not perturbed by external influences such as mechanical deformation or irradiation. The diffusion of point defects leads to mass transport and possibly changes of phase in the solid state. For example diffusion in pure crystals usually occurs through the migration of vacancies, as they are sites in the crystal that can undergo a change of occupation through a host atom jumping into one of them. This is analogous to electronic conduction in a p-type semiconductor which occurs through electrons jumping into holes at the top of the valence band.

Point defects may be created in concentrations far in excess of those prescribed by thermodynamic equilibrium through processes such as plastic deformation and irradiation by high energy particles. When this happens the point defects may aggregate and form interstitial and vacancy clusters that can be very mobile, especially when they are small. Intrinsic point defects such as vacancies may be attracted to foreign atoms and form new combined defects called complexes. The variety of point defects can be huge.

In this chapter we will focus on how point defects interact with each other through their elastic fields. We begin with the model of a point defect as a misfitting sphere. We will go on to consider models of point defects that capture the symmetry of the atomic environment of the point defect.

5.2 The misfitting sphere model of a point defect

The simplest model of a point defect is a misfitting sphere[2] due to Bilby[3]. Cut out a spherical hole of radius r_h in an elastic continuum. Insert into the hole a sphere of radius $r_s \neq r_h$ and weld the interface. The sphere may be larger or smaller than the hole. We consider first the case where the inserted sphere is *rigid* and the continuum is infinite in extent. All the deformation occurs in the continuum surrounding the sphere.

Let $\mathbf{u}^\infty(\mathbf{r})$ be the displacement field in the continuum surrounding the sphere at radius r. The superscipt ∞ is to remind us that the continuum is assumed to be infinite in extent. In an isotropic continuum the displacement field will be purely radial so that $\mathbf{u}^\infty(\mathbf{r}) = \mathbf{u}^\infty(|\mathbf{r}|) = \mathbf{u}^\infty(r)$. In the isotropic approximation the displacement field must

[2]Bilby, B A, Proc. Phys. Soc. A **63**, 191 (1950).

[3]Bruce Alexander Bilby FRS 1922-2013, British materials physicist.

satisfy Navier's equation, eq 4.4, which in the absence of body forces becomes:

$$\mu\nabla^2\mathbf{u}^\infty + (\lambda+\mu)\nabla(\nabla\cdot\mathbf{u}^\infty) = 0. \tag{5.1}$$

We have to be careful how we interpret this equation in spherical polar coordinates. If we naively substitute the usual expressions for the Laplacian, gradient and divergence in spherical polar coordinates (r,θ,ϕ) we will go wrong. When we rotate a cartesian coordinate system (x_1,x_2,x_3) into another cartesian coordinate system (x_1',x_2',x_3') the transformation is the same at all points in space. But when we use spherical polar coordinates (or other curvilinear coordinates) the transformation varies with position because the coordinates r,θ and ϕ vary with position. Consequently, $(\mathrm{d}u_r, \mathrm{d}u_\theta, \mathrm{d}u_\phi)$ does not form a vector. The solution is to use tensor calculus, following Sokolnikoff[4]. Then it is found that the strain tensor in spherical polar coordinates is as follows:

$$\begin{aligned}
e_{rr} &= \frac{\partial u_r}{\partial r}\\
e_{\theta\theta} &= \frac{1}{r}\frac{\partial u_\theta}{\partial\theta} + \frac{u_r}{r}\\
e_{\phi\phi} &= \frac{1}{r\sin\theta}\frac{\partial u_\phi}{\partial\phi} + \frac{u_r}{r} + u_\theta\frac{\cot\theta}{r}\\
e_{r\phi} &= \frac{1}{2}\left(\frac{1}{r\sin\theta}\frac{\partial u_r}{\partial\phi} - \frac{u_\phi}{r} + \frac{\partial u_\phi}{\partial r}\right)\\
e_{r\theta} &= \frac{1}{2}\left(\frac{1}{r}\frac{\partial u_r}{\partial\theta} - \frac{u_\theta}{r} + \frac{\partial u_\theta}{\partial r}\right)\\
e_{\phi\theta} &= \frac{1}{2}\left(\frac{1}{r}\frac{\partial u_\phi}{\partial\theta} - \frac{u_\phi\cot\theta}{r} + \frac{1}{r\sin\theta}\frac{\partial u_\theta}{\partial\phi}\right)
\end{aligned}$$

Hooke's law in isotropic elasticity in spherical polar coordinates is as follows:

$$\begin{aligned}
\sigma_{rr} &= 2\mu e_{rr} + \lambda(e_{rr} + e_{\theta\theta} + e_{\phi\phi})\\
\sigma_{\theta\theta} &= 2\mu e_{\theta\theta} + \lambda(e_{rr} + e_{\theta\theta} + e_{\phi\phi})\\
\sigma_{\phi\phi} &= 2\mu e_{\phi\phi} + \lambda(e_{rr} + e_{\theta\theta} + e_{\phi\phi})\\
\sigma_{r\theta} &= 2\mu e_{r\theta}\\
\sigma_{r\phi} &= 2\mu e_{r\phi}\\
\sigma_{\theta\phi} &= 2\mu e_{\theta\phi}
\end{aligned}$$

[4]Sokolnikoff, I S, *Mathematical theory of elasticity*, 2nd edn., section 48, McGraw-Hill Book Company Inc.: New York (1956). ISBN 978-0070596290. Ivan Stephan Sokolnikoff 1901-1976 Russian-born, US mathematician.

The equations of equilibrium in isotropic elasticity in spherical polar coordinates are as follows:

$$
\begin{aligned}
\frac{\partial \sigma_{rr}}{\partial r} + \frac{1}{r\sin\theta}\frac{\partial \sigma_{r\phi}}{\partial \phi} + \frac{1}{r}\frac{\partial \sigma_{r\theta}}{\partial \theta} + \frac{2\sigma_{rr} - \sigma_{\theta\theta} - \sigma_{\phi\phi} + \sigma_{r\theta}\cot\theta}{r} + f_r &= 0 \\
\frac{\partial \sigma_{\phi r}}{\partial r} + \frac{1}{r\sin\theta}\frac{\partial \sigma_{\phi\phi}}{\partial \phi} + \frac{1}{r}\frac{\partial \sigma_{\phi\theta}}{\partial \theta} + \frac{3\sigma_{\phi r} + 2\sigma_{\phi\theta}\cot\theta}{r} + f_\phi &= 0 \\
\frac{\partial \sigma_{\theta r}}{\partial r} + \frac{1}{r\sin\theta}\frac{\partial \sigma_{\theta\phi}}{\partial \phi} + \frac{1}{r}\frac{\partial \sigma_{\theta\theta}}{\partial \theta} + \frac{3\sigma_{\theta r} + (\sigma_{\theta\theta} - \sigma_{\phi\phi})\cot\theta}{r} + f_\theta &= 0
\end{aligned}
$$

where (f_r, f_θ, f_ϕ) is the body force at (r, θ, ϕ).

Exercise 5.1 Consider a radially symmetric elastic displacement field. In spherical polar coordinates the displacement field $\mathbf{u}$ is then purely radial, $\mathbf{u} = [u_r, 0, 0]$ and u_r is a function of r only. Using the equations above for the strain tensor, Hooke's law and the equations of equilibrium in spherical polar coordinates show that the equations of equilibrium become just one equation:

$$
\frac{\mathrm{d}\sigma_{rr}}{\mathrm{d}r} + \frac{2\sigma_{rr} - \sigma_{\theta\theta} - \sigma_{\phi\phi}}{r} + f_r = 0
$$

Hence derive a differential equation for u_r in isotropic elasticity, in the absence of body forces, and show that the general solution is:

$$
u_r(r) = Ar + D/r^2, \tag{5.2}
$$

where A and D are arbitrary constants. ■

Solution For $\mathbf{u} = [u_r, 0, 0]$ in spherical polar coordinates we have:

$$
e_{rr} = \frac{\mathrm{d}u_r}{\mathrm{d}r},\ e_{\theta\theta} = \frac{u_r}{r},\ e_{\phi\phi} = \frac{u_r}{r},\ e_{r\theta} = e_{\theta\phi} = e_{\phi r} = 0.
$$

Therefore, Hooke's law becomes:

$$
\begin{aligned}
\sigma_{rr} &= 2\mu e_{rr} + \lambda(e_{rr} + e_{\theta\theta} + e_{\phi\phi}) = (2\mu + \lambda)\frac{\mathrm{d}u_r}{\mathrm{d}r} + 2\lambda\frac{u_r}{r} \\
\sigma_{\theta\theta} &= 2\mu e_{\theta\theta} + \lambda(e_{rr} + e_{\theta\theta} + e_{\phi\phi}) = (2\mu + 2\lambda)\frac{u_r}{r} + \lambda\frac{\mathrm{d}u_r}{\mathrm{d}r}
\end{aligned}
$$

$$\begin{aligned}\sigma_{\phi\phi} &= 2\mu e_{\phi\phi} + \lambda(e_{rr} + e_{\theta\theta} + e_{\phi\phi}) = (2\mu + 2\lambda)\frac{u_r}{r} + \lambda\frac{\mathrm{d}u_r}{\mathrm{d}r} \\ \sigma_{r\theta} &= 0 \\ \sigma_{\theta\phi} &= 0 \\ \sigma_{\phi r} &= 0.\end{aligned}$$

Two of the three equations of mechanical equilibrium are satisfied identically. The third is as follows:

$$\frac{\mathrm{d}\sigma_{rr}}{\mathrm{d}r} + \frac{2\sigma_{rr} - \sigma_{\theta\theta} - \sigma_{\phi\phi}}{r} + f_r = 0.$$

Substituting Hooke's law into this equation with $f_r = 0$ we obtain:

$$(2\mu + \lambda)\frac{\mathrm{d}^2 u_r}{\mathrm{d}r^2} + 2(2\mu + \lambda)\frac{1}{r}\frac{\mathrm{d}u_r}{\mathrm{d}r} - 2(2\mu + \lambda)\frac{u_r}{r^2} = 0.$$

Cancelling the factor of $(2\mu + \lambda)$ we obtain the following differential equation for u_r:

$$\frac{\mathrm{d}^2 u_r}{\mathrm{d}r^2} + \frac{2}{r}\frac{\mathrm{d}u_r}{\mathrm{d}r} - 2\frac{u_r}{r^2} = 0.$$

This equation is homogeneous. A suitable trial solution is $u_r = Ar^n$. Substituting this trail solution into the differential equation we find $n^2 + n - 2 = 0$, for which $n = -2$ or $n = 1$. Therefore the general solution is:

$$u_r(r) = \frac{A}{r^2} + Br$$

where A and B are arbitrary constants determined by boundary conditions on the solution for each problem, such as the required behaviour of the solution at $r = 0$ and $r = \infty$.

Returning to the misfitting sphere in an infinite medium we note that since $\mathbf{u}^\infty$ cannot be infinite at $r = \infty$ we choose $u^\infty(r) = D/r^2$. The constant D is determined by the boundary condition at the surface of the hole: $u^\infty(r_h) = r_s - r_h$. Therefore $D = (r_s - r_h)r_h^2$.

One of the surprising features of this model is that the dilation is zero everywhere[5]. Inside the sphere there is no elastic strain because the sphere is assumed to be rigid. Outside the sphere the strain in the radial direction is $e_{rr} = \mathrm{d}u_r^\infty/\mathrm{d}r = -2D/r^3$, while $e_{\theta\theta} = e_{\phi\phi} = u_r/r = D/r^3$, so that the trace of the strain tensor is zero. But

[5] This is a special case of a more general result proved in problem 8.4.

there is an overall volume change ΔV^{∞} and it is all located in the misfitting sphere: $\Delta V^{\infty} = (4\pi/3)(r_s^3 - r_h^3) \approx 4\pi r_h^2(r_s - r_h) = 4\pi D$.

In reality all bodies are finite. Consider a finite body with a free surface $\mathcal{S}$. The solution for the misfitting sphere in an infinite medium has to be corrected by applying surface tractions $-\sigma_{ij}^{\infty} n_j$ over the surface $\mathcal{S}$, where $\sigma_{ij}^{\infty} = 2\mu \partial u_i^{\infty}/\partial x_j$, and in cartesian coordinates $u_i^{\infty} = Dx_i/r^3$. These tractions cancel those that would exist on the surface if the infinite solution were not corrected. To calculate the change ΔV^I in the volume of the body caused by this distribution of surface tractions consider the integral:

$$\int_S \sigma_{ij}^{\infty} x_i n_j \mathrm{d}S = \int_V (\sigma_{ij}^{\infty} x_i)_{,j} \mathrm{d}V = \int_V \sigma_{ij,j}^{\infty} x_i + \sigma_{ij}^{\infty}\delta_{ij} \mathrm{d}V = \int_V -f_i x_i \mathrm{d}V + V\langle\sigma_{ii}\rangle,$$

where the divergence theorem and the equilibrium condition have been used. There is no net body force in V due to the point defect. The average trace of the stress tensor, $\langle\sigma_{ii}\rangle$, within the body due to the cancelling distribution of surface tractions, $-\sigma_{ij}^{\infty} n_j$, is then as follows:

$$\begin{aligned} V\langle\sigma_{ii}\rangle &= -\int_S \sigma_{ij}^{\infty} x_i n_j \mathrm{d}S \\ &= -\int_S 2\mu D\left(\frac{\delta_{ij}}{r^3} - \frac{3x_i x_j}{r^5}\right) x_i n_j \mathrm{d}S \\ &= 4\mu D \int_S \frac{x_i n_i}{r^3} \mathrm{d}S, \end{aligned}$$

where $(x_i n_i/r^3)\mathrm{d}S$ is an element of solid angle, and the final surface integral is therefore 4π. The change in volume, ΔV^I, caused by the elimination of the surface tractions is given by $3B\Delta V^I = \langle\sigma_{ii}\rangle V = 16\pi\mu D = 4\mu\Delta V^{\infty}$. The total volume change caused by the misfitting sphere in a finite body is given by:

$$\Delta V = \Delta V^{\infty} + \Delta V^I = 4\pi D\left(\frac{3B + 4\mu}{3B}\right) = 3\Delta V^{\infty}\left(\frac{1-\nu}{1+\nu}\right), \tag{5.3}$$

where we have used $B = 2\mu(1+\nu)/(3(1-2\nu))$, and ν is Poisson's ratio. For $\nu = 1/3$, which is typical of many metals, the total volume change caused by the misfitting sphere is approximately 50% greater in a finite body than in an infinite body. It is remarkable that this result does not depend on the size of the body, or its shape, or where the defect is located within it. It is an example of an insight from a simple model into a universal property of point defects that would be virtually impossible to obtain from a purely atomistic model.

Exercise 5.2 Consider a misfitting *deformable* sphere at the centre of a much larger sphere of radius R on the surface of which there are no tractions. The radius of the hole into which the misfitting sphere is inserted is r_h. The radius of the unconstrained (i.e. stress-free) deformable sphere is r_s. The bulk modulus and the shear modulus of the large sphere of radius R are B and μ respectively, and the bulk modulus of the misfitting sphere is B'.

To first order in the difference $r_s - r_h$ show that the volume change of the sphere of radius R is

$$\Delta V = \left(\frac{1 + \frac{4\mu}{3B}}{1 + \frac{4\mu}{3B'}} \right) \delta v$$

where $\delta v = 4\pi r_h^2 (r_s - r_h)$ is the misfit volume.

Show that the dilation in the large sphere is

$$e_{ii} = \frac{\frac{4\mu}{3B}}{1 + \frac{4\mu}{3B'}} \frac{\delta v}{V},$$

where $V = 4\pi R^3/3$ is the volume of the sphere of radius R.

■

Solution The radial solution between $r = r_h$ and $r = R$ has the general form $u_r^{\text{out}} = Ar + D/r^2$. Between $r = 0$ and $r = r_h$ the solution has the form $u_r^{\text{in}} = Cr$, because it must not blow up at $r = 0$. The superscripts 'out' and 'in' label the fields *out*side and *in*side the misfitting sphere. There are three unknowns: A, D and C. They are determined by three boundary conditions:

1. To ensure there are no holes of overlapping material at the interface between the inner and outer spheres we must have:

$$r_h + u_r^{\text{out}}(r_h) = r_s + u_r^{\text{in}}(r_s).$$

2. The tractions at the interface between the spheres must be continuous. The spherical symmetry requires they are purely radial, and therefore:

$$\sigma_{rr}^{\text{in}}(r_s) = \sigma_{rr}^{\text{out}}(r_h).$$

3. The tractions on the external surface of the outer sphere must be zero. This requires:

$$\sigma_{rr}^{\text{out}}(R) = 0.$$

These three boundary conditions generate the following simultaneous equations for A, D and C:

$$\begin{aligned}
r_h + A r_h + \frac{D}{r_h^2} &= r_s + C r_s \\
3B'C &= 3BA - 4\frac{\mu D}{r_h^3} \\
3BA - 4\frac{\mu D}{R^3} &= 0,
\end{aligned}$$

where B and B' are the bulk moduli of the large outer sphere of radius R and the inner sphere of radius r_s respectively. Using these equations and $R \gg r_h, r_s$ and retaining terms only to first order in $(r_s - r_h)/r_h$ we obtain:

$$\begin{aligned}
A &= \frac{1}{\frac{R^3}{r_h^3}\left(1 + \frac{3B'}{4\mu}\right)} \frac{B'}{B} \left(\frac{r_s - r_h}{r_h}\right) \\
C &= -\frac{1}{\left(1 + \frac{3B'}{4\mu}\right)} \left(\frac{r_s - r_h}{r_h}\right) \\
D &= \frac{\frac{3B'}{4\mu}}{1 + \frac{3B'}{4\mu}} r_h^3 \left(\frac{r_s - r_h}{r_h}\right).
\end{aligned}$$

As a check we note that D has the dimensions of cubic length as required, and A and C are dimensionless as required.

To first order in $u_r^{(out)}/R$ the volume change of the large sphere is:

$$\begin{aligned}
\Delta V &= 4\pi R^2 u_r^{(out)}(R) \\
&= 4\pi r_h^2 (r_s - r_h) \frac{B'}{B} \frac{\left(1 + \frac{3B}{4\mu}\right)}{\left(1 + \frac{3B'}{4\mu}\right)}
\end{aligned}$$

$$= \delta v \frac{\left(1+\frac{4\mu}{3B}\right)}{\left(1+\frac{4\mu}{3B'}\right)}.$$

The dilation in the large sphere is:

$$\begin{aligned}
\frac{\mathrm{d}u_r^{(out)}}{\mathrm{d}r} + 2\frac{u_r^{(out)}}{r} &= 3A \\
&= \frac{3}{\frac{R^3}{r_h^3}\left(1+\frac{3B'}{4\mu}\right)} \frac{B'}{B}\left(\frac{r_s - r_h}{r_h}\right) \\
&= \frac{B'}{B} \frac{1}{\left(1+\frac{3B'}{4\mu}\right)} \frac{4\pi r_h^2(r_s - r_h)}{4\pi R^3/3} \\
&= \frac{\frac{4\mu}{3B}}{1+\frac{4\mu}{3B'}} \frac{\delta v}{V}.
\end{aligned}$$

Comment

As $B' \to \infty$ the misfitting sphere becomes rigid. Then $\Delta V \to (1+4\mu/(3B))\delta v$, which reproduces the result for a rigid misfitting sphere embedded in a finite medium (see eq 5.3). Also, as $B' \to 0$ the 'misfitting sphere' is an empty hole, and $\Delta V \to 0$, as expected.

5.3 Interaction energies

The misfitting sphere model has been used widely owing to its attractive simplicity. But its central assumption that the defect has spherical symmetry limits its applicability principally to substitutional defects at sites of cubic symmetry. It is less applicable to interstitial defects which may display highly non-spherical symmetries, such as split interstitials and crowdion defects. We saw in section 4.5 how the multipole expansion may be applied to derive the displacement field of a substitutional point defect. It may also be applied to the displacement field of an interstitial defect, with the sole difference that the defect forces are the entire forces, rather than the excess forces, exerted on the neighbours. The example of a dumb-bell interstitial defect was treated in problem 4.2. The defect forces have the symmetry of the site occupied by the defect. In this section we consider the interaction energy between two point defects using the multipole expansion. This is a multiscale approach using interatomic forces to describe the sources of the elastic fields and elasticity theory to describe the interactions over distances much larger than the atomic scale. The analysis of this section follows

Siems[6].

We begin by thinking about the problem entirely from an atomistic viewpoint. In eq 3.28 we wrote down the harmonic expansion of the potential energy of the crystal. Here we write it slightly differently:

$$E = -\sum_n f_i^{(n)} u_i^{(n)} + \frac{1}{2}\sum_n \sum_p \frac{\partial^2 E}{\partial u_i^{(n)} \partial u_j^{(p)}} u_i^{(n)} u_j^{(p)}, \tag{5.4}$$

where we interpret the forces $f_i^{(n)}$ as being forces due to an existing defect which create the displacement field $u_i^{(n)}$. Let the vector of defect forces be **f**, the vector of atomic displacements be **u** and the matrix of second derivatives be **S**. Then we can rewrite eq 5.4 for the potential energy of the defect as follows:

$$E = -\mathbf{f}\cdot\mathbf{u} + \frac{1}{2}\mathbf{uSu}.$$

E is minimised when $\mathbf{f} = \mathbf{Su}$. At the minimum, $E = -\frac{1}{2}\mathbf{f}\cdot\mathbf{u} = -\frac{1}{2}\mathbf{uSu}$. Notice that at the minimum E is negative. It is the energy of *relaxing* the system of defect forces and harmonic atomic interactions. The defect forces do work when the atoms on which they act are displaced thereby reducing the potential energy by $-\mathbf{f}\cdot\mathbf{u}$. These displacements induce further displacements that extend throughout the crystal and which raise the potential energy by $\frac{1}{2}\mathbf{uSu}$. A balance between these two competing terms is established when $\mathbf{f} = \mathbf{Su}$, and the reduction in the potential energy is then only half of $-\mathbf{f}\cdot\mathbf{u}$.

If there is a second defect, with defect force vector $\tilde{\mathbf{f}}$ and vector of atomic displacements $\tilde{\mathbf{u}}$, with $\tilde{\mathbf{f}} = \mathbf{S}\tilde{\mathbf{u}}$ and $\tilde{E} = -\frac{1}{2}\tilde{\mathbf{f}}\cdot\tilde{\mathbf{u}}$, we may use the linear superposition principle to write the total potential energy as follows:

$$\begin{aligned} E^T &= -(\mathbf{f}+\tilde{\mathbf{f}})\cdot(\mathbf{u}+\tilde{\mathbf{u}}) + \frac{1}{2}(\mathbf{u}+\tilde{\mathbf{u}})\mathbf{S}(\mathbf{u}+\tilde{\mathbf{u}}) \\ &= E + \tilde{E} + E_{int} \end{aligned}$$

where the interaction energy E_{int} is given by:

$$\begin{aligned} E_{int} &= -(\mathbf{f}\cdot\tilde{\mathbf{u}} + \tilde{\mathbf{f}}\cdot\mathbf{u}) + \frac{1}{2}(\mathbf{uS}\tilde{\mathbf{u}} + \tilde{\mathbf{u}}\mathbf{Su}) \\ &= -\mathbf{f}\cdot\tilde{\mathbf{u}} = -\tilde{\mathbf{f}}\cdot\mathbf{u}. \end{aligned} \tag{5.5}$$

The last equality follows from Maxwell's reciprocity theorem. This result for the interaction energy may be understood in the following way. Suppose the first defect already exists and we introduce the second defect. The atoms on which the defect

[6]Siems, R Phys. Stat. Sol. **30**, 645-658 (1968). Rolf Siems 1930-, German theoretical physicist.

forces $\mathbf{f}$ of the first defect act are displaced further by the displacement field $\tilde{\mathbf{u}}$ of the second defect. The additional work done by the defect forces $\mathbf{f}$ is $-\mathbf{f}\cdot\tilde{\mathbf{u}}$. If we repeat the argument but with the second defect existing before the first the work done by the defect forces $\tilde{\mathbf{f}}$ is $-\tilde{\mathbf{f}}\cdot\mathbf{u}$.

To describe the interaction energy over large separations between the defects we may use defect forces from an atomistic calculation and elasticity theory to evaluate the displacement field. That is why we call this a multiscale approach.

With the first point defect at the origin of the cartesian coordinate system the interaction energy with a second point defect at $\mathbf{x}$ is

$$E_{int} = -\sum_n \tilde{f}_i(\mathbf{R}^{(n)})u_i(\mathbf{x}+\mathbf{R}^{(n)}) \tag{5.6}$$

where $\tilde{f}_i(\mathbf{R}^{(n)})$ is the force exerted by the defect at $\mathbf{x}$ on its neighbour at the relative position $\mathbf{R}^{(n)}$ and $u_i(\mathbf{x}+\mathbf{R}^{(n)})$ is the displacement of this neighbour due to the defect at the origin. Provided $|\mathbf{x}| \gg |\mathbf{R}^{(n)}|$ we may expand $u_i(\mathbf{x}+\mathbf{R}^{(n)})$ about $u_i(\mathbf{x})$:

$$u_i(\mathbf{x}+\mathbf{R}^{(n)}) = u_i(\mathbf{x})+R_j^{(n)}u_{i,j}(\mathbf{x})+\frac{1}{2}R_j^{(n)}R_k^{(n)}u_{i,jk}(\mathbf{x})+\frac{1}{6}R_j^{(n)}R_k^{(n)}R_l^{(n)}u_{i,jkl}(\mathbf{x})+\ldots \tag{5.7}$$

Substituting this Taylor expansion into the interaction energy, eq 5.6, we obtain:

$$E_{int} = -\left(\sum_n \tilde{f}_i(\mathbf{R}^{(n)})\right)u_i(\mathbf{x}) - \tilde{p}_{ij}u_{i,j}(\mathbf{x}) - \frac{1}{2}\tilde{q}_{ijk}u_{i,jk}(\mathbf{x}) - \frac{1}{6}\tilde{o}_{ijkl}u_{i,jkl}(\mathbf{x}) - \ldots \tag{5.8}$$

where the sum of the defect forces is zero provided the defect at $\mathbf{x}$ is relaxed, and $\tilde{p}_{ij}$, $\tilde{q}_{ijk}$ and $\tilde{o}_{ijkl}$ are the dipole, quadrupole and octupole moments respectively of the forces exerted by the defect at $\mathbf{x}$.

We can now use the multipole expansion, eq 4.21, to expand the displacement field due to the defect at the origin:

$$\begin{aligned} u_i(\mathbf{x}) &= \sum_m G_{ia}(\mathbf{x}-\mathbf{R}^{(m)})f_a(\mathbf{R}^{(m)}) \\ &= \sum_m \Bigg(G_{ia}(\mathbf{x}) - R_b^{(m)}G_{ia,b}(\mathbf{x}) + \frac{1}{2!}R_b^{(m)}R_c^{(m)}G_{ia,bc}(\mathbf{x}) \\ &\quad -\frac{1}{3!}R_b^{(m)}R_c^{(m)}R_d^{(m)}G_{ia,bcd}(\mathbf{x})+\ldots\Bigg)f_a(\mathbf{R}^{(m)}) \\ &= G_{ia}(\mathbf{x})\sum_m f_a(\mathbf{R}^{(m)}) - G_{ia,b}(\mathbf{x})p_{ab} + \frac{1}{2}G_{ia,bc}(\mathbf{x})q_{abc} \\ &\quad -\frac{1}{6}G_{ia,bcd}(\mathbf{x})o_{abcd}+\ldots \end{aligned} \tag{5.9}$$

The sum of the forces exerted by the defect at the origin on its neighbours is zero provided the defect is relaxed. The dipole, quadrupole and octupole moments of the forces exerted by the defect at the origin are ρ_{ab}, q_{abc}, o_{abcd} respectively. Inserting the displacement field $u_i(\mathbf{x})$ into the interaction energy of eq 5.8, and collecting terms involving the same order of differentiation of the Green's function we obtain finally:

$$\begin{aligned} E_{int} &= \tilde{\rho}_{ij} G_{ia,bj}(\mathbf{x}) \rho_{ab} \\ &+ \frac{1}{2}\left(\tilde{q}_{ijk} G_{ia,bjk}(\mathbf{x}) \rho_{ab} - \tilde{\rho}_{ij} G_{ia,bcj}(\mathbf{x}) q_{abc}\right) \\ &+ \frac{1}{12}\Big(2\tilde{o}_{ijkl} G_{ia,bjkl}(\mathbf{x}) \rho_{ab} + 2\tilde{\rho}_{ij} G_{ia,bcdj}(\mathbf{x}) o_{abcd} \\ &\qquad -3\tilde{q}_{ijk} G_{ia,bcjk}(\mathbf{x}) q_{abc}\Big) + \ldots \end{aligned} \tag{5.10}$$

The first line is the dipole-dipole interaction and it decays as $1/|\mathbf{x}|^3$. The second line is the dipole-quadrupole interaction and it decays as $1/|\mathbf{x}|^4$. The third and fourth lines describe interactions that depend on the fourth derivative of the Green's function, which decays as $1/|\mathbf{x}|^5$, and involve quadrupole-quadrupole interactions and dipole-octupole interactions.

5.4 The λ-tensor

In the previous section we saw that a point defect at $\mathbf{x}$ has an interaction energy with another elastic field according to eq 5.8. Thus, if the strain field with which it interacts is homogeneous then the interaction energy involves only the dipole tensor, a strain gradient at $\mathbf{x}$ will interact with the quadrupole tensor, the second derivative of the strain at $\mathbf{x}$ will interact with the octupole tensor of the defect, and so on. If there is a dilute solution of these point defects their dipole tensors can drive a change of shape of the crystal. If c is the atomic concentration of the point defect then the λ-tensor describes the rate of change of the spontaneous homogeneous strain of the crystal with c in the absence of any forces applied to the surface of the crystal:

$$\lambda_{kl} = \frac{\mathrm{d}e^h_{kl}}{\mathrm{d}c} \tag{5.11}$$

To expose the relationship between the λ-tensor and the dipole tensor consider the elastic energy of the crystal containing a solution of the defects. It is assumed that the defects are sufficiently far apart that their defect forces do not overlap, and therefore the dipole tensors are the same as those for an isolated defect. In a volume V let there be N defects. If the atomic volume of the host atoms is Ω then $c = N\Omega/V$ in the dilute limit.

If the crystal undergoes a homogeneous strain e^h_{ij} the elastic energy of the volume V becomes:

$$E = -N\rho_{ij}e^h_{ij} + \frac{V}{2}c_{ijkl}e^h_{ij}e^h_{kl}.$$

Minimising this elastic energy with respect to the homogeneous strain we obtain:

$$N\rho_{ij} = Vc_{ijkl}e^h_{kl}$$

Substituting $N = Vc/\Omega$ and differentiating the resulting expression with respect to the concentration c we obtain:

$$\rho_{ij} = \Omega c_{ijkl}\lambda_{kl} \tag{5.12}$$

This is a very useful relationship because the λ-tensor is experimentally measurable using X-ray diffraction for example. By measuring the dependence of the homogeneous strain on concentration c of the solute atoms, in the limit of small concentrations, the λ-tensor may be deduced using eq 5.11. If the elastic constants are also known then the dipole tensor can be obtained from eq 5.12.

5.5 Problems

Problem 5.1 Consider two point defects occupying sites of cubic point group symmetry in an infinite, slightly anisotropic cubic crystal, such as aluminium for which the anisotropy ratio is 1.2. Their elastic dipole tensors are $\tilde{\rho}_{ij} = \tilde{d}\delta_{ij}$ and $\rho_{ij} = \delta_{ij}d$. The elastic interaction energy at large separations is given by:

$$E_{int} = \tilde{\rho}_{ij}G_{ia,jb}(\mathbf{x})\rho_{ab} = \tilde{d}dG_{ij,ji}(\mathbf{x}).$$

Hence show that in an infinite isotropic medium the interaction energy is zero.

Using eq 4.43 for the first order correction to the Fourier transform of the Green's function in an infinite cubic crystal show that

$$\begin{aligned} G_{ij,ji}(\mathbf{x}) &= -\left(\frac{C_{11}-C_{12}-2C_{44}}{(C_{12}+2C_{44})^2}\right)\frac{1}{(2\pi)^3}\int\frac{k_1^4+k_2^4+k_3^4}{k^4}e^{-i\mathbf{k}\cdot\mathbf{x}}\mathrm{d}^3k \\ &= -\left(\frac{3}{8\pi}\right)\left(\frac{C_{11}-C_{12}-2C_{44}}{(C_{12}+2C_{44})^2}\right)\frac{1}{x^3}\left[5\frac{x_1^4+x_2^4+x_3^4}{x^4}-3\right]. \end{aligned} \tag{5.13}$$

Solution Starting with the following expression for the isotropic elastic Green's function

$$G_{ij}(\mathbf{x}) = \frac{1}{16\pi\mu(1-\nu)}\left(\frac{3-4\nu}{x}\delta_{ij} + \frac{x_i x_j}{x^3}\right)$$

we find its divergence is

$$G_{ij,j}(\mathbf{x}) = -\frac{1-2\nu}{8\pi\mu(1-\nu)}\frac{x_i}{x^3}.$$

Differentiating this expression with respect to x_i we obtain:

$$G_{ij,ji} = -\frac{1-2\nu}{8\pi\mu(1-\nu)}\left(\frac{3}{x^3} - \frac{3x_i x_i}{x^5}\right) = -\frac{1-2\nu}{8\pi\mu(1-\nu)}\left(\frac{3}{x^3} - \frac{3}{x^3}\right) = 0.$$

It follows that in an infinite isotropic elastic medium two spherically symmetric point defects do not interact. In reality two spherically symmetric point defects do interact (a) because the medium is not elastically isotropic (although tungsten is very nearly isotropic) and (b) all bodies are finite with free surfaces which give rise to image interactions.

From problem 4.3 we have the following expression for the correction to the Fourier transform of the Green's function in a cubic crystal to first order in D/C_{44}:

$$\begin{aligned}\Delta\tilde{G}_{ij}(\mathbf{k}) &= \frac{D}{\mu}\frac{1}{\mu}\frac{(\delta_{i1}k_1^2\delta_{j1} + \delta_{i2}k_2^2\delta_{j2} + \delta_{i3}k_3^2\delta_{j3})}{k^4} \\ &\quad -\frac{D}{\mu}\frac{1}{\mu}\left(\frac{\lambda+\mu}{\lambda+2\mu}\right)\times \\ &\qquad \frac{\left(k_1^3(\delta_{i1}k_j + k_i\delta_{j1}) + k_2^3(\delta_{i2}k_j + k_i\delta_{j2}) + k_3^3(\delta_{i3}k_j + k_i\delta_{j3})\right)}{k^6} \\ &\quad +\frac{D}{\mu}\frac{1}{\mu}\left(\frac{\lambda+\mu}{\lambda+2\mu}\right)^2 k_i k_j \frac{(k_1^4 + k_2^4 + k_3^4)}{k^8}\end{aligned}$$

where $D = C_{11} - C_{12} - 2C_{44}$.

The Fourier transform of $\Delta G_{ij,ji}(\mathbf{x})$ is $-\Delta\tilde{G}_{ij}(\mathbf{k})k_i k_j$. Using the above expression for $\Delta\tilde{G}_{ij}(\mathbf{k})$ we obtain:

$$\Delta\tilde{G}_{ij}(\mathbf{k})k_i k_j = \frac{D}{(\lambda+2\mu)^2}\frac{k_1^4+k_2^4+k_3^4}{k^4} = \frac{(C_{11}-C_{12}-2C_{44})}{(C_{12}+2C_{44})^2}\frac{k_1^4+k_2^4+k_3^4}{k^4}.$$

Using contour integration (see problem 4.3) we obtain

$$\frac{1}{(2\pi)^3}\int \frac{1}{k^4}e^{-i\mathbf{k}\cdot\mathbf{x}}\,\mathrm{d}^3k = -\frac{x}{8\pi}.$$

Therefore,

$$\frac{1}{(2\pi)^4}\int \frac{k_1^4}{k^4}e^{-i\mathbf{k}\cdot\mathbf{x}}\,\mathrm{d}^3k = \frac{1}{(2\pi)^4}\frac{\partial^4}{\partial x_1^4}\int \frac{1}{k^4}e^{-i\mathbf{k}\cdot\mathbf{x}}\,\mathrm{d}^3k = -\frac{1}{8\pi}\frac{\partial^4 x}{\partial x_1^4}$$

But,

$$\frac{\partial^4 x}{\partial x_1^4} = -\frac{3}{x^3} + 18\frac{x_1^2}{x^5} - 15\frac{x_1^4}{x^7}.$$

Therefore,

$$\Delta G_{ij,ji}(\mathbf{x}) = \frac{3}{8\pi}\frac{(C_{12}+2C_{44}-C_{11})}{(C_{12}+2C_{44})^2}\frac{1}{x^3}\left(5\frac{x_1^4+x_2^4+x_3^4}{x^4}-3\right)$$

Comment

To first order in D/C_{44}, the radial dependence of the interaction energy between two point defects in an infinite cubic crystal, occupying sites of cubic symmetry, varies as the inverse cube of their separation. The angular dependence, which is the term in large brackets, has cubic symmetry as expected in a cubic crystal.

Problem 5.2 Consider a substitutional point defect located at the origin of a cartesian coordinate system in an infinite simple cubic crystal structure with lattice constant a. The forces exerted by the defect on each of the six nearest neighbours have magnitude f, and they are directed along the bonds. Show that the displacement field $u_i(\mathbf{x})$ at $\mathbf{x}$ is approximately $-2afG_{ij,j}(\mathbf{x})$, and state the nature of the approximation.

Show that in the isotropic elastic approximation:

$$u_i(\mathbf{x}) = \frac{(1-2\nu)af}{4\pi\mu(1-\nu)}\frac{x_i}{x^3}.$$

Hence calculate the volume change ΔV^∞ associated with the point defect in an infinite medium. By comparing your answer with ΔV^∞ in the model of a point defect as a misfitting sphere in an infinite medium obtain an expression for $D = \Delta V^\infty/(4\pi)$ in the misfitting sphere model in terms of the bond length a and defect force f.

Solution The displacement field at $\mathbf{x}$ is found as follows:

$$\begin{aligned}
u_i(\mathbf{x}) &= \sum_{n=1}^{6} G_{ij}\left(\mathbf{x}-\mathbf{R}^{(n)}\right) f_j\left(\mathbf{R}^{(n)}\right) \\
&= G_{i1}(\mathbf{x}-a[1,0,0])f - G_{i1}(\mathbf{x}-a[\bar{1},0,0])f \\
&\quad +G_{i2}(\mathbf{x}-a[0,1,0])f - G_{i2}(\mathbf{x}-a[0,\bar{1},0])f \\
&\quad +G_{i3}(\mathbf{x}-a[0,0,1])f - G_{i3}(\mathbf{x}-a[0,0,\bar{1}])f \\
&\approx -2af\left(\frac{\partial G_{i1}(\mathbf{x})}{\partial x_1} + \frac{\partial G_{i2}(\mathbf{x})}{\partial x_2} + \frac{\partial G_{i3}(\mathbf{x})}{\partial x_3}\right) \\
&= -2afG_{ij,j}(\mathbf{x}).
\end{aligned}$$

The approximation is that $x \gg a$ so that only the first order terms in the Taylor expansions are included.

In isotropic elasticity the divergence of the Green's function is:

$$G_{ij,j}(\mathbf{x}) = -\frac{1-2\nu}{8\pi\mu(1-\nu)}\frac{x_i}{x^3}.$$

Therefore,

$$u_i(\mathbf{x}) = \frac{(1-2\nu)}{4\pi\mu(1-\nu)}af\frac{x_i}{x^3}.$$

The volume change ΔV^∞ associated with the point defect is:

$$\Delta V^\infty = 4\pi \lim_{x\to\infty} x^2|\mathbf{u}(\mathbf{x})| = \frac{(1-2\nu)}{\mu(1-\nu)}af$$

Equating ΔV^{∞} to the volume change of a misfitting sphere in an infinite medium, $4\pi D$, we obtain an expression for D in terms of a and f:

$$D = \frac{(1-2\nu)}{4\pi\mu(1-\nu)} af.$$

Comment

For $\nu \approx \frac{1}{3}$, $\mu \approx 100\text{GPa}$, $a \approx 2\text{Å}$, $4\pi D \approx 0.1\text{Å}^3$ we obtain $f \approx 0.1\,\text{nN} \approx 0.06\text{eV/Å}$. This is a credible interatomic force for a point defect of this size misfit to exert on neighbouring host atoms.

Problem 5.3 Consider two vacancies in tungsten, which has a body centred cubic crystal structure, and its elastic properties are well approximated as isotropic[7]. Aligning the coordinate axes with the sides of the cubic unit cell show that the dipole tensor for each vacancy has the form $\rho_{ij} = p\delta_{ij}$.

In problem 5.1 it was shown that the dipole-dipole interaction energy is zero because $G_{ij,ji} = 0$. The symmetry of atomic sites in the bcc crystal dictates that the quadrupole tensor is zero. Show that there are only two types of non-zero independent elements of the octupole tensor: o_{1111} and o_{1122}. Note that $o_{1122} = o_{1212} = o_{1221}$. Hence show that the octupole tensor may be expressed as:

$$o_{ijkl} = o_{1111}\delta_{ij}\delta_{jk}\delta_{kl} + o_{1122}\left(\delta_{ij}\delta_{kl} + \delta_{ik}\delta_{jl} + \delta_{il}\delta_{jk} - 3\delta_{ij}\delta_{jk}\delta_{kl}\right).$$

Show that the first non-zero term in the interaction energy between the defects is:

$$E_{int} = \frac{1}{6}G_{ij,jiii}\left[p(\tilde{o}_{1111} - 3\tilde{o}_{1122}) + \tilde{p}(o_{1111} - 3o_{1122})\right] + \frac{1}{2}G_{ij,jikk}(p\tilde{o}_{1122} + \tilde{p}o_{1122}),$$

where $G_{ij,jiii}$ means $G_{1j,j111} + G_{2j,j222} + G_{3j,j333}$.

Hence show that

$$E_{int} = \frac{7(1-2\nu)}{8\pi\mu(1-\nu)x^5}\left[p(\tilde{o}_{1111} - 3\tilde{o}_{1122})\right]\left(\frac{5(x_1^4 + x_2^4 + x_3^4)}{x^4} - 3\right)$$

Thus, in an infinite isotropic cubic crystal two vacancies interact through the dipole-octupole interaction. This interaction energy displays the same angular dependence[8]

[7]This problem is inspired by section 23.2 of Teodosiu's book

[8]The angular dependence is the cubic harmonic $K_{4,1} = (\sqrt{21}/4)[5((x_1/x)^4 + (x_2/x)^4 + (x_3/x)^4) - 3]$,

as was found in problem 5.1 for the dipole-dipole interaction energy between two point defects occupying sites of cubic symmetry in a weakly anisotropic cubic crystal. However, the dipole-dipole interaction energy in a cubic crystal varies with separation x as $1/x^3$, as compared with $1/x^5$ for the variation of the dipole-octupole interaction energy in an isotropic crystal. Therefore, whenever there is a departure of the anisotropy ratio from unity in a cubic crystal the dipole-dipole interaction will dominate at long-range.

Solution Since the vacancies are identical $\rho_{ij} = \tilde{\rho}_{ij}$ and $o_{ijkl} = \tilde{o}_{ijkl}$. Since the vacancies occupy sites of cubic symmetry the dipole tensor must be of the form $\rho_{ij} = p\delta_{ij}$. This follows from the transformation properties of any physical property represented by a second rank tensor in the presence of cubic symmetry.

In cubic symmetry only those components of the multipolar moments with pairs of equal indices are non-zero. This immediately dictates that all elements of the quadrupole tensor are zero, since the quadrupole tensor has an odd number of indices. The non-zero components of the octupole tensor are:

$$
\begin{aligned}
o_{1111} &= o_{2222} = o_{3333} \\
o_{1122} &= o_{2211} = o_{1212} = o_{1221} = o_{2112} = o_{2121} \\
&= o_{2233} = o_{3322} = o_{2323} = o_{2332} = o_{3223} = o_{3232} \\
&= o_{3311} = o_{1133} = o_{3131} = o_{3113} = o_{1331} = o_{1313}.
\end{aligned}
$$

Therefore,

$$
o_{ijkl} = \tilde{o}_{ijkl} = o_{1111}\delta_{ij}\delta_{jk}\delta_{kl} + o_{1122}\left(\delta_{ij}\delta_{kl} + \delta_{ik}\delta_{jl} + \delta_{il}\delta_{jk} - 3\delta_{ij}\delta_{jk}\delta_{kl}\right).
$$

The dipole-octupole interaction energy is as follows:

$$
\begin{aligned}
E_{int} &= \frac{1}{6}\left(\tilde{o}_{ijkl}G_{ia,bjkl}\rho_{ab} + \tilde{\rho}_{ij}G_{ia,bcdj}o_{abcd}\right) \\
&= \frac{1}{6}\left\{(\tilde{o}_{1111} - 3\tilde{o}_{1122})\delta_{ij}\delta_{jk}\delta_{kl} + \tilde{o}_{1122}\left(\delta_{ij}\delta_{kl} + \delta_{ik}\delta_{jl} + \delta_{il}\delta_{jk}\right)\right\} \times \\
&\quad G_{ia,bjkl}p\delta_{ab}
\end{aligned}
$$

which is normalised as follows: $\int (K_{4,1})^2 \mathrm{d}\Omega = 4\pi$ where $\mathrm{d}\Omega = \sin\theta \mathrm{d}\theta \mathrm{d}\phi$ is an element of solid angle. Cubic harmonics are combinations of spherical harmonics with cubic symmetry. See Fehlner, W R and Vosko, S H, Can. J. Phys. **54**, 2159-2169 (1976).

$$+\frac{1}{6}\tilde{p}\delta_{ij}G_{ia,bcdj}\Big\{(o_{1111}-3o_{1122})\delta_{ab}\delta_{bc}\delta_{cd}$$

$$+o_{1122}(\delta_{ab}\delta_{cd}+\delta_{ac}\delta_{bd}+\delta_{ad}\delta_{bc})\Big\}$$

$$=\frac{1}{6}\Big\{(\tilde{o}_{1111}-3\tilde{o}_{1122})p+(o_{1111}-3o_{1122})\tilde{p}\Big\}\sum_{a,i=1}^{3}G_{ia,aiii}$$

$$+\frac{1}{2}(\tilde{o}_{1122}p+o_{1122}\tilde{p})\sum_{a,i,j=1}^{3}G_{ai,iajj}.$$

Since there is a potential confusion arising from a subscript repeated three times we have dropped the summation convention and made the sums explicit. Now,

$$\sum_{a,i,j=1}^{3}G_{ai,iajj}=0$$

because we have already shown that $\sum_{a,i}G_{ai,ia}=0$ in isotropic elasticity. Therefore, we need only to evaluate $\sum_{a,i}G_{ia,aiii}$. We have seen already that the divergence of the isotropic Green's function is

$$\sum_{a=1}^{3}G_{ia,a}=-\frac{(1-2\nu)}{8\pi\mu(1-\nu)}\frac{x_i}{x^3}.$$

Let us evaluate the third partial derivative of x_1/x^3 with respect to x_1:

$$\frac{\partial}{\partial x_1}\left(\frac{x_1}{x^3}\right)=\frac{1}{x^3}-\frac{3x_1^2}{x^5}$$

$$\frac{\partial^2}{\partial x_1^2}\left(\frac{x_1}{x^3}\right)=-\frac{9x_1}{x^5}+\frac{15x_1^3}{x^7}$$

$$\frac{\partial^3}{\partial x_1^3}\left(\frac{x_1}{x^3}\right)=-\frac{9}{x^5}+\frac{90x_1^2}{x^7}-\frac{105x_1^4}{x^9}.$$

Therefore,

$$\sum_{a,i=1}^{3}G_{ia,aiii}=-\frac{(1-2\nu)}{8\pi\mu(1-\nu)}\left(63\frac{1}{x^5}-105\frac{x_1^4+x_2^4+x_3^4}{x^9}\right)$$

$$= -\frac{(1-2\nu)}{8\pi\mu(1-\nu)}\frac{21}{x^5}\left(3-5\frac{x_1^4+x_2^4+x_3^4}{x^4}\right).$$

Inserting this expression into the equation for the interaction energy we get:

$$\begin{aligned} E_{int} &= \frac{(1-2\nu)}{8\pi\mu(1-\nu)}\frac{1}{6}\left\{(\tilde{o}_{1111}-3\tilde{o}_{1122})\,p+(o_{1111}-3o_{1122})\,\tilde{p}\right\}\times \\ &\quad \frac{21}{x^5}\left(5\frac{x_1^4+x_2^4+x_3^4}{x^4}-3\right) \\ &= \frac{7(1-2\nu)}{8\pi\mu(1-\nu)}p\,(\tilde{o}_{1111}-3\tilde{o}_{1122})\,\frac{1}{x^5}\left(5\frac{x_1^4+x_2^4+x_3^4}{x^4}-3\right), \end{aligned}$$

where in the last line we have used $(\tilde{o}_{1111}-3\tilde{o}_{1122})\,p=(o_{1111}-3o_{1122})\,\tilde{p}$.

6. Dislocations

.1 Introduction

When crystals are deformed there is a limit to which the deformation remains elastic or reversible. Further deformation is irreversible or *plastic*. It is irreversible because the crystal changes shape permanently as a result of shearing processes. It is a characteristic property of many metals and alloys that they undergo plastic deformation before they fracture, which enables them to be extruded, pressed, rolled and forged into everyday objects from car bodies to drink cans and from furniture to girders for bridges and skyscrapers.

In the early twentieth century there was a great deal of interest to discover what happened inside a crystal when it ceased to deform elastically and began to deform plastically. It was a mystery why some crystals such as copper and gold were extremely ductile whereas others such as diamond were brittle, at least at room temperature. Some ductile metals and alloys, including some steels, become brittle when the temperature is lowered. Beginning to address these questions brought about the birth of the modern science of materials in the 1930s. Today our understanding of all these phenomena and processes is based on dislocations and how they interact with other defects in the material. It is the movement of dislocations that leads to plastic deformation of crystalline matter, and in general this happens more readily in metals than non-metals. During the 1950s transmission electron microscopy made it possible to observe dislocations moving inside very thin crystals undergoing plastic deformation. The complexity of many processes involved in plastic deformation, or 'plasticity', is so great that our understanding is still far from complete, and we shall return to some of these issues in Chapter 11.

.2 Dislocations as the agents of plastic deformation

To a very good approximation crystals deform plastically at constant volume. Permanent changes of shape are brought about by shearing processes in which planes of atoms slide over each other. The sliding creates steps at the surface of the crystal.

In 1926 Frenkel[1] produced a startling back-of-an-envelope calculation. He showed that the stress required to slide a plane of atoms *en masse* over another was orders of magnitude larger than the stresses observed to initiate plastic deformation of many metals. He considered two adjacent planes of atoms in a crystal spaced d apart. Let b be the spacing of atoms in the plane in the direction of sliding. Frenkel sought to estimate the maximum stress required to slide one entire plane *en masse* over the other.

Let x be a relative, rigid displacement of the two atomic planes in the direction of sliding. Let $\gamma(x)$ be the energy per unit area of the plane associated with this relative displacement. We know $\gamma(x)$ is periodic with a period of b. Therefore, we can write

$$\gamma(x) = A_0 + \sum_{n=1}^{\infty} A_n \cos\left(\frac{2n\pi x}{b}\right) + B_n \sin\left(\frac{2n\pi x}{b}\right), \tag{6.1}$$

which is just a Fourier expansion of $\gamma(x)$. If we make the simplest approximation and assume $\gamma(x)$ is an even function, taking just the first term of the cosine series we obtain $\gamma(x) = A\sin^2(\pi x/b)$, where A is a positive constant with the dimensions of energy per unit area. The minima of $\gamma(x)$ are at $x = nb$, where n is an integer. The slope of $\gamma(x)$ is the stress required to sustain the relative displacement x. Frenkel argued that as $x \longrightarrow 0$ this slope is determined by the elastic shear modulus μ. Thus, $\mathrm{d}\gamma/\mathrm{d}x = (A\pi/b)\sin(2\pi x/b) \longrightarrow A2\pi^2 x/b^2$ is equated to $\mu x/d$. Therefore, $A = (\mu b^2)/(2\pi^2 d)$. The maximum stress is reached when $x = b/4$, when it is $\mu b/(2\pi d) \approx \mu/10$.

The stress required to transition from elastic to plastic deformation is called the yield stress. Frenkel's estimate is that the yield stress is of order $\mu/10$. Experimentally observed values of the yield stress are typically between 3 and 5 orders of magnitude less than this. Frenkel's simple analysis can be improved by using more realistic descriptions of $\gamma(x)$. However such improvements do not reduce the estimated yield stress by orders of magnitude. We are forced to conclude that entire planes of atoms do not slide *en masse* over each other during plastic deformation.

The resolution of the paradox came with the independent but almost simultaneous insights of Orowan[2] in Budapest, Polanyi[3] in Manchester (UK) and Taylor[4] in Cambridge (UK) in 1934. They suggested that planes *slip* past each other through the movement of linear defects called dislocations. Taylor's paper is particularly

[1]Jakov I Frenkel 1894-1952, Soviet condensed matter physicist.

[2]Egon Orowan FRS 1902 -1989, Hungarian/British/US physicist and metallurgist, in Z. Physik **89**, 634 (1934) (in German).

[3]Michael Polanyi FRS 1891-1976, Hungarian/British polymath, in Z. Physik **89**, 660 (1934) (in German).

[4]Sir Geoffrey Ingram Taylor FRS OM 1891-1976, British physicist and mathematician, in Proc. R. Soc. A **145**, 388 (1934).

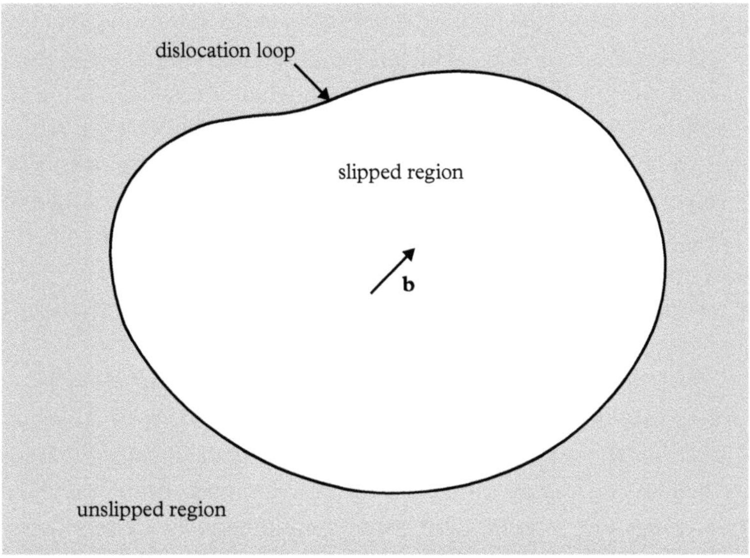

Figure 6.1: A dislocation loop lying in a slip plane separating slipped and unslipped regions. The dislocation is the boundary of the slipped region. Inside the loop the material beneath the slip plane has been translated with respect to material above it by the Burgers vector **b**.

remarkable as it includes a detailed discussion of the crystallographic nature of slip, namely the observation that it occurs on particular planes in particular directions, and of the hardening of crystals that occurs with increasing plastic deformation, which is called work hardening, and elastic interactions between dislocations leading to stable and unstable configurations of dislocations. Taylor drew on the earlier mathematical treatment of dislocations by Volterra[5]. Volterra's interest was more mathematical and centred on the elastic equilibrium of bodies rendered multiply connected through the existence of dislocations[6].

The sliding of one crystal plane over another, which is called slip, begins with the nucleation of a small dislocation loop. The plane where slip occurs is called the slip plane. Inside the loop the planes have slipped past each other by a vector **b**, which is called the Burgers vector. Outside the loop the planes have not slipped. The dislocation is the line separating the slipped and unslipped regions of the slip plane - see Fig 6.1. We will show that under the influence of an applied shear stress on the slip plane in the direction of **b** the loop expands converting more of the unslipped region into the

[5]Vito Volterra 1860-1940, Italian mathematician and physicist.

[6]Volterra, V, Annales scientifiques de l'École Normale Supérieure **24** 401-517, (1907) (in French).

slipped region. This enables the applied shear stress to do more work thereby reducing the potential energy of the system, which includes an external loading mechanism if there is one. If the loop is in a single crystal then when it reaches the surfaces of the crystal the entire crystal on one side of the slip plane has slipped by **b** with respect to the other side. On a surface with unit normal $\hat{\mathbf{n}}$ there is then a step of height $(\mathbf{b} \cdot \hat{\mathbf{n}})$. If the Burgers vector **b** is a crystal lattice vector in the slip plane then as the dislocation moves the same crystal structure is created in its wake. In principle **b** may be any lattice vector in the slip plane, but we shall see that the elastic energy of the dislocation varies as $|\mathbf{b}|^2$ so that usually **b** is one of the smallest lattice vectors. The smallest lattice vectors tend to occur in planes with the largest spacing.

The key physical insight of Orowan, Polanyi and Taylor in 1934 is that slip by dislocation motion localises the inevitable bond breaking and making, when one plane slides over another, to the very much smaller region where the dislocation line is located. In contrast, when an entire plane slides *en masse* over another the bond breaking and making occurs everywhere in the slip plane simultaneously. Therefore, the stress required to move a dislocation is orders of magnitude less than that required to slide a plane of a macroscopic crystal over another *en masse*. This was a giant step forward. But fundamental questions remained, such as why some crystals seem to undergo very limited slip, if any, before they fracture, why some are brittle at low temperatures and ductile at higher temperatures, why small concentration of impurities can have a seemingly disproportionate effect on the ease of slip, what determines the selection of slip plane and Burgers vector, the stress to move a dislocation, how dislocations are created, and even whether they do in fact exist. Addressing these questions over the intervening years has led to some of the most interesting experimental, theoretical and computational condensed matter physics, with profound consequences for engineering and technology.

6.3 Characterisation of dislocations: the Burgers circuit

Consider a dislocation loop lying in a slip plane. Within the loop there is a relative displacement of the crystals below and above the slip plane equal to the Burgers vector. Because the relative displacement is constant throughout the slipped region inside the loop it does not vary with the direction of the dislocation line. Where the direction of the dislocation line is perpendicular to the Burgers vector the dislocation is said to have *edge* character. Perhaps the reason for this name is that the dislocation line is then along the edge of a terminating half-plane, as shown in Fig 6.2. An edge dislocation is a long straight dislocation where the dislocation line is everywhere perpendicular to the Burgers vector. The edge dislocation appeared in Taylor's paper of 1934. When the dislocation line is parallel to the Burgers vector the dislocation is said to have *screw*

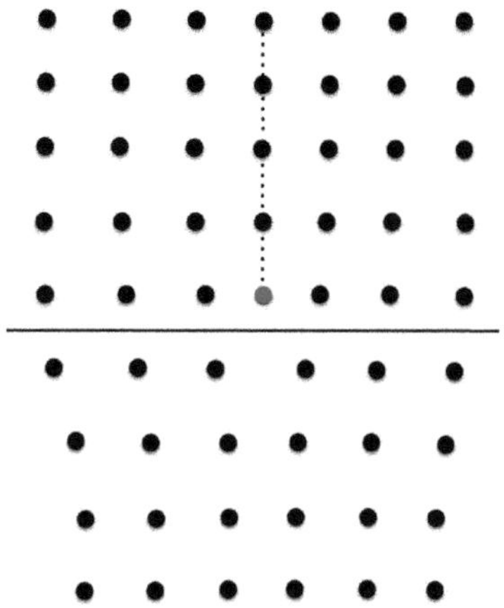

Figure 6.2: A schematic illustration of an edge dislocation viewed along the dislocation line. Each dot represent a column of atoms normal to the page. The horizontal line is the trace of the slip plane. Above the slip plane there is an extra half plane shown by dotted lines. The edge dislocation is located at the termination of this extra half plane, which is shown by the red column of atoms.

character. A screw dislocation is a long straight dislocation where the dislocation line is everywhere parallel to the Burgers vector. The screw dislocation was introduced by J M Burgers[7] in 1939[8]. The reason for this name is that the crystal lattice planes normal to the dislocation line become a helicoidal surface like the thread of a screw, as shown in Fig 6.3. If the angle between the line direction and the Burgers vector is ϕ then the dislocation may be regarded as a superposition of an edge dislocation with Burgers vector $\mathbf{b}\sin\phi$ and a screw dislocation with Burgers vector $\mathbf{b}\cos\phi$. Such a dislocation is said to have *mixed* character, and its atomic structure changes from screw to edge type as ϕ varies from 0 to $\pi/2$.

Suppose we see some defect in a crystal. How do we know whether it is a dislocation or some other defect? If it is a dislocation how do we determine its Burgers vector? The answer to both questions is the Burgers circuit construction. This is a geometrical construction that was introduced by Frank[9] in a paper[10] which defined rigorously many of the terms in use today when discussing dislocations and plasticity of crystals. The construction is illustrated in Fig 6.4 for an edge dislocation. We draw a closed circuit in

[7]Johannes Martinus Burgers 1895-1981, Dutch physicist, whose brother, Wilhelm Gerard Burgers 1897-1988, also a Dutch physicist, also worked on dislocations.

[8]Burgers, J M, Koninklijke Nederlandsche Akademie van Wetenschappen, **42**, 293-325 (1939).

[9]Sir (Frederick) Charles Frank FRS 1911-1998, British physicist.

[10]Frank, F C, Phil. Mag. **42**, 809 (1951).

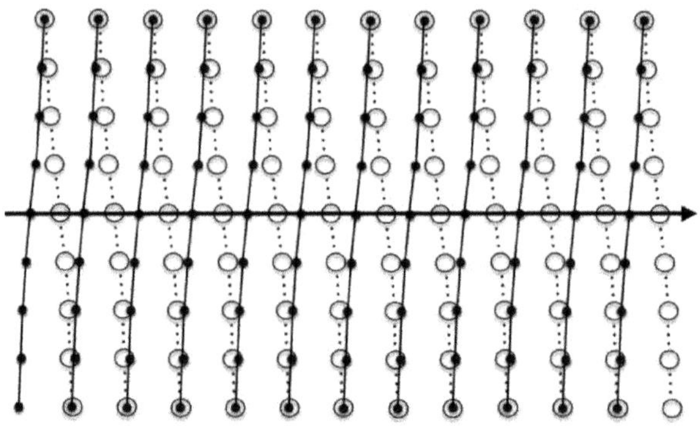

Figure 6.3: A schematic illustration of a screw dislocation. The dislocation line is shown by the horizontal arrow. Solid circles and solid lines are below the page. Open circles and dotted lines are above the page. Planes of atoms are converted into a continuous spiral by the dislocation. This drawing was adapted from Hull, D, and Bacon, D J, *Introduction to Dislocations*, Pergamon Press: Oxford, 3rd edn. (1984). ISBN 0-08-028720-4.

the elastically distorted crystal surrounding the dislocation which begins and ends at the same site. The circuit comprises steps between lattice sites of the elastically distorted crystal. It is important that the circuit passes through material that is recognisable as perfect crystal that has only been elastically strained, and that it does not go near the core of the dislocation where the local atomic environment cannot be mapped onto the perfect crystal. It does not matter how far the circuit goes from the dislocation as long as it does not enclose any other dislocations. The circuit is then mapped onto a perfect lattice, as shown in Fig 6.4. If the defect is a dislocation the circuit mapped onto the perfect lattice will not close. By convention, the closure failure, from the finish to the start of the circuit mapped onto the perfect crystal, is the Burgers vector when the circuit is taken in a clockwise sense looking along the dislocation line, i.e. into the page in Fig 6.4. If the circuit is started somewhere else or follows a different closed path in the dislocated crystal the Burgers vector will always be the same provided the circuit does not enclose any other dislocations and that it does not pass through the dislocation core. The Burgers vector is an invariant property of the dislocation. The existence of the closure failure when the circuit is mapped into the perfect crystal is the defining property of a dislocation[11]. There is no closure failure if the circuit encloses a point defect only.

In a continuum there is no lattice to define a Burgers circuit in the manner of Fig

[11]The Burgers circuit is an example of anholonomy which arises elsewhere in physics including the Aharonov-Bohm effect, Foucault's pendulum and the Berry phase.

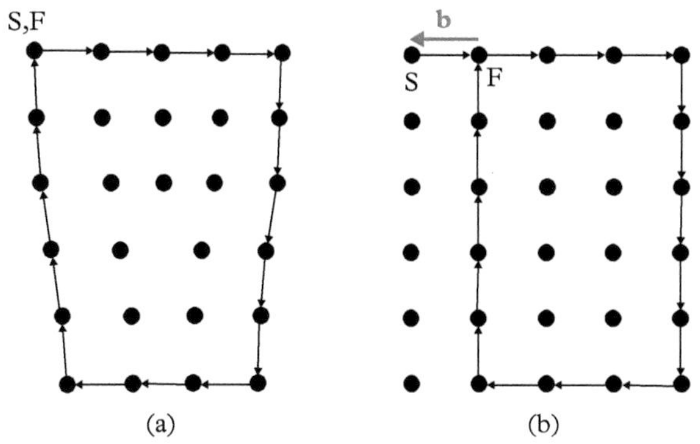

Figure 6.4: Illustration of the Burgers circuit construction for an edge dislocation. The line sense of the edge dislocation in (a) is defined to be positive looking into the page. A right handed circuit is drawn around the dislocation line, starting at S and finishing at the same site F. The circuit is mapped, step by step, onto the perfect crystal in (b). It is found that S and F are no longer coincident, and therefore the circuit has a closure failure. The vector joining F to S, shown by the arrow in red, is defined by the FS/RH convention as the Burgers vector **b** of the dislocation.

6.4. If **u** is the elastic displacement field of the dislocation then the Burgers vector is defined in a continuum by the line integral:

$$b_i = \oint\limits_C \frac{\partial u_i}{\partial x_k} \mathrm{d}x_k, \tag{6.2}$$

where C is any clockwise circuit taken around the dislocation line viewed along its positive direction. A word of caution: this convention is followed by many authors but some define the Burgers vector as the negative of this integral.

The *sign* of a dislocation is defined by two vectors: the direction of its line $\hat{\mathbf{t}}$ and its Burgers vector **b**. If either $\hat{\mathbf{t}}$ or **b** changes sign then the sign of the dislocation changes. If both $\hat{\mathbf{t}}$ and **b** change sign then the dislocation retains the same sign. Dislocations with the same sign repel while dislocations of opposite sign attract each other. At all points of a dislocation loop the Burgers vector is the same but the line direction changes. Dislocation segments in the loop with opposite line directions have opposite signs and if they come together they annihilate. We will see an example of this in the operation of a Frank-Read source of dislocation loops.

A bit more terminology: when the Burgers vector is a lattice vector the dislocation is said to be a *perfect* dislocation. When the Burgers vector is a fraction of a lattice vector the dislocation is called a *partial* dislocation, or sometimes an *imperfect* dislocation.

6.4 Glide, climb and cross slip

When dislocations move in their slip plane no diffusion of atoms is required. This kind of motion is called *glide* or conservative because the number of atoms involved in the motion of the dislocation is conserved. The normal to the slip plane of an edge dislocation is $\hat{\mathbf{n}} = \hat{\mathbf{b}} \times \hat{\mathbf{t}}$, where $\hat{\mathbf{b}}$ is a unit vector parallel to the Burgers vector and $\hat{\mathbf{t}}$ is a unit vector along the line direction. As long as an edge dislocation moves within its slip plane the number of atoms in the extra half-plane associated with the dislocation does not change, and this is why such motion is conservative.

If an edge dislocation moves out of its slip plane then the extra half plane either grows or shrinks requiring atoms to be added or removed from it. This is called *climb* or non-conservative motion. Because diffusion is involved the speed of climb is generally much less than the speed of glide. Climb may enable an edge dislocation to overcome an obstacle blocking glide on its slip plane, but since diffusion is involved it is a thermally activated process. The extra half plane of an edge dislocation grows when atoms are added to it, which is equivalent to saying that vacancies are emitted from it. In this way edge dislocations are sources and sinks for vacancies and self-interstitials, and thus they can regulate the populations of these point defects when the populations deviate from those in thermal equilibrium, e.g. due to irradiation.

In contrast to an edge dislocation the slip plane of a screw dislocation is not uniquely defined because $\mathbf{b}$ and $\hat{\mathbf{t}}$ are parallel or anti-parallel to each other. Consequently a screw dislocation gliding on one plane may switch to gliding on another plane, which is a process called cross slip. For example, a screw dislocation in a body-centred cubic (b.c.c.) crystal with Burgers vector $\mathbf{b} = 1/2[111]$ gliding on a $(1\bar{1}0)$ plane may cross slip onto the planes $(10\bar{1})$ or $(0\bar{1}1)$. As a result screw dislocations move only conservatively. Cross slip provides a mechanism for screw dislocations to overcome barriers to glide on a slip plane by gliding on an inclined plane over or under the obstacle. Although cross slip is a process that happens during glide it is thermally activated. In fcc crystals this is because perfect dislocations may be dissociated into partial dislocations separated by stacking faults (see Chapter 7), and before cross slip can take place the partial dislocations have to be recombined which is a process requiring energy. But even when perfect dislocations cross slip the atomic structure of the dislocation core has to adjust from that on one slip plane to that on another, and this is also a process requiring energy.

Exercise 6.1 During irradiation with high energy neutrons a highly non-equilibrium population of vacancies is created in a metal. In a b.c.c. crystal some of the vacancies cluster together on a $\{111\}$-type plane forming hexagonal dislocation loops with sides along $\langle 1\bar{1}0 \rangle$ directions. The Burgers vector of the dislocation loop is $1/2\langle 111 \rangle$

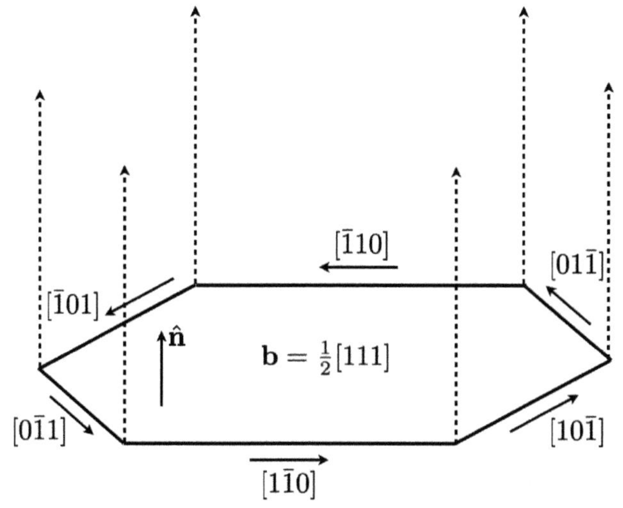

Figure 6.5: Sketch of a hexagonal prismatic loop with Burgers vector $\mathbf{b} = \frac{1}{2}[111]$. The solid arrows show the positive directions of each segment of the dislocation line. The positive normal to the loop is $\hat{\mathbf{n}}$ points upwards in accord with right-hand screw rule applied to the positive line directions of the dislocation segments. This is also the direction of the Burgers vector $\mathbf{b}$, so that $\hat{\mathbf{n}} \cdot \mathbf{b} > 0$ since the loop is a vacancy loop. The broken lines outline the positions of the vertices of the loop as it glides along the loop normal, outlining a hexagonal prism.

type. Show that such a loop may move in a conservative manner in the direction of its Burgers vector, tracing out a hexagonal prism as it moves. Such a loop is called *prismatic* for this reason. This example illustrates how individual defects, vacancies in this case, may come together to form a new defect, a prismatic dislocation loop in this case, and the mechanism of their motion changes from diffusion of individual vacancies to glide of the loop as a whole involving no diffusion at all. ▪

Solution With reference to Fig 6.5 the Burgers vector of the loop, $\mathbf{b} = \frac{1}{2}[111]$, is parallel to the positive loop normal. Each of the six segments of the loop is perpendicular to the Burgers vector, so the dislocation bounding the loop is an edge dislocation. As the loop glides along [111] its area is conserved. Therefore the number of vacancies the loop contains is also conserved, so no diffusion is required. If the loop were created by condensation of self-interstitials the Burgers vector

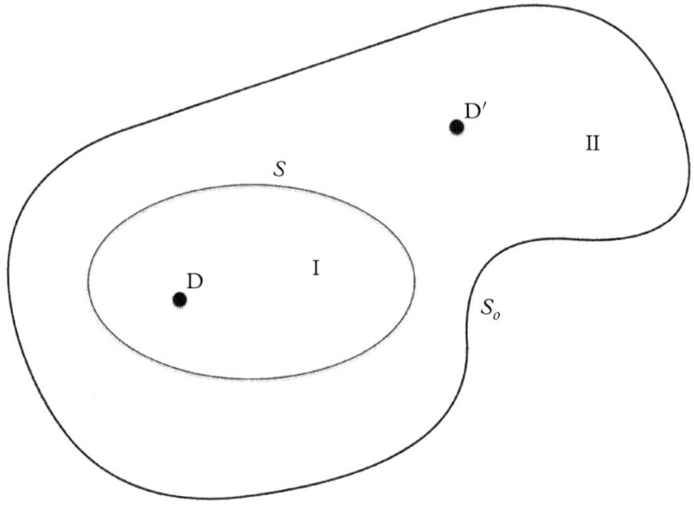

Figure 6.6: A schematic illustration of a body with an external surface S_o containing a dislocation D and other defects represented by D′. The internal closed surface S separates D from the other defects in the body, and divides the body into region I containing D only and region II which is the rest of the body.

would be $\mathbf{b} = \frac{1}{2}[\bar{1}1\bar{1}]$ and it would also be able to glide in a conservative manner along the loop normal.

6.5 Interaction energy between a dislocation and a source of stress

This section and the following section are based on the analysis presented in section 5 of a review article by Eshelby[12]. Consider a dislocation D in a body with external surface S_o. There may be constant tractions acting on the surface of the body, and other defects D′ inside the body creating internal stresses. Define a closed surface S inside the body enclosing D only so that it separates D from other sources of internal stress and the external load. The position of S is otherwise arbitrary. The internal surface S divides the body into two regions: region I contains D and region II is the rest of the body, as illustrated in Fig 6.6.

Let the elastic fields created by D be u_i^D, e_{ij}^D and σ_{ij}^D. Let u_i^A, e_{ij}^A and σ_{ij}^A be the elastic fields created by the constant loads applied to the external surface and by sources of internal stress other than D. Using linear superposition the total elastic fields in the body are $\sigma_{ij} = \sigma_{ij}^D + \sigma_{ij}^A$, $e_{ij} = e_{ij}^D + e_{ij}^A$ and $u_i = u_i^D + u_i^A$. The total elastic energy

[12]Eshelby, J D, Solid State Physics **3**, 79 (1956).

density is $\frac{1}{2}\sigma_{ij}e_{ij} = \frac{1}{2}(\sigma^D_{ij} + \sigma^A_{ij})(e^D_{ij} + e^A_{ij})$. The elastic interaction energy density involves only the cross terms between D and A, so that the total interaction energy is the following integral:

$$E_{int} = \frac{1}{2}\int_V \left(\sigma^D_{ij} e^A_{ij} + \sigma^A_{ij} e^D_{ij}\right) \mathrm{d}V. \tag{6.3}$$

In region I u^D_i cannot be defined everywhere owing to the elastic singularity D, but u^A_i is well defined throughout region I. Similarly u^A_i cannot be defined everywhere in region II, but u^D_i is well defined throughout region II. It follows from Hooke's law that $\sigma^A_{ij} e^D_{ij} = \sigma^D_{ij} e^A_{ij}$. Therefore eq 6.3 can be rewritten as:

$$E_{int} = \int_I \sigma^D_{ij} u^A_{i,j} \mathrm{d}V + \int_{II} \sigma^A_{ij} u^D_{i,j} \mathrm{d}V. \tag{6.4}$$

Furthermore, $\sigma^D_{ij,j} = 0$ and $\sigma^A_{ij,j} = 0$ in the absence of resultant body forces and therefore $\sigma^D_{ij} u^A_{i,j} = (\sigma^D_{ij} u^A_i)_{,j}$ and $\sigma^A_{ij} u^D_{i,j} = (\sigma^A_{ij} u^D_i)_{,j}$. The divergence theorem then transforms eq 6.4 into the following surface integrals:

$$E_{int} = \int_S \sigma^D_{ij} u^A_i n_j \mathrm{d}S + \int_{S_o} \sigma^A_{ij} u^D_i n_j \,\mathrm{d}S - \int_S \sigma^A_{ij} u^D_i n_j \,\mathrm{d}S. \tag{6.5}$$

Suppose D moves by an infinitesimal amount. The displacement field u^D_i will change by δu^D_i. The external load will do additional work given by $\int_{S_o} \sigma^A_{ij} \delta u^D_i n_j \,\mathrm{d}S$. This is represented by the change in the second term on the right of eq 6.5. The potential energy of the external loading mechanism will change by $-\int_{S_o} \sigma^A_{ij} \delta u^D_i n_j \mathrm{d}S$ because it has expended this amount of energy to do the work. Provided we include the potential energy of the external loading mechanism in the change of the total interaction energy these two terms cancel. They amount to nothing more than a redistribution of energy within the system comprising the body and the external loading mechanism. The integral over the external surface S_o of the body in eq 6.5 does not contribute to changes in the total interaction energy, regardless of whether or not there is an external load applied to the body. Thus, the total interaction energy, including the potential energy of the external loading mechanism, may be expressed as follows:

$$E^T_{int} = E_{int} + E_{ext} = \int_S \left(\sigma^D_{ij} u^A_i - \sigma^A_{ij} u^D_i\right) n_j \,\mathrm{d}S. \tag{6.6}$$

As it stands eq 6.6 has limited utility because σ^D_{ij} and u^D_i are the stress and displacement field of the defect D that satisfy the boundary conditions at the surface S_o

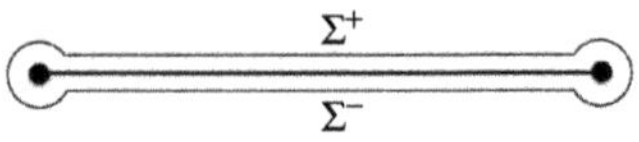

Figure 6.7: Illustration of the surface of the integral in eq 6.6 for a dislocation loop. The loop is viewed edge on and the plane of the loop is the thicker horizontal line. The positive sense of the dislocation line is coming out of the page on the left and into the page on the right. The positive loop normal points upwards. The surface S comprises the surfaces Σ^+ and Σ^- above and below the plane of the loop and a tube surrounding the dislocation. The positive loop normal points from the negative surface Σ^- to the positive surface Σ^+.

of the body. But suppose we replace σ_{ij}^D and u_i^D with $\sigma_{ij}^D + \sigma_{ij}^W$ and $u_i^D + u_i^W$ where σ_{ij}^W and u_i^W are any stress and displacement fields that have no singularities within S. Then a straightforward application of the divergence theorem shows that the additional terms do not change the integral in eq 6.6. It follows that σ_{ij}^D and u_i^D may be replaced by any stress and displacement fields that have the same singularities within S. In particular they may be replaced by $\sigma_{ij}^{D\infty}$ and $u_i^{D\infty}$ where $\sigma_{ij}^{D\infty}$ and $u_i^{D\infty}$ are the stress and displacement fields of D in an *infinite* medium. With this change eq 6.6 is very useful indeed.

If D is a dislocation loop enclosing a surface Σ then Σ may be taken as the cut on either side of which the displacement by the Burgers vector **b** exists. Then the surface S may be taken as the positive and negative sides of Σ together with a tube surrounding the dislocation line, as shown in Fig 6.7. The positive side of the loop is defined by the right-hand screw rule: progressing around the loop along the positive direction of the dislocation line advances a right-hand screw in the direction of the positive loop normal. As the radius of the tube shrinks to zero it contributes nothing to the integral in eq 6.6 because the dislocation is not associated with a resultant line of force. The term $\sigma_{ij}^D u_i^A$ is the same at opposing points on Σ^+ and Σ^-, and therefore this term contributes nothing to the integral. But there is a discontinuity in the displacement u_i^D on either side of the cut with u_i^D on Σ^+ minus u_i^D on Σ^- equal to $-b_i$. Therefore the total interaction energy becomes:

$$E_{int}^T = +b_i \int_{\Sigma^+} \sigma_{ij}^A n_j \, \mathrm{d}S. \tag{6.7}$$

This equation has a clear physical interpretation. If we imagine the dislocation loop is created in the presence of the stress field σ_{ij}^A then $-E_{int}^T$ represents the work done by the stress field σ_{ij}^A when the surface Σ^- is translated by the Burgers vector with respect to the surface Σ^+.

Exercise 6.2 Using Maxwell's reciprocity theorem, eq 4.7, show that the interaction energy between two dislocation loops A and B is as follows:

$$E_{int} = b_i^A \int_{\Sigma_A} \sigma_{ij}^B n_j \, \mathrm{d}S = b_i^B \int_{\Sigma_B} \sigma_{ij}^A n_j \, \mathrm{d}S. \tag{6.8}$$

■

Solution Suppose loop A already exists. Introduce loop B in the presence of loop A. Then the work done by the stress field σ_{ij}^A of loop A when the negative side of loop B is displaced by b_i^B with respect to the positive side is:

$$W = -b_i^B \int_{\Sigma_B^+} \sigma_{ij}^A n_j \, \mathrm{d}S.$$

Therefore, the interaction energy between the loops is:

$$E_{int} = +b_i^B \int_{\Sigma_B^+} \sigma_{ij}^A n_j \, \mathrm{d}S.$$

Repeating this but with A and B interchanged we find the interaction energy is also:

$$E_{int} = +b_i^A \int_{\Sigma_A^+} \sigma_{ij}^B n_j \, \mathrm{d}S.$$

Therefore,

$$E_{int} = +b_i^A \int_{\Sigma_A^+} \sigma_{ij}^B n_j \, \mathrm{d}S = +b_i^B \int_{\Sigma_B^+} \sigma_{ij}^A n_j \, \mathrm{d}S.$$

6.6 The Peach-Koehler force on a dislocation

Now that we have the interaction energy in eq 6.7 we may derive a general expression for the force on a dislocation due to an applied stress field. This force was first derived[13]

[13] Peach, M O and Koehler, J S, Phys. Rev. **80**, 436 (1950).

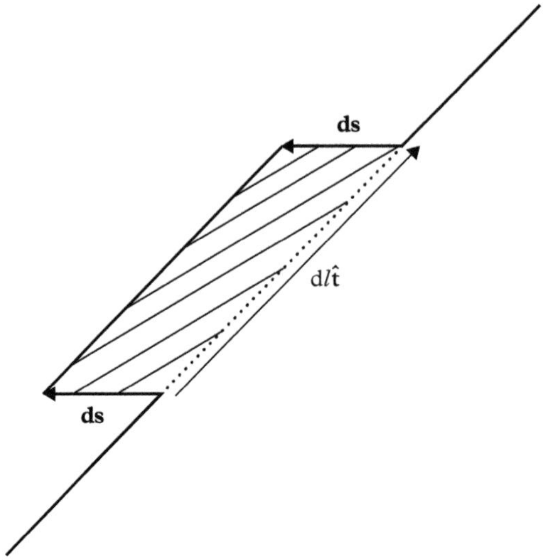

Figure 6.8: A segment of dislocation of length dl moves by the vector **ds** sweeping the area shaded.

by Peach and Koehler[14] and it is known as the Peach-Koehler force.

Consider a dislocation where the local direction of the dislocation line is $\hat{\mathbf{t}}$. Suppose an infinitesimal segment dl of the dislocation line is displaced by a vector **ds**, as shown in Fig 6.8. Bearing in mind the convention for defining the positive loop normal, with the shaded area in Fig 6.8 in the plane of the page, the positive loop normal points into the page. Then the vector area swept by the displaced segment is positive if it is defined as $\mathrm{d}\mathbf{A} = \mathrm{d}\mathbf{s} \times \mathrm{d}l\,\hat{\mathbf{t}}$. The displacement of the dislocation segment changes the surface Σ in the surface integral of eq 6.7. As in the previous section let σ^A_{ij} be the stress field acting locally on the dislocation segment, where this stress is the resultant of the stresses due to an external loading mechanism and sources of internal stresses. The work done by the applied stress is given by the change of $-E^T_{int}$ in eq 6.7:

$$\mathrm{d}w = -\mathrm{d}E^T_{int} = -b_i \sigma^A_{ij} \mathrm{d}l\, \varepsilon_{jpq} \mathrm{d}s_p t_q \qquad (6.9)$$

The force acting on the dislocation segment is $\partial w / \partial s_q = -\partial E^T_{int} / \partial s_q = F_q \mathrm{d}l$, where F_q is the force per unit length:

$$F_q = b_i \sigma^A_{ij} \varepsilon_{jpq} t_p = \varepsilon_{qjp} \left(\sigma^A_{ji} b_i \right) t_p. \qquad (6.10)$$

[14]This work formed part of the doctoral thesis of M O Peach under the supervision of James Stark Koehler 1914-2006, undertaken at Carnegie Institute of Technology (which merged in 1967 with Mellon Institute of Industrial Research to create Carnegie-Mellon University).

In vector notation this is $\mathbf{F} = (\sigma\mathbf{b}) \times \hat{\mathbf{t}}$. This is the Peach-Koehler force. The force is always perpendicular to the dislocation line. For edge dislocations, where the slip plane is uniquely defined, it includes components that promote glide and climb of dislocations. If $\hat{\mathbf{n}}$ is the normal to the slip plane the climb component is $\mathbf{F}_c = (\mathbf{F} \cdot \hat{\mathbf{n}})\hat{\mathbf{n}}$ and the glide component is $\mathbf{F}_g = \mathbf{F} - \mathbf{F}_c = \hat{\mathbf{n}} \times (\mathbf{F} \times \hat{\mathbf{n}})$.

Exercise 6.3 Show that the glide component of the Peach-Koehler force per unit length on any dislocation is τb where τ is the shear stress resolved on the slip plane in the direction of the Burgers vector and b is the magnitude of the Burgers vector.

Show that the climb component of the Peach-Koehler force on an edge dislocation is σb where σ is the resolved normal stress in the direction of the Burgers vector. ■

Solution The Peach-Koehler force per unit length on a dislocation is:

$$\mathbf{F} = (\sigma\mathbf{b}) \times \hat{\boldsymbol{\xi}},$$

where $\hat{\boldsymbol{\xi}}$ is the line direction and $\boldsymbol{\sigma}$ is the stress field acting on the dislocation. If $\hat{\mathbf{n}}$ is the normal to the slip plane then the glide force is

$$\mathbf{F}_{glide} = \mathbf{F} - (\mathbf{F} \cdot \hat{\mathbf{n}})\hat{\mathbf{n}} = \hat{\mathbf{n}} \times (\mathbf{F} \times \hat{\mathbf{n}}).$$

But

$$\mathbf{F} \times \hat{\mathbf{n}} = -\hat{\mathbf{n}} \times \mathbf{F} = -\hat{\mathbf{n}} \times \left((\sigma\mathbf{b}) \times \hat{\boldsymbol{\xi}}\right) = -(\hat{\mathbf{n}} \cdot \hat{\boldsymbol{\xi}})(\sigma\mathbf{b}) + (\hat{\mathbf{n}}\sigma\mathbf{b})\hat{\boldsymbol{\xi}}.$$

$\hat{\mathbf{n}} \cdot \hat{\boldsymbol{\xi}} = 0$ because the dislocation line is in the slip plane. Therefore,

$$\hat{\mathbf{n}} \times (\mathbf{F} \times \hat{\mathbf{n}}) = (\hat{\mathbf{n}}\sigma\mathbf{b})(\hat{\mathbf{n}} \times \hat{\boldsymbol{\xi}})$$

$(\hat{\mathbf{n}}\sigma\mathbf{b}) = \tau b$, where τ is the resolved stress on the slip plane in the direction of the Burgers vector and b is the magnitude of the Burgers vector. Since the Burgers vector is in the slip plane, τ is a shear stress. The direction of the force is perpendicular to the dislocation line in the slip plane.

The climb component of the Peach-Koehler force is

$$\mathbf{F}_{climb} = (\mathbf{F} \cdot \hat{\mathbf{n}})\hat{\mathbf{n}} = \left((\sigma\mathbf{b}) \times \hat{\boldsymbol{\xi}}\right) \cdot \hat{\mathbf{n}} = (\sigma\mathbf{b}) \cdot (\hat{\boldsymbol{\xi}} \times \hat{\mathbf{n}}) = (\hat{\boldsymbol{\xi}} \times \hat{\mathbf{n}}) \cdot (\sigma\mathbf{b})$$

For an edge dislocation $\hat{\boldsymbol{\xi}} \times \hat{\mathbf{n}}$ is a unit vector parallel to $\mathbf{b}$. Therefore, $|\mathbf{F}_{climb}| = (\hat{b}_i \sigma_{ij} \hat{b}_j) b$, where $\hat{b}_i \sigma_{ij} \hat{b}_j$ is the resolved normal stress in the direction of $\mathbf{b}$.

6.7 Volterra's formula

We come now to the derivation of a general expression for the displacement field of a dislocation loop in anisotropic elasticity. This expression is useful because, as shown[15] by Nabarro[16], the displacement field for an infinitesimal loop may be integrated to obtain solutions for loops of arbitrary shapes and sizes, which may then be differentiated to give strain fields and stress fields using Hooke's law. Assign a line sense to the dislocation loop. As before, the positive loop normal $\hat{\mathbf{n}}$ is defined by the right-hand screw rule applied to the positive line direction of the dislocation loop. Looking along the positive direction of the dislocation line draw a right-handed circuit around the dislocation to define the Burgers vector $\mathbf{b}$, as in eq 6.2. The surface inside the loop may be taken as the cut on either side of which the displacement field changes discontinuously by the Burgers vector. More precisely, the displacement field on the *negative* side of this cut minus the displacement field on the *positive* side of this cut is equal to the Burgers vector. The third line of eq 4.12 gives the displacement field in an infinite medium when displacements are prescribed on a surface $\mathcal{S}$. We may use this relation to write down the displacement field of a dislocation loop $\mathcal{L}$ with a displacement discontinuity equal to the Burgers vector between the negative and positive sides of a cut surface $\mathcal{S}$ bounded by $\mathcal{L}$. The result is known as Volterra's formula:

$$u_j(\mathbf{x}) = -c_{mikp}\, b_k \int_{\mathcal{S}^+} G_{ij,m'}(\mathbf{x}-\mathbf{x}')\, n_{p'} \mathrm{d}S' = c_{mikp}\, b_k \int_{\mathcal{S}^+} G_{ij,m}(\mathbf{x}-\mathbf{x}')\, n_{p'} \mathrm{d}S', \quad (6.11)$$

where the integration is carried out on the positive side $\mathcal{S}^+$ of the cut only.

A dislocation loop created in this way is called a Volterra dislocation. It is a mathematical simplification of a real dislocation in the sense that it is a line in the mathematical sense where the displacement discontinuity equal to the Burgers vector appears infinitely abruptly. In other words a Volterra dislocation has no width, and the displacement by the Burgers vector appears as a step function at the dislocation line. Real dislocations in crystals have finite widths, which are called dislocation cores, where the relative displacement by the Burgers vector accumulates over several interatomic bond lengths. Nevertheless the concept of a Volterra dislocation describes

[15] Nabarro, F R N, Phil. Mag. **42**, 1224 (1951).

[16] Frank Reginald Nunes Nabarro FRS 1916-2006, South African physicist born and educated in England

the elastic field further from the core quite accurately. In Chapter 7 we will meet more realistic models of dislocations.

The stress and strain fields of the loop are more significant physically than the displacement field because they determine the energy of the loop and its interaction with other defects. In contrast to the stress and strain fields of the loop, the displacement field is inevitably dependent on the choice of the cut where the displacement by the Burgers vector is introduced. Differentiation of the displacement field obtained with Volterra's formula yields the strain field, which must be independent of the location of cut. Later in this chapter we will derive Mura's formula for the strain field of the dislocation which depends only on the configuration of the dislocation *line*, and not on the cut bounded by it.

We have implied in this section that the choice of the cut plane is arbitrary. In an elastic continuum that is true. The dislocation is defined by the line where the cut terminates inside the medium. The location of the cut does affect the displacement field because it has to show the discontinuity by the Burgers vector on crossing the cut. But it does not affect the strain field and hence the stress field of the dislocation. For a crystal dislocation the location of the cut is defined by the slip plane of the dislocation when it glides. For an edge crystal dislocation the slip plane is uniquely defined by the cross product between the Burgers vector and the dislocation line direction. For a crystal screw dislocation the slip plane may change, as in cross slip, but there is still only a finite number of choices of cut plane. Only in an elastic continuum is the cut plane truly arbitrary, and there it does not even need to be flat.

6.8 The infinitesimal loop

Consider a loop of infinitesimal area δA and unit normal n_p located at $\mathbf{x}'$. Volterra's formula, eq 6.11, provides the displacement field at $\mathbf{x}$:

$$\delta u_j(\mathbf{x}) = \delta A \, c_{mikp} \, b_k \, n_p \, G_{ij,m}(\mathbf{x}-\mathbf{x}'). \tag{6.12}$$

This expression is exact within linear anisotropic elasticity. If the loop is finite, with a characteristic size L, then it remains a good approximation provided $|\mathbf{x}-\mathbf{x}'|$ is more than about 2L. To see that this is true consider a finite, planar, centrosymmetric loop, with its centre located at $\mathbf{R}$, where we take the cut to be the plane of area A bounded by the loop. Writing $\mathbf{x}' = \mathbf{R}+\rho$ we have the Taylor expansion:

$$G_{ij,m}(\mathbf{x}-\mathbf{R}-\rho) = G_{ij,m}(\mathbf{x}-\mathbf{R}) - \rho_p \, G_{ij,mp}(\mathbf{x}-\mathbf{R}) + \frac{1}{2}\rho_p \, \rho_q \, G_{ij,mpq}(\mathbf{x}-\mathbf{R}) - \ldots$$

Inserting this into Volterra's formula, eq 6.11, we obtain an expansion for the displacement field of the loop in terms of its areal moments:

$$u_j(\mathbf{x}) = c_{mikp}\, b_k n_p \left[G_{ij,m}(\mathbf{x}-\mathbf{R})\, A - G_{ij,mp}(\mathbf{x}-\mathbf{R}) \int_S \rho_p \,\mathrm{d}S + \frac{1}{2} G_{ij,mpq}(\mathbf{x}-\mathbf{R}) \int_S \rho_p \rho_q \,\mathrm{d}S - \ldots \right] \quad (6.13)$$

For a centrosymmetric loop, such as a circle, ellipse, rectangle, hexagon etc., only the even moments are non-zero. The first correction to eq 6.12 comes from second moments of the form $G_{ij,mpp}(\mathbf{x}-\mathbf{R}) \int_S \rho_p^2 \mathrm{d}S$, which is of order $L^4/|\mathbf{x}-\mathbf{R}|^4$. Thus, provided the distance from a finite centrosymmetric loop is more than about twice its size, eq 6.12 is remarkably accurate.

In isotropic elasticity eq 6.12 becomes:

$$\delta u_j(\mathbf{x}) = -\frac{\delta A\, b_k n_p}{8\pi(1-\nu)} \left[(1-2\nu) \frac{\delta_{kj} X_p + \delta_{pj} X_k - \delta_{kp} X_j}{X^3} + 3 \frac{X_k X_p X_j}{X^5} \right], \quad (6.14)$$

where $\mathbf{X} = \mathbf{x} - \mathbf{x}'$. By integrating this expression the displacement fields of finite loops may be derived in isotropic elasticity. The displacement fields of infinite straight dislocations may be found by considering loops closed at infinity.

6.9 The dipole tensor of an infinitesimal loop

The leading term in the interaction energy between two point defects in the multipole expansion of eq 5.10 involves their dipole tensors. Since an infinitesimal dislocation loop is also a point defect it should be possible to define a dipole tensor for it too. One way to derive the dipole tensor for an infinitesimal loop is to consider the interaction energy between two infinitesimal loops A and B using eq 6.8. If loop A is at the origin and loop B is at **x** then the interaction energy between them is as follows:

$$\delta E_{int} = b_f^B n_g^B (\delta A^B)\, \delta\sigma_{fg}^A(\mathbf{x}), \quad (6.15)$$

where δA^B is the area of loop B, and $\delta\sigma_{fg}^A(\mathbf{x})$ is the stress field of the infinitesimal loop A evaluated at loop B. Taking only the first term of the moment expansion of the displacement field of a loop in eq 6.13, and differentiating it to obtain the strain field we obtain:

$$\delta\sigma_{fg}^A(\mathbf{x}) = c_{fgjl} c_{mikp} b_k^A n_p^A \left(\delta A^A\right) G_{ij,ml}(\mathbf{x}). \quad (6.16)$$

Inserting this stress field into eq 6.15 we obtain an expression for the elastic interaction energy in terms of the dipole tensors $\delta\rho$ of the infinitesimal loops:

$$\delta E_{int} = \delta\rho^B_{jl} G_{ij,ml}(\mathbf{x})\delta\rho^A_{mi}, \tag{6.17}$$

where

$$\begin{aligned} \delta\rho^A_{mi} &= c_{mikp} b^A_k n^A_p (\delta A^A) \\ \delta\rho^B_{jl} &= c_{jlfg} b^B_f n^B_g (\delta A^B). \end{aligned} \tag{6.18}$$

As with the interaction energy between two point defects eq 6.17 separates properties of the loops themselves, which are contained in the dipole tensors, from the radial and angular dependences of their interaction energy, which are contained in the second derivatives of the Green's function. The dipole tensor of a loop in eq 6.18 has a simple physical interpretation. As first discussed[17] by Kroupa[18] a dislocation loop may be thought of as a region that has undergone a transformation strain[19] $e^T_{kp} = \frac{1}{2}(b_k n_p + b_p n_k)/\Delta$ where Δ is the thickness of the transformed region in the direction of the loop normal. We will take the limit $\Delta \to 0$. The stress associated with the transformation strain through Hooke's law is $\sigma^T_{im} = c_{imkp} e^T_{kp} = c_{imkp} b_k n_p/\Delta$. The surfaces of the loop are separated by the vector $\hat{\mathbf{n}}\Delta$. These surfaces are subjected to equal and opposite forces per unit area (tractions) equal to $\sigma^T_{im} n_m$. The displacement at $\mathbf{x}$ is then given by:

$$u_j(\mathbf{x}) = \int_{S'} G_{ji}(\mathbf{x} - \mathbf{x}')\sigma^T_{im} n_m \mathrm{d}S', \tag{6.19}$$

where the integration is over the surface of the transformed region constituting the loop. Inside the loop σ^T_{im} is constant. Applying the divergence theorem we obtain:

$$u_j(\mathbf{x}) = \sigma^T_{im} \int_{V'} G_{ji,m}(\mathbf{x} - \mathbf{x}')\mathrm{d}V', \tag{6.20}$$

where the integral is over the volume of the transformed region[20]. In the limit of an infinitesimal loop, and in the limit of $\Delta \to 0$, the volume integral becomes

[17]Kroupa, F in *Theory of crystal defects*, Proceedings of the Summer School held in Hrazany in September 1964, Academia Publishing House of the Czechoslovak Academy of Sciences (1966), pp 275-316.

[18]František Kroupa 1925-2009, Czech physicist.

[19]see section 8.1 and problem 8.3

[20]This equation is identical to eq 8.164 for the displacement field far from an inclusion with volume $(\delta A)\Delta$

$G_{ji,m}(\mathbf{x})(\delta A)\Delta$. The displacement field of the infinitesimal loop is therefore as follows:

$$u_j(\mathbf{x}) = c_{imkp}\frac{b_k n_p}{\Delta}\Delta(\delta A)G_{ji,m}(\mathbf{x}) = c_{imkp}b_k n_p(\delta A)G_{ji,m}(\mathbf{x}) = \rho_{im}G_{ji,m}(\mathbf{x}). \quad (6.21)$$

As discussed by Landau and Lifshitz[21] and the paper[22] by Burridge and Knopoff[23] the displacement by the Burgers vector at a dislocation may be thought of as the response of the medium to a dipolar distribution of fictitious forces across the cut. These forces are fictitious in the sense that they are the forces required by linear elasticity to generate the relative displacement across the cut. They are equivalent to Kanzaki forces in harmonic lattice theory, which are the forces required to create a defect within a model of harmonic atomic interactions. The real forces are what we have called *defect forces*, and they are the true forces acting between atoms, as determined by quantum mechanics. But in a linear elastic theory it is the fictitious forces we need to generate a dislocation and they appear in the dipole tensor for the dislocation loop.

Volterra's formula, eq 6.11, follows directly from eq 6.20. That is because $\sigma^T_{im} = c_{imkp}b_k n_p/\Delta$ and $dV' = dS'\Delta$. This way of deriving Volterra's formula is arguably more satisfying than the rather formal presentation in section 6.7. A dislocation loop may thus be viewed in two equivalent ways. First, as discussed in section 6.7 the dislocation delineates a region in a plane which has undergone a displacement by the Burgers vector **b**. Second, the dislocation delineates a region which has undergone a transformation strain $e^T_{kp} = \frac{1}{2}(b_k n_p + b_p n_k)/\Delta$, where Δ is the thickness of the transformed region in the direction of the loop normal and is very small compared to the diameter of the loop except in nanoscale loops. The transformed region inside the loop sets up closely spaced dipolar sheets of surface tractions inside the loop which generate the elastic field of the loop. This second way of viewing dislocations is particularly useful in the modelling of cracks as distributions of dislocations, as we shall see in Chapter 10.

Does the picture of terminating dipolar sheets of forces at a dislocation apply in an atomistic model? As a dislocation glides atoms on either side of the slip plane experience forces that establish the relative displacement by the Burgers vector and enlarge the slipped region. Those forces generate vibrations of the crystal lattice, and as discussed in section 11.3.3 this radiation is a principal source of drag on the dislocation, and it gives rise to acoustic emission which can be detected experimentally. The forces are real and they have observable consequences. Once the dislocation has moved on

[21]Landau, L D and Lifshitz, E M, *Theory of Elasticity*, Pergamon Press: Oxford, 3rd edition (1986), section 27, p.111. ISBN 978-0750626330.

[22]Burridge, R and Knopoff, L, Bulletin of the Seismological Society of America **54**, 1875-1888 (1964).

[23]Leon Knopoff 1925-2011 US geophysicist and musicologist.

the forces return to zero if the perfect crystal structure is recreated in the wake of the dislocation[24]. Since there is no crystal structure in a continuum the forces required to shear the medium to establish the relative displacement by the Burgers vector persist even after the dislocation has moved on. However, they produce no stress or strain field in the limit $\Delta \to 0$, and only the relative displacement by the Burgers vector on either side of the slip plane[25]. Another difference is that the minimum value of Δ in a crystal is the spacing of atomic planes parallel to the slip plane, whereas in a continuum the limit $\Delta \to 0$ is usually taken.

Exercise 6.4 For a screw dislocation along the x_3-axis, with Burgers vector $b_k = b\delta_{k3}$, with the sheets occupying the half spaces $x_1 > 0, x_2 = \Delta/2$ and $x_1 > 0, x_2 = -\Delta/2$ show that in isotropic elasticity:

$$\sigma^T_{im} = \frac{\mu b}{\Delta}(\delta_{i3}\delta_{m2} + \delta_{i2}\delta_{m3}).$$

An edge dislocation along the x_3 axis may be created with $b_k = b\delta_{k1}$ and the same location of the sheets of force as in the screw dislocation. Show that for an edge dislocation in isotropic elasticity:

$$\sigma^T_{im} = \frac{\mu b}{\Delta}(\delta_{i1}\delta_{m2} + \delta_{i2}\delta_{m1}).$$

■

Solution For a screw dislocation with Burgers vector $b_k = b\delta_{k3}$ and $n_p = \delta_{p2}$ in isotropic elasticity we have:

$$\begin{aligned}
\sigma^T_{im} &= c_{imkp} b_k n_p/\Delta \\
&= \left[\lambda\delta_{im}\delta_{kp} + \mu(\delta_{ik}\delta_{mp} + \delta_{ip}\delta_{mk})\right] b\delta_{k3}\delta_{p2}/\Delta \\
&= \mu(\delta_{i3}\delta_{m2} + \delta_{i2}\delta_{m3})b/\Delta.
\end{aligned}$$

[24]This is always true for a perfect dislocation. More generally the forces return to zero for any dislocation separating regions of the slip plane that correspond to local minima in the γ-surface.

[25]This is why the stress and strain fields of a dislocation are independent of the choice of cut on either side of which the relative displacement by the Burgers vector exists.

For an edge dislocation with Burgers vector $b_k = b\delta_{k1}$ and $n_p = \delta_{p2}$ in isotropic elasticity we have:

$$\begin{aligned}\sigma^T_{im} &= c_{imkp}b_k n_p/\Delta \\ &= \left[\lambda\delta_{im}\delta_{kp} + \mu(\delta_{ik}\delta_{mp} + \delta_{ip}\delta_{mk})\right] b\delta_{k1}\delta_{p2}/\Delta \\ &= \mu(\delta_{i1}\delta_{m2} + \delta_{i2}\delta_{m1})b/\Delta.\end{aligned}$$

6.10 The infinitesimal loop in isotropic elasticity

The stress field of an infinitesimal loop is readily obtained in isotropic elasticity by differentiating the displacement field in eq 6.14 to get the displacement gradient and then applying Hooke's law:

$$\begin{aligned}\sigma_{ij}(\mathbf{x}) &= -\frac{\mu(\delta A)b}{4\pi(1-\nu)x^3}\left\{\left[3(1-2\nu)(\hat{\mathbf{b}}\cdot\hat{\mathbf{x}})(\hat{\mathbf{n}}\cdot\hat{\mathbf{x}}) + (4\nu-1)(\hat{\mathbf{b}}\cdot\hat{\mathbf{n}})\right]\delta_{ij}\right. \\ &\quad + (1-2\nu)(\hat{b}_i n_j + n_i\hat{b}_j) + 3\nu\left[(\hat{\mathbf{b}}\cdot\hat{\mathbf{x}})(n_i\hat{x}_j + \hat{x}_i n_j) + (\hat{\mathbf{n}}\cdot\hat{\mathbf{x}})(\hat{b}_i\hat{x}_j + \hat{x}_i\hat{b}_j)\right] \\ &\quad \left. + 3(1-2\nu)(\hat{\mathbf{b}}\cdot\hat{\mathbf{n}})\hat{x}_i\hat{x}_j - 15(\hat{\mathbf{b}}\cdot\hat{\mathbf{x}})(\hat{\mathbf{n}}\cdot\hat{\mathbf{x}})\hat{x}_i\hat{x}_j\right\}. \qquad (6.22)\end{aligned}$$

where b is the magnitude of the Burgers vector, $\hat{\mathbf{b}} = \mathbf{b}/b$ is the unit vector parallel to $\mathbf{b}$, and $\hat{\mathbf{x}} = \mathbf{x}/x$.

In isotropic elasticity the second derivative of the Green's function is as follows:

$$\begin{aligned}G_{ik,jl}(\mathbf{x}) &= \frac{1}{16\pi\mu(1-\nu)x^3}\left\{(3-4\nu)\delta_{ik}(3\hat{x}_l\hat{x}_j - \delta_{lj})\right. \\ &\quad + 15\hat{x}_i\hat{x}_j\hat{x}_k\hat{x}_l - 3(\delta_{ij}\hat{x}_k\hat{x}_l + \delta_{il}\hat{x}_j\hat{x}_k + \delta_{jl}\hat{x}_i\hat{x}_k + \delta_{kj}\hat{x}_i\hat{x}_l + \delta_{kl}\hat{x}_i\hat{x}_j) \\ &\quad \left. + (\delta_{il}\delta_{kj} + \delta_{kl}\delta_{ij})\right\}. \qquad (6.23)\end{aligned}$$

When this expression is inserted in eq 6.17 we obtain, after some tedious algebra, the following expression for the elastic interaction energy between two infinitesimal loops, A and B, in isotropic elasticity[26]:

$$E^{(AB)}_{int} = \frac{\mu b^A b^B\left(\delta A^A\right)\left(\delta A^B\right)}{4\pi(1-\nu)x^3}\Big[$$

[26]Dudarev, S L and Sutton, A P, Acta Materialia **125**, 425-430 (2017). Sergei Lvovich Dudarev 1960-, British materials physicist born in Belarus.

$$\begin{aligned}
&15(\hat{\mathbf{b}}^A\cdot\hat{\mathbf{x}})(\hat{\mathbf{b}}^B\cdot\hat{\mathbf{x}})(\hat{\mathbf{n}}^A\cdot\hat{\mathbf{x}})(\hat{\mathbf{n}}^B\cdot\hat{\mathbf{x}})\\
&-3\nu\Big\{(\hat{\mathbf{b}}^A\cdot\hat{\mathbf{b}}^B)(\hat{\mathbf{n}}^A\cdot\hat{\mathbf{x}})(\hat{\mathbf{n}}^B\cdot\hat{\mathbf{x}})+(\hat{\mathbf{b}}^A\cdot\hat{\mathbf{n}}^B)(\hat{\mathbf{n}}^A\cdot\hat{\mathbf{x}})(\hat{\mathbf{b}}^B\cdot\hat{\mathbf{x}})\\
&\qquad+(\hat{\mathbf{n}}^A\cdot\hat{\mathbf{b}}^B)(\hat{\mathbf{n}}^B\cdot\hat{\mathbf{x}})(\hat{\mathbf{b}}^A\cdot\hat{\mathbf{x}})+(\hat{\mathbf{n}}^A\cdot\hat{\mathbf{n}}^B)(\hat{\mathbf{b}}^A\cdot\hat{\mathbf{x}})(\hat{\mathbf{b}}^B\cdot\hat{\mathbf{x}})\Big\}\\
&-3(1-2\nu)(\hat{\mathbf{b}}^A\cdot\hat{\mathbf{n}}^A)(\hat{\mathbf{b}}^B\cdot\hat{\mathbf{x}})(\hat{\mathbf{n}}^B\cdot\hat{\mathbf{x}})\\
&-3(1-2\nu)(\hat{\mathbf{b}}^B\cdot\hat{\mathbf{n}}^B)(\hat{\mathbf{b}}^A\cdot\hat{\mathbf{x}})(\hat{\mathbf{n}}^A\cdot\hat{\mathbf{x}})\\
&-(1-2\nu)\Big\{(\hat{\mathbf{b}}^A\cdot\hat{\mathbf{b}}^B)(\hat{\mathbf{n}}^A\cdot\hat{\mathbf{n}}^B)+(\hat{\mathbf{b}}^A\cdot\hat{\mathbf{n}}^B)(\hat{\mathbf{n}}^A\cdot\hat{\mathbf{b}}^B)\Big\}\\
&-(4\nu-1)(\hat{\mathbf{b}}^A\cdot\hat{\mathbf{n}}^A)(\hat{\mathbf{b}}^B\cdot\hat{\mathbf{n}}^B)\Big].
\end{aligned}\tag{6.24}$$

In this expression the loops may be prismatic ($\hat{\mathbf{b}}\cdot\hat{\mathbf{n}}=1$ for a vacancy loop and $\hat{\mathbf{b}}\cdot\hat{\mathbf{n}}=-1$ for an interstitial loop), shear ($\hat{\mathbf{b}}\cdot\hat{\mathbf{n}}=0$) or a mixture ($0<|\hat{\mathbf{b}}\cdot\hat{\mathbf{n}}|<1$). The interaction energy separates into an inverse cube dependence on the separation between the loops and a term that depends on no less than *ten* angles. They are all the angles between the five unit vectors $\hat{\mathbf{x}},\hat{\mathbf{b}}^A,\hat{\mathbf{b}}^B,\hat{\mathbf{n}}^A$ and $\hat{\mathbf{n}}^B$, taken in pairs. Together with the magnitudes of the Burgers vectors, the loop areas and the distance between the loop centres, the interaction energy is a function of fifteen variables, and it is remarkable that this function has a closed form. Equation 6.24 may be applied to finite-sized planar loops provided the separation between them is more than about twice their size.

6.11 Mura's formula

We have already noted that the displacement field of a dislocation depends on the choice of the cut across which the relative displacement by the Burgers vector is introduced. This is explicit in Volterra's formula, eq 6.11. However, the strain and stress fields cannot depend on the choice of cut. Nevertheless, we can differentiate Volterra's formula to get the displacement gradient field:

$$u_{j,g}(\mathbf{x})=c_{mikp}b_k\int_{S^+}G_{ij,mg}(\mathbf{x}-\mathbf{x}')n_{p'}\mathrm{d}S',\tag{6.25}$$

but this also involves an integral over the cut surface. The displacement gradient field should be independent of the choice of the cut and dependent only on the configuration of the dislocation line. Mura's formula achieves this by transforming the surface integral in eq 6.25 into a line integral along the dislocation line using a version of Stokes' theorem.

A familiar form of Stokes' theorem is:

$$\int_S \varepsilon_{ijk} V_{k,j}(\mathbf{x}-\mathbf{x}')n_i \,\mathrm{d}S(\mathbf{x}) = \oint_L V_p(\mathbf{x}-\mathbf{x}')\,\mathrm{d}x_p, \tag{6.26}$$

where L is the closed line where the surface S terminates. In this equation $\mathbf{x}'$ is constant and $\mathbf{x}$ ranges over the surface S and around the line L. In the following we revert to our usual notation $\mathrm{d}S(\mathbf{x}) = \mathrm{d}S$. To obtain the version of Stokes' theorem needed here let $V_k(\mathbf{x}-\mathbf{x}') = f(\mathbf{x}-\mathbf{x}')\delta_{kq}$ so that the vector function $\mathbf{V}(\mathbf{x}-\mathbf{x}')$ has only one non-zero component and that is $V_q(\mathbf{x}-\mathbf{x}') = f(\mathbf{x}-\mathbf{x}')$. Then Stokes' theorem becomes:

$$\varepsilon_{qij}\int_S f_{,j}(\mathbf{x}-\mathbf{x}')n_i\,\mathrm{d}S = \oint_L f(\mathbf{x}-\mathbf{x}')\,\mathrm{d}x_q.$$

Multiplying both sides of this equation by ε_{qpg} we obtain the following:

$$\begin{aligned}
\varepsilon_{qpg}\varepsilon_{qij}\int_S f_{,j}(\mathbf{x}-\mathbf{x}')n_i\,\mathrm{d}S &= \varepsilon_{qpg}\oint_L f(\mathbf{x}-\mathbf{x}')\,\mathrm{d}x_q \\
(\delta_{ip}\delta_{jg}-\delta_{jp}\delta_{ig})\int_S f_{,j}(\mathbf{x}-\mathbf{x}')n_i\,\mathrm{d}S &= \varepsilon_{qpg}\oint_L f(\mathbf{x}-\mathbf{x}')\,\mathrm{d}x_q \\
\int_S f_{,g}(\mathbf{x}-\mathbf{x}')n_p\,\mathrm{d}S &= \int_S f_{,p}(\mathbf{x}-\mathbf{x}')n_g\,\mathrm{d}S + \varepsilon_{qpg}\oint_L f(\mathbf{x}-\mathbf{x}')\,\mathrm{d}x_q.
\end{aligned} \tag{6.27}$$

Now we switch the integration variable from $\mathbf{x}$ to $\mathbf{x}'$, treating $\mathbf{x}$ as constant. Writing $\mathrm{d}S'$ for $\mathrm{d}S(\mathbf{x}')$ we obtain:

$$\int_S f_{,g'}(\mathbf{x}-\mathbf{x}')n'_p\,\mathrm{d}S' = \int_S f_{,p'}(\mathbf{x}-\mathbf{x}')n'_g\,\mathrm{d}S' + \varepsilon_{qpg}\oint_L f(\mathbf{x}-\mathbf{x}')\,\mathrm{d}x'_q. \tag{6.28}$$

If we set $f(\mathbf{x}-\mathbf{x}') = G_{ij,m'}(\mathbf{x}-\mathbf{x}')$ then we obtain:

$$\int_S G_{ij,m'g'}(\mathbf{x}-\mathbf{x}')n'_p\,\mathrm{d}S' = \int_S G_{ij,m'p'}(\mathbf{x}-\mathbf{x}')n'_g\,\mathrm{d}S' + \varepsilon_{qpg}\oint_L G_{ij,m'}(\mathbf{x}-\mathbf{x}')\,\mathrm{d}x'_q. \tag{6.29}$$

Multiplying both sides by c_{mikp} we obtain:

$$c_{mikp}\int_S G_{ij,m'g'}(\mathbf{x}-\mathbf{x}')n'_p\,\mathrm{d}S' = c_{mikp}\int_S G_{ij,m'p'}(\mathbf{x}-\mathbf{x}')n'_g\,\mathrm{d}S'$$

$$+c_{mikp}\,\varepsilon_{qpg}\oint_L G_{ij,m'}(\mathbf{x}-\mathbf{x}')\,\mathrm{d}x'_q. \quad (6.30)$$

Recalling that $G_{ij,m'p'} = G_{ij,mp}$, it follows from the defining equation, (eq 4.8), for the Green's function that $c_{mikp}G_{ij,m'p'} = 0$ at all points except $\mathbf{x}' = \mathbf{x}$. Since the surface S can always be chosen to avoid $\mathbf{x}$ the surface integral on the right hand side is zero. Inserting this result in eq 6.25 we obtain Mura's formula for the displacement gradient:

$$u_{j,g}(\mathbf{x}) = \varepsilon_{qpg}\,c_{mikp}b_k\oint_{L'} G_{ij,m'}(\mathbf{x}-\mathbf{x}')\,\mathrm{d}x_{q'}, \quad (6.31)$$

The stress field of the dislocation follows from Hooke's law: $\sigma_{ab}(\mathbf{x}) = c_{abjg}u_{j,g}(\mathbf{x})$. In isotropic elasticity this becomes:

$$\sigma_{ab}(\mathbf{x}) = c_{abjg}\frac{\varepsilon_{qpg}b_k}{8\pi(1-\nu)}\oint_{L'}\left[(1-2\nu)\frac{\delta_{kj}X_p+\delta_{pj}X_k-\delta_{kp}X_j}{X^3}+3\frac{X_kX_pX_j}{X^5}\right]\mathrm{d}x_{q'},$$

where $\mathbf{X} = \mathbf{x}-\mathbf{x}'$. Defining the line integral I_{kjpq} as

$$I_{kjpq} = \oint_{L'}\left[(1-2\nu)\frac{\delta_{kj}X_p+\delta_{pj}X_k-\delta_{kp}X_j}{X^3}+3\frac{X_kX_pX_j}{X^5}\right]\mathrm{d}x_{q'}$$

the stress field may be conveniently expressed as follows:

$$\sigma_{ab}(\mathbf{x}) = \frac{\mu b_k}{8\pi(1-\nu)}\left[\frac{2\nu}{1-2\nu}\delta_{ab}\varepsilon_{qpj}I_{kjpq}+\varepsilon_{qpb}I_{kapq}+\varepsilon_{qpa}I_{kbpq}\right]. \quad (6.32)$$

6.12 The stress field of an edge dislocation in isotropic elasticity

To illustrate the application of Mura's formula the stress field of an infinitely long edge dislocation along the x_3-axis with $\mathbf{b} = [b,0,0]$ will be derived. In eq 6.32 we have $b_k = b\delta_{k1}$ and $q = 3$. We obtain:

$$\sigma_{11} = \frac{\mu b}{4\pi(1-\nu)(1-2\nu)}(\nu I_{1213}-(1-\nu)I_{1123})$$

$$\sigma_{12} = \sigma_{21} = \frac{\mu b}{8\pi(1-\nu)}(I_{1113}-I_{1223})$$

$$\sigma_{13} = \sigma_{31} = -\frac{\mu b}{8\pi(1-\nu)}I_{1323}$$

$$\sigma_{22} = \frac{\mu b}{4\pi(1-\nu)(1-2\nu)}((1-\nu)I_{1213}-\nu I_{1123})$$

$$\sigma_{23} = \sigma_{32} = \frac{\mu b}{8\pi(1-\nu)} I_{1313}$$

$$\sigma_{33} = \frac{\mu b}{4\pi(1-\nu)(1-2\nu)} \nu (I_{1213} - I_{1123}) = \nu\,(\sigma_{11} + \sigma_{22}) \,.$$

It is straightforward to evaluate the integrals:

$$\begin{aligned}
I_{1213} &= -\frac{2(1-2\nu)x_2}{(x_1^2+x_2^2)} + \frac{4x_1^2 x_2}{(x_1^2+x_2^2)^2} \\
I_{1123} &= \frac{2(1-2\nu)x_2}{(x_1^2+x_2^2)} + \frac{4x_1^2 x_2}{(x_1^2+x_2^2)^2} \\
I_{1113} &= \frac{2(1-2\nu)x_1}{(x_1^2+x_2^2)} + \frac{4x_1^3}{(x_1^2+x_2^2)^2} \\
I_{1223} &= \frac{2(1-2\nu)x_1}{(x_1^2+x_2^2)} + \frac{4x_1 x_2^2}{(x_1^2+x_2^2)^2} \\
I_{1313} &= I_{1323} = 0
\end{aligned}$$

We obtain the following stress components for an edge dislocation in isotropic elasticity:

$$\begin{aligned}
\sigma_{11} &= -\frac{\mu b}{2\pi(1-\nu)} x_2 \frac{3x_1^2+x_2^2}{(x_1^2+x_2^2)^2} \\
\sigma_{12} &= \frac{\mu b}{2\pi(1-\nu)} x_1 \frac{x_1^2-x_2^2}{(x_1^2+x_2^2)^2} \\
\sigma_{13} &= 0 \\
\sigma_{22} &= \frac{\mu b}{2\pi(1-\nu)} x_2 \frac{x_1^2-x_2^2}{(x_1^2+x_2^2)^2} \\
\sigma_{23} &= 0 \\
\sigma_{33} &= -\frac{\mu b \nu}{\pi(1-\nu)} \frac{x_2}{(x_1^2+x_2^2)}
\end{aligned} \qquad (6.33)$$

We see in eq 6.33 that the stress field is inversely proportional to the distance from the dislocation line. This is a general feature of straight dislocations. The stress diverges as the dislocation line is approached, which is a consequence of the assumption that the dislocation is a line of no width. In more realistic models the dislocation has a finite width and the stress does not become infinite.

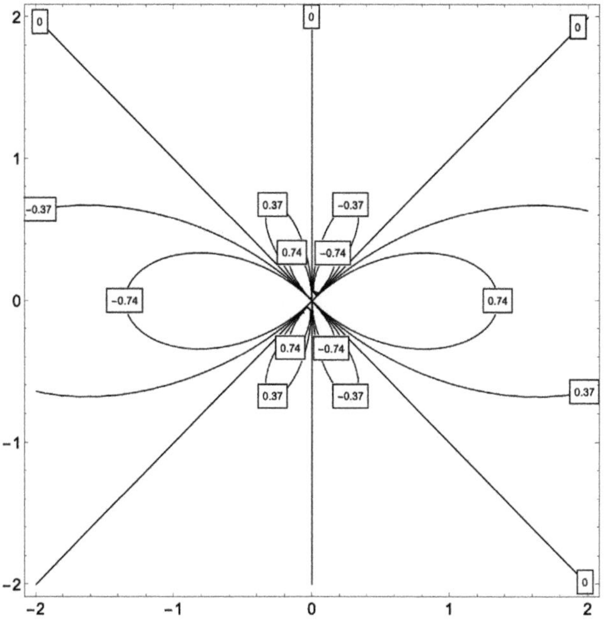

Figure 6.9: Contour plot of the function $x_1(x_1^2 - x_2^2)/(x_1^2 + x_2^2)^2$ for $-2 \leq x_1, x_2 \leq +2$. The stress field σ_{12} of an edge dislocation at $(0,0)$ with $\mathbf{b} = (b,0,0)$ is directly proportional to this function.

Exercise 6.5 Sketch the stress component σ_{12} of eq 6.33 in the $x_1 - x_2$ plane. Sketch the hydrostatic stress $\frac{1}{3}\mathrm{Tr}\,\sigma$ for an edge dislocation in the $x_1 - x_2$ plane. Edge dislocations attract misfitting atoms. Where would a larger misfitting atom tend to locate itself to reduce the elastic energy?

■

Solution The dependence of σ_{12} on x_1 and x_2 is illustrated in Fig 6.9.
The hydrostatic stress is:

$$\frac{1}{3}\mathrm{Tr}\,\sigma = -\frac{\mu b(1+\nu)}{3(1-\nu)} \frac{x_2}{(x_1^2 + x_2^2)}.$$

Its dependence on $-x_2/(x_1^2 + x_2^2)$ is illustrated in Fig 6.10. A large misfitting atom will lower the total elastic energy if it sits beneath the extra half plane where the hydrostatic stress is positive and hence tensile.

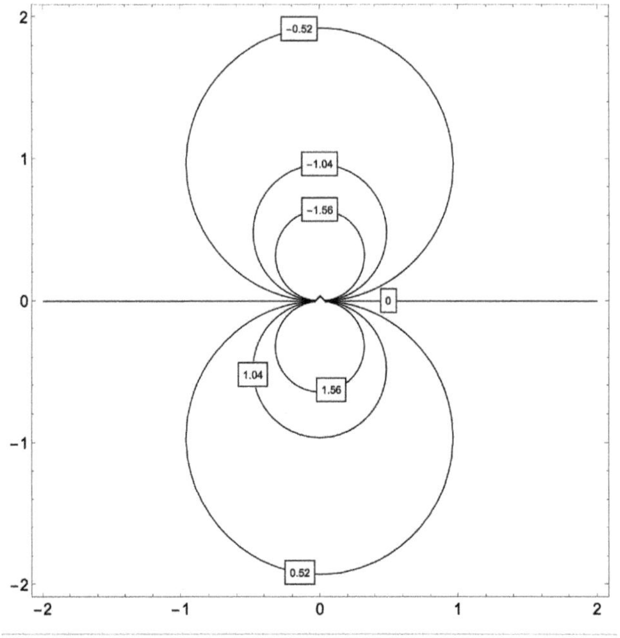

Figure 6.10: Contour plot of the function $-x_2/(x_1^2+x_2^2)$ for $-2 \leq x_1, x_2 \leq +2$. The hydrostatic stress field, $\frac{1}{3}\mathrm{Tr}\,\sigma$, of an edge dislocation at $(0,0)$ with $\mathbf{b} = (b,0,0)$ is directly proportional to this function.

6.13 The elastic energy of a dislocation

The elastic energy of a dislocation is just the volume integral of the elastic energy density it creates:

$$E_{el} = \frac{1}{2}\int_V \sigma_{ij} u_{i,j} \mathrm{d}V.$$

It would be quite laborious to evaluate the elastic energy in this way as there are usually many non-zero stress and strain components. However, the task can be simplified significantly by applying the divergence theorem.

Consider an infinite straight dislocation parallel to the x_3-axis in an anisotropic elastic medium. Let the cut surface coincide with the half-plane $x_2 = 0$ where $x_1 \geq 0$, as shown in Fig 6.11. Applying the divergence theorem to the elastic strain energy per

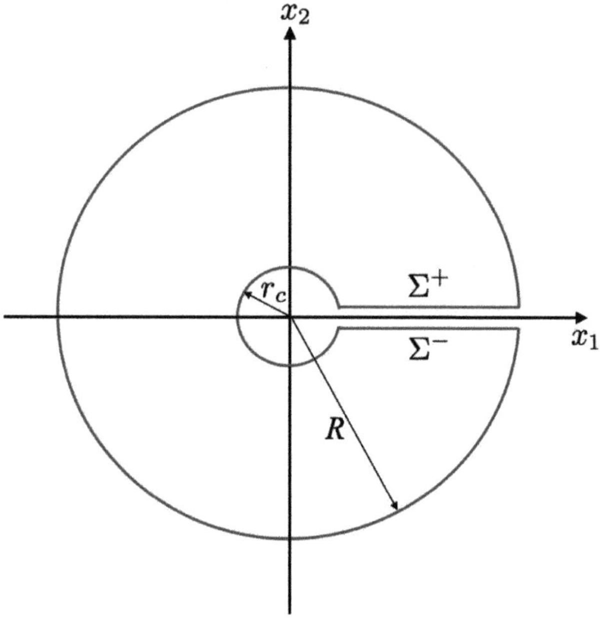

Figure 6.11: To illustrate the surface C (in red) of the surface integral in eq 6.34 for a straight dislocation lying along the x_3-axis. The cut of the dislocation is along the positive x_1-axis, across which the displacement changes discontinuously by the Burgers vector.

unit length along x_3, and recalling that $\sigma_{ij,j} = 0$, we obtain:

$$E_{el} = \frac{1}{2} \int_C \sigma_{ij} u_i n_j \, \mathrm{d}S, \tag{6.34}$$

where the surface C comprises four segments of unit length along x_3. The first is the almost complete small cylinder of radius r_c around the dislocation line. In the limit the separation of the surfaces on either of the cut becomes infinitesimal the cylinder of radius r_c becomes complete. The integral of $\sigma_{ij} n_j$ around the cylinder is then zero because there is no resultant force associated with the dislocation line. The same argument applies to the segment around the outer cylinder of radius R, and it too is zero. That leaves the two contributions from the segments Σ^+ and Σ^- on either side of the cut. As the separation of the segments shrinks to zero the energy of the elastic field per unit length of dislocation is:

$$E_{el} = \frac{1}{2}\int_{r_c}^{R} \sigma_{i2}(x_1, 0)\, b_i\, \mathrm{d}x_1 \tag{6.35}$$

For example, for an edge dislocation $b_i = b\delta_{i1}$, and the only stress component contributing to this integral is $\sigma_{12}(x_1, 0)$. Thus we obtain the following elastic energy per unit length of an edge dislocation in isotropic elasticity:

$$E_{el}^{edge} = \frac{1}{2}\int_{r_c}^{R} \frac{\mu b^2}{2\pi(1-\nu)}\frac{\mathrm{d}x_1}{x_1} = \frac{\mu b^2}{4\pi(1-\nu)}\ln(R/r_c). \tag{6.36}$$

For a screw dislocation $b_i = b\delta_{i3}$, and the only stress component contributing to the integral in eq 6.35 is $\sigma_{32}(x_1, 0)$. In the isotropic elastic approximation the elastic energy per unit length of a screw dislocation is as follows:

$$E_{el}^{screw} = \frac{\mu b^2}{4\pi}\ln(R/r_c) \tag{6.37}$$

The logarithmic divergence of the elastic energy of a straight dislocation is common to other line singularities in physics, e.g. the electrostatic energy per unit length of an infinitely long thin line of charge, or the energy per unit length of the magnetic field of an infinitely long thin wire carrying a constant current. Of course dislocations exist in crystals of finite size, so their energy never becomes infinite. More significantly, dislocations often lower their elastic energies by organising themselves into configurations where their elastic fields have much smaller spatial extent than they would have if each dislocation were isolated. A classic example of such self-organisation is the elastic field of a grain boundary comprising an array of dislocations with a spacing d. It is found (see problem 6.7) that the elastic field decays exponentially from the grain boundary with a decay length of $d/(2\pi)$.

The surface C for the integral in eq 6.34 has to avoid the singularity at $x_1 = x_2 = 0$ where the stress tensor diverges. r_c is often called the core radius, but as we have argued the core has no radius in this theory. The meaning of r_c is more sensibly described as the radius at which the elastic solution provides an acceptable description of the actual stresses near the dislocation. It is of order 1 nm in most crystalline materials. However, it is not necessarily the circular cross-section that has been assumed here and many other places. For example, in fcc metals it is often found that the core is spread on the slip plane, so that it is much longer in that direction than it is normal to the slip plane by as much as a factor of ten. The total energy per unit length of the dislocation is the sum of the elastic energy and the core energy per unit length.

The surface tension of a liquid arises because the surface of the liquid has an energy per unit area. If the surface is stretched more surface area is created by diffusion of atoms from beneath the surface, which raises the energy of the system. The increase in energy may be thought of as the work done against the surface tension. Similarly a line tension may be associated with the energy per unit length of a dislocation: in isotropic elasticity if there are no other forces acting on them dislocations will always seek to minimise their lengths. But this is not necessarily true in elastic anisotropy because the energies per unit length of some line directions may be significantly less than those of other directions. In that case the total elastic energy may decrease even though there is an overall increase in the length of dislocation line.

The logarithmic singularity in the elastic energy per unit length of a dislocation is weak. Increasing R/r_c to $(R/r_c)^n$, where $n > 1$, changes the elastic energy per unit length by only a factor of n. Sometimes we need a rough estimate of the energy of a dislocation per unit length for back-of-an-envelope calculations, and then we use $\mu b^2/2$. This is also a useful approximation for the line tension of a dislocation. It is of order 0.1 to 1 $\text{eV} \cdot \text{Å}^{-1}$, which is equivalent to 0.16 to 1.6 nN. This is also the range of energies per unit length of the core.

The energy of a dislocation loop is also given by eq 6.34 where the surface C is that shown in Fig 6.7. The energy of the loop is then:

$$E_{el}^{loop} = \frac{1}{2} \int_{\Sigma^+} \sigma_{ij} b_i n_j \mathrm{d}S \tag{6.38}$$

where the integral is taken over the positive side of the cut surface inside the loop. This expression may be understood as the work done against the self-stress of the loop when it is created by increasing its Burgers vector from zero to the final b_i. By the conservation of energy this work becomes elastic energy stored in the medium.

6.14 The Frank-Read source

For many years after it was suggested that dislocations are the agents of plastic deformation in crystals it was not clear how they are created. As the Frenkel analysis showed, to create a dislocation loop by homogeneous nucleation requires stresses approaching the theoretical shear strength, which is of order $\mu/10$ and much higher than the yield stresses observed at normal strain rates. A mechanism of generating dislocations was needed to explain these much smaller yield stresses. One mechanism was proposed by Frank and Read[27] before it was observed, and is known as a Frank-Read

[27]Frank, F C, and Read, W T, Phys. Rev. **79**, 722 (1950). W Thornton Read was a US physicist.

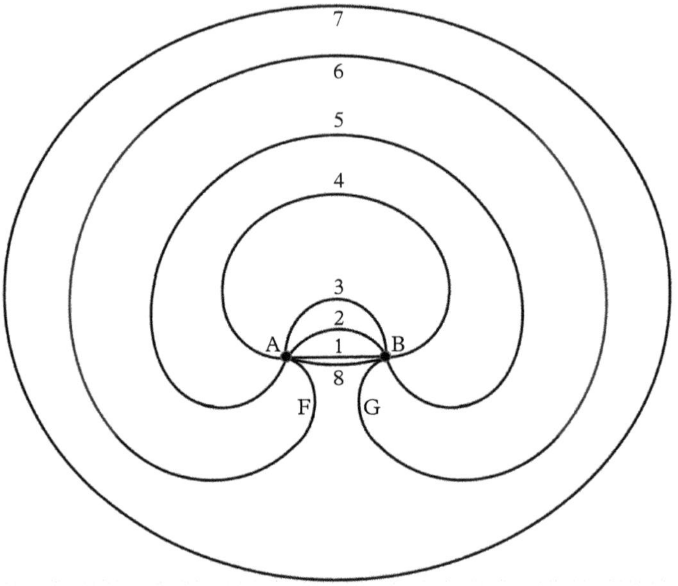

Figure 6.12: Schematic illustration of the operation of a Frank-Read source. A segment of dislocation line is pinned at A and B (configuration 1). Under the action of an applied shear stress on the slip plane in the direction of the Burgers vector the dislocation segment bows (configuration 2) against the line tension tending to keep it straight. Eventually a critical configuration is reached, configuration 3, where the radius of curvature of the dislocation segment is a minimum. The segment continues to expand (configurations 4, 5). Segments F and G have opposite line senses and attract each other. When they meet, a loop is liberated and expands under the action of the applied stress, configuration 7. A segment is left behind, configuration 8, and returns to AB and the process starts again.

source[28].

Fig 6.12 illustrates the operation of a Frank-Read source. We consider a segment of a dislocation pinned at A and B. There are many ways the dislocation could be pinned at two points. In the paper by Frank and Read they considered a rectangular prismatic loop ABCD with Burgers vector normal to the plane of the loop, see Fig 6.13. A pure shear stress on the plane containing the Burgers vector and the line segment AB will make segments AB and CD move in opposite directions, but the segments AD and

[28]There is an intriguing account by Sir Charles Frank of how he and Thornton Read developed the idea of their dislocation source independently and precisely simultaneously in Proc. R. Soc. Lond. A **371**, 136 (1980).

BC will experience no Peach-Koehler force and remain static. The result is that the segments AB and CD are pinned at their ends.

Let τ be the resolved shear stress on the slip plane in the direction of the Burgers vector of the segment AB. The Peach-Koehler force acting on the segment is then τb where, as usual, b is the magnitude of the Burgers vector. This force acts in a direction normal to the line and the dislocation bows out between the pinning points (see configuration 2 in Fig 6.12). As a result the dislocation line length increases and this is opposed by the line tension, T. As the radius of curvature of the bowed segment decreases the stress required to make it bow further increases. Eventually a critical configuration is reached (configuration 3) where the radius of curvature is a minimum and the bowed segment expands with no further increase in the applied stress required. The segments F and G (see Fig 6.12) have opposite line directions and they attract each other and annihilate. This reaction liberates a loop (configuration 7) which then expands freely under the influence of the applied stress leaving behind a segment (configuration 8) which returns to the configuration AB at the beginning of the operation of the source. The process then repeats sending out a succession of loops into the slip plane. Eventually these loops meet obstacles such as grain boundaries where they may form a *pileup*. The dislocations in the pileup exert a stress on the source, called a *back stress*, which eventually reduces the total stress acting on the source to zero. Further operation of the source then requires the applied stress to be increased.

It is instructive to derive an approximate expression for the stress required to operate a Frank-Read source. This is another example of a useful back-of-an-envelope estimate that provides insight. Let the length of the pinned segment AB be L. We assume that when the dislocation starts to bow it forms the arc of a circle with a radius R. An element $\mathrm{d}l$ of the arc experiences an outward force due to the applied stress equal to $\tau b \mathrm{d}l$. It also experiences an inward force due to the line tension, given by $T\mathrm{d}l/R$ where $T \approx \mu b^2/2$. At equilibrium $\tau \approx \mu b/(2R)$. As the segment bows further, R decreases and τ increases until the critical configuration is reached where R is a minimum. This happens when the bowed segment is a semi-circle, with $R = L/2$. Thus, the minimum stress required to operate the source is $\tau_{min} \approx \mu b/L$.

Efforts have been made to improve on this rough estimate. They include replacing the line tension approximation with an evaluation of the dislocation self-interactions when it bows, using anisotropic elasticity and including a friction stress that must be overcome before a dislocation will move. These are all possible but much more difficult than our simple estimate.

The relationship $\tau \approx \mu b/L$ is known as the Orowan flow stress[29]. It provides insight in a variety of contexts of strengthening mechanisms, and it is one of the most useful

[29]The flow stress is the stress required to sustain the glide of dislocations when they encounter obstacles.

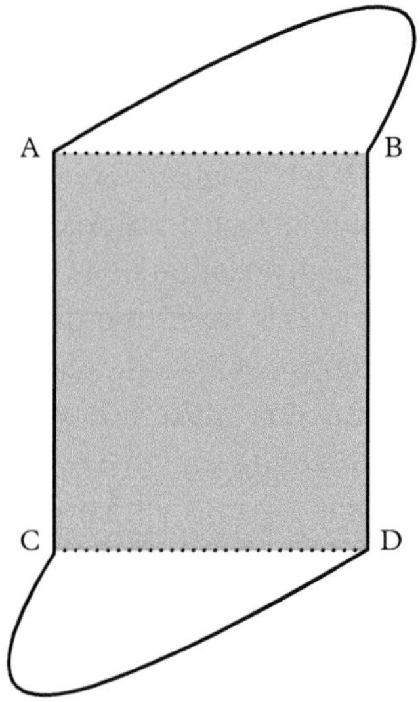

Figure 6.13: A rectangular prismatic loop ABCD with Burgers vector normal to the page initially occupies the shaded area. A shear stress is applied in the direction of the Burgers vector on planes containing the Burgers vector and the lines AB and CD. Segments AB and CD bow under the influence of this applied shear stress. However, segments AC and BD experience no resolved shear stress and remain static. Segments AB and CD may then operate as Frank-Read sources of loops, as illustrated in Fig 6.12. Jaehyun Cho has uploaded a movie of such a pair of Frank-Read sources operating at `https://www.youtube.com/watch?v=jwK-TF7o2Oo`.

formulae in the theory of dislocations. Here are three examples of its usefulness:

- *Precipitation hardening*. One way metals are made stronger is by alloying them with elements that result in a dispersion of second phase particles. Dislocations moving in their slip planes encounter these particles. If the particles are sufficiently large that dislocations cannot cut through them they bow out between them and proceed on the slip plane after leaving a loop around each precipitate. The critical stress required for this process is of order $\mu b/L$ where L is the average separation of the precipitates. This is a key relationship in designing age-hardened alloys.

- *Work hardening*. Another way metals are made stronger is by deforming them plastically. This is called work hardening and it is discussed in Chapter 11. It arises because dislocations moving on one slip plane encounter dislocations moving on inclined slip planes, as a result of which a variety of obstacles may form that impede slip on both slip planes. Further slip then proceeds by dislocations bowing out between these obstacles. If L is the separation between the obstacles then the stress required for further slip is again of order $\mu b/L$. The spacing L is related to the dislocation density ρ, which is defined as the number of dislocations crossing unit area or equivalently the total dislocation length per unit volume: $\rho \approx 1/L^2$. Therefore the stress required for dislocations to glide varies as $\mu b/L \approx \mu b\sqrt{\rho}$. This relationship was first proposed by G I Taylor in his paper of 1934 where he proposed dislocations as the agents of plasticity.

- *Plasticity in nanocrystals.* We saw at the beginning of this chapter that dislocations are the agents of plasticity because the stress required to slide an entire plane of atoms within a crystal over an adjacent plane is far too large. But if the area of the plane where slip occurs is very small, there might be a transition from slip mediated by dislocations to *block* slip where an entire plane of atoms *can* slide over another. If a dislocation is introduced into a cubic crystal of side L it will tend to be pinned by the surfaces. The stress required to make it bow out will be of order $\mu b/L$. Thus if $L \approx 10-100b$ the stress required to move the dislocation becomes comparable to that required for block slip.

6.15 Problems

Problem 6.1 In eq 4.13 the displacement field of an infinite planar fault was derived:

$$u_i(\mathbf{x}) = -\int_{fault} G_{ij,l'}(\mathbf{x}-\mathbf{x}')c_{jlmp}\, t_m n_p\, \mathrm{d}^2x', \tag{6.39}$$

where $\mathbf{t}$ is the translation of the medium on the negative side of the fault relative to that on the positive side. In isotropic elasticity we have:

$$c_{jlmp}G_{ij,l'}(\mathbf{X}) = \frac{1}{8\pi(1-\nu)}\left[(1-2\nu)\frac{\delta_{mi}X_p + \delta_{pi}X_m - \delta_{mp}X_i}{X^3} + 3\frac{X_m X_p X_i}{X^5}\right], \tag{6.40}$$

where $\mathbf{X} = \mathbf{x} - \mathbf{x}'$.

If the fault is in the plane $x_3 = 0$ verify that eq 6.39 yields $u_i(\mathbf{x}) = -\frac{1}{2}\,\mathrm{sgn}(x_3)t_i$.

Solution Using the expression for $c_{jlmp}G_{ij,l'}(\mathbf{X})$ in isotropic elasticity, and with $n_p = \delta_{p3}$, the displacement field of the fault at $\mathbf{x}$ becomes:

$$u_i(\mathbf{x}) = -\frac{t_m}{8\pi(1-\nu)} \int_{-\infty}^{\infty} \mathrm{d}x_1' \int_{-\infty}^{\infty} \mathrm{d}x_2' \left\{ (1-2\nu)\left[\frac{\delta_{mi}X_3 + \delta_{i3}X_m - \delta_{m3}X_i}{X^3}\right] + \frac{3X_m X_i X_3}{X^5} \right\}$$

The translational symmetry of the fault enables us to choose $x_1 = 0$ and $x_2 = 0$. The fault is in the plane $x_3' = 0$, so $X_3 = x_3$. Then $X_1 = -x_1'$, $X_2 = -x_2'$, $X_m t_m = -x_1' t_1 - x_2' t_2 + x_3 t_3$ and $X^2 = x_1'^2 + x_2'^2 + x_3^2$. The displacement field is then as follows:

$$u_i(\mathbf{x}) = \frac{-1}{8\pi(1-\nu)} \int_{-\infty}^{\infty} \mathrm{d}x_1' \int_{-\infty}^{\infty} \mathrm{d}x_2' \Bigg\{$$

$$(1-2\nu)\left[\frac{t_i x_3 + \delta_{i3}(-x_1' t_1 - x_2' t_2 + x_3 t_3) - t_3(-x_1'\delta_{i1} - x_2'\delta_{i2} + x_3\delta_{i3})}{(x_1'^2 + x_2'^2 + x_3^2)^{\frac{3}{2}}}\right]$$

$$+ \frac{3(-x_1' t_1 - x_2' t_2 + x_3 t_3)(-x_1'\delta_{i1} - x_2'\delta_{i2} + x_3\delta_{i3})x_3}{(x_1'^2 + x_2'^2 + x_3^2)^{\frac{5}{2}}}\Bigg\}.$$

Consider $u_1(\mathbf{x})$, for which $i = 1$:

$$\begin{aligned} u_1(\mathbf{x}) &= \frac{-1}{8\pi(1-\nu)} \int_{-\infty}^{\infty} \mathrm{d}x_1' \int_{-\infty}^{\infty} \mathrm{d}x_2' \Bigg\{ \frac{(1-2\nu)}{(x_1'^2 + x_2'^2 + x_3^2)^{\frac{3}{2}}} \left[t_1 x_3 + t_3 \cancel{x_1'}\right] \\ &\quad + \frac{3(x_1'^2 t_1 + \cancel{x_1'} x_2' t_2 - x_3 t_3 \cancel{x_1'})x_3}{(x_1'^2 + x_2'^2 + x_3^2)^{\frac{5}{2}}} \Bigg\}, \end{aligned}$$

where the terms crossed out cancel because they are odd in x_1'. That leaves:

$$u_1(\mathbf{x}) = \frac{-t_1 x_3}{8\pi(1-\nu)} \int_{-\infty}^{\infty} \mathrm{d}x_1' \int_{-\infty}^{\infty} \mathrm{d}x_2' \left\{ \frac{(1-2\nu)}{(x_1'^2 + x_2'^2 + x_3^2)^{\frac{3}{2}}} + \frac{3x_1'^2}{(x_1'^2 + x_2'^2 + x_3^2)^{\frac{5}{2}}} \right\}.$$

The integrals may be evaluated by transforming the integrands to circular polar coordinates, $x'_1 = r\cos\theta$ and $x'_2 = r\sin\theta$, yielding:

$$\int_{-\infty}^{\infty} dx'_1 \int_{-\infty}^{\infty} dx'_2 \frac{1}{(x_1'^2 + x_2'^2 + x_3^2)^{\frac{3}{2}}} = \int_0^{\infty} dr\, r \int_0^{2\pi} d\theta \frac{1}{(r^2 + x_3^2)^{\frac{3}{2}}} = \frac{2\pi}{|x_3|}$$

$$\int_{-\infty}^{\infty} dx'_1 \int_{-\infty}^{\infty} dx'_2 \frac{x_1'^2}{(x_1'^2 + x_2'^2 + x_3^2)^{\frac{5}{2}}} = \int_0^{\infty} dr\, r \int_0^{2\pi} d\theta \frac{r^2\cos^2\theta}{(r^2 + x_3^2)^{\frac{5}{2}}} = \frac{2\pi}{3\,|x_3|}$$

Therefore,

$$\begin{aligned} u_1(\mathbf{x}) &= -\frac{t_1 x_3}{8\pi(1-\nu)}\left((1-2\nu)\frac{2\pi}{|x_3|} + \cancel{3}\frac{2\pi}{\cancel{3}\,|x_3|}\right) \\ &= -\frac{t_1}{2}\frac{x_3}{|x_3|} = -\frac{t_1}{2}\operatorname{sgn}(x_3). \end{aligned}$$

The same set of steps applies to $u_2(\mathbf{x}) = -(t_2/2)\operatorname{sgn}(x_3)$. But $u_3(\mathbf{x})$ is different:

$$u_3(\mathbf{x}) = -\frac{t_3 x_3}{8\pi(1-\nu)} \int_{-\infty}^{\infty} dx'_1 \int_{-\infty}^{\infty} dx'_2 \left\{ \frac{(1-2\nu)}{(x_1'^2 + x_2'^2 + x_3^2)^{\frac{3}{2}}} + \frac{3x_3^2}{(x_1'^2 + x_2'^2 + x_3^2)^{\frac{5}{2}}} \right\}$$

The second integral is easily evaluated using circular polars:

$$\int_{-\infty}^{\infty} dx'_1 \int_{-\infty}^{\infty} dx'_2 \frac{1}{(x_1'^2 + x_2'^2 + x_3^2)^{\frac{5}{2}}} = \frac{2\pi}{3}\frac{1}{|x_3|^3}.$$

Therefore,

$$u_3(\mathbf{x}) = -\frac{2\pi t_3}{8\pi(1-\nu)}\left((1-2\nu)\frac{x_3}{|x_3|} + \frac{x_3^3}{|x_3|^3}\right) = -\frac{t_3}{2}\operatorname{sgn}(x_3).$$

This completes the derivation.

Problem 6.2 Using Volterra's formula for the displacement field of a dislocation in the form:

$$u_i(\mathbf{x}) = -c_{jlmp}\, b_m \int_{S^+} G_{ij,l'}(\mathbf{x}-\mathbf{x}')\, n'_p\, dS',$$

and using eq 6.40 show that, with the cut in the half-plane $x_2 = 0, x_1 \geq 0$, the displacement field of a screw dislocation in isotropic elasticity is as follows:

$$\begin{aligned} u_1(x_1,x_2) &= 0 \\ u_2(x_1,x_2) &= 0 \\ u_3(x_1,x_2) &= \frac{b}{2\pi}\left[\tan^{-1}\left(\frac{x_2}{x_1}\right) - \pi\right]. \end{aligned}$$

N.B. For $x_1 > 0$ the arctangent function varies from 0 to 2π as x_2 varies from just above the cut to just below the cut. Therefore, $u_3 = -b/2$ just above the cut and $u_3 = b/2$ just below the cut.

Solution For $b_m = b\delta_{m3}$ and $n'_p = \delta_{p2}$ Volterra's formula becomes:

$$u_i(\mathbf{x}) = -\frac{b}{8\pi(1-\nu)}\int_0^{\infty} \mathrm{d}x'_1 \int_{-\infty}^{\infty} \mathrm{d}x'_3 \left\{(1-2\nu)\left(\frac{\delta_{mi}X_p + \delta_{pi}X_m - \delta_{mp}X_i}{X^3}\right) + 3\frac{X_m X_p X_i}{X^5}\right\}\delta_{m3}\delta_{p2}$$

Translational symmetry along the x_3-axis permits the arbitrary choice $x_3 = 0$, so that $X_3 = -x'_3$. Since $x'_2 = 0$ in the cut plane, $X_2 = x_2$. Finally, $X_1 = x_1 - x'_1$. Therefore,

$$u_i(\mathbf{x}) = -\frac{b}{8\pi(1-\nu)}\int_0^{\infty} \mathrm{d}x'_1 \int_{-\infty}^{\infty} \mathrm{d}x'_3$$

$$\left\{(1-2\nu)\left(\frac{\delta_{i3}x_2 - \delta_{i2}x'_3}{\left((x_1-x'_1)^2 + x_2^2 + x'^2_3\right)^{\frac{3}{2}}}\right) - 3x'_3 x_2\left(\frac{(x_1-x'_1)\delta_{i1} + x_2\delta_{i2} - x'_3\delta_{i3}}{\left((x_1-x'_1)^2 + x_2^2 + x'^2_3\right)^{\frac{5}{2}}}\right)\right\}.$$

It follows that $u_1(\mathbf{x}) = 0$ and $u_2(\mathbf{x}) = 0$ because the relevant integrals are odd in x'_3. That leaves $u_3(\mathbf{x})$:

$$u_3(\mathbf{x}) = -\frac{b}{8\pi(1-\nu)}\int_0^{\infty} \mathrm{d}x'_1 \int_{-\infty}^{\infty} \mathrm{d}x'_3 \left\{\frac{(1-2\nu)x_2}{\left((x_1-x'_1)^2 + x_2^2 + x'^2_3\right)^{\frac{3}{2}}} + \frac{3x'^2_3 x_2}{\left((x_1-x'_1)^2 + x_2^2 + x'^2_3\right)^{\frac{5}{2}}}\right\}.$$

It is straightforward to derive the following integrals:

$$\int_{-\infty}^{\infty} \frac{dx}{(a^2+x^2)^{\frac{3}{2}}} = \frac{2}{a^2}$$

$$\int_{-\infty}^{\infty} \frac{x^2 dx}{(a^2+x^2)^{\frac{5}{2}}} = \frac{2}{3a^2}$$

Therefore,

$$\begin{aligned}
u_3(\mathbf{x}) &= -\frac{b}{8\pi(1-\nu)} \int_0^{\infty} dx_1' \left\{ \frac{2(1-2\nu)x_2}{(x_1-x_1')^2+x_2^2} + \cancel{3}\frac{2x_2}{\cancel{3}\left((x_1-x_1')^2+x_2^2\right)} \right\} \\
&= -\frac{b}{2\pi} \int_0^{\infty} dx_1' \frac{x_2}{(x_1'-x_1)^2+x_2^2} \\
&= -\frac{b}{2\pi} \left[\tan^{-1}\left(\frac{x_1'-x_1}{x_2} \right) \right]_{x_1'=0}^{\infty} \\
&= -\frac{b}{2\pi} \left(\frac{\pi}{2} + \tan^{-1}\left(\frac{x_1}{x_2} \right) \right)
\end{aligned}$$

But $\tan^{-1}(x_1/x_2) + \tan^{-1}(x_2/x_1) = \pi/2$. Therefore,

$$u_3(\mathbf{x}) = \frac{b}{2\pi} \left(\tan^{-1}\left(\frac{x_2}{x_1} \right) - \pi \right).$$

Problem 6.3 Using Mura's formula, eq 6.31, in isotropic elasticity show that the strain and stress fields of a screw dislocation lying along the x_3-axis with Burgers vector $\mathbf{b} = [0,0,b]$ are as follows:

$$\begin{aligned}
e_{11} &= e_{22} = e_{33} = e_{12} = 0 \\
e_{13} &= -\frac{b}{4\pi} \frac{x_2}{x_1^2+x_2^2} \\
e_{23} &= \frac{b}{4\pi} \frac{x_1}{x_1^2+x_2^2}
\end{aligned}$$

$$\begin{aligned}\sigma_{11} &= \sigma_{22} = \sigma_{33} = \sigma_{12} = 0 \\ \sigma_{13} &= -\frac{\mu b}{2\pi}\frac{x_2}{x_1^2+x_2^2} \\ \sigma_{23} &= \frac{\mu b}{2\pi}\frac{x_1}{x_1^2+x_2^2}\end{aligned} \tag{6.41}$$

Verify that your answer is consistent with the stress and strain fields derived from the displacement field of the screw dislocation obtained in the previous problem.

Solution Mura's formula is

$$u_{j,g}(\mathbf{x}) = \varepsilon_{qpg}c_{mikp}b_k \oint_{L'} G_{ij,m'}(\mathbf{x}-\mathbf{x}')\,\mathrm{d}x_{q'}.$$

For a screw dislocation along x_3, $b_k = b\delta_{k3}$ and $q = 3$. We may choose $x_3 = 0$ owing to the translational symmetry along x_3. Therefore, $x_1' = x_2' = 0$, and $x_1 - x_{1'} = x_1$, $x_2 - x_{2'} = x_2$ and $x_3 - x_{3'} = -x_{3'}$.

Substituting the isotropic elastic constant tensor $c_{mi3p} = \lambda\delta_{mi}\delta_{3p} + \mu(\delta_{m3}\delta_{ip} + \delta_{mp}\delta_{i3})$ we obtain:

$$\begin{aligned}u_{j,g}(\mathbf{x}) &= \varepsilon_{3pg}\mu(\delta_{m3}\delta_{ip} + \delta_{mp}\delta_{i3})b\oint_{L'} G_{ij,m'}(\mathbf{x}-\mathbf{x}')\,\mathrm{d}x_{3'} \\ &= \mu b\int_{-\infty}^{\infty}\mathrm{d}x_{3'}\Big\{\delta_{g2}G_{1j,3'}(\mathbf{x}-\mathbf{x}') - \delta_{g1}G_{2j,3'}(\mathbf{x}-\mathbf{x}') \\ &\quad + \delta_{g2}G_{3j,1'}(\mathbf{x}-\mathbf{x}') - \delta_{g1}G_{3j,2'}(\mathbf{x}-\mathbf{x}')\Big\}\end{aligned}$$

where the term involving λ is zero because $\varepsilon_{3pg}\delta_{3p} = 0$. Since $g = 3$ never occurs we must have $u_{1,3} = u_{2,3} = u_{3,3} = 0$. The remaining displacement gradients are as follows:

$$u_{1,1}(\mathbf{x}) = -\mu b\int_{-\infty}^{\infty}\mathrm{d}x_{3'}\,\big(G_{21,3'}(\mathbf{x}-\mathbf{x}') + G_{31,2'}(\mathbf{x}-\mathbf{x}')\big)$$

$$
\begin{aligned}
u_{1,2}(\mathbf{x}) &= \mu b \int_{-\infty}^{\infty} \mathrm{d}x_{3'} \left(G_{11,3'}(\mathbf{x}-\mathbf{x}') + G_{31,1'}(\mathbf{x}-\mathbf{x}') \right) \\
u_{2,1}(\mathbf{x}) &= -\mu b \int_{-\infty}^{\infty} \mathrm{d}x_{3'} \left(G_{22,3'}(\mathbf{x}-\mathbf{x}') + G_{32,2'}(\mathbf{x}-\mathbf{x}') \right) \\
u_{2,2}(\mathbf{x}) &= \mu b \int_{-\infty}^{\infty} \mathrm{d}x_{3'} \left(G_{12,3'}(\mathbf{x}-\mathbf{x}') + G_{32,1'}(\mathbf{x}-\mathbf{x}') \right) \\
u_{3,1}(\mathbf{x}) &= -\mu b \int_{-\infty}^{\infty} \mathrm{d}x_{3'} \left(G_{23,3'}(\mathbf{x}-\mathbf{x}') + G_{33,2'}(\mathbf{x}-\mathbf{x}') \right) \\
u_{3,2}(\mathbf{x}) &= \mu b \int_{-\infty}^{\infty} \mathrm{d}x_{3'} \left(G_{13,3'}(\mathbf{x}-\mathbf{x}') + G_{33,1'}(\mathbf{x}-\mathbf{x}') \right)
\end{aligned}
$$

The derivative of the isotropic elastic Green's function is as follows:

$$
G_{ij,l'}(\mathbf{x}-\mathbf{x}') = \frac{1}{16\pi\mu(1-\nu)} \left\{ \delta_{ij} \frac{(3-4\nu)(x_l - x_{l'})}{|\mathbf{x}-\mathbf{x}'|^3} - \frac{\delta_{il'}(x_j - x_{j'}) + \delta_{jl'}(x_i - x_{i'})}{|\mathbf{x}-\mathbf{x}'|^3} \right.
$$
$$
\left. + 3\frac{(x_i - x_{i'})(x_j - x_{j'})(x_l - x_{l'})}{|\mathbf{x}-\mathbf{x}'|^5} \right\}
$$

Any derivative of a Green's function matrix element that is odd in $(x_3 - x_{3'})$ integrates to zero. Therefore, the following eight Green's function derivatives integrate to zero:

$$
G_{21,3'},\ G_{31,2'},\ G_{11,3'},\ G_{31,1'},\ G_{22,3'},\ G_{32,3'},\ G_{12,3'},\ G_{32,1'}.
$$

Only $u_{3,1}$ and $u_{3,2}$ are non-zero:

$$
u_{3,1}(\mathbf{x}) = -\frac{\mu b}{16\pi\mu(1-\nu)} \int_{-\infty}^{\infty} \mathrm{d}x_{3'} \left\{ \frac{(2-4\nu)x_2}{\left(x_1^2 + x_2^2 + x_3'^2\right)^{\frac{3}{2}}} + \frac{6x_2 x_3'^2}{\left(x_1^2 + x_2^2 + x_3'^2\right)^{\frac{5}{2}}} \right\}
$$

$$= -\frac{b}{2\pi}\frac{x_2}{x_1^2+x_2^2},$$

where we have used the integrals:

$$\int_{-\infty}^{\infty} dx_{3'} \frac{1}{\left(x_1^2+x_2^2+x_3'^2\right)^{\frac{3}{2}}} = \frac{2}{x_1^2+x_2^2}$$

$$\int_{-\infty}^{\infty} dx_{3'} \frac{x_3'^2}{\left(x_1^2+x_2^2+x_3'^2\right)^{\frac{5}{2}}} = \frac{2}{3}\frac{1}{x_1^2+x_2^2}.$$

Since $u_{1,3} = 0$ the strain component $e_{13} = e_{31}$ is as follows:

$$e_{13} = -\frac{b}{4\pi}\frac{x_2}{x_1^2+x_2^2},$$

and the stress component $\sigma_{13} = \sigma_{31}$ is then:

$$\sigma_{13} = 2\mu e_{13} = -\frac{\mu b}{2\pi}\frac{x_2}{x_1^2+x_2^2}.$$

Similarly, $u_{3,2}(\mathbf{x})$ is evaluated as follows:

$$u_{3,2}(\mathbf{x}) = \frac{\mu b}{16\pi\mu(1-\nu)}\int_{-\infty}^{\infty} dx_{3'}\left\{\frac{(2-4\nu)x_1}{\left(x_1^2+x_2^2+x_3'^2\right)^{\frac{3}{2}}} + \frac{6x_1x_3'^2}{\left(x_1^2+x_2^2+x_3'^2\right)^{\frac{5}{2}}}\right\}$$

$$= \frac{b}{2\pi}\frac{x_1}{x_1^2+x_2^2},$$

Therefore,

$$e_{23} = \frac{b}{4\pi}\frac{x_1}{x_1^2+x_2^2}$$

and

$$\sigma_{23} = 2\mu e_{23} = \frac{\mu b}{2\pi}\frac{x_1}{x_1^2+x_2^2}.$$

In the previous question we derived the displacement field:

$$u_3(\mathbf{x}) = \frac{b}{2\pi}\left(\tan^{-1}\left(\frac{x_2}{x_1}\right) - \pi\right).$$

Differentiating this expressions with respect to x_1 and x_2 we obtain:

$$u_{3,1}(\mathbf{x}) = -\frac{b}{2\pi}\frac{x_2}{x_1^2 + x_2^2}$$

$$u_{3,2}(\mathbf{x}) = \frac{b}{2\pi}\frac{x_1}{x_1^2 + x_2^2}$$

These displacement gradients are identical to those obtained from Mura's formula.

Problem 6.4 Calculate the elastic energy of a screw dislocation in isotropic elasticity in two ways:
(i) by integrating the elastic energy density $\frac{1}{2}\sigma_{ij}e_{ij}$ using the stress and strain tensors derived in the previous question.
(ii) by using eq 6.35.

Solution (i) The elastic energy density for the screw dislocation is as follows:

$$\frac{1}{2}\sigma_{ij}e_{ij} = \frac{1}{2}(\sigma_{13}e_{13} + \sigma_{31}e_{31} + \sigma_{23}e_{23} + \sigma_{32}e_{32}) = \frac{\mu b^2}{8\pi^2}\frac{1}{x_1^2 + x_2^2}.$$

In circular polar coordinates $x_1 = r\cos\theta$ and $x_2 = r\sin\theta$ the elastic energy density becomes $\mu b^2/(8\pi^2 r^2)$. Therefore, the elastic energy per unit length of the dislocation is as follows:

$$E_{el} = \int_0^{2\pi} \mathrm{d}\theta \int_{r_c}^{R} \mathrm{d}r\, r\frac{\mu b^2}{8\pi^2 r^2} = \frac{\mu b^2}{4\pi}\ln\left(\frac{R}{r_c}\right).$$

(ii) With the dislocation line along x_3 and with the cut occupying $x_1 \geq 0, x_2 = 0$, we have $n_j = \delta j2$. For

$$u_i(\mathbf{x}) = \delta_{i3}\frac{b}{2\pi}\left(\tan^{-1}\left(\frac{x_2}{x_1}\right) - \pi\right).$$

Just beneath the cut $u_3(x_1,0^-,x_3) = -b/2$ and just above the cut $u_3(x_1,0^+,x_3) = b/2$. The only stress component appearing in the integral along the cut is $\sigma_{32}(\mathbf{x}) = \mu b/(2\pi x_1)$, which is continuous across the cut. Therefore the elastic energy per unit length of dislocation line is:

$$E_{el} = \frac{1}{2}\left(\frac{-b}{2}\right)\int_{\infty}^{r_c} \mathrm{d}x_1 \frac{\mu b}{2\pi x_1} + \frac{1}{2}\left(\frac{b}{2}\right)\int_{r_c}^{\infty} \mathrm{d}x_1 \frac{\mu b}{2\pi x_1} = \frac{\mu b^2}{4\pi}\ln\left(\frac{R}{r_c}\right),$$

which is the same as in part (i).

Problem 6.5 An isotropic elastic crystal contains edge dislocations with Burgers vectors $\pm\mathbf{b}$ gliding on a set of parallel slip planes. Let the dislocations lie along the x_3-axis and let the normal to the slip planes be $[0,1,0]$. An edge dislocation with Burgers vector $\mathbf{b} = [b,0,0]$ is pinned at the origin of the coordinate system. A second edge dislocation with Burgers vector $\mathbf{b} = [b,0,0]$ is gliding on a parallel slip plane. The distance between the slip planes of the two dislocations is D. Show that the position of stable equilibrium of the second dislocation is at $x_1 = 0, x_2 = D$. If now the Burgers vector of the second dislocation is $\mathbf{b} = -[b,0,0]$ show that it has two positions of stable equilibrium at $x_1 = \pm D, x_2 = D$.

Solution The Peach-Koehler force on the dislocation at (x_1, D) is as follows:

$$\mathbf{F} = \begin{pmatrix} \sigma_{11} & \sigma_{12} & 0 \\ \sigma_{21} & \sigma_{22} & 0 \\ 0 & 0 & \sigma_{33} \end{pmatrix}\begin{pmatrix} b \\ 0 \\ 0 \end{pmatrix} \times \begin{pmatrix} 0 \\ 0 \\ 1 \end{pmatrix} = \begin{pmatrix} \sigma_{12}b \\ -\sigma_{11}b \\ 0 \end{pmatrix}$$

where the stress tensor is that of the edge dislocation at the origin. The glide force is $\sigma_{12}(x_1, D)b$, where :

$$\sigma_{12}(x_1, D)b = \frac{\mu b^2}{2\pi(1-\nu)} x_1 \frac{x_1^2 - D^2}{(x_1^2 + D^2)^2}.$$

Fig 6.14 is a plot of this force of interaction. We see that the glide force is zero at $x_1 = 0$ and at $x_1 = \pm D$. Only $x_1 = 0$ is a position of stable equilibrium because if the position of the dislocation deviates a small amount along $+x_1$ the force is negative bringing it back to $x_1 = 0$. If its position deviates a small amount along $-x_1$ the force is positive bringing it back to $x_1 = 0$. Small deviations from the

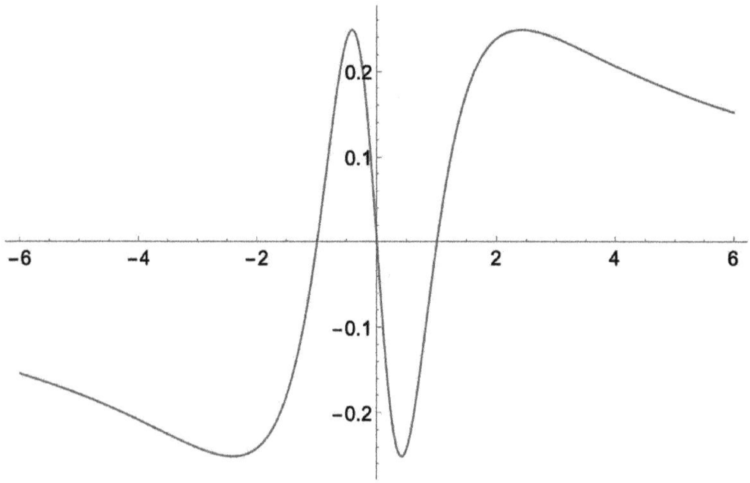

Figure 6.14: Peach Koehler force of interaction between two parallel edge dislocations on parallel slip planes separated by $D = 1$. The vertical axis is in units of $\mu b^2/(2\pi(1-\nu))$.

points $x_1 = \pm D$ grow through the force of interaction. If the second dislocation is at $|x_1| > D$ it will be repelled to $\pm\infty$. If it is at $|x_1| < D$ it will be attracted to $x_1 = 0$.

If the Burgers vectors of the two edge dislocations have opposite signs then the force of interaction changes sign. Again the glide force is zero at $x_1 = 0$ and at $x_1 = \pm D$. But now we find $x_1 = 0$ is a position of unstable equilibrium while $x_1 = \pm D$ are positions of stable equilibrium. If $x_1 > 0$ the mobile dislocation is attracted to $x_1 = D$. If $x_1 < 0$ it is attracted to $x_1 = -D$.

Problem 6.6 *Stress field of a symmetric tilt boundary.* This question illustrates how dislocations may organise themselves into configurations where they mutually screen their elastic fields, reducing the overall elastic energy.

A small angle symmetric tilt grain boundary in the plane $x_1 = 0$ comprises an infinite array of edge dislocations with Burgers vector $[b, 0, 0]$. The dislocation lines are parallel to the x_3-axis and their positions along the x_2-axis are $x_2 = 0, \pm p, \pm 2p, \pm 3p, \ldots, \pm\infty$. Using the components of the stress tensor for an edge dislocation given in eq 6.33 show that the non-zero stress components of the grain boundary are:

$$\sigma_{11}(X_1, X_2) = -\frac{\mu b}{2\pi(1-\nu)p}\sum_{n=-\infty}^{\infty}(X_2 - n)\left(\frac{3X_1^2 + (X_2-n)^2}{(X_1^2 + (X_2-n)^2)^2}\right)$$

$$\sigma_{12}(X_1, X_2) = \frac{\mu b}{2\pi(1-\nu)p} \sum_{n=-\infty}^{\infty} X_1 \left(\frac{X_1^2 - (X_2 - n)^2}{\left(X_1^2 + (X_2 - n)^2\right)^2} \right)$$

$$\sigma_{22}(X_1, X_2) = \frac{\mu b}{2\pi(1-\nu)p} \sum_{n=-\infty}^{\infty} (X_2 - n) \left(\frac{X_1^2 - (X_2 - n)^2}{\left(X_1^2 + (X_2 - n)^2\right)^2} \right)$$

$$\sigma_{33}(X_1, X_2) = -\frac{\mu b \nu}{\pi(1-\nu)p} \sum_{n=-\infty}^{\infty} \frac{X_2 - n}{X_1^2 + (X_2 - n)^2}$$

where $X_1 = x_1/p$ and $X_2 = x_2/p$.

The sums may be evaluated as contour integrals using the Sommerfeld-Watson transformation. For example,

$$\sum_{n=-\infty}^{\infty} \frac{1}{n+a} = \frac{1}{2\pi i} \oint_C \pi \cot(\pi z) \frac{1}{z+a} \mathrm{d}z = \pi \cot \pi a,$$

where the contour includes all the poles of $\cot(\pi z)$ along the real axis where $z = n$, but excludes the pole at $z = -a$ and is closed by a circle at infinity. The result holds for real or complex a. We may use this result to derive all the sums we need to evaluate the stress components for the grain boundary. For example, since

$$\frac{1}{n + X_2 + iX_1} + \frac{1}{n + X_2 - iX_1} = \frac{2(n + X_2)}{(n + X_2)^2 + X_1^2}$$

we deduce that

$$\sum_{n=-\infty}^{\infty} \frac{n + X_2}{(n + X_2)^2 + X_1^2} = \frac{\pi \sin(2\pi X_2)}{\cosh(2\pi X_1) - \cos(2\pi X_2)}.$$

Hence show that the non-zero components of the stress tensor associated with the grain boundary are as follows:

$$\sigma_{11} = -\frac{\mu b}{2(1-\nu)p} \frac{\sin(2\pi X_2)\left[\cosh(2\pi X_1) - \cos(2\pi X_2) + 2\pi X_1 \sinh(2\pi X_1)\right]}{(\cosh(2\pi X_1) - \cos(2\pi X_2))^2}$$

$$\sigma_{12} = \frac{\pi \mu b}{(1-\nu)p} \frac{X_1\left[\cosh(2\pi X_1)\cos(2\pi X_2) - 1\right]}{(\cosh(2\pi X_1) - \cos(2\pi X_2))^2}$$

$$\sigma_{22} = -\frac{\mu b}{2(1-\nu)p} \frac{\sin(2\pi X_2)\left[\cosh(2\pi X_1) - \cos(2\pi X_2) - 2\pi X_1 \sinh(2\pi X_1)\right]}{(\cosh(2\pi X_1) - \cos(2\pi X_2))^2}$$

$$\sigma_{33} = -\frac{\mu b \nu}{(1-\nu)p} \frac{\sin(2\pi X_2)}{(\cosh(2\pi X_1) - \cos(2\pi X_2))}$$

When $|x_1| > p$ show that these stress components are approximately proportional to $e^{-2\pi|x_1|/p}$.

Show that in the limit $p \rightarrow \infty$ these stress components become those of an isolated dislocation, given in eq 6.33.

Solution In the grain boundary there are dislocations at $(0, np)$ where n is an integer. Therefore,

$$\sigma_{11}(x_1, x_2) = -\frac{\mu b}{2\pi(1-\nu)} \sum_{n=-\infty}^{\infty} (x_2 - np) \left(\frac{3x_1^2 + (x_2 - np)^2}{\left(x_1^2 + (x_2 - np)^2\right)^2} \right)$$

Dividing through by p we obtain:

$$\sigma_{11}(X_1, X_2) = -\frac{\mu b}{2\pi(1-\nu)p} \sum_{n=-\infty}^{\infty} (X_2 - n) \left(\frac{3X_1^2 + (X_2 - n)^2}{\left(X_1^2 + (X_2 - n)^2\right)^2} \right)$$

The sums for $\sigma_{12}(X_1, X_2)$, $\sigma_{22}(X_1, X_2)$ and $\sigma_{33}(X_1, X_2)$ are derived in the same way. It is straightforward to derive the following sums:

$$S_1 = \sum_{n=-\infty}^{\infty} \frac{1}{\left(X_1^2 + (X_2 - n)^2\right)} = \frac{\pi}{X_1} \frac{\sinh(2\pi X_1)}{(\cosh(2\pi X_1) - \cos(2\pi X_2))}$$

$$S_2 = \sum_{n=-\infty}^{\infty} \frac{(X_2 - n)}{\left(X_1^2 + (X_2 - n)^2\right)} = \pi \frac{\sin(2\pi X_2)}{(\cosh(2\pi X_1) - \cos(2\pi X_2))}$$

$$S_3 = \sum_{n=-\infty}^{\infty} \frac{1}{\left(X_1^2 + (X_2 - n)^2\right)^2}$$

$$= \frac{\pi}{2X_1^3} \frac{\sinh(2\pi X_1)}{(\cosh(2\pi X_1) - \cos(2\pi X_2))} + \frac{\pi^2}{X_1^2} \frac{\cosh(2\pi X_1)\cos(2\pi X_2) - 1}{(\cosh(2\pi X_1) - \cos(2\pi X_2))^2}$$

$$S_4 = \sum_{n=-\infty}^{\infty} \frac{(X_2-n)}{\left(X_1^2+(X_2-n)^2\right)^2} = \frac{\pi^2}{X_1}\frac{\sinh(2\pi X_1)\sin(2\pi X_2)}{(\cosh(2\pi X_1)-\cos(2\pi X_2))^2}$$

$$S_5 = \sum_{n=-\infty}^{\infty} \frac{(X_2-n)^2}{\left(X_1^2+(X_2-n)^2\right)^2}$$

$$= \frac{\pi}{2X_1}\frac{\sinh(2\pi X_1)}{(\cosh(2\pi X_1)-\cos(2\pi X_2))} - \pi^2\frac{\cosh(2\pi X_1)\cos(2\pi X_2)-1}{(\cosh(2\pi X_1)-\cos(2\pi X_2))^2}$$

$$S_6 = \sum_{n=-\infty}^{\infty} \frac{(X_2-n)^3}{\left(X_1^2+(X_2-n)^2\right)^2}$$

$$= \pi\frac{\sin(2\pi X_2)}{(\cosh(2\pi X_1)-\cos(2\pi X_2))} - \pi^2 X_1\frac{\sinh(2\pi X_1)\sin(2\pi X_2)}{(\cosh(2\pi X_1)-\cos(2\pi X_2))^2}$$

Using these sums we can sum the series involved in the components of the stress tensor:

$$\sigma_{11} = -\frac{\mu b}{2\pi(1-\nu)p}(3X_1^2 S_4 + S_6)$$

$$= -\frac{\mu b}{2(1-\nu)p}\frac{\sin(2\pi X_2)\left[\cosh(2\pi X_1)-\cos(2\pi X_2)+2\pi X_1\sinh(2\pi X_1)\right]}{\left[\cosh(2\pi X_1)-\cos(2\pi X_2)\right]^2}$$

$$\sigma_{12} = \frac{\mu b}{2\pi(1-\nu)p}(X_1^3 S_3 - X_1 S_5)$$

$$= \frac{\pi\mu b}{(1-\nu)p}\frac{X_1\left[\cosh(2\pi X_1)\cos(2\pi X_2)-1\right]}{\left[\cosh(2\pi X_1)-\cos(2\pi X_2)\right]^2}$$

$$\sigma_{22} = \frac{\mu b}{2\pi(1-\nu)p}(X_1^2 S_4 - S_6)$$

$$= -\frac{\mu b}{2(1-\nu)p}\frac{\sin(2\pi X_2)\left[\cosh(2\pi X_1)-\cos(2\pi X_2)-2\pi X_1\sinh(2\pi X_1)\right]}{\left[\cosh(2\pi X_1)-\cos(2\pi X_2)\right]^2}$$

$$\sigma_{33} = -\frac{\mu b\nu}{\pi(1-\nu)p}S_2$$

$$= -\frac{\mu b \nu}{(1-\nu)p} \frac{\sin(2\pi X_2)}{\cosh(2\pi X_1) - \cos(2\pi X_2)}$$

When $|X_1| > 1$ all non-zero stress components are approximately proportional to $\operatorname{sech}(2\pi X_1)$, so that the stress field decays exponentially as $\exp(-2\pi |x_1| / p)$.

It is a useful check on these expressions for the components of the stress tensor that they revert to those of an isolated dislocation at the origin as $p \to \infty$. Retaining terms only to second order in $1/p$ we have:

$$\sin(2\pi X_2) \to \frac{2\pi x_2}{p} \text{ and } \cos(2\pi X_2) \to 1 - \frac{2\pi^2 x_2^2}{p^2}$$

$$\sinh(2\pi X_1) \to \frac{2\pi x_1}{p} \text{ and } \cosh(2\pi X_1) \to 1 + \frac{2\pi^2 x_1^2}{p^2}$$

Then we find:

$$\sigma_{11}(x_1/p, x_2/p) \quad \to \quad -\frac{\mu b}{2(1-\nu)p} \frac{\frac{2\pi x_2}{p}\left(\frac{2\pi^2 x_1^2}{p^2} + \frac{2\pi^2 x_2^2}{p^2} + \frac{4\pi^2 x_1^2}{p^2}\right)}{\left(\frac{2\pi^2 x_1^2}{p^2} + \frac{2\pi^2 x_2^2}{p^2}\right)^2}$$

$$= \quad -\frac{\mu b}{2\pi(1-\nu)} \frac{x_2(3x_1^2 + x_2^2)}{(x_1^2 + x_2^2)^2}$$

$$\sigma_{12}(x_1/p, x_2/p) \quad \to \quad \frac{\pi \mu b}{(1-\nu)p} \frac{x_1}{p} \frac{\left(\left(1 + \frac{2\pi^2 x_1^2}{p^2}\right)\left(1 - \frac{2\pi^2 x_2^2}{p^2}\right) - 1\right)}{\left(\frac{2\pi^2 x_1^2}{p^2} + \frac{2\pi^2 x_2^2}{p^2}\right)^2}$$

$$= \quad \frac{\mu b}{2\pi(1-\nu)} \frac{x_1(x_1^2 - x_2^2)}{(x_1^2 + x_2^2)^2}$$

$$\sigma_{22}(x_1/p, x_2/p) \quad \to \quad -\frac{\mu b}{2(1-\nu)p} \frac{2\pi x_2}{p} \frac{\left(1 + \frac{2\pi^2 x_1^2}{p^2} - 1 + \frac{2\pi^2 x_2^2}{p^2} - \frac{2\pi x_1}{p}\frac{2\pi x_1}{p}\right)}{\left(\frac{2\pi^2 x_1^2}{p^2} + \frac{2\pi^2 x_2^2}{p^2}\right)^2}$$

$$= -\frac{\mu b}{2\pi(1-\nu)}\frac{x_2(x_2^2-x_1^2)}{(x_1^2+x_2^2)^2}$$

$$\sigma_{33}(x_1/p, x_2/p) \rightarrow -\frac{\mu b \nu}{(1-\nu)p}\frac{2\pi x_2/p}{\left(\frac{2\pi^2 x_1^2}{p^2}+\frac{2\pi^2 x_2^2}{p^2}\right)}$$

$$= -\frac{\mu b \nu}{\pi(1-\nu)}\frac{x_2}{x_1^2+x_2^2}$$

This confirms that the stress field of the grain boundary has the correct limit when $p \to \infty$.

Comment

Another interesting limit to consider is $b \to 0$ and $p \to 0$ such that b/p remains constant. This amounts to turning the discrete dislocations, with their finite Burgers vectors, into a continuous distribution of dislocations, with infinitesimal Burgers vectors, without altering the misorientation between the crystals. It is not difficult to see that all stress components then become zero at all x_1 and x_2 except in the boundary plane, $x_1 = 0$, where σ_{11}, σ_{22} and σ_{33} are proportional to $\tan(\pi x_2/p)$. However, in any finite strip of the boundary, with infinite length along x_3 and finite width along x_2, the average values of σ_{11}, σ_{22} and σ_{33} are zero. All components of the stress tensor are then zero everywhere.

Problem 6.7 Having derived the stress component $\sigma_{12}(x_1, x_2)$ for the symmetrical tilt grain boundary in the previous question we may use it to calculate the energy of the boundary. Let the cut associated with each edge dislocation at $x_1 = 0, x_2 = np$ be in the half-plane $x_2 = np, x_1 \geq 0$. Following the same argument using the divergence theorem to derive eq 6.35, where the contour C now involves an infinite number of circuits around the cuts and the dislocations in the grain boundary, show that the elastic energy of the grain boundary per unit area is as follows:

$$E_{GB} = \frac{1}{p}\left[\frac{1}{2}\int_{r_c}^{R}\sigma_{12}(x_1, 0)\, b\, \mathrm{d}x_1 + E_c\right]$$

where E_c is the energy per unit length of the material inside the radius r_c of each dislocation that cannot be described with elasticity. It is assumed that E_c does not vary with the misorientation angle θ. This assumption is reasonable provided θ is small. In that case $\theta = b/p$ is also a good approximation.

Given the standard integral

$$\int \frac{u}{\sinh^2 u} du = \ln(\sinh u) - u \coth u$$

show that

$$E_{GB} = \theta \left[\left\{ \frac{E_c}{b} + \frac{\mu b}{4\pi(1-\nu)} \ln\left(\frac{eb}{2\pi r_c}\right) \right\} - \frac{\mu b}{4\pi(1-\nu)} \ln\theta \right]$$

This equation has the form $E_{GB} = \theta(A - B\ln\theta)$ where A and B are constants:

$$A = \frac{E_c}{b} + \frac{\mu b}{4\pi(1-\nu)} \ln\left(\frac{eb}{2\pi r_c}\right)$$

$$B = \frac{\mu b}{4\pi(1-\nu)}$$

It is known as the Read-Shockley[30] formula[31].

Solution Using the divergence theorem the elastic energy of the array of edge dislocations may be written in the equivalent forms:

$$E_{el} = \frac{1}{2} \int_{\text{all space}} \sigma_{ij} u_{i,j} \mathrm{d}V = \frac{1}{2} \int_S \sigma_{ij} u_i n_j \mathrm{d}S,$$

where σ_{ij} is the stress field of the array of edge dislocations, u_i is the displacement field of the array of edge dislocations, and the surface S is shown in red in Fig 6.15. Provided no tractions are applied to the surface of the body containing the grain boundary there is no contribution to the surface integral over the external surface. Since each dislocation is not associated with a net body force there is also no contribution to the surface integral from the small cylinder surrounding each dislocation line. Therefore, the only contribution to the surface integral comes from just beneath each cut, where $u_i = -\delta_{i1} b/2$ and $n_j = -\delta_{j2}$, and above each cut, where $u_i = \delta_{i1} b/2$ and $n_j = \delta_{j2}$. It follows that only the stress component $\sigma_{12}(x_1, np)$ appears in the surface integral. Unlike the displacement field the stress field is continuous across the cut. Since the dislocations are all equivalent the energy of the grain boundary per unit area is the number of dislocations per unit length along

[30] William Bradford Shockley 1910-1989, US Nobel prize winning physicist

[31] Read, W T, and Shockley, W, Phys. Rev. **78**, 275 (1950).

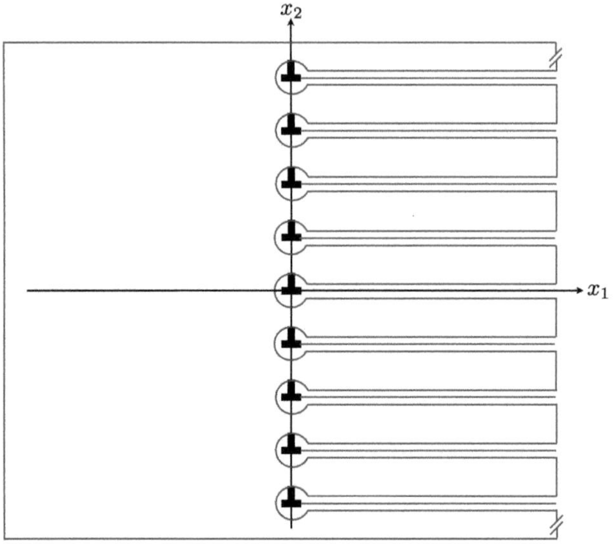

Figure 6.15: Illustration of the surface (in red) of the surface integral used to calculate the elastic energy of a symmetric tilt grain boundary. The boundary comprises an infinite stack of edge dislocations at $x_2 = np$, where n is an integer from $-\infty$ to $+\infty$. The dislocation lines are normal to the page along x_3. The blue lines are planar cuts in the half planes $x_2 = np$, $x_1 > 0$, viewed edge on. On traversing each cut there is an abrupt change in the displacement along x_1 equal to the Burgers vector $[b,0,0]$. The cuts extend to infinity along the positive x_1-axis. The surface integral includes the surface of the body, which is free of tractions.

x_2, which is $1/p$, multiplied by the contribution to the surface integral from one dislocation:

$$E_{el} = \frac{1}{2p}\int_{r_c}^{\infty} \sigma_{12}(x_1, np)\{b/2 - (-b/2)\}\mathrm{d}x_1 = \frac{1}{2p}\int_{r_c}^{\infty} \sigma_{12}(x_1, np)b\mathrm{d}x_1,$$

where r_c is the radius of the dislocation core, within which elasticity provides an inadequate description of the structure and energy of the dislocation.

Using the expression derived for $\sigma_{12}(x_1,x_2)$ in problem 6.6 we find:

$$\sigma_{12}(x_1,np) = \frac{\pi\mu b}{(1-\nu)p}\frac{x_1}{p}\frac{1}{\left(\cosh\left(\frac{2\pi x_1}{p}\right)-1\right)} = \frac{\mu b}{2(1-\nu)p}\frac{\pi x_1/p}{\sinh^2(\pi x_1/p)}.$$

Therefore, the elastic energy of the grain boundary is as follows:

$$\begin{aligned} E_{el} &= \frac{\mu b^2}{4(1-\nu)p^2}\lim_{R\to\infty}\int_{r_c}^{R}\frac{\pi x_1/p}{\sinh^2(\pi x_1/p)}\mathrm{d}x_1 \\ &= \frac{\mu b^2}{4\pi(1-\nu)p}\lim_{R\to\infty}\int_{\pi r_c/p}^{\pi R/p}\frac{u}{\sinh^2 u}\mathrm{d}u \end{aligned}$$

Using the given standard integral we get:

$$\begin{aligned} E_{el} &= \frac{\mu b^2}{4\pi(1-\nu)p}\lim_{R\to\infty}\Big[\ln(\sinh u) - u\coth u\Big]_{\pi r_c/p}^{\pi R/p} \\ &= \frac{\mu b^2}{4\pi(1-\nu)p}\lim_{R\to\infty}\left\{\ln\sinh\left(\frac{\pi R}{p}\right) - \frac{\pi R}{p}\coth\left(\frac{\pi R}{p}\right) - \ln\sinh\left(\frac{\pi r_c}{p}\right)\right. \\ &\quad \left. + \frac{\pi r_c}{p}\coth\left(\frac{\pi r_c}{p}\right)\right\} \end{aligned}$$

As $R\to\infty$ we have $\ln\sinh(\pi R/p)\to \pi R/p - \ln 2$ and $\coth(\pi R/p)\to 1$. Assuming $r_c \ll p$ then $\sinh(\pi r_c/p)\to \pi r_c/p$ and $\coth(\pi r_c/p)\to p/(\pi r_c)$. Then, we get:

$$E_{el} = -\frac{\mu b^2}{4\pi(1-\nu)p}\ln\left(\frac{2\pi r_c}{ep}\right).$$

The boundary misorientation is approximately $\theta = b/p$. Therefore, the elastic energy may be expressed in terms of θ as follows:

$$E_{el} = -\frac{\mu b\theta}{4\pi(1-\nu)}\ln\left(\frac{2\pi r_c\theta}{eb}\right) = -\frac{\mu b\theta}{4\pi(1-\nu)}\left[\ln\left(\frac{2\pi r_c}{eb}\right) + \ln\theta\right]$$

The energy per unit area of the grain boundary, E_{GB}, is E_{el} plus the energies of the dislocation cores, which we have not been able to treat with elasticity theory. Since we have assumed $r_c \ll p$ it is reasonable to assume the core energies are independent of p. The contribution to E_{GB} from the core energy is therefore $E_c/p = E_c\theta/b$, where E_c is the energy per unit length of one dislocation core. Therefore,

$$E_{GB} = \theta\Bigg(\underbrace{\frac{E_c}{b} + \frac{\mu b}{4\pi(1-\nu)}\ln\left(\frac{eb}{2\pi r_c}\right)}_{A} - \underbrace{\frac{\mu b}{4\pi(1-\nu)}}_{B}\ln\theta\Bigg),$$

where A and B are constants.

Comment

The small angle symmetric tilt grain boundary is an example of self-organisation of dislocations to reduce their elastic energy. There are other examples of small angle grain boundaries comprising arrays of dislocations. One is an asymmetric tilt grain boundary which is made of two sets of edge dislocations with differing Burgers vectors. Another is a twist grain boundary which comprises a planar network of screw dislocations.

Problem 6.8 In this question we repeat the calculation of the energy of a small angle symmetrical tilt grain boundary in a more heuristic but insightful way. The insight is to exploit the mutual screening of the elastic fields of the edge dislocations it contains. The radius R in eq 6.36 for the energy of an isolated edge dislocation may be taken as half the spacing p of the dislocations in the boundary. Show that the energy per unit area of the grain boundary is then

$$E_{GB} \approx \frac{1}{p}\left[\frac{\mu b^2}{4\pi(1-\nu)}\ln\left(\frac{p}{2r_c}\right) + E_c\right]$$

where E_c is the energy of the material inside the radius r_c which cannot be described with elasticity. As before, the angle θ of misorientation across the boundary is approximately b/p. Show that E_{GB} has the form

$$E_{GB} \approx \theta(A - B\ln\theta) \tag{6.42}$$

where A and B are constants with the dimensions of energy per unit area:

$$A = \frac{E_c}{b} + \frac{\mu b}{4\pi(1-\nu)}\ln\left(\frac{b}{2r_c}\right)$$

$$B = \frac{\mu b}{4\pi(1-\nu)}.$$

Solution As seen in problem 6.7, there are $1/p$ dislocations per unit length along x_2. Setting $R = p/2$ we add the energies of unit lengths of the $1/p$ dislocations to get:

$$
\begin{aligned}
E_{GB} &= \frac{1}{p}\left[\frac{\mu b^2}{4\pi(1-\nu)}\ln\left(\frac{p}{2r_c}\right)+E_c\right] \\
&= \theta\left[\frac{\mu b}{4\pi(1-\nu)}\ln\left(\frac{b}{2r_c}\frac{1}{\theta}\right)+\frac{E_c}{b}\right] \\
&= \theta\left[\frac{\mu b}{4\pi(1-\nu)}\ln\left(\frac{b}{2r_c}\right)+\frac{E_c}{b}-\frac{\mu b}{4\pi(1-\nu)}\ln\theta\right].
\end{aligned}
$$

Comment

This is remarkably close to the more rigorous solution in problem 6.7, which supports the idea of mutual screening of the elastic fields.

Problem 6.9 One application of eq 6.24 is to the evolution of prismatic loops in tungsten created by irradiation with high energy neutrons. Tungsten is very close to being elastically isotropic, and the distribution of loops evolves in time driven in part by their elastic interactions. Suppose the prismatic loops are such that $\hat{\mathbf{b}}\cdot\hat{\mathbf{n}} = 1$. Show that eq 6.24 simplifies as follows, and discuss its validity:

$$
\begin{aligned}
E_{int}^{(AB)} &= \frac{\mu b^A b^B (\delta A^A)(\delta A^B)}{4\pi(1-\nu)x^3}\Big\{ \\
&\quad 15(\hat{\mathbf{n}}^A\cdot\hat{\mathbf{x}})^2(\hat{\mathbf{n}}^B\cdot\hat{\mathbf{x}})^2-(4\nu-1)-12\nu(\hat{\mathbf{n}}^A\cdot\hat{\mathbf{n}}^B)(\hat{\mathbf{n}}^A\cdot\hat{\mathbf{x}})(\hat{\mathbf{n}}^B\cdot\hat{\mathbf{x}}) \\
&\quad -(1-2\nu)\left[3(\hat{\mathbf{n}}^B\cdot\hat{\mathbf{x}})^2+3(\hat{\mathbf{n}}^A\cdot\hat{\mathbf{x}})^2+2(\hat{\mathbf{n}}^A\cdot\hat{\mathbf{n}}^B)^2\right]\Big\}. \qquad (6.43)
\end{aligned}
$$

Note the angular dependence has been reduced from ten to just three angles between the unit vectors $\hat{\mathbf{x}}$, $\hat{\mathbf{n}}^A$, $\hat{\mathbf{n}}^B$ taken in pairs.

Solution Setting $\hat{\mathbf{n}}^A = \hat{\mathbf{b}}^A$ and $\hat{\mathbf{n}}^B = \hat{\mathbf{b}}^B$ in eq 6.24, we obtain eq 6.43. The key assumption is that the distance between the loops is significantly larger than the sizes of the loops. If the distance between the loops becomes comparable to, or smaller than, their sizes then the variation of the stress field of one loop within the other has to be taken into account.

7. Multiscale models of dislocations

7.1 Introduction

The treatment of dislocations in an elastic continuum in Chapter 6 has a number of weaknesses which are addressed in this chapter by taking into consideration the discrete atomic structure of the crystal. In the continuum theory of dislocations the Burgers vector is arbitrary. But in a crystal to maintain equivalent structures in slipped and unslipped regions of a slip plane the Burgers vector must be a crystal lattice translation vector. The stress associated with a Volterra dislocation becomes infinite at its centre because the dislocation is treated as a line singularity separating slipped and unslipped regions. Infinite stresses cannot be sustained. In reality the relative displacement across the slip plane by the Burgers vector is accumulated over a finite number of interatomic distances in the dislocation core, giving the core a finite width within which the Burgers vector is distributed. By including the energy of distorted bonds between atoms on either side of the slip plane the Frenkel-Kontorova and Peierls-Nabarro models predict a finite core width, and stresses remain finite in the dislocation core. The Peierls-Nabarro model also shows there is a finite stress required to move a dislocation, which is a direct result of its discrete atomic structure. This leads to a mechanism of dislocation motion involving kinks on dislocations, which is thermally activated, and which provides an explanation the strain rate dependence of the stress at which dislocations move in pure crystals.

A key concept in this chapter is the γ-surface introduced in section 7.3. The γ-surface provides the restoring force tending to keep the dislocation core as narrow as possible. The γ-surface also indicates the possible existence of metastable planar faults in the crystal which may enable dislocations with Burgers vectors equal to lattice translation vectors to split into two or more partial dislocations separated by the metastable fault(s), as discussed in section 7.4.

The Peierls-Nabarro model of a dislocation is multiscale. It couples the elastic energy of the defect extending throughout the body, to interatomic bonding in the dislocation core. It is a paragon of multiscale models of defects in crystals. The Frenkel-Kontorova model is a simpler treatment of the interplay between elastic energy and interatomic bonding confined to the slip plane of the dislocation.

7.2 Frenkel-Kontorova model

The Frenkel-Kontorova model[1] is a development of a model introduced[2] ten years earlier by Dehlinger[3]. The original model considers a chain of atoms connected by identical harmonic springs, with spring constant κ and natural length a, placed on a rigid substrate characterised by a sinusoidally varying potential $V\sin^2(\pi x/a)$. In the ground state the position of each atom along the x-axis is $x = na$, where n is an integer, and since all springs have their natural length a and all atoms sit in the minima of the substrate potential the total energy is zero. If u_n is the displacement of atom n from its ground state position $x = na$ the total energy of the system is as follows:

$$E = \sum_n \frac{1}{2}\kappa(u_n - u_{n-1})^2 + V\sin^2(\pi u_n/a) \tag{7.1}$$

Now suppose the chain is stretched by a. If the chain were not in the presence of the substrate potential the stretch would be taken up equally by all the springs in the chain. But in the presence of the substrate potential this is unlikely because the constant strain would put too many atoms at or near the maximum in the potential. The substrate potential tends to localise the strain to a region where the strain is correspondingly large, so that the stretch displaces a smaller number of atoms from the minima in the potential. Minimisation of the total energy in eq 7.1 leads to a balance between the elastic energy, tending to reduce the strain by spreading it over a large region, and the substrate potential, tending to confine the strain to a small region.

The relationship of this one-dimensional model to an edge dislocation is illustrated in Fig 7.1. It is assumed the dislocation core is planar lying in the slip plane, and corresponds to the atoms shown in red in Fig 7.1. But rather than there being one chain of atoms separated by harmonic springs sitting in a rigid periodic potential there are two harmonic chains, which interact with each other. Far from the location of the extra half plane of the dislocation each harmonic chain provides a periodic potential for the

[1]Frenkel, J, and Kontorova, T, J. Phys. Acad. Sci. USSR **1**, 137-149 (1939), in Russian. Tatiana Abramovna Kontorova 1911-1977. Soviet theoretical physicist.

[2]Dehlinger, U, Ann. Phys. **394**, 749-793 (1929), in German.

[3]Ulrich Dehlinger 1901-1981, German theoretical physicist at the Technische Hochschule Stuttgart (now University of Stuttgart).

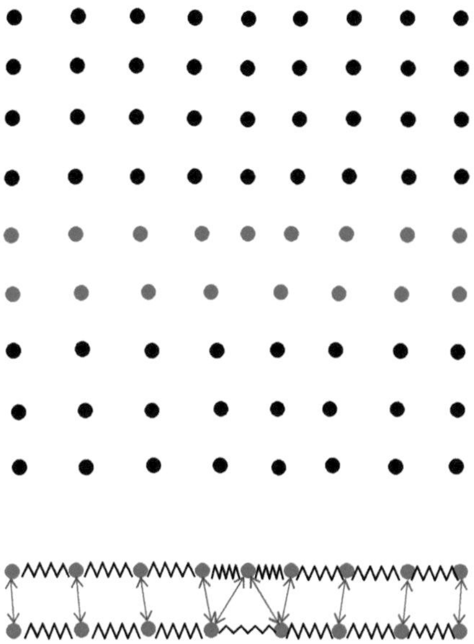

Figure 7.1: A sketch to illustrate the relationship between the Frenkel-Kontorova model and an edge dislocation in a 2D square lattice. The upper figure is a sketch of the atomic structure of an edge dislocation. The atoms coloured red are on either side of the horizontal slip plane. These atoms are reproduced in the lower figure, which is a sketch of the Frenkel-Kontorova model. In each chain of atoms nearest neighbours are linked together by harmonic springs. The disregistry between nearest neighbour atoms in either chain is shown by blue arrows. When there is no disregistry the arrows are vertical. The disregistry is seen to increase to a maximum where the extra half plane of the dislocation terminates.

other, as in the original Frenkel-Kontorova model. But in the dislocation core both chains distort, with one being compressed by $a/2$ and the other stretched by $a/2$, to accommodate the Burgers vector with magnitude equal to a, as illustrated in the lower sketch in Fig 7.1.

It is straightforward to modify the original Frenkel-Kontorova model to have two interacting harmonic chains. If v_n describes the deviation of atom n in the second harmonic chain we may express the total energy of both chains as follows:

$$E = \sum_n \frac{1}{2}\kappa\,(u_n - u_{n-1})^2 + \frac{1}{2}\kappa\,(v_n - v_{n-1})^2 + V\sin^2\left(\frac{\pi(u_n - v_n)}{a}\right) \tag{7.2}$$

The first two terms are the elastic energies of the harmonic chains. The third term is the energy associated with the disregistry $u_n - v_n$ between atoms n in the two chains.

Provided the displacement fields u_n and v_n are varying slowly we may replace the discrete sums in eq 7.2 with integrals. In that limit $\sum_n \to \int \frac{dx}{a}$, $u_n \to u(x)$, $u_{n+1} \to u(x+a) \to u(x) + a\mathrm{d}u/\mathrm{d}x$ and similarly for v. We can then write eq 7.2 as follows:

$$E = \int_{-\infty}^{\infty} \left\{ \frac{\kappa a^2}{2} \left[\left(\frac{\mathrm{d}u}{\mathrm{d}x}\right)^2 + \left(\frac{\mathrm{d}v}{\mathrm{d}x}\right)^2 \right] + V \sin^2\left(\frac{\pi(u-v)}{a}\right) \right\} \frac{\mathrm{d}x}{a} \tag{7.3}$$

Minimisation of the integral in eq 7.3 with respect to variations δu and δv leads to the following coupled Euler-Lagrange equations:

$$\kappa a \frac{\mathrm{d}^2 u}{\mathrm{d}x^2} = \frac{\pi V}{a^2} \sin\left(\frac{2\pi(u-v)}{a}\right) \tag{7.4}$$

$$\kappa a \frac{\mathrm{d}^2 v}{\mathrm{d}x^2} = -\frac{\pi V}{a^2} \sin\left(\frac{2\pi(u-v)}{a}\right) \tag{7.5}$$

By subtracting and adding eqs 7.4 and 7.5 we obtain:

$$\frac{\mathrm{d}^2(u-v)}{\mathrm{d}x^2} = \frac{2\pi V}{\kappa a^3} \sin\left(\frac{2\pi(u-v)}{a}\right) \tag{7.6}$$

$$\frac{\mathrm{d}^2(u+v)}{\mathrm{d}x^2} = 0. \tag{7.7}$$

To model a dislocation with a Burgers vector of magnitude a we impose the boundary conditions $u(-\infty) = v(-\infty) = 0$ and $u(+\infty) = a/2$, $v(+\infty) = -a/2$. It follows from these boundary conditions and eq 7.7 that $u(x) = -v(x)$ for all x. It then follows from eq 7.6 that $u(x)$ is determined by the sine-Gordon equation:

$$\frac{\mathrm{d}^2 u}{\mathrm{d}x^2} = \frac{\pi V}{\kappa a^3} \sin\left(\frac{4\pi u}{a}\right). \tag{7.8}$$

The solution to this equation satisfying the boundary conditions for $u(x)$ at $\pm\infty$ is as follows:

$$u = \frac{a}{\pi} \tan^{-1}\left[\exp\left\{\sqrt{\frac{V}{\kappa a^2}} \frac{2\pi x}{a}\right\}\right] \tag{7.9}$$

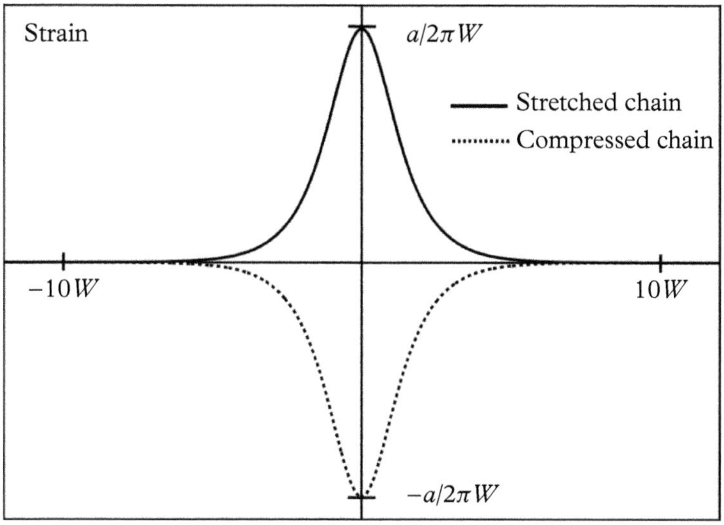

Figure 7.2: Plot of the strains in a pair of interacting chains stretched (solid line) and compressed (broken line) by $a/2$. Reproduced from Hammad, A, Swinburne, T D, Hasan, H, Del Rosso, S, Iannucci, L and Sutton, A P, Proc. R. Soc. A **471** 20150171 (2015).

The strains in the two chains are equal and opposite:

$$\frac{\mathrm{d}u}{\mathrm{d}x} = -\frac{\mathrm{d}v}{\mathrm{d}x} = \sqrt{\frac{V}{\kappa a^2}}\,\mathrm{sech}\left(\sqrt{\frac{V}{\kappa a^2}}\frac{2\pi x}{a}\right). \tag{7.10}$$

These strains are plotted in Fig 7.2, where it is seen they are localised around the origin $x = 0$. The width of the strain profiles may be defined by:

$$W = \frac{a}{2\pi}\sqrt{\frac{\kappa a^2}{V}} \tag{7.11}$$

The physics of the Frenkel-Kontorova model is clear. The disregistry associated with the Burgers vector between atoms on either side of the slip plane is accommodated in a finite width characterised by W. The magnitude of W increases with increasing spring stiffness κ and decreasing amplitude V of the interaction potential between the chains. When $V \to \infty$ the curves in Fig 7.2 become two delta functions at $x = 0$ of opposite sign, and the relative displacement of the chains by the Burgers vector becomes a step function at the origin: the core of a Volterra dislocation has no width. In the next section the Peierls-Nabarro model also predicts a finite core width, and goes further by treating the crystals above and below the red atoms in Fig 7.1.

Exercise 7.1 Derive the solution, eq 7.9, to the sine-Gordon equation, eq 7.8, with the boundary conditions stated above.

Hint: multiply both sides of eq 7.8 by $\mathrm{d}u/\mathrm{d}x$ and integrate. Then use an arctangent substitution. ■

Solution Multiplying both sides of eq 7.8 by $\mathrm{d}u/\mathrm{d}x$ we get:

$$\frac{1}{2}\frac{\mathrm{d}}{\mathrm{d}x}\left(\frac{\mathrm{d}u}{\mathrm{d}x}\right)^2 = -\frac{V}{4\kappa a^2}\frac{\mathrm{d}}{\mathrm{d}x}\cos\left(\frac{4\pi u}{a}\right).$$

Therefore,

$$\left(\frac{\mathrm{d}u}{\mathrm{d}x}\right)^2 = -\frac{V}{2\kappa a^2}\cos\left(\frac{4\pi u}{a}\right) + C_1,$$

where C_1 is a constant determined by the boundary conditions. As $x \to \pm\infty$, $\mathrm{d}u/\mathrm{d}x \to 0$. At $x = -\infty$ the displacement u is zero, so that $\cos(4\pi u/a) = 1$. At $x = +\infty$ the displacement u is $a/2$, so that again $\cos(4\pi u/a) = 1$. To satisfy these boundary conditions the constant C_1 must be $V/(2\kappa a^2)$. It follows that:

$$\frac{\mathrm{d}u}{\mathrm{d}x} = \sqrt{\frac{V}{\kappa a^2}}\sin\left(\frac{2\pi u}{a}\right). \tag{7.12}$$

To solve this differential equation we make the following change of variable:

$$u = \frac{a}{\pi}\tan^{-1} z.$$

Then we find,

$$\frac{\mathrm{d}u}{\mathrm{d}x} = \frac{a}{\pi}\frac{1}{1+z^2}\frac{\mathrm{d}z}{\mathrm{d}x}, \quad \text{and} \quad \sin\left(\frac{2\pi u}{a}\right) = \sin\left(2\tan^{-1} z\right) = \frac{2z}{1+z^2}.$$

The differential equation for $u(x)$ becomes the following differential equation for $z(x)$:

$$\frac{\mathrm{d}z}{\mathrm{d}x} = \frac{2\pi}{a}\sqrt{\frac{V}{\kappa a^2}}\,z.$$

The solution is:

$$\ln z = \sqrt{\frac{V}{\kappa a^2}}\frac{2\pi x}{a} + C_2,$$

where C_2 is another constant. Expressing the solution in terms of the displacement u:

$$\ln\left[\tan\left(\frac{\pi u}{a}\right)\right] = \sqrt{\frac{V}{\kappa a^2}}\frac{2\pi x}{a} + C_2.$$

The symmetry of the problem dictates that $u = a/4$ when $x = 0$. Therefore, $C_2 = 0$, and

$$u = \frac{a}{\pi}\tan^{-1}\left\{\exp\left(\sqrt{\frac{V}{\kappa a^2}}\frac{2\pi x}{a}\right)\right\}.$$

Exercise 7.2 Show that in the ground state there is equipartition of the total energy between the elastic energy of the harmonic springs (i.e. the first two terms in eq 7.3) and the misfit energy (i.e. the third term in eq 7.3) and that each is equal to $2VW/a = \kappa a^3/(2\pi^2 W)$. ■

Solution The elastic energy of the harmonic springs is:

$$E_{el} = \int_{-\infty}^{\infty}\frac{\kappa a^2}{2}\left\{\left(\frac{\mathrm{d}u}{\mathrm{d}x}\right)^2 + \left(\frac{\mathrm{d}v}{\mathrm{d}x}\right)^2\right\}\frac{\mathrm{d}x}{a}.$$

From eq 7.12:

$$\frac{\mathrm{d}u}{\mathrm{d}x} = -\frac{\mathrm{d}v}{\mathrm{d}x} = \sqrt{\frac{V}{\kappa a^2}}\sin\left(\frac{2\pi u}{a}\right).$$

Therefore,

$$E_{el} = \kappa a\int_{-\infty}^{\infty}\left(\frac{\mathrm{d}u}{\mathrm{d}x}\right)^2\mathrm{d}x$$

$$= \frac{V}{a}\int_{-\infty}^{\infty} \sin^2\left(\frac{2\pi u}{a}\right) \mathrm{d}x$$

$$= \frac{V}{a}\int_{-\infty}^{\infty} \sin^2\left(\frac{\pi(u-v)}{a}\right) \mathrm{d}x$$

$$= E_c$$

where we used $u = -v$. Hence the total energy is partitioned equally between the elastic and misfit energies.

Using eq 7.10 the elastic energy is evaluated as follows:

$$E_{el} = \kappa a \int_{-\infty}^{\infty} \left(\frac{\mathrm{d}u}{\mathrm{d}x}\right)^2 \mathrm{d}x = \kappa a \int_{-\infty}^{\infty} \frac{V}{\kappa a^2} \operatorname{sech}^2\left\{\sqrt{\frac{V}{\kappa a^2}}\frac{2\pi x}{a}\right\} \mathrm{d}x.$$

Let $y = \sqrt{V/(\kappa a^2)}(2\pi x/a)$. Then

$$E_{el} = \underbrace{\frac{2V}{a}\frac{a}{2\pi}\sqrt{\frac{\kappa a^2}{V}}}_{=W} \underbrace{\int_0^{\infty} \operatorname{sech}^2 y \, \mathrm{d}y}_{=1} = \frac{2VW}{a} = \frac{\kappa a^3}{2\pi^2 W},$$

where we have used eq 7.11. The integral $\int_0^{\infty} \operatorname{sech}^2 y \, \mathrm{d}y = 1$ is easily shown by making the substitution $y = \ln s$.

Comment

In problem 7.2 it is shown that this is an example of the equipartition of the total energy between the elastic energy and the misfit energy in the Frenkel-Kontorova model for *any* functional form of the misfit energy.

7.3 Peierls-Nabarro model

Although the physics of the Frenkel-Kontorova model is appealing it treats the interaction between only two linear chains of atoms. It does not address the elastic energy stored in the semi-infinite crystals on either side of the slip plane. The Peierls[4]-Nabarro model treats both the elastic energy of the dislocation and the misfit energy in the slip plane, albeit in a simplified manner. We will develop the model for an edge dislocation,

[4]Sir Rudolf Ernst Peierls, FRS 1907-1995 German born British physicist.

following Eshelby's approach[5] of treating the dislocation as a continuous distribution of edge dislocations along the slip plane with infinitesimal Burgers vectors[6].

We regard the edge dislocation shown in Fig 7.1 as a projection of its atomic structure viewed along the dislocation line. The atoms coloured red are in (010) planes, seen edge on, on either side of the geometrical slip plane, which is midway between them. As in the Frenkel-Kontorova model the disregistry between these atoms will be treated, but as the relative translation along x between columns of atoms on either side of the slip plane rather than between single atoms in linear chains.

Vitek[7] introduced the γ-surface in 1968[8]. There is a γ-surface for every plane within a crystal. Each γ-surface describes the energy γ associated with the imposition of a relative translation by $\mathbf{t}$ of one crystal half with respect to the other on either side of the plane. The relative translation vector is in the plane. The planar defect formed as a result of the relative translation is called a *generalised fault.* The γ-surface is an atomic-scale property of the crystal, and it is calculated increasingly using electronic density functional theory. An example is shown if Fig 7.3.

The relative translation $\mathbf{t}$ parallel to the chosen crystal plane is imposed as a step function on atoms on either side of the plane. Atoms are allowed to move only along the plane normal to positions where they experience no normal force. $\gamma(\mathbf{t})$ is the excess energy per unit area of the whole crystal containing the fault in this partially relaxed and constrained state. When the plane contains two non-parallel sets of lattice vectors $\gamma(\mathbf{t})$ is periodic in two dimensions. If $\mathbf{t} = \mathbf{0}$ coincides with the perfect crystal configuration there is a global energy minimum at $\mathbf{t} = \mathbf{0}$ and at all translation vectors equal to lattice vectors within the plane. If there are n atoms associated with each lattice site of the crystal then there are up to n γ-surfaces on each plane. We will return to the γ surface in section 7.4. The purpose of introducing it here is to put the Peierls-Nabarro model, in particular eq 7.13, in the context of current thinking.

In the Frenkel-Kontorova and Peierls-Nabarro models we consider relative translations of the two crystal halves on either side of the slip plane, where the translations are parallel to the Burgers vector of the dislocation. The energy associated with these relative translations is a section through the γ-surface on the slip plane. Equation 6.1 is a general expression for the energy per unit area of the generalised faults created by these relative translations along one direction in the γ-surface. The Peierls-Nabarro and Frenkel-Kontorova models simplify eq 6.1 by taking just the first harmonic of the

[5]Eshelby, J D, Phil. Mag. **40**, 903-912 (1949).

[6]We have chosen to present the model for an edge dislocation because it is most easily visualised. The model is equally applicable to screw and mixed edge and screw dislocations, although in all cases the core is usually assumed to be planar.

[7]Vaclav Vitek FRS 1940- Czech born British and US materials physicist.

[8]Vitek, V, Phil. Mag. **18**, 773-786 (1968).

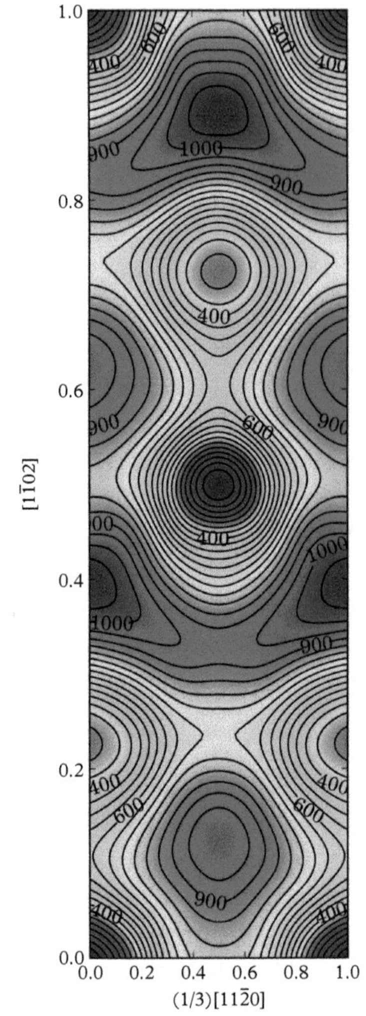

Figure 7.3: γ-surface for widely spaced $(1\bar{1}0\bar{1})$ atomic planes in titanium computed by density functional theory. The energy surface is viewed from above like a map showing hills in red and valleys in blue with contours separated by 50 mJ $\cdot$ m^{-2}. The periodic cell contains two lattice sites - one at the four corners and one in the centre - where the energy is zero. There is a local minimum at [0.00, 0.22] in fractional coordinates which is seen most clearly at [0.50, 0.72], where the energy is 279 mJ $\cdot$ m^{-2}. Reproduced with permission by Taylor & Francis (`www.tandfonline.com`) from Ready, A J, Haynes, P D, Rugg, D, and Sutton, A P, Phil. Mag. **97**, 1129-1143 (2017).

Fourier expansion.

To illustrate the Peierls-Nabarro model consider an edge dislocation in a simple cubic crystal, with just one atom at each lattice site. The slip plane is assumed to be (010) and the Burgers vector $\mathbf{b} = [a, 0, 0]$ where a is the lattice constant of the simple cubic crystal. Let w be the relative translation in the [100] direction of the crystal halves on either side of the (010) slip plane. As with the Frenkel-Kontorova model $\gamma(w)$ is approximated as follows:

$$\gamma(w) = V \sin^2\left(\frac{\pi w}{a}\right), \tag{7.13}$$

where V is the amplitude of the energy variation with relative translation w.

The dislocation creates a relative displacement parallel to [100] between atomic columns along [001] that were directly opposite each other on either side of the slip plane (010) before the dislocation was introduced. This relative displacement is the disregistry, w, between the atomic columns, and inside the dislocation core it varies with position: $w = w(x)$. Provided the disregistry does not vary too rapidly with position we may define the misfit energy, E_c, of the dislocation per unit length of the dislocation line by the following integral:

$$E_c = \int_{-\infty}^{\infty} \gamma(w(x))\, \mathrm{d}x. \tag{7.14}$$

This is the energy associated with the misfitting atomic bonds across the slip plane. Far from the dislocation core $\gamma(w)$ is zero. Although the limits of the integral extend from minus infinity to plus infinity the dominant contribution to E_c is confined to the region where $w(x)$ is changing, which is in the dislocation core. It is worth noting that Peierls and Nabarro did not allow any displacements of atoms normal to the slip plane. This has two consequences. The first is that their $\gamma(w)$ differs from a section of the γ-surface as defined above, where relaxations normal to the plane are allowed. Secondly, the misfit energy of eq 7.14 would be reduced if normal relaxations were allowed.

So far the Peierls-Nabarro model does not differ much from the Frenkel-Kontorova model. It is the treatment of the elastic energy where they diverge, because it is non-local in the Peierls-Nabarro model. The dislocation is now viewed as a continuous distribution of dislocations with infinitesimal Burgers vectors. In the case of a Volterra dislocation of Burgers vector a at the origin the distribution is $a\delta(x)$. The δ-function is smeared out by writing the distribution of the Burgers vector as $af(x)$ where the integral $\int_{-\infty}^{\infty} f(x)\mathrm{d}x = 1$. Then $af(x)\mathrm{d}x$ is the infinitesimal Burgers vector of a dislocation between x and $x + \mathrm{d}x$. Note that $f(x)$ has the dimensions of one over length. It is clear that when $f(x) \neq 0$ the disregistry $w(x)$ is changing. More precisely, $af(x)\mathrm{d}x = \mathrm{d}w$, which leads to the following relation:

$$\frac{\mathrm{d}w}{\mathrm{d}x} = af(x). \tag{7.15}$$

The Peach-Koehler force on an edge dislocation with Burgers vector b_1 at x due to another edge dislocation with Burgers vector b_2 at x_2 in isotropic elasticity is as follows:

$$F = \frac{\mu b_1 b_2}{2\pi(1-\nu)} \frac{1}{(x-x_2)}.$$

Integrating $-F$ between $x = x_2 + r_c$, where r_c is the radius of the core of each dislocation[9], and $x = x_1$ we obtain their interaction energy:

$$E_{int} = -\frac{\mu b_1 b_2}{2\pi(1-\nu)} \ln\left(\frac{|x_1 - x_2|}{r_c}\right) \tag{7.16}$$

It follows that for the continuous distribution of dislocations comprising the dislocation with Burgers vector a the elastic energy is as follows:

$$E_{el} = -\frac{1}{2}\frac{\mu a^2}{2\pi(1-\nu)} \int_{-\infty}^{\infty} \mathrm{d}x_1 \int_{-\infty}^{\infty} \mathrm{d}x_2\, f(x_1) f(x_2) \ln\left(\frac{|x_1 - x_2|}{r_c}\right) \tag{7.17}$$

where the factor of one half is to correct for the double counting in the double integral. Combining eqs 7.14, 7.15 and 7.17 Christian[10] and Vitek obtain[11] the following expression for the total energy of the dislocation per unit length in the Peierls-Nabarro model:

$$E_{PN} = \int_{-\infty}^{\infty} V \sin^2\left(\frac{\pi w(x)}{a}\right) \mathrm{d}x - \frac{\mu}{4\pi(1-\nu)} \int_{-\infty}^{\infty} \mathrm{d}x_1 \int_{-\infty}^{\infty} \mathrm{d}x_2 \left(\frac{\mathrm{d}w}{\mathrm{d}x}\right)_{x=x_1} \left(\frac{\mathrm{d}w}{\mathrm{d}x}\right)_{x=x_2} \ln\left(\frac{|x_1 - x_2|}{r_c}\right). \tag{7.18}$$

This equation is the analogue of eq 7.3 in the Frenkel-Kontorova model, where the disregistry is $u(x) - v(x)$. Minimising the integral in eq 7.18 with respect to variations $\delta w(x)$ we obtain the following Euler-Lagrange equation:

$$\frac{\mu}{2\pi(1-\nu)} P \int_{-\infty}^{\infty} \mathrm{d}x_1 \frac{(\mathrm{d}w/\mathrm{d}x)_{x=x_1}}{x - x_1} = -\frac{\mathrm{d}\gamma}{\mathrm{d}w(x)} = -\frac{V\pi}{a} \sin\left(\frac{2\pi w(x)}{a}\right). \tag{7.19}$$

[9]It has to be non-zero because the force becomes infinite when the dislocations are coincident. It is not present in the final equation, so its value is unimportant.

[10]John Wyrill Christian FRS 1926-2001. British materials scientist.

[11]Christian, J W, and Vitek, V, Reports on Progress in Physics **33**, 307 (1970).

Although mathematically this is a daunting non-linear, singular, integro-differential equation, physically it is just a balance of shear stresses at each point on the slip plane: the left hand side is the elastic shear stress at x arising from the continuous distribution of dislocations, and the right hand side is the shear stress at x arising from the $\gamma(w)$. The P in front of the integral signifies it is a Cauchy principal value integral because each dislocation in the distribution interacts with all the other dislocations in the distribution, but not with itself. As with the Frenkel-Kontorova model, the elastic energy is tending to make the distribution of dislocations as wide as possible, so that the gradient of the disregistry is minimised. But $\gamma(w)$ is tending to make the distribution of dislocations as narrow as possible. The actual distribution is determined by a balance between these two opposing tendencies.

The boundary conditions on the disregistry are that $w(-\infty) = 0$ and $w(+\infty) = a$. Peierls[12] guessed the following solution and showed that it solved eq 7.19:

$$w(x) = (a/\pi)\tan^{-1}(x/\zeta) + a/2 \tag{7.20}$$

since it satisfies the boundary conditions. The derivative of $w(x)$ shows that the distribution $f(x)$ of the Burgers vector is a Lorentzian:

$$f(x) = \frac{1}{\pi}\frac{\zeta}{x^2+\zeta^2}. \tag{7.21}$$

The full width at half maximum of this distribution is 2ζ. Therefore, we may define 2ζ as the width of the dislocation core.

Exercise 7.3 Verify that eq 7.20 solves eq 7.19 provided

$$\zeta = \frac{\mu a^2}{4\pi^2 V(1-\nu)}. \tag{7.22}$$

Hint: differentiate eq 7.20 and substitute the derivative dw/dx into the principal value integral, which may be evaluated using contour integration. When the expression for $w(x)$ is substituted into the right hand-side of eq 7.19 it is found the equation is satisfied provided eq 7.22 holds. It is customary to invoke the Frenkel argument of section 6.2 to express the amplitude as $V = \mu a^2/(2\pi^2 d)$, where d is the spacing of atomic planes parallel to the slip plane. Then we obtain:

$$\zeta = \frac{d}{2(1-\nu)}.$$

[12] Peierls, R, Proc. Phys. Soc. **52**, 34 (1940).

Show that the misfit energy in eq 7.14 is given by:

$$E_c = \pi V \zeta = \frac{\mu a^2}{4\pi(1-\nu)}. \tag{7.23}$$

■

Solution Differentiating eq 7.20 we get

$$\frac{\mathrm{d}w}{\mathrm{d}x} = \frac{a}{\pi}\frac{\zeta}{\zeta^2+x^2}.$$

Substituting this derivative into the integral in eq 7.19 we obtain:

$$I = \frac{a\zeta}{\pi}\int_{-\infty}^{\infty}\mathrm{d}x'\frac{1}{(x-x')}\frac{1}{(x'+i\zeta)}\frac{1}{(x'-i\zeta)}.$$

To evaluate this integral consider the following contour integral in the complex plane:

$$J = \oint_C \mathrm{d}z\frac{1}{(z-x)}\frac{1}{(z+i\zeta)}\frac{1}{(z-i\zeta)},$$

where the contour C includes the real axis from $-\infty$ to $+\infty$ and is closed by a semi-circle in the upper half-plane. There are simple poles at $z = \pm i\zeta$ on the imaginary axis and at $z = x$ on the real axis. There is a small semi-circle over the pole at $z = x$, so that the pole is outside the contour. Only the pole at $z = i\zeta$ lies within the contour. The contribution to the contour integral from the large semi-circle is zero. The contribution from the small semi-circle over the pole on the real axis at $z = x$ is $-\pi i/(x^2+\zeta^2)$. Invoking Cauchy's theorem we obtain:

$$\begin{aligned} -I - \frac{\pi i}{x^2+\zeta^2} &= 2\pi i\frac{1}{2i\zeta}\frac{1}{(i\zeta-x)} \\ I &= \frac{\pi}{\zeta}\frac{x}{x^2+\zeta^2}. \end{aligned}$$

The left hand side of eq 7.19 is then:

$$\frac{\mu a}{2\pi(1-\nu)}\frac{x}{x^2+\zeta^2}.$$

Substituting $w(x)$ from eq 7.20 into the right hand side of eq 7.19 we obtain:

$$2\pi V \frac{\zeta}{a} \frac{x}{x^2+\zeta^2}.$$

The two sides of eq 7.20 are equal provided

$$\zeta = \frac{1}{4\pi^2(1-\nu)} \frac{\mu a^2}{V}.$$

For $w(x)$ defined by eq 7.20 we find:

$$\sin\left(\frac{\pi w(x)}{a}\right) = \frac{\zeta}{\sqrt{\zeta^2+x^2}}.$$

The total misfit energy per unit length of dislocation is then:

$$E_c = V\zeta \int_{-\infty}^{\infty} \mathrm{d}x \frac{\zeta}{x^2+\zeta^2} = \pi V \zeta = \frac{\mu a^2}{4\pi(1-\nu)}.$$

Having replaced a Volterra edge dislocation with its singular core by a more realistic dislocation with a finite core width, the stress and strain fields of the dislocation no longer become infinite at the centre of the dislocation. This may be seen by convolving the components of the stress tensor for a Volterra edge dislocation in eq 6.33 with the Burgers vector distribution function $af(x)$ given by eq 7.21:

$$\sigma_{11}(x_1,x_2) = -\frac{\mu a}{2\pi(1-\nu)} \int_{-\infty}^{\infty} \mathrm{d}x_1' \frac{x_2\left(3(x_1-x_1')^2+x_2^2\right)}{\left((x_1-x_1')^2+x_2^2\right)^2} \frac{\zeta}{\pi\left(\zeta^2+(x_1')^2\right)}$$

$$\sigma_{22}(x_1,x_2) = \frac{\mu a}{2\pi(1-\nu)} \int_{-\infty}^{\infty} \mathrm{d}x_1' \frac{x_2\left((x_1-x_1')^2-x_2^2\right)}{\left((x_1-x_1')^2+x_2^2\right)^2} \frac{\zeta}{\pi\left(\zeta^2+(x_1')^2\right)}$$

$$\sigma_{12}(x_1,x_2) = \frac{\mu a}{2\pi(1-\nu)} \int_{-\infty}^{\infty} \mathrm{d}x_1' \frac{(x_1-x_1')\left((x_1-x_1')^2-x_2^2\right)}{\left((x_1-x_1')^2+x_2^2\right)^2} \frac{\zeta}{\pi\left(\zeta^2+(x_1')^2\right)},$$

These integrals may be evaluated using contour integration to yield the following stress components:

$$
\begin{aligned}
\sigma_{11}(x_1,x_2) &= -\frac{\mu a}{2\pi(1-\nu)}\left(\frac{3x_2+2\zeta}{x_1^2+(x_2+\zeta)^2}-\frac{2x_2(x_2+\zeta)^2}{\left(x_1^2+(x_2+\zeta)^2\right)^2}\right)\\
\sigma_{22}(x_1,x_2) &= \frac{\mu a}{2\pi(1-\nu)}\left(\frac{2x_1^2x_2}{\left(x_1^2+(x_2+\zeta)^2\right)^2}-\frac{x_2}{x_1^2+(x_2+\zeta)^2}\right)\\
\sigma_{12}(x_1,x_2) &= \frac{\mu a}{2\pi(1-\nu)}\left(\frac{x_1}{x_1^2+(x_2+\zeta)^2}-\frac{2x_1x_2(x_2+\zeta)}{\left(x_1^2+(x_2+\zeta)^2\right)^2}\right)\\
\sigma_{33}(x_1,x_2) &= \nu(\sigma_{11}(x_1,x_2)+\sigma_{22}(x_1,x_2))
\end{aligned}
\tag{7.24}
$$

The presence of the finite value of ζ in the denominators of these expressions removes the singularity, which is present in the stress field of a Volterra dislocation as $x_1^2+x_2^2 \to 0$. Notice also that as $\zeta \to 0$ these stress components converge to those of the Volterra dislocation in eq 6.33.

7.3.1 Comments on the Peierls-Nabarro model

Like its predecessor, the Frenkel-Kontorova model, the Peierls-Nabarro model captures the physics of the balance between the misfit energy tending to make the dislocation as narrow as possible and the elastic energy tending to make it as wide as possible. In both cases this is achieved by introducing new physics into the Volterra treatment of a dislocation, namely atomic interactions across the slip plane, and a length scale, namely the atomic spacing. Whereas the Frenkel-Kontorova model involves just two interacting linear chains of atoms, the Peierls-Nabarro model treats the dislocation in a three dimensional medium. In both models the misfit energy arising from the distortion of bonds across the slip plane is approximated by a simple sinusoidal form. This is undoubtedly a gross simplification, as atomistic calculations of γ-surfaces have shown for a variety of slip planes in different crystal structures. The small value of the calculated width 2ζ of the Burgers vector distribution leads to strains near the dislocation centre that are far too large to be described by linear elasticity. It also raises serious doubts about the neglect of gradient terms in eq 7.14 for the misfit energy. But the most serious weakness of both the Frenkel-Kontorova and Peierls-Nabarro models is the presumption that the dislocation core is *planar*. As Vitek[13] has stressed, in any given crystal there may be some orientations of the dislocation line where the cores are planar, but there may be other orientations where the cores are non-planar. Orientations

[13]Vitek, V, private communication (2019).

of dislocations with non-planar cores govern plastic deformation because they require higher stresses to make them glide. For example, in bcc metals screw dislocations have non-planar cores, whereas edge dislocations have planar cores. Plastic deformation in these metals is limited by the motion of screw dislocations[14].

The genesis of the Peierls-Nabarro model has been described by Peierls[15]. He explained that the key idea of embedding a layer in the slip plane, where the misfit is treated atomistically, between elastic continua, was conceived by neither Peierls nor Nabarro, but by Orowan. Orowan approached Peierls for help with the mathematical formulation of the model. Peierls derived the integral equation, eq 7.19, but not by the method followed here. Seeing its non-linear form he thought it was probably insoluble. He guessed the arctan solution, eq 7.20, and when he inserted it into the integral equation he was amazed to discover it was *the* solution. He introduced an error of a factor of two in the algebra associated with the stress required to move the dislocation, which appears in a large exponent and therefore had dramatic consequences. The error was corrected by Nabarro seven years later[16], and thereafter it was known as the Peierls-Nabarro model. Peierls had tried to persuade Orowan to publish the 1940 paper under Orowan's name alone, or as a joint paper, because Orowan had conceived the model. But Orowan refused. Peierls' concern about the authorship of the 1940 paper grew as its fame increased, and he felt he should have insisted Orowan was at least a coauthor, if not the sole author.

7.4 Stacking faults and partial dislocations

Sometimes there are local minima in the γ-surface in addition to global minima. Suppose there is a local minimum at $\mathbf{t} = \mathbf{t}_0$ relative to the nearest global energy minimum. A metastable planar fault in the crystal may arise at which there is a relative translation of the crystals on either side of the fault equal to $\mathbf{t}_0$. It sometimes happens that the fault may also be created through a local change in the normal sequence of the stacking of atomic planes parallel to the fault. It is then called a stacking fault. When the energy associated with these faults is relatively small they may enable dislocations to reduce their elastic energy by dissociating into two or more partial dislocations separated by one or more faults.

We will illustrate these ideas with the example of the dissociation of crystal lattice

[14]Vitek, V, Progress in Materials Science **56**, 577-585 (2011).

[15]Peierls, R E, Proc. R. Soc. London A **371**, 28-38 (1980).
A slightly more complete account by Peierls is given on pages xiii-xiv of *Dislocation Dynamics*, eds. A R Rosenfield, G T Hahn, A L Bement Jr and R I Jaffee. McGraw-Hill: New York (1968). Peierls' original derivation of the integral equation and his solution is transcribed on pages xvii to xx of this book.

[16]Nabarro, F R N, Proc. Phys. Soc. **59**, 256 (1947).

dislocations into partial dislocations in face centred cubic lattices. Slip takes place in these lattices on $\{111\}$ planes with Burgers vectors $\frac{1}{2}\langle 110\rangle$, where we have set the lattice constant to 1. In the γ-surface of the (111) plane there are global minima at the three lattice vectors $\frac{1}{2}[1\bar{1}0], \frac{1}{2}[10\bar{1}]$ and $\frac{1}{2}[01\bar{1}]$ and integer combinations of these vectors, where there is no misfit and the energy γ may be set to zero. Relative to each of these lattice vectors there are three local minima at $\mathbf{t} = \frac{1}{6}[2\bar{1}\bar{1}], \frac{1}{6}[\bar{1}2\bar{1}]$ and $\frac{1}{6}[\bar{1}\bar{1}2]$, where the energy $\gamma(\mathbf{t}) = \gamma_{SF} > 0$. Therefore, faults may exist in these lattices with $\frac{1}{6}\langle 211\rangle$ fault vectors on $\{111\}$ planes. In addition to being created by these relative translations they may also be created by altering the stacking sequence of $\{111\}$ planes by removing a plane, so that the perfect crystal sequence $\ldots ABCABCABC\ldots$ becomes $\ldots ABCBCABCA\ldots$. These particular faults are therefore *stacking faults*.

Consider a dislocation with Burgers vector $\frac{1}{2}[1\bar{1}0]$ on the (111) slip plane. This dislocation may dissociate into two partial dislocations, called Shockley partials, as follows:

$$\frac{1}{2}[1\bar{1}0] \to \frac{1}{6}[2\bar{1}\bar{1}] + \frac{1}{6}[1\bar{2}1]. \tag{7.25}$$

Thus, a dislocation with Burgers vector $\frac{1}{2}[1\bar{1}0]$ becomes two dislocations with Burgers vectors $\frac{1}{6}[2\bar{1}\bar{1}]$ and $\frac{1}{6}[1\bar{2}1]$ separated by a ribbon of stacking fault. The driving force for the dissociation is the reduction of the elastic energy of the dislocation which depends on the square of the magnitude of the Burgers vector. The separation of the partial dislocations is determined by a balance between the Peach-Koehler force of repulsion between them and the magnitude of the stacking fault energy γ_{SF}. In face centred cubic crystals with large stacking fault energies, such as aluminium, the separation of the partials, if it exists, may be too small to resolve experimentally. In silver the stacking fault energy is relatively small and the partial dislocations are clearly resolved in the transmission electron microscope.

In isotropic elasticity the force per unit length acting between two parallel, straight dislocations with Burgers vectors $\mathbf{b}_1$ and $\mathbf{b}_2$ and line direction $\hat{\boldsymbol{\xi}}$, with a separation S, is as follows[17]:

$$F_{el} = \frac{\mu}{2\pi S}(\mathbf{b}_1 \cdot \hat{\boldsymbol{\xi}})(\mathbf{b}_2 \cdot \hat{\boldsymbol{\xi}}) + \frac{\mu}{2\pi(1-\nu)S}\left[(\mathbf{b}_1 \times \hat{\boldsymbol{\xi}}) \cdot (\mathbf{b}_2 \times \hat{\boldsymbol{\xi}})\right]. \tag{7.26}$$

If the separation increases by $\mathrm{d}S$ the change in the energy per unit length of the elastic interaction between the dislocations is $-F_{el}\mathrm{d}S$. At the same time the change in the energy of the stacking fault, per unit length of dislocation line, is $\gamma_{SF}\mathrm{d}S$. At equilibrium

[17]Anderson, P M, Hirth, J P and Lothe, J, *Theory of dislocations*, 3rd edn., Cambridge University Press: Cambridge and New York (2017), p.110. ISBN 978-0-521-86436-7. Peter M Anderson, US materials engineer. John Price Hirth 1930- , US materials engineer. Jens Lothe 1931-2016, Norweigan physicist.

these two changes cancel exactly, so that $\gamma_{SF} = F_{el}$. Thus, a measurement of S enables the stacking fault energy to be determined. The tails of the Burgers vector distributions associated with each partial dislocation extend into the stacking fault slightly, which introduces an error that diminishes with increasing S/ζ.

The existence of partial dislocations separated by stacking faults reduces the ability of screw dislocations to cross slip because the partials have to be forced to form a constriction before they can cross slip[18]. The ease of cross slip has consequences for macroscopic plasticity because it enables dislocations to overcome obstacles and it may also lead to new slip systems being activated. It is remarkable that the origin of these macroscopic properties can be traced to local minima in the γ-surfaces of slip planes, which in turn are determined by the quantum mechanical behaviour of electrons in the distorted bonding environments sampled at each point of the γ-surface.

7.5 The static friction stress on a dislocation

Peierls and Nabarro calculated the minimum stress required to initiate glide of an edge dislocation on its slip plane. This is a static friction stress originating from the discrete atomic structure of the crystal, and it is often called a Peierls stress or lattice friction stress. In section 11.3 we will consider sources of dynamic friction a dislocation has to overcome to keep gliding once it has overcome the Peierls stress.

Consider the glide of the edge dislocation shown in Fig 7.1. As the dislocation centre moves from one atomic row to the next along the slip plane its elastic energy does not change if we assume the disregistry $w(x)$ is rigidly translated with the dislocation with no change to its functional form. But the misfit energy E_c changes as bonds are stretched across the slip plane, switch from one neighbour to the next, and return to their original length. The maximum slope of this periodic function is the stress required to initiate dislocation glide.

It had been presumed since 1934, when dislocations were proposed as the agents of plastic deformation, that the stress required to initiate movement dislocations in metals was several orders of magnitude less than the shear modulus. But it was not demonstrated theoretically until Peierls calculated it. To evaluate the Peierls stress we have to evaluate the misfit energy E_c as a discrete sum rather than a continuous integral, as the dislocation centre moves along the slip plane.

To evaluate the misfit energy E_c as a sum of discrete interactions between atoms we assume eq 7.20 for the disregistry $w(x)$ remains valid as the dislocation moves. The calculation proceeds by evaluating the energy per unit length of each row of atoms on either side of the slip plane. The misfit energy $\gamma(w)$ per unit area given by eq 7.13

[18]see section 2.4 of L P Kubin, *Dislocations, mesoscale simulations and plastic flow*, Oxford University Press: Oxford (2013). ISBN 978-0-19-852501-1. Ladislas P Kubin, French materials scientist, 1943-2022.

is shared by all atoms in the two atomic planes on either side of the slip plane. The energy per unit length per row of atoms on either side of the slip plane is then given by:

$$\begin{aligned}
E_{row} &= \frac{aV}{2}\sin^2\left(\frac{\pi w(x)}{a}\right) \\
&= \frac{\mu a^3}{4\pi^2 d}\sin^2\left(\frac{\pi w(x)}{a}\right) \\
&= \frac{\mu a^3}{8\pi^2 d}\left[1-\cos\left(\frac{2\pi}{a}\left(\frac{a}{\pi}\tan^{-1}\left(\frac{x}{\zeta}\right)+\frac{a}{2}\right)\right)\right] \\
&= \frac{\mu a^3}{8\pi^2 d}\left(1+\cos\left(2\tan^{-1}\left(\frac{x}{\zeta}\right)\right)\right) \\
&= \frac{\mu a^3}{4\pi^2 d}\frac{\zeta^2}{\zeta^2+x^2}.
\end{aligned} \tag{7.27}$$

Before the dislocation is introduced the positions of atomic rows on say the lower side of the slip plane are $x_n = na$ and on the upper side they are $x_n = (n+\frac{1}{2})a$. Assuming the disregistry $w(x)$ is accommodated equally by the two crystal halves, the positions of atomic rows on the lower side of the slip plane become $x_n = na + \beta a/2$ and on the upper side $x_n = (n+\frac{1}{2})a - \beta a/2$, where $0 \leq \beta \leq 1$. The misfit energy then becomes:

$$\begin{aligned}
E_c &= \frac{\mu a^3}{4\pi^2 d}\frac{\zeta^2}{a^2}\left(\sum_{n=-\infty}^{\infty}\frac{1}{\frac{\zeta^2}{a^2}+\left(n+\frac{1}{2}-\frac{\beta}{2}\right)^2}+\frac{1}{\frac{\zeta^2}{a^2}+\left(n+\frac{\beta}{2}\right)^2}\right) \\
&= \frac{\mu a^3}{4\pi^2 d}\frac{\zeta^2}{a^2}\frac{\pi a}{\zeta}\left(\frac{\sinh(2\pi\zeta/a)}{\cosh(2\pi\zeta/a)+\cos(\pi\beta)}+\frac{\sinh(2\pi\zeta/a)}{\cosh(2\pi\zeta/a)-\cos(\pi\beta)}\right) \\
&= \frac{\mu a^2}{4\pi(1-\nu)}\frac{\sinh(4\pi\zeta/a)}{\cosh(4\pi\zeta/a)-\cos(2\pi\beta)} \\
&\approx \frac{\mu a^2}{4\pi(1-\nu)}\left(1+2\cos(2\pi\beta)\exp(-4\pi\zeta/a)\right).
\end{aligned} \tag{7.28}$$

As expected the energy is a periodic function of β with the periodicity of the crystal lattice. The periodic variations in the energy is called the Peierls potential and its amplitude is equal to $[\mu a^2/(2\pi(1-\nu))]e^{-4\pi\zeta/a}$. The stress required to initiate glide is determined by the maximum value of $(1/a^2)\partial E_c/\partial\beta$, which is called the Peierls stress,

σ_P:

$$\sigma_P = \frac{\mu}{(1-\nu)} e^{-4\pi\zeta/a}. \tag{7.29}$$

Provided a dislocation has a planar core eq 7.29 explains why the resolved shear stress required to initiate glide in single crystals of pure metals is so much less than the shear modulus. The Peierls stress is smaller for dislocations with smaller Burgers vectors in slip planes with larger interplanar spacings. These predictions are broadly consistent with the selection of slip systems found experimentally in a wide range of crystals.

The details of the Peierls analysis may be criticised for many reasons, but the central ideas have been seminal. The existence of a minimum stress required to initiate glide over the periodic energy surface, E_c in eq 7.28, is a key distinction between dislocation motion in a crystal lattice as compared to a continuum. The maxima and minima in this energy surface are often referred to as Peierls barriers and Peierls valleys respectively. The undulations in this energy surface reflect the changes in bonding between atoms on either side of the slip plane as the dislocation glides[19]. When the core is very narrow a small number of atomic rows undergo a short sequence of relatively large shear displacements on either side of the slip plane. With a wider core a larger number of atomic rows undergo a longer sequence of smaller shear displacements, which require a smaller stress σ_P to bring about.

In crystals with strong bonding the Peierls stress of dislocations even with planar cores can be very high. For example, diamond has a face centred cubic lattice with the same slip systems as copper, but dislocation motion in diamond at room temperature is much more difficult because the Peierls barriers are much larger owing to the strong covalent bonding. In this case dislocation motion also can occur only through a kink mechanism, which requires thermal activation.

Vitek and Paidar[20] have shown[21] that non-planar dislocation cores are common in crystalline materials, even when atomic bonding is not directional. The modern approach to Peierls barriers is to use atomistic modelling of dislocation motion with a variety of models of atomic interactions from empirical potentials to density functional theory. Since bond breaking and making are always involved in dislocation motion, methods that treat the quantum mechanical nature of bonding explicitly should be more

[19]Note that unlike the γ-surface the Peierls energy surface is not a property of the slip plane alone because it also depends on the orientation of the dislocation line: there are different Peierls barriers and valleys for dislocations with different line directions in the same slip plane.

[20]Vaclav Paidar 1946 -, Czech materials physicist.

[21]Vitek, V and Paidar, V, *Dislocations in solids* ed. F R N Nabarro, **14**, 439-514 (2008). Elsevier: Amsterdam. ISBN 9780444531667.

(a)
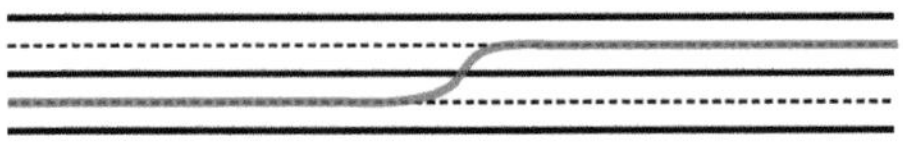
(b)
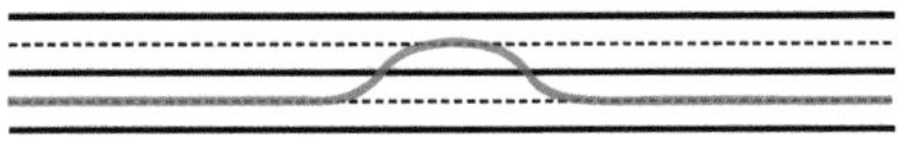
(c)
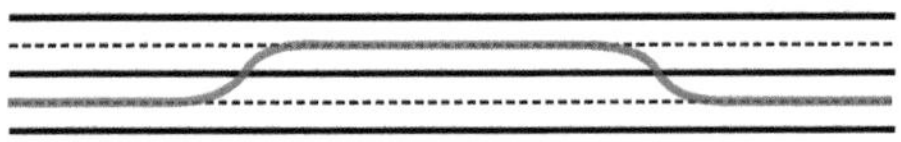

Figure 7.4: Schematic illustrations of kinks. Black solid lines are Peierls barriers. Black broken lines are Peierls valleys. The dislocation line is shown in red. If the single kink in (a) moves the left(right) the dislocation line moves up(down). In (b) a kink pair is nucleated and separates in (c) enabling a segment of the dislocation line to move up.

reliable, *provided* the size of the system modelled is sufficient for the calculation to be credible.

7.6 Dislocation motion by a kink mechanism

Dislocations are the agents of slip because they localise in the dislocation core the far more extensive bond breaking and making that would occur if one plane of atoms were to slide *en masse* over another. But when the Peierls barrier is too large for the dislocation line itself to move *en masse* over the barrier the bond breaking and making is further localised to a small region, called a kink, where the dislocation crosses the Peierls barrier. The motion of the dislocation line over the Peierls barrier is then effected by the sideways propagation of the kink along the Peierls barrier, as illustrated in Fig 7.4a. The energy of the kink along the line is also periodic and the maxima are called secondary Peierls barriers. The further localisation of the bond breaking and making associated with the motion of a kink reduces the barrier to dislocation motion significantly. Movement of the kink over the secondary Peierls barriers may be thermally activated, with an activation Gibbs free energy of migration, G_m.

Kinks are geometrically necessary when the direction of a dislocation line is inclined to the Peierls valleys. Changes in direction of the dislocation line are effected through variations in the number and sense of kinks per unit length of dislocation line. Kink pairs may be introduced into a straight dislocation by a nucleation process as illustrated in Fig 7.4b and c.

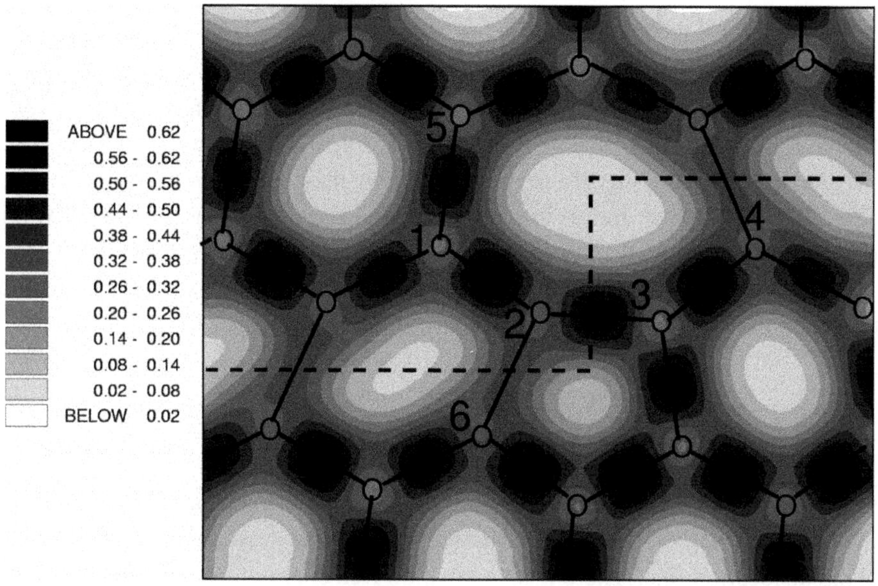

Figure 7.5: Valence electronic charge density (in electrons·Å^{-3}), viewed in the slip plane, of a relaxed kink on a Shockley partial edge dislocation in silicon calculated quantum mechanically with density functional theory. The broken line is the centre of the dislocation. Atoms (circles) lie either slightly above or below the plane of the figure. Solid lines signify bonds between atoms. Reprinted with permission from Valladares, A, White, J A and Sutton, A P, Phys. Rev. Letts., **81**, 4903-4906 (1998). Copyright 1997 by the American Physical Society.

As an example of a calculation using current computational techniques, Fig 7.5 shows the relaxed structure of a kink on a Shockley partial dislocation in silicon calculated with density functional theory. The bonds within the kink are fully reconstructed and comparable in strength to those in the crystal and along the dislocation line. For the kink to move one crystal period to the left the bond between atoms 1 and 2 has to be rotated into the orientation of the bond between atoms 3 and 4. When this rotation was applied in nine equal steps, using a constrained energy minimisation procedure, the activation energy for migration of the kink was found to be 1.1 eV, in agreement with experiment.

Since kinks are defects on linear objects they are point defects, and the thermodynamics of points defects applies to them. The lowest energy state of a dislocation at absolute zero is achieved when it lies in a Peierls valley along its entire length. At a finite temperature kink pairs may be nucleated through thermal fluctuations along the dislocation. Let the Gibbs free energy of formation of a kink pair be $G^f_{kp} = H^f_{kp} - TS^f_{kp}$, where H^f_{kp} is the enthalpy of formation and S^f_{kp} is the vibrational entropy of formation of a kink pair. Then if λ is the period of the atomic structure along the dislocation line in the Peierls valley there are $1/\lambda$ possible sites for the nucleation of a kink pair per unit length of dislocation line. In each kink pair we may designate one kink as positive and the other as negative. Let c_+ and c_- be the number of positive and negative kinks per unit length of dislocation line. In thermal equilibrium the rate at which positive and negative kinks annihilate each other is equal to the rate at which they are generated. This leads to the following law of mass action[22]:

$$c_+c_- = \frac{1}{\lambda^2}\exp(-G_{kp}/k_BT) \tag{7.30}$$

Once the separation between the positive and negative kinks of a pair exceeds a few atomic spacings they escape their mutual attraction and they perform a random walk along the dislocation line. There is an activation free energy for migration of the kinks, which in general is not the same for positive and negative kinks. In the presence of a resolved shear stress on the slip plane in the slip direction the nucleation and migration of the kinks is biased and the dislocation acquires a drift velocity, which is proportional to the resolved shear stress[23]. Since both the formation and migration of kinks are thermally activated plasticity becomes easier with increasing temperature.

The macroscopic strain rate achieved by dislocation motion depends on the average velocity of mobile dislocations, which varies linearly in the kink mechanism with stress. Conversely if a strain rate is imposed on the material the stress required to maintain the strain rate will increase as the strain rate increases. Here we have an explanation for the dependence of the stress required to achieve plastic deformation on the rate at which it is applied.

7.7 Problems

Problem 7.1 In this question we see how the energy for the two-chain Frenkel-Kontorova model in eq 7.2 arises from a description of atomic interactions. To be specific we

[22]It may be helpful to compare this with the concentrations, n and p, of electrons and holes in an intrinsic semiconductor with a band gap of E_g, where $np = \text{constant} \times \exp(-E_g/k_BT)$. In an n-type semiconductor this relation still holds but $n \gg p$. The analogue of doping in the case of kinks is to tilt the dislocation line to some angle to the Peierls valley. If it is tilted in a positive sense then $c_+ \gg c_-$

[23]for a detailed discussion see Anderson, P M, Hirth, J P and Lothe, J, *Theory of dislocations*, 3rd edn., Cambridge University Press: Cambridge and New York (2017), section 15.2. ISBN 978-0-521-86436-7.

assume atoms interact through Lennard-Jones pair potentials:

$$E_{LJ} = \frac{1}{2}\sum_i \sum_{j \neq i} \varepsilon \left[\left(\frac{r_0}{r_{ij}}\right)^{12} - 2\left(\frac{r_0}{r_{ij}}\right)^{6} \right] \tag{7.31}$$

where ε and r_0 are parameters with the dimensions of energy and length respectively. i and j label atoms in the two chains, and r_{ij} is the distance between atoms i and j. Since the Frenkel-Kontorova model assumes harmonic springs exist only between neighbouring atoms in either chain we also assume that atoms in the same chain interact only with their nearest neighbours. In that case show that r_0 must equal a, and $\kappa = 72\varepsilon/a^2$.

Let the separation of the chains be ρ. Let atoms in the first chain be at $x = na$, where n is an integer. Let the atoms in the second chain be at $x = na + \Delta$ where Δ is a constant. The distance between the atom at $x = 0$ in the first chain and atom n in the second chain is then $\left(\rho^2 + (na+\Delta)^2\right)^{1/2}$. The energy of interaction between the atom at $x = 0$ in the first chain and all atoms in the second chain is then:

$$E_{int} = \varepsilon \sum_{n=-\infty}^{\infty} \left\{ \left(\frac{a^{12}}{\left[\rho^2 + (na+\Delta)^2\right]^6} \right) - 2\left(\frac{a^6}{\left[\rho^2 + (na+\Delta)^2\right]^3} \right) \right\} \tag{7.32}$$

The sum in this equation is taken over all atoms n in the second chain to ensure that the interaction energy is a smooth, periodic function of Δ. These sums may be evaluated exactly using the Sommerfeld-Watson transformation. A simpler approach is to note that the interaction energy is an even periodic function of Δ, and therefore it can be expanded as a Fourier cosine series:

$$E_{int} = c_0 + \sum_{m=1}^{\infty} c_m \cos\left(\frac{2m\pi\Delta}{a}\right) \tag{7.33}$$

The constant c_0 is of no significance. Show that for $m \neq 0$, c_m is given by the Fourier integral:

$$c_m = 2\varepsilon \int_{-\infty}^{\infty} \left[\frac{1}{\left(\zeta^2 + z^2\right)^6} - \frac{2}{\left(\zeta^2 + z^2\right)^3} \right] e^{i2m\pi z}\, \mathrm{d}z, \tag{7.34}$$

where $\zeta = \rho/a \approx 1$. There are third and sixth order poles at $z = i\zeta$. Without evaluating the integral in detail it is clear[24] it is going to be a polynomial multiplied by $\exp(-2m\pi\zeta)$.

[24]For example,

$$I_6 = \int_{-\infty}^{\infty} \frac{e^{i2m\pi z}}{(z^2+\zeta^2)^3} \mathrm{d}z = \frac{4\pi^2 m^2 \zeta^2 + 6\pi m\zeta + 3}{8\zeta^5}\, \pi\, e^{-2m\pi\zeta}.$$

Therefore, if we assume that terms with $m > 1$ in eq 7.33 are negligible then E_{int} may be expressed approximately as $V\sin^2(\pi\Delta/a)$, ignoring the unimportant term which is independent of Δ. Equation 7.2 then follows if we make a local approximation for Δ, i.e. if we ignore the contributions to the interaction energy arising from gradients in Δ.

Solution Since r_0 is the separation at which the LJ potential is a minimum, and interactions between atoms in the same chain are limited to nearest neighbours only, we must equate r_0 to a.

The spring constant κ is the second derivative of the LJ potential evaluated at the minimum, $r = r_0 = a$:

$$\left(\frac{\mathrm{d}^2 V_{LJ}}{\mathrm{d}r^2}\right)_{r=a} = \varepsilon\left(\frac{156a^{12}}{a^{14}} - \frac{84a^6}{a^8}\right) = 72\frac{\varepsilon}{a^2}.$$

The Fourier coefficient c_m is evaluated as follows:

$$c_m = \frac{2\varepsilon}{a}\int_0^a \mathrm{d}\Delta \sum_{n=-\infty}^{\infty}\left\{\frac{a^{12}}{\left(\rho^2+(na+\Delta)^2\right)^6} - 2\frac{a^6}{\left(\rho^2+(na+\Delta)^2\right)^3}\right\}\cos\left(\frac{2m\pi\Delta}{a}\right)$$

Let $na + \Delta = s$. Then we get

$$\begin{aligned} c_m &= \frac{2\varepsilon}{a}\sum_{n=-\infty}^{\infty}\int_{na}^{(n+1)a}\mathrm{d}s\left\{\frac{a^{12}}{\left(\rho^2+s^2\right)^6} - 2\frac{a^6}{\left(\rho^2+s^2\right)^3}\right\}\cos\left(\frac{2m\pi s}{a}\right) \\ &= 2\varepsilon\int_{-\infty}^{\infty}\mathrm{d}z\left\{\frac{1}{\left(z^2+\zeta^2\right)^6} - 2\frac{1}{\left(z^2+\zeta^2\right)^3}\right\}e^{i2m\pi z}, \end{aligned}$$

where $z = s/a$ and $\zeta = \rho/a$.

Problem 7.2 In the Frenkel-Kontorova model let $\gamma(w(x))$ be the misfit energy associated with the disregistry at x, where $w(x) = u(x) - v(x)$. In the continuum approximation the

---

and

$$I_{12} = \int_{-\infty}^{\infty}\frac{e^{i2m\pi z}}{(z^2+\zeta^2)^6}\mathrm{d}z = -\frac{1}{60}\frac{\mathrm{d}^3 I_6}{\mathrm{d}(\zeta^2)^3}.$$

total energy of the two interacting linear chains in eq 7.3 becomes:

$$E = \int_{-\infty}^{\infty} \left\{ \frac{\kappa a^2}{2} \left[\left(\frac{\mathrm{d}u}{\mathrm{d}x} \right)^2 + \left(\frac{\mathrm{d}v}{\mathrm{d}x} \right)^2 \right] + \gamma(w(x)) \right\} \frac{\mathrm{d}x}{a}$$

By minimising the integral with respect to variations in $u(x)$ and $v(x)$ and applying the same boundary conditions as in section 7.2 show that:

$$\frac{\kappa a^2}{2} \frac{\mathrm{d}^2 w}{\mathrm{d}x^2} = \frac{\mathrm{d}\gamma}{\mathrm{d}w}. \tag{7.35}$$

Show that the misfit energy is equal to the elastic energy stored in the chains, i.e.:

$$\int_{-\infty}^{\infty} \gamma(w(x)) \frac{\mathrm{d}x}{a} = \int_{-\infty}^{\infty} \frac{\kappa a^2}{2} \left[\left(\frac{\mathrm{d}u}{\mathrm{d}x} \right)^2 + \left(\frac{\mathrm{d}v}{\mathrm{d}x} \right)^2 \right] \frac{\mathrm{d}x}{a}.$$

Hint: Integrate the left hand side by parts and use eq 7.35.

Solution Consider the independent variations δu and δv, for which $\delta w = \delta u - \delta v$. The variation in the total energy is as follows:

$$\begin{aligned}
\delta E &= \int_{-\infty}^{\infty} \frac{\mathrm{d}x}{a} \kappa a^2 \left(\frac{\mathrm{d}u}{\mathrm{d}x} \frac{\mathrm{d}(\delta u)}{\mathrm{d}x} + \frac{\mathrm{d}v}{\mathrm{d}x} \frac{\mathrm{d}(\delta v)}{\mathrm{d}x} \right) + \int_{-\infty}^{\infty} \frac{\mathrm{d}x}{a} \frac{\mathrm{d}\gamma(w)}{\mathrm{d}w} (\delta u - \delta v) \\
&= \kappa a \cancelto{0}{\left[\frac{\mathrm{d}u}{\mathrm{d}x} \delta u + \frac{\mathrm{d}v}{\mathrm{d}x} \delta v \right]_{-\infty}^{\infty}} - \kappa a^2 \int_{-\infty}^{\infty} \frac{\mathrm{d}x}{a} \left(\frac{\mathrm{d}^2 u}{\mathrm{d}x^2} \delta u + \frac{\mathrm{d}^2 v}{\mathrm{d}x^2} \delta v \right) \\
&\quad + \int_{-\infty}^{\infty} \frac{\mathrm{d}x}{a} \frac{\mathrm{d}\gamma(w)}{\mathrm{d}w} (\delta u - \delta v),
\end{aligned}$$

where we have used $\mathrm{d}u/\mathrm{d}x$ and $\mathrm{d}v/\mathrm{d}x \to 0$ as $x \to \pm\infty$. Since the variations δu and δv are independent we must have

$$\begin{aligned}
\kappa a^2 \frac{\mathrm{d}^2 u}{\mathrm{d}x^2} &= \frac{\mathrm{d}\gamma(w)}{\mathrm{d}w} \\
\kappa a^2 \frac{\mathrm{d}^2 v}{\mathrm{d}x^2} &= -\frac{\mathrm{d}\gamma(w)}{\mathrm{d}w}
\end{aligned}$$

Subtracting these equations we get:

$$\frac{\kappa a^2}{2}\frac{\mathrm{d}^2 w}{\mathrm{d}x^2} = \frac{\mathrm{d}\gamma(w)}{\mathrm{d}w}. \tag{7.36}$$

The misfit energy E_c is as follows:

$$E_c = \int_{-\infty}^{\infty} \mathrm{d}x\, \gamma(w) = \cancelto{0}{\left[x\gamma(w)\right]_{x=-\infty}^{\infty}} - \int_{-\infty}^{\infty} \mathrm{d}x\, x \frac{\mathrm{d}\gamma(w)}{\mathrm{d}w}\frac{\mathrm{d}w}{\mathrm{d}x},$$

where $\gamma(w) \to 0$ as $x \to \pm\infty$ because perfect registry is restored infinitely far from the centre of the dislocation. Using eq 7.36 we obtain:

$$\begin{aligned}
E_c &= -\frac{\kappa a^2}{2}\int_{-\infty}^{\infty} \mathrm{d}x\, x \frac{\mathrm{d}^2 w}{\mathrm{d}x^2}\frac{\mathrm{d}w}{\mathrm{d}x} \\
&= -\frac{\kappa a^2}{4}\int_{-\infty}^{\infty} \mathrm{d}x\, x \frac{\mathrm{d}}{\mathrm{d}x}\left(\frac{\mathrm{d}w}{\mathrm{d}x}\right)^2 \\
&= -\frac{\kappa a^2}{4}\cancelto{0}{\left[x\left(\frac{\mathrm{d}w}{\mathrm{d}x}\right)^2\right]_{x=-\infty}^{\infty}} + \frac{\kappa a^2}{4}\int_{-\infty}^{\infty} \mathrm{d}x \left(\frac{\mathrm{d}w}{\mathrm{d}x}\right)^2
\end{aligned}$$

where we have used $\mathrm{d}w/\mathrm{d}x \to 0$ as $x \to \pm\infty$. Therefore,

$$\begin{aligned}
E_c &= \frac{\kappa a^2}{4}\int_{-\infty}^{\infty} \mathrm{d}x \left\{\left(\frac{\mathrm{d}u}{\mathrm{d}x}\right)^2 + \left(\frac{\mathrm{d}v}{\mathrm{d}x}\right)^2 - 2\left(\frac{\mathrm{d}u}{\mathrm{d}x}\right)\left(\frac{\mathrm{d}v}{\mathrm{d}x}\right)\right\} \\
&= \frac{\kappa a^2}{2}\int_{-\infty}^{\infty} \mathrm{d}x \left\{\left(\frac{\mathrm{d}u}{\mathrm{d}x}\right)^2 + \left(\frac{\mathrm{d}v}{\mathrm{d}x}\right)^2\right\} \\
&= E_{el}
\end{aligned}$$

where we have used $w = u - v$ and $u = -v$.

Problem 7.3 Prove that for *any* function $\gamma(w)$ satisfying the integral equation in eq 7.19, the misfit energy of eq 7.14 is equal to $\mu a^2/[4\pi(1-\nu)]$. This surprising result, due originally to A J E Foreman[25], demonstrates that the total misfit energy is independent of the functional form of the γ-surface in the Peierls-Nabarro model.

Solution As in problem 7.2 the core energy may be expressed as follows:

$$E_c = \int_{-\infty}^{\infty} \mathrm{d}x\,\gamma(w) = \left[x\gamma(w)\right]_{x=-\infty}^{\infty} \rightarrow 0 \quad - \int_{-\infty}^{\infty} \mathrm{d}x\, x \frac{\mathrm{d}\gamma(w)}{\mathrm{d}w}\frac{\mathrm{d}w}{\mathrm{d}x} = - \int_{-\infty}^{\infty} \mathrm{d}x\, x \frac{\mathrm{d}\gamma(w)}{\mathrm{d}w}\frac{\mathrm{d}w}{\mathrm{d}x}.$$

It follows from the integral equation, eq 7.19, that:

$$-\frac{\mathrm{d}\gamma(w)}{\mathrm{d}w} = \frac{\mu}{2\pi(1-\nu)} P\int_{-\infty}^{\infty} \mathrm{d}x' \,\frac{(\mathrm{d}w/\mathrm{d}x)_{x=x'}}{x-x'}.$$

Inserting this equation into the expression for the misfit energy E_c we obtain:

$$\begin{aligned} E_c &= \frac{\mu}{2\pi(1-\nu)} \int_{-\infty}^{\infty} \mathrm{d}x\, P\int_{-\infty}^{\infty} \mathrm{d}x' \,\frac{x}{x-x'}\left(\frac{\mathrm{d}w}{\mathrm{d}x}\right)\left(\frac{\mathrm{d}w}{\mathrm{d}x'}\right) \\ &= \frac{\mu}{2\pi(1-\nu)} \int_{-\infty}^{\infty} \mathrm{d}x\, P\int_{-\infty}^{\infty} \mathrm{d}x' \left(1-\frac{x'}{x'-x}\right)\left(\frac{\mathrm{d}w}{\mathrm{d}x}\right)\left(\frac{\mathrm{d}w}{\mathrm{d}x'}\right) \end{aligned}$$

But swopping the integration variables x and x' the misfit energy may also be expressed as:

$$E_c = \frac{\mu}{2\pi(1-\nu)} \int_{-\infty}^{\infty} \mathrm{d}x'\, P\int_{-\infty}^{\infty} \mathrm{d}x \,\frac{x'}{x'-x}\left(\frac{\mathrm{d}w}{\mathrm{d}x'}\right)\left(\frac{\mathrm{d}w}{\mathrm{d}x}\right)$$

Adding the last two equations for the misfit energy we get:

$$E_c = \frac{1}{2}\frac{\mu}{2\pi(1-\nu)} \int_{-\infty}^{\infty} \mathrm{d}x \int_{-\infty}^{\infty} \mathrm{d}x' \left(\frac{\mathrm{d}w}{\mathrm{d}x}\right)\left(\frac{\mathrm{d}w}{\mathrm{d}x'}\right) = \frac{\mu b^2}{4\pi(1-\nu)}.$$

[25]see Nabarro, F R N, Adv. Phys. **1**, 269-394 (1952), p.360.

Comment

It is interesting to note that the elastic energy per unit length of the Peierls-Nabarro edge dislocation in a cylinder of radius $R \gg \zeta$ is :

$$E_{el} = \frac{\mu b^2}{4\pi(1-\nu)} \ln\left(\frac{R}{2\zeta}\right).$$

It follows that the elastic energy still diverges with increasing R even with a core of finite size.

Provided $R \gg \zeta$ the total energy per unit length of the Peierls-Nabarro dislocation is:

$$E_{tot} = \frac{\mu b^2}{4\pi(1-\nu)} \ln\left(\frac{eR}{2\zeta}\right).$$

8. The ellipsoidal inclusion

The summation convention is suspended in this chapter

8.1 Introduction

So far we have considered zero – , one – and two – dimensional defects, namely point defects, dislocations, planar faults and grain boundaries[1]. In this chapter we will treat a three – dimensional defect, namely an ellipsoidal inclusion. An ellipsoidal region in an infinite, homogeneous, isotropic, elastic medium undergoes a transformation which changes its shape and or size, and which establishes an elastic field in the surrounding medium. The transformed region is an ellipsoidal inclusion. During the transformation the elastic displacements and tractions are assumed to remain continuous at the interface of the inclusion. If the region were not constrained by the surrounding medium it would undergo a homogeneous, affine, *transformation strain* e^T_{ij}. The transformation strain is sometimes called an *eigenstrain* or *stress-free strain*. But since the inclusion *is* constrained by the surrounding medium, owing to the continuity of elastic displacements and tractions at its surface, what is the final elastic state of the inclusion and the surrounding medium?

Eshelby formulated a set of operations to define the problem clearly and rigorously, and solved the elastic field inside[2] and outside[3] an ellipsoidal inclusion. He assumed the inclusion and the surrounding medium have the same isotropic elastic constants.

[1]A grain boundary modelled as an array of dislocations was treated in problem 6.7.

[2]Eshelby, J D, Proc. R. Soc. A, **241**, 376-396 (1957). In the preface to *The Collected Works of J D Eshelby*, published in 2006, Xanthippi Markenscoff wrote that this paper *is the most cited reference in solid mechanics in the last fifty years and has been called by many the elasticity solution of the century.*

[3]Eshelby, J D, Proc. R. Soc. A **252**, 561-569 (1959). Eshelby wrote a review article about his work on inclusions and inhomogeneities in Progress in Solid Mechanics **2**, 89-140 (1961), which appears to be unavailable online. It is included in *The Collected Works of J D Eshelby*.

Applications include mechanical twinning and slip bands[4], where the *inclusion* is a region of the host material that undergoes a simple shear in response to an applied shear stress. It has also been applied to martensitic transformations where a region undergoes a spontaneous phase change involving a shape change with shear and dilation components in general. But perhaps its most widespread applications have been to composite materials, where fibre, sheet and particulate reinforcements are modelled as ellipsoidal inclusions. The versatility of Eshelby's analysis stems in part from the wide range of shapes that can be described by ellipsoids, including needles, discs, prolate and oblate spheroids and spheres. Mura has solved the inclusion problem in anisotropic elasticity[5].

Eshelby's elegant thought experiment to formulate the problem may be summarised as follows. Cut out and remove from the medium the ellipsoidal region $\mathcal{R}$ that is to transform. Allow it to undergo the homogeneous transformation strain e_{ij}^T without any constraint applied to its surface $\mathcal{S}$. At this point there is no stress in the inclusion or the surrounding medium. The region $\mathcal{R}$ no longer fits back into the hole from which it was removed. Return the transformed region to its original shape by applying tractions $-\sum_j \sigma_{ij}^T n_j$ to the surface $\mathcal{S}$, where $\sigma_{ij}^T = 2\mu e_{ij}^T + \lambda\delta_{ij}\sum_k e_{kk}^T$. The ellipsoidal region $\mathcal{R}$ is then inserted back into the hole and the bonds across the interface are reformed. At this point there is no strain in the medium or inclusion, but there are tractions $-\sum_j \sigma_{ij}^T n_j$ at the interface. They are annihilated by applying an equal and opposite distribution of tractions $+\sum_j \sigma_{ij}^T n_j$, which produce the constrained displacements u_i^C in the inclusion and the surrounding medium, from which strains e_{ij}^C and stresses σ_{ij}^C may be calculated. The final constrained stresses in the surrounding medium are σ_{ij}^C. But since the inclusion was already stressed by $-\sigma_{ij}^T$ the final stresses in the inclusion are $\sigma_{ij}^I = \sigma_{ij}^C - \sigma_{ij}^T$.

The approach we will take here is slightly different from Eshelby's. It centres on the explicit evaluation of two potentials both inside and outside the inclusion. This enables us to find the elastic fields in a consistent way throughout the body. We shall derive the displacement, stress and strain fields inside *and* outside the inclusion, reproducing and extending Eshelby's results. Along the way we will derive results that Eshelby quotes but does not derive. We will see there is a connection between the mathematics of the ellipsoidal inclusion and research, spanning more than two centuries after Newton, on the equilibrium shapes of celestial bodies and their gravitational interactions.

A consequence of making this chapter self-contained is that it is rather long. Recognising that not all readers will want to work through all the mathematical detail,

---

[4]The field of an ellipsoidal slip band is solved in problem 8.7

[5]Mura, T, *Micromechanics of defects in solids*, Chapter 3, Kluwer Academic Publishers: Dordrecht (1991). ISBN 90-247-3256-5.

some of it is consigned to three appendices at the end of the chapter.

8.2 The harmonic and biharmonic potentials

To calculate the constrained displacement field u_i^C created by the distribution of tractions $\sum_k \sigma_{jk}^T n_k$ at the surface of the ellipsoid, we use the isotropic elastic Green's function of eq 4.19:

$$u_i^C(\mathbf{x}) = \sum_{j,k} \int_S G_{ij}(\mathbf{x}-\mathbf{x}')\sigma_{jk}^T n_k \mathrm{d}S', \tag{8.1}$$

where the integral is over the surface of the inclusion and the prime on dS' signifies the integration variable is $\mathbf{x}'$. The field point $\mathbf{x}$ may be inside or outside the inclusion.

Using the divergence theorem the surface integral in eq 8.1 may be transformed into a volume integral where the integration variable $\mathbf{x}'$ ranges over the interior $\mathcal{R}$ of the inclusion:

$$\begin{aligned} u_i^C(\mathbf{x}) &= \sum_{j,k} \int_{\mathcal{R}} G_{ij,k'}(\mathbf{x}-\mathbf{x}')\sigma_{jk}^T \mathrm{d}V' \\ &= -\sum_{j,k} \int_{\mathcal{R}} G_{ij,k}(\mathbf{x}-\mathbf{x}')\sigma_{jk}^T \mathrm{d}V' \end{aligned} \tag{8.2}$$

In the first line, the Green's function $G_{ij}(\mathbf{x}-\mathbf{x}')$ is differentiated with respect to x_k'. In the second line it is differentiated with respect to x_k, which introduces the minus sign.

Rewriting the isotropic elastic Green's function as:

$$G_{ij}(\mathbf{x}-\mathbf{x}') = \frac{1}{4\pi\mu}\frac{\delta_{ij}}{|\mathbf{x}-\mathbf{x}'|} - \frac{1}{16\pi\mu(1-\nu)}\frac{\partial}{\partial x_i}\frac{\partial}{\partial x_j}|\mathbf{x}-\mathbf{x}'|, \tag{8.3}$$

the constrained displacement field of eq 8.2 becomes:

$$u_i^C(\mathbf{x}) = -\sum_{j,k} \sigma_{jk}^T \left\{ \frac{1}{4\pi\mu}\delta_{ij}\Phi_{,k} - \frac{1}{16\pi\mu(1-\nu)}\Psi_{,ijk} \right\}, \tag{8.4}$$

where $\Phi(\mathbf{x})$ is called the *harmonic potential*[6] and $\Psi(\mathbf{x})$ is called the *biharmonic potential.* They are defined as follows:

$$\Phi(\mathbf{x}) = \int_{\mathcal{R}} \frac{1}{|\mathbf{x}-\mathbf{x}'|}\mathrm{d}V' \tag{8.5}$$

[6] It is sometimes called the *Newtonian potential* owing to its relation to the gravitational potential.

$$\Psi(\mathbf{x}) = \int_{\mathcal{R}} |\mathbf{x}-\mathbf{x}'|\,\mathrm{d}V' \tag{8.6}$$

The integrals are over the volume $\mathcal{R}$ of the ellipsoid. We shall obtain explicit expressions for these potentials inside and outside the ellipsoid. As noted by Chandrasekhar[7], for more than two centuries after Newton introduced the universal law of gravitation, the determination of the harmonic potential inside and outside ellipsoids attracted a glittering array of mathematicians and astronomers including Laplace, Maclaurin, d'Alembert, Cayley, Legendre, Gauss, Jacobi, Rodrigues, Poisson, Ferrers, Dyson (F W), Lyapunov, Poincaré, Dirichlet and others.

Exercise 8.1 Show that

$$\nabla^2\Psi(\mathbf{x}) = 2\Phi(\mathbf{x}) \tag{8.7}$$

$$\nabla^2\Phi(\mathbf{x}) = \begin{cases} -4\pi & \text{if } \mathbf{x} \text{ is in } \mathcal{R} \\ 0 & \text{otherwise.} \end{cases} \tag{8.8}$$

Using eqs 8.4, 8.7 and Hooke's law show that $u_i^C(\mathbf{x})$ may be expressed in terms of the transformation strain as follows:

$$u_i^C(\mathbf{x}) = \frac{1}{8\pi(1-\nu)}\sum_{j,k}\left(\Psi_{,ijk} - 2\nu\Phi_{,i}\delta_{jk} - 4(1-\nu)\delta_{ij}\Phi_{,k}\right)e_{jk}^T. \tag{8.9}$$

■

Solution Expressing eq 8.6 in Cartesian coordinates we have:

$$\Psi(\mathbf{x}) = \int_{\mathcal{R}} \sqrt{(x_1-x_1')^2+(x_2-x_2')^2+(x_3-x_3')^2}\,\mathrm{d}V'.$$

The derivative $\Psi_{,1}(\mathbf{x})$ is as follows:

$$\Psi_{,1}(\mathbf{x}) = \int_{\mathcal{R}} \frac{(x_1-x_1')}{\sqrt{(x_1-x_1')^2+(x_2-x_2')^2+(x_3-x_3')^2}}\,\mathrm{d}V',$$

[7]Chandrasekhar, S, *Ellipsoidal figures of equilibrium*, Dover Publications: New York (1987). ISBN 978-0486652580. Subrahmanyan Chandrasekhar FRS 1910-1995, Nobel Prize winning Indian-US physicist.

and the second derivative $\Psi_{,11}(\mathbf{x})$ is as follows:

$$\begin{aligned}\Psi_{,11}(\mathbf{x}) &= \int_{\mathcal{R}} \frac{1}{\sqrt{(x_1-x_1')^2+(x_2-x_2')^2+(x_3-x_3')^2}}\,\mathrm{d}V' \\ &- \int_{\mathcal{R}} \frac{(x_1-x_1')^2}{\left[(x_1-x_1')^2+(x_2-x_2')^2+(x_3-x_3')^2\right]^{\frac{3}{2}}}\,\mathrm{d}V'\end{aligned}$$

Therefore,

$$\nabla^2\Psi(\mathbf{x}) = 3\Phi(\mathbf{x}) - \Phi(\mathbf{x}) = 2\Phi(\mathbf{x}).$$

Using

$$\nabla^2 \frac{1}{|\mathbf{x}-\mathbf{x}'|} = -4\pi\delta(\mathbf{x}-\mathbf{x}')$$

we find

$$\nabla^2\Phi(\mathbf{x}) = -4\pi\int_{\mathcal{R}} \delta(\mathbf{x}-\mathbf{x}')\,\mathrm{d}V' = \begin{cases} -4\pi & \text{if } \mathbf{x} \text{ is in } \mathcal{R} \\ 0 & \text{otherwise.}\end{cases}$$

Using eqs 8.4, 8.7 and Hooke's law (with $\lambda = 2\mu\nu/(1-2\nu)$), the constrained displacement field may be written as:

$$\begin{aligned} u_i^C &= -\sum_{j,k}\left(2\mu e_{jk}^T + \frac{2\nu\mu}{1-2\nu}\delta_{jk}\sum_m e_{mm}^T\right)\left(\frac{1}{4\pi\mu}\delta_{ij}\Phi_{,k} - \frac{1}{16\pi\mu(1-\nu)}\Psi_{,ijk}\right) \\ &= -\sum_{j,k}\frac{1}{2\pi}\left(\delta_{ij}\Phi_{,k} - \frac{\Psi_{,ijk}}{4(1-\nu)} + \frac{\nu}{(1-2\nu)}\Phi_{,i}\delta_{jk} - \frac{\nu}{(1-2\nu)}\frac{1}{2(1-\nu)}\Phi_{,i}\delta_{jk}\right)e_{jk}^T \\ &= \sum_{j,k}\frac{1}{8\pi(1-\nu)}\left(\Psi_{,ijk} - 4(1-\nu)\delta_{ij}\Phi_{,k} - 2\nu\Phi_{,i}\delta_{jk}\right)e_{jk}^T \end{aligned}$$

The strain e_{in}^C is obtained by differentiating the displacement field u_i^C in the usual way:

$$e_{in}^C(\mathbf{x}) = \frac{1}{2}(u_{i,n}^C + u_{n,i}^C)$$

$$= \sum_{j,k} S_{injk} e^T_{jk} \tag{8.10}$$

where

$$S_{injk} = \frac{1}{8\pi(1-\nu)} \left(\Psi_{,injk} - 2\nu\Phi_{,in}\delta_{jk} - 2(1-\nu)\left(\delta_{ij}\Phi_{,nk} + \delta_{nj}\Phi_{,ik}\right)\right) \tag{8.11}$$

is called the Eshelby tensor. Equations 8.10 and 8.11 apply inside and outside the ellipsoid, although the Eshelby tensors are different inside and outside.

In sections 8.3 and 8.4 we will see that S_{injk} is homogeneous inside the ellipsoid. Since the transformation strain is also homogeneous inside the ellipsoid the constrained stress and strain in the ellipsoid are homogeneous. Markenscoff[8] has proved[9] in isotropic and anisotropic elasticity, that an ellipsoid is the *only* shape of an inclusion within which the constrained stress and strain fields are homogeneous.

Viewed from afar the inclusion resembles a point defect with a volume V, and its remote elastic field is then easily evaluated. With the origin of the coordinate system at the centre of the inclusion, we have $|\mathbf{x}-\mathbf{x}'| \approx |\mathbf{x}|$ for all $\mathbf{x}'$ inside the inclusion. The displacement field of the inclusion is then:

$$u_i(\mathbf{x}) = -V \sum_{j,k} G_{ij,k}(\mathbf{x})\sigma^T_{jk}, \tag{8.12}$$

This limiting form is valid for any shape of inclusion, not only ellipsoidal, provided $|\mathbf{x}|$ is sufficiently large. Using eq 4.42 for $G_{ij,k}(\mathbf{x})$ the elastic displacement field far from the inclusion is as follows:

$$\begin{aligned} u_i(\mathbf{x}) &= \frac{V}{16\pi\mu(1-\nu)} \sum_{j,k} \left(2(1-2\nu)\delta_{ij}\frac{x_k}{x^3} - \delta_{jk}\frac{x_i}{x^3} + 3\frac{x_i x_j x_k}{x^5}\right)\sigma^T_{jk} \\ &= \frac{V}{8\pi(1-\nu)} \sum_{j,k} \left(2(1-2\nu)\delta_{ij}\frac{x_k}{x^3} - (1-2\nu)\delta_{jk}\frac{x_i}{x^3} + 3\frac{x_i x_j x_k}{x^5}\right)e^T_{jk}, \end{aligned} \tag{8.13}$$

where $x = |\mathbf{x}|$. By differentiating this displacement field and using Hooke's law we may derive the following limiting forms for the strain and stress fields far from the inclusion:

$$\begin{aligned} e_{ij} &= \frac{V}{8\pi(1-\nu)} \sum_{k,l} \left\{(1-2\nu)\left(\frac{2\delta_{il}\delta_{jk} - \delta_{kl}\delta_{ij}}{x^3}\right) + 6\nu\left(\frac{\delta_{il}x_j x_k + \delta_{jl}x_i x_k - \delta_{kl}x_i x_j}{x^5}\right)\right. \\ &\quad \left. + \frac{3\delta_{kl}x_i x_j + 3\delta_{ij}x_k x_l}{x^5} - 15\frac{x_i x_j x_k x_l}{x^7}\right\} e^T_{kl}. \end{aligned} \tag{8.14}$$

[8]Xanthippi Markenscoff 1947-, Greek born, US theoretical materials scientist and engineer.
[9]Markenscoff, X, J Elasticity **49**, 163-166 (1998).

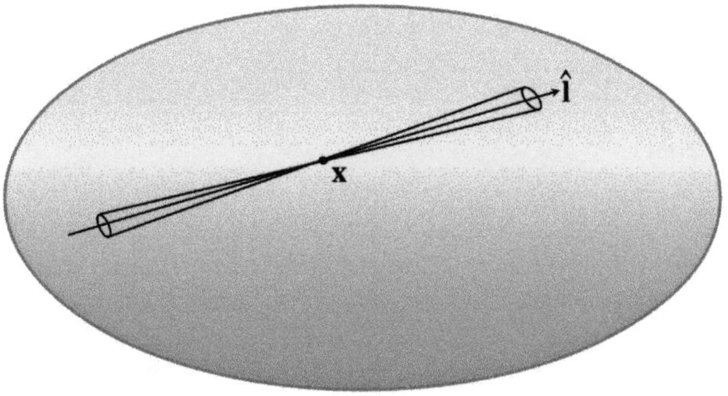

Figure 8.1: An ellipsoidal inclusion in which there are two elementary cones with vertices at $\mathbf{x}$ and axes parallel to the unit vector $\hat{\mathbf{l}}$. The bases of the cones are on the surface of the ellipsoid.

$$\sigma_{ij} = \frac{\mu V}{4\pi(1-\nu)} \sum_{k,l} \left\{ (1-2\nu)\left(\frac{2\delta_{il}\delta_{jk} - \delta_{kl}\delta_{ij}}{x^3}\right) + 2\nu\left(\frac{\delta_{ij}\delta_{kl}}{x^3} + \frac{3\delta_{il}x_j x_k + 3\delta_{jl}x_i x_k}{x^5}\right) \right.$$
$$\left. +3(1-2\nu)\frac{\delta_{kl}x_i x_j + \delta_{ij}x_k x_l}{x^5} - 15\frac{x_i x_j x_k x_l}{x^7} \right\} e^T_{kl} \quad (8.15)$$

Notice the displacement field varies as $1/x^2$ and the stresses and strains vary as $1/x^3$ when x is large. In problem 8.5 we show these limiting forms are reproduced by the exact expressions we shall derive for the elastic fields outside the inclusion when x is large compared to the inclusion.

8.3 The harmonic potential inside a homogeneous ellipsoid

We consider an ellipsoidal inclusion with semi-axes $a_1 > a_2 > a_3$, defined by the equation:

$$\frac{x_1^2}{a_1^2} + \frac{x_2^2}{a_2^2} + \frac{x_3^2}{a_3^2} = 1. \quad (8.16)$$

The origin of the cartesian coordinate system is at the centre of the ellipsoid and its axes are aligned with the principal axes of the ellipsoid. We construct two elementary cones of solid angle $d\omega$ with vertices at $\mathbf{x}$ inside the ellipsoid. Their bases are on the surface of the ellipsoid and their axes are parallel to the unit vector $\hat{\mathbf{l}}$, as shown in Fig 8.1.

The contribution of the cones to $\Phi(\mathbf{x})$ is:

$$\begin{aligned}
\mathrm{d}\Phi(\mathbf{x}) &= \int_{cones} \frac{\mathrm{d}V}{\rho} \\
&= \mathrm{d}\omega \left\{ \int_0^{R_1} \frac{\rho^2}{\rho} \mathrm{d}\rho + \int_0^{R_2} \frac{\rho^2}{\rho} \mathrm{d}\rho \right\} = \frac{1}{2} \mathrm{d}\omega \left(R_1^2 + R_2^2 \right), \qquad (8.17)
\end{aligned}$$

where ρ is the distance from $\mathbf{x}$ along the axis $\hat{\mathbf{l}}$ and $\mathrm{d}V = \rho^2 \mathrm{d}\rho \mathrm{d}\omega$. R_1 and R_2 are the distances along $\hat{\mathbf{l}}$ from $\mathbf{x}$ to the surface of the ellipsoid. Integrating over all solid angles we obtain twice the total potential at $\mathbf{x}$ because each cone is counted twice:

$$\Phi(\mathbf{x}) = \frac{1}{4} \int_S (R_1^2 + R_2^2) \mathrm{d}\omega. \qquad (8.18)$$

The coordinates of the points at which the axis of the cones intersects the surface of the ellipsoid are $(x_1 + l_1 R_1, x_2 + l_2 R_1, x_3 + l_3 R_1)$ and $(x_1 - l_1 R_2, x_2 - l_2 R_2, x_3 - l_3 R_2)$, where $R_1, R_2 \geq 0$. Since these points are on the surface of the ellipsoid they satisfy eq 8.16. It follows that R_1 and R_2 are the roots of the quadratic equation:

$$\frac{(x_1 + l_1 R)^2}{a_1^2} + \frac{(x_2 + l_2 R)^2}{a_2^2} + \frac{(x_3 + l_3 R)^2}{a_3^2} = 1. \qquad (8.19)$$

Expanding, we obtain:

$$\left(\frac{l_1^2}{a_1^2} + \frac{l_2^2}{a_2^2} + \frac{l_3^2}{a_3^2} \right) R^2 + 2 \left(\frac{l_1 x_1}{a_1^2} + \frac{l_2 x_2}{a_2^2} + \frac{l_3 x_3}{a_3^2} \right) R + \left(\frac{x_1^2}{a_1^2} + \frac{x_2^2}{a_2^2} + \frac{x_3^2}{a_3^2} - 1 \right) = 0. \qquad (8.20)$$

The coefficient of R^2 is equal to $1/r^2$, where r is the distance from the centre of the ellipsoid to the surface along the direction $\hat{\mathbf{l}}$. That is because if $(X_1, X_2, X_3) = r(l_1, l_2, l_3)$ are the coordinates of a point on the surface of the ellipsoid, relative to the centre of the ellipsoid, then $r^2 \left(l_1^2/a_1^2 + l_2^2/a_2^2 + l_3^2/a_3^2 \right) = 1$. Using this relation the sum of the roots of eq 8.20 is as follows:

$$R_1 + R_2 = -2r^2 \left(l_1 x_1/a_1^2 + l_2 x_2/a_2^2 + l_3 x_3/a_3^2 \right).$$

Therefore,

$$\frac{R_1^2 + R_2^2}{r^2} = -2 \left(\frac{l_1 x_1}{a_1^2} + \frac{l_2 x_2}{a_2^2} + \frac{l_3 x_3}{a_3^2} \right) (R_1 + R_2) + 2 \left(1 - \frac{x_1^2}{a_1^2} - \frac{x_2^2}{a_2^2} - \frac{x_3^2}{a_3^2} \right)$$

$$R_1^2 + R_2^2 \;=\; 4r^4\left(\frac{l_1x_1}{a_1^2} + \frac{l_2x_2}{a_2^2} + \frac{l_3x_3}{a_3^2}\right)^2 + 2r^2\left(1 - \frac{x_1^2}{a_1^2} - \frac{x_2^2}{a_2^2} - \frac{x_3^2}{a_3^2}\right). \qquad (8.21)$$

Substituting this expression for $R_1^2 + R_2^2$ into eq 8.18 we obtain:

$$\begin{aligned}\Phi(\mathbf{x}) &= \int_S r^4\left(\frac{l_1x_1}{a_1^2} + \frac{l_2x_2}{a_2^2} + \frac{l_3x_3}{a_3^2}\right)^2 + \frac{1}{2}r^2\left(1 - \frac{x_1^2}{a_1^2} - \frac{x_2^2}{a_2^2} - \frac{x_3^2}{a_3^2}\right)\mathrm{d}\omega \\ &= \int_S r^4\left(\frac{l_1^2x_1^2}{a_1^4} + \frac{l_2^2x_2^2}{a_2^4} + \frac{l_3^2x_3^2}{a_3^4}\right) + \frac{1}{2}r^2\left(1 - \frac{x_1^2}{a_1^2} - \frac{x_2^2}{a_2^2} - \frac{x_3^2}{a_3^2}\right)\mathrm{d}\omega. \qquad (8.22)\end{aligned}$$

In the second line cross terms, such as l_1l_2, are dropped because they integrate to zero, owing to the symmetry of the ellipsoid.

There are two integrals to be evaluated:

$$I_1 = \int_S r^2\mathrm{d}\omega \qquad (8.23)$$

$$I_{2i} = \frac{1}{a_i^4}\int_S r^4 l_i^2\mathrm{d}\omega, \text{ where } i = 1,2,3. \qquad (8.24)$$

To evaluate I_1 we introduce spherical polar coordinates:

$$\begin{aligned}x_1 &= r\sin\theta\cos\phi \\ x_2 &= r\sin\theta\sin\phi \\ x_3 &= r\cos\theta, \qquad (8.25)\end{aligned}$$

where $0 \le \theta \le \pi$ and $0 \le \phi \le 2\pi$. Invoking eq 8.16 we find the following relation between r, θ and ϕ:

$$r^2 = \frac{1}{\frac{\sin^2\theta\cos^2\phi}{a_1^2} + \frac{\sin^2\theta\sin^2\phi}{a_2^2} + \frac{\cos^2\theta}{a_3^2}} \qquad (8.26)$$

Using this expression for r^2 and $\mathrm{d}\omega = \sin\theta\mathrm{d}\theta\mathrm{d}\phi$, we obtain:

$$I_1 = \int_0^\pi \mathrm{d}\theta \int_0^{2\pi} \mathrm{d}\phi \, \frac{\sin\theta}{\frac{\sin^2\theta\cos^2\phi}{a_1^2} + \frac{\sin^2\theta\sin^2\phi}{a_2^2} + \frac{\cos^2\theta}{a_3^2}}$$

$$= 8\int_0^{\pi/2} d\theta \int_0^{\pi/2} d\phi \, \frac{\sin\theta}{\frac{\sin^2\theta\cos^2\phi}{a_1^2} + \frac{\sin^2\theta\sin^2\phi}{a_2^2} + \frac{\cos^2\theta}{a_3^2}}. \tag{8.27}$$

In the second line we have used the symmetry of the ellipsoid to reduce the integration to one octant. Making the substitution $t = \tan\phi$, I_1 becomes:

$$\begin{aligned}
I_1 &= 8\int_0^{\pi/2} d\theta \sin\theta \int_0^{\infty} dt \, \frac{1}{\left(\frac{\sin^2\theta}{a_1^2} + \frac{\cos^2\theta}{a_3^2}\right) + \left(\frac{\sin^2\theta}{a_2^2} + \frac{\cos^2\theta}{a_3^2}\right)t^2} \\
&= 8\int_0^{\pi/2} d\theta \sin\theta \frac{1}{\sqrt{\left(\frac{\sin^2\theta}{a_1^2} + \frac{\cos^2\theta}{a_3^2}\right)\left(\frac{\sin^2\theta}{a_2^2} + \frac{\cos^2\theta}{a_3^2}\right)}} \\
&\quad \times \left[\tan^{-1}\left\{\frac{\sqrt{\left(\frac{\sin^2\theta}{a_2^2} + \frac{\cos^2\theta}{a_3^2}\right)}}{\sqrt{\left(\frac{\sin^2\theta}{a_1^2} + \frac{\cos^2\theta}{a_3^2}\right)}} t\right\}\right]_{t=0}^{\infty} \\
&= 4\pi\int_0^{\pi/2} d\theta \sin\theta \frac{1}{\sqrt{\left(\frac{\sin^2\theta}{a_1^2} + \frac{\cos^2\theta}{a_3^2}\right)\left(\frac{\sin^2\theta}{a_2^2} + \frac{\cos^2\theta}{a_3^2}\right)}} \\
&= 4\pi a_1 a_2 a_3^2 \int_0^{\pi/2} d\theta \frac{\sin\theta\sec^2\theta}{\sqrt{(a_1^2 + a_3^2\tan^2\theta)(a_2^2 + a_3^2\tan^2\theta)}}
\end{aligned} \tag{8.28}$$

Substituting $u = a_3^2\tan^2\theta$ then $\cos\theta = a_3\left(a_3^2 + u\right)^{-1/2}$ and $du = 2a_3^2\sin\theta\sec^3\theta d\theta$. I_1 becomes:

$$I_1 = 2\pi a_1 a_2 a_3 \int_0^{\infty} \frac{du}{\Delta(u)}, \tag{8.29}$$

where

$$\Delta(u) = \sqrt{(a_1^2 + u)(a_2^2 + u)(a_3^2 + u)}. \tag{8.30}$$

The integral I_{2i} is obtained by differentiating I_1 with respect to a_i:

$$\frac{\partial I_1}{\partial a_i} = \int_S 2r\frac{\partial r}{\partial a_i}\mathrm{d}\omega$$

$$\frac{\partial r}{\partial a_i} = \frac{r^3}{a_i^3}l_i^2$$

$$\frac{\partial I_1}{\partial a_i} = \frac{2}{a_i^3}\int_S r^4 l_i^2 \mathrm{d}\omega = 2a_i I_{2i}, \tag{8.31}$$

where the second line is obtained from $r^2\left(l_1^2/a_1^2 + l_2^2/a_2^2 + l_3^2/a_3^2\right) = 1$. Therefore,

$$I_{2i} = \frac{1}{2a_i}\frac{\partial I_1}{\partial a_i} \tag{8.32}$$

$$= \frac{I_1}{2a_i^2} - \pi a_1 a_2 a_3 \int_0^\infty \frac{\mathrm{d}u}{(a_i^2+u)\Delta(u)}. \tag{8.33}$$

Defining

$$A_i = \pi a_1 a_2 a_3 \int_0^\infty \frac{\mathrm{d}u}{(a_i^2+u)\Delta(u)} \tag{8.34}$$

and using eqs 8.22, 8.23, 8.24 and 8.33 we obtain the following formula for the harmonic potential in a homogeneous ellipsoid:

$$\Phi(\mathbf{x}) = \frac{I_1}{2} - \sum_i A_i x_i^2. \tag{8.35}$$

This is sometimes written as follows:

$$\Phi(\mathbf{x}) = \pi a_1 a_2 a_3 \int_0^\infty \frac{1}{\Delta(u)}\left(1 - \sum_i \frac{x_i^2}{a_i^2+u}\right)\mathrm{d}u \tag{8.36}$$

The integrals may be expressed in terms of incomplete elliptic integrals of the first and second kinds:

$$\text{first kind: } F(\phi, k^2) = \int_0^\phi \frac{\mathrm{d}\theta}{\sqrt{1-k^2\sin^2\theta}}$$

$$\text{second kind: } E(\phi,k^2) = \int_0^{\phi} \sqrt{1-k^2\sin^2\theta}\,\mathrm{d}\theta. \tag{8.37}$$

With the substitution $(a_1^2+u)\sin^2\theta = (a_1^2-a_3^2)$, and setting $k^2 = (a_1^2-a_2^2)/(a_1^2-a_3^2)$ and $\phi = \cos^{-1}(a_3/a_1)$ the integral I_1 of eq 8.29 is expressed in terms of the following elliptic integral of the first kind:

$$I_1 = \frac{4\pi a_1a_2a_3}{\sqrt{a_1^2-a_3^2}} F(\phi,k^2). \tag{8.38}$$

With the same substitution the three integrals of eq 8.34 become:

$$\begin{aligned} A_1 &= \frac{2\pi a_1a_2a_3}{(a_1^2-a_3^2)^{3/2}k^2}\left[F(\phi,k^2)-E(\phi,k^2)\right] \\ A_2 &= \frac{2\pi a_1a_2a_3}{k^2(1-k^2)(a_1^2-a_3^2)^{3/2}}\left[E(\phi,k^2)-(1-k^2)F(\phi,k^2)-\frac{k^2\sin\phi\cos\phi}{(1-k^2\sin^2\phi)^{1/2}}\right] \\ A_3 &= \frac{2\pi a_1a_2a_3}{(1-k^2)(a_1^2-a_3^2)^{3/2}}\left[(\tan\phi)(1-k^2\sin^2\phi)^{1/2}-E(\phi,k^2)\right] \end{aligned} \tag{8.39}$$

The expressions for I_1, A_1, A_2, A_3 have to be evaluated only once to obtain the harmonic potential throughout the interior of the ellipsoid.

Only two of the integrals I_1, A_1, A_2 and A_3 are independent because there are two relations between them:

$$A_1a_1^2+A_2a_2^2+A_3a_3^2 = I_1/2 \tag{8.40}$$

$$A_1+A_2+A_3 = 2\pi. \tag{8.41}$$

The first of these relations may be proved as follows. Using eqs 8.23, 8.29, 8.34 and $\cos^2\theta = a_3^2/(a_3^2+u)$ we have:

$$\begin{aligned} \int_S r^2 l_3^2\,\mathrm{d}\omega &= 2\pi a_1a_2a_3\int_0^{\infty}\frac{a_3^2}{a_3^2+u}\frac{\mathrm{d}u}{\Delta(u)} \\ &= 2A_3a_3^2. \end{aligned} \tag{8.42}$$

It follows by symmetry that $\int_S r^2l_1^2\mathrm{d}\omega = 2A_1a_1^2$ and $\int_S r^2l_2^2\mathrm{d}\omega = 2A_2a_2^2$. Since $(l_1^2+l_2^2+l_3^2) = 1$, eq 8.40 follows.

To derive eq 8.41 we observe that:

$$\frac{1}{\Delta(u)}\frac{\mathrm{d}\Delta(u)}{\mathrm{d}u} = \frac{1}{2}\sum_j\frac{1}{a_j^2+u}. \tag{8.43}$$

Therefore,

$$\begin{aligned}
\sum_{j=1}^{3} A_j &= \pi a_1 a_2 a_3 \int_0^\infty \left(\sum_j \frac{1}{a_j^2+u} \right) \frac{\mathrm{d}u}{\Delta(u)} \\
&= 2\pi a_1 a_2 a_3 \int_0^\infty \frac{1}{\Delta^2(u)} \frac{\mathrm{d}\Delta(u)}{\mathrm{d}u} \mathrm{d}u \\
&= -2\pi a_1 a_2 a_3 \int_0^\infty \frac{\mathrm{d}}{\mathrm{d}u} \left(\frac{1}{\Delta(u)} \right) \mathrm{d}u \\
&= -2\pi a_1 a_2 a_3 \left[\frac{1}{\Delta(u)} \right]_{u=0}^{\infty} \\
&= 2\pi. \qquad (8.44)
\end{aligned}$$

It follows that eq 8.35 satisfies $\nabla^2\Phi(\mathbf{x}) = -4\pi$, as required for $\mathbf{x}$ inside the ellipsoid (see eq 8.8). Eqs 8.40 and 8.41 are also satisfied by the expressions for I_1 in eq 8.38 and A_1, A_2 and A_3 in eq 8.39.

8.4 The biharmonic potential inside a homogeneous ellipsoid

Repeating the steps that led to eq 8.18 we find:

$$\Psi(\mathbf{x}) = \frac{1}{8} \int_S \left(R_1^4 + R_2^4 \right) \mathrm{d}\omega. \qquad (8.45)$$

Using eqs 8.20 and 8.21 and ignoring all terms that are odd in any of the direction cosines we obtain:

$$\begin{aligned}
\Psi(\mathbf{x}) = \int_S \Bigg\{ & 2r^8 \left[\frac{x_1^4}{a_1^8} l_1^4 + \frac{x_2^4}{a_2^8} l_2^4 + \frac{x_3^4}{a_3^8} l_3^4 + 6 \left(\frac{x_1^2 x_2^2}{a_1^4 a_2^4} l_1^2 l_2^2 + \frac{x_2^2 x_3^2}{a_2^4 a_3^4} l_2^2 l_3^2 + \frac{x_3^2 x_1^2}{a_3^4 a_1^4} l_3^2 l_1^2 \right) \right] \\
& + 2g r^6 \left(\frac{x_1^2}{a_1^4} l_1^2 + \frac{x_2^2}{a_2^4} l_2^2 + \frac{x_3^2}{a_3^4} l_3^2 \right) + \frac{1}{4} g^2 r^4 \Bigg\} \mathrm{d}\omega, \qquad (8.46)
\end{aligned}$$

where $g = 1 - x_1^2/a_1^2 - x_2^2/a_2^2 - x_3^2/a_3^2$. Therefore, $\Psi(\mathbf{x})$ is an even quartic polynomial in the coordinates of $\mathbf{x}$. Since $\Phi(\mathbf{x})$ is a quadratic polynomial in the coordinates of

x it follows from eqs 8.10 and 8.11 that the constrained strain inside the ellipsoid is homogeneous.

The following integrals are needed:

$$I_2 = \int_S r^4 \mathrm{d}\omega \tag{8.47}$$

$$I_{3i} = \frac{1}{a_i^4}\int_S r^6 l_i^2 \mathrm{d}\omega, \text{ where } i = 1,2,3 \tag{8.48}$$

$$I_{4ij} = \frac{1}{a_i^4 a_j^4}\int_S r^8 l_i^2 l_j^2 \mathrm{d}\omega \text{ where } i,j = 1,2,3, \tag{8.49}$$

They are evaluated in section 8.10. Here we proceed to use the integrals to construct the biharmonic potential inside the ellipsoid.

$\Psi(\mathbf{x})$ may be rewritten as follows:

$$\Psi(\mathbf{x}) = \sum_{i,j} \xi_{ij} x_i^2 x_j^2 + \sum_i \zeta_i x_i^2 + I_2/4, \tag{8.50}$$

where $\xi_{ij} = \xi_{ji}$. The coefficient of x_1^4 is as follows:

$$\xi_{11} = 2I_{411} - 2\frac{I_{31}}{a_1^2} + \frac{I_2}{4a_1^4} = -\frac{1}{12}\left(B_{12}a_2^2 + B_{13}a_3^2\right) = \frac{1}{4}\left(B_{11}a_1^2 - A_1\right). \tag{8.51}$$

The coefficients of x_2^4 and x_3^4 are obtained by cyclic permutation of the subscripts.

We have introduced the integrals:

$$B_{ij} = \pi a_1 a_2 a_3 \int_0^\infty \frac{\mathrm{d}u}{(a_i^2+u)(a_j^2+u)\Delta(u)}. \tag{8.52}$$

For $i \neq j$ we may split the integrand into partial fractions to obtain:

$$\begin{aligned} B_{ij} &= \frac{\pi a_1 a_2 a_3}{(a_j^2 - a_i^2)}\int_0^\infty \frac{\mathrm{d}u}{(a_i^2+u)\Delta(u)} + \frac{\pi a_1 a_2 a_3}{(a_i^2 - a_j^2)}\int_0^\infty \frac{\mathrm{d}u}{(a_j^2+u)\Delta(u)} \\ &= \frac{A_i - A_j}{(a_j^2 - a_i^2)} \end{aligned} \tag{8.53}$$

For B_{ii} we observe that the integrals B_{i1}, B_{i2}, B_{i3} are related as a consequence of eq 8.43:

$$\begin{aligned}
\sum_j B_{ij} &= \pi a_1 a_2 a_3 \int_0^\infty \frac{\mathrm{d}u}{(a_i^2+u)\Delta(u)} \sum_j \frac{1}{(a_j^2+u)} \\
&= \pi a_1 a_2 a_3 \int_0^\infty \frac{\mathrm{d}u}{(a_i^2+u)} \frac{2}{\Delta^2(u)} \frac{\mathrm{d}\Delta(u)}{\mathrm{d}u} \\
&= -2\pi a_1 a_2 a_3 \int_0^\infty \frac{\mathrm{d}u}{(a_i^2+u)} \frac{\mathrm{d}}{\mathrm{d}u} \frac{1}{\Delta(u)} \\
&= 2\pi a_1 a_2 a_3 \left(\frac{1}{a_i^2} \frac{1}{a_1 a_2 a_3} - \int_0^\infty \frac{\mathrm{d}u}{(a_i^2+u)^2} \frac{1}{\Delta(u)} \right) \\
&= \frac{2\pi}{a_i^2} - 2B_{ii}.
\end{aligned} \tag{8.54}$$

Once the 3 off diagonal integrals are determined using eq 8.53, the 3 diagonal components follow from eq 8.54. For example,

$$3B_{11} = 2\pi/a_1^2 - B_{12} - B_{13}. \tag{8.55}$$

Exercise 8.2 Using eqs 8.41, 8.53 and 8.54 show that:

$$B_{kk}a_k^2 - A_k = -\frac{1}{3} \sum_{i \neq k} B_{ki} a_i^2. \tag{8.56}$$

■

Solution From eq 8.54 we have

$$B_{kk}a_k^2 = \frac{1}{3}\left(2\pi - \sum_{j\neq k} B_{jk} a_k^2\right).$$

Using eq 8.41 this may be rewritten as:

$$B_{kk}a_k^2 = \frac{1}{3}\left(A_k + \sum_{j\neq k} A_j - B_{jk} a_k^2\right)$$

Using eq 8.53 this may be recast as follows:

$$\begin{aligned}
B_{kk}a_k^2 &= \frac{1}{3}\left(A_k + \sum_{j\neq k} A_j - \frac{(A_j - A_k)}{(a_k^2 - a_j^2)}a_k^2\right) \\
&= \frac{1}{3}\left(A_k - \sum_{j\neq k} \frac{A_j a_j^2 - A_k a_k^2}{a_k^2 - a_j^2}\right)
\end{aligned}$$

$$\begin{aligned}
B_{kk}a_k^2 &= \frac{1}{3}\left(A_k - \sum_{j\neq k} \frac{A_j a_j^2 - A_k(a_k^2 - a_j^2) - A_k a_j^2}{a_k^2 - a_j^2}\right) \\
&= \frac{1}{3}\left(3A_k - \sum_{j\neq k} \frac{(A_j - A_k)a_j^2}{a_k^2 - a_j^2}\right) \\
&= A_k - \frac{1}{3}\sum_{j\neq k} B_{kj}a_j^2.
\end{aligned}$$

Therefore,

$$B_{kk}a_k^2 - A_k = -\frac{1}{3}\sum_{j\neq k} B_{kj}a_j^2.$$

The coefficient ξ_{12} is as follows:

$$\xi_{12} = 6I_{412} - \frac{I_{32}}{a_1^2} - \frac{I_{31}}{a_2^2} + \frac{I_2}{4a_1^2a_2^2} = \frac{1}{4}\left(B_{12}a_2^2 - A_1\right) = \frac{1}{4}\left(B_{12}a_1^2 - A_2\right). \tag{8.57}$$

The coefficients ξ_{23} and ξ_{31} are obtained by cyclic permutation.

We arrive at the following remarkably simple result for ξ_{ij}:

$$\xi_{ij} = \frac{1}{4}\left(B_{ij}a_j^2 - A_i\right) = \frac{1}{4}\left(B_{ji}a_i^2 - A_j\right). \tag{8.58}$$

Notice that ξ_{ij} is symmetric.

The coefficient of x_1^2 is as follows:

$$\zeta_1 = 2I_{31} - \frac{I_2}{2a_1^2} = \frac{1}{2}(A_2a_2^2 + A_3a_3^2) = \frac{1}{2}(I_1/2 - A_1a_1^2). \tag{8.59}$$

The coefficients of x_2^2 and x_3^2 are obtained by cyclic permutation.

Finally, the constant $I_2/4$ is as follows:

$$I_2/4 = \frac{1}{4}\left\{(2\pi - A_1)a_2^2a_3^2 + (2\pi - A_2)a_3^2a_1^2 + (2\pi - A_3)a_1^2a_2^2\right\} \tag{8.60}$$

Exercise 8.3 Show that:

$$\Psi_{,i}(\mathbf{x}) = x_i\left(\Phi(\mathbf{x}) - a_i^2\Phi_i(\mathbf{x})\right) \tag{8.61}$$

where $\Phi(\mathbf{x})$ is the harmonic potential of eq 8.35 and $\Phi_i(\mathbf{x})$ is defined as follows:

$$\Phi_i(\mathbf{x}) = A_i - \sum_j B_{ij}x_j^2. \tag{8.62}$$

Show that:

$$\Psi_{,ij}(\mathbf{x}) = \delta_{ij}\left(\Phi(\mathbf{x}) - a_i^2\Phi_i(\mathbf{x})\right) + 2x_ix_j\left(B_{ij}a_j^2 - A_j\right). \tag{8.63}$$

Verify that $\nabla^2\Psi = 2\Phi$.

Show that:

$$\Psi_{,ijk}(\mathbf{x}) = 2\left\{\delta_{jk}\left(B_{ij}a_j^2 - A_i\right)x_i + \delta_{ki}\left(B_{jk}a_k^2 - A_j\right)x_j + \delta_{ij}\left(B_{ki}a_i^2 - A_k\right)x_k\right\}. \tag{8.64}$$

and hence:

$$\Psi_{,ijkl}(\mathbf{x}) = 2\left\{\delta_{jk}\delta_{il}\left(B_{ij}a_j^2 - A_i\right) + \delta_{ki}\delta_{jl}\left(B_{jk}a_k^2 - A_j\right) + \delta_{ij}\delta_{kl}\left(B_{ki}a_i^2 - A_k\right)\right\}. \tag{8.65}$$

■

Solution From eq 8.50 we have:

$$\Psi(\mathbf{x}) = \sum_{i,j}\xi_{ij}x_i^2x_j^2 + \sum_i \zeta_i x_i^2 + I_2/4,$$

where ξ_{ij}, ζ_i and I_2 are independent of $\mathbf{x}$. Therefore, the first derivative is as follows:

$$\begin{aligned}
\Psi_{,k} &= \sum_j 4\xi_{kj}x_j^2 x_k + 2\zeta_k x_k \\
&= \sum_j \left(B_{jk}a_k^2 - A_j\right) x_j^2 x_k + \left(I_1/2 - A_k a_k^2\right) x_k \\
&= \left(I_1/2 - \sum_j A_j x_j^2\right) x_k - a_k^2 \left(A_k - \sum_j B_{kj} x_j^2\right) x_k \\
&= \left(\Phi(\mathbf{x}) - a_k^2 \Phi_k(\mathbf{x})\right) x_k
\end{aligned}$$

Therefore, the second derivative $\Psi_{,ij}$ is a follows:

$$\begin{aligned}
\Psi_{,ij} &= \left(\Phi - a_i^2\Phi_i\right)\delta_{ij} - 2A_j x_j x_i + 2x_i x_j B_{ij} a_i^2 \\
&= \left(\Phi - a_i^2\Phi_i\right)\delta_{ij} + 2x_i x_j \left(B_{ij}a_j^2 - A_i\right),
\end{aligned}$$

where in the last line we used $B_{ij}a_i^2 - A_j = B_{ij}a_j^2 - A_i$.

Therefore,

$$\Psi_{,11} = \Phi - a_1^2 \left(A_1 - \sum_j B_{1j} x_j^2\right) + 2x_1^2 \left(B_{11}a_1^2 - A_1\right).$$

and hence

$$\begin{aligned}
\nabla^2\Psi &= 3\Phi - \underbrace{\sum_i A_i a_i^2}_{=I_1/2} + \sum_{i,j} a_i^2 B_{ij} x_j^2 + 2\sum_i \underbrace{(B_{ii}a_i^2 - A_i)}_{=-\frac{1}{3}\sum_{j\neq i} B_{ij}a_j^2} x_i^2 \\
&= 3\Phi - \frac{I_1}{2} + \sum_i a_i^2 B_{ii} x_i^2 + \frac{1}{3}\sum_j \underbrace{\left(\sum_{i\neq j} a_i^2 B_{ij}\right)}_{3A_j - 3B_{jj}a_j^2} x_j^2 \\
&= 3\Phi - \left(\frac{I_1}{2} - \sum_j A_j x_j^2\right) \\
&= 2\Phi
\end{aligned}$$

where we have used eq 8.56 in the first and second lines.

The third derivative $\Psi_{,ijk}$ is evaluated as follows:

$$\begin{aligned}\Psi_{,ijk} &= 2\delta_{ij}\left(-A_k x_k + B_{ik}a_i^2 x_k\right) + 2\delta_{ik}x_j\left(B_{ij}a_j^2 - A_i\right) + 2\delta_{jk}x_i\left(B_{ij}a_j^2 - A_i\right)\\ &= 2\left\{\delta_{jk}\left(B_{ij}a_j^2 - A_i\right)x_i + \delta_{ki}\left(B_{jk}a_k^2 - A_j\right)x_j + \delta_{ij}\left(B_{ki}a_i^2 - A_k\right)x_k\right\},\end{aligned}$$

where we have used $B_{ij} = B_{ji}$ and $B_{ij}a_j^2 - A_i = B_{ji}a_i^2 - A_j$.

The fourth derivative $\Psi_{,ijkl}$ is then as follows:

$$\Psi_{,ijkl} = 2\left\{\delta_{il}\delta_{jk}\left(B_{ij}a_j^2 - A_i\right) + \delta_{ki}\delta_{jl}\left(B_{jk}a_k^2 - A_j\right) + \delta_{ij}\delta_{kl}\left(B_{ki}a_i^2 - A_k\right)\right\}.$$

8.5 The elastic fields inside the inclusion

The elastic displacement field inside the inclusion follows from eqs 8.9, 8.35 and 8.64:

$$\begin{aligned}4\pi(1-\nu)u_i^C(\mathbf{x}) &= \sum_{j,k}\left\{\delta_{jk}\left(B_{ij}a_j^2 - A_i\right)x_i + \delta_{ki}\left(B_{jk}a_k^2 - A_j\right)x_j\right.\\ &\quad \left. + \delta_{ij}\left(B_{ki}a_i^2 - A_k\right)x_k\right\}e_{jk}^T\\ &\quad + \sum_{j,k}\left(2\nu\delta_{jk}A_i x_i + 4(1-\nu)\delta_{ij}A_k x_k\right)e_{jk}^T. \end{aligned} \tag{8.66}$$

The elastic strain field inside the inclusion is then as follows:

$$\begin{aligned}8\pi(1-\nu)e_{il}^C &= \delta_{il}\left(\sum_k\left(B_{ki}a_i^2 + B_{kl}a_l^2 - (1-2\nu)(A_i + A_l)\right)e_{kk}^T\right)\\ &\quad + \left((B_{li}a_i^2 + B_{il}a_l^2) + (1-2\nu)(A_i + A_l)\right)(e_{il}^T + e_{li}^T). \end{aligned} \tag{8.67}$$

Therefore the Eshelby tensor S_{injk} of eq 8.11 is as follows:

$$\begin{aligned}8\pi(1-\nu)S_{injk} &= \left(\delta_{ij}\delta_{nk} + \delta_{ik}\delta_{nj}\right)\left[B_{ni}a_i^2 + B_{in}a_n^2 + (1-2\nu)(A_i + A_n)\right]\\ &\quad + \delta_{in}\delta_{jk}\left[B_{ki}a_i^2 + B_{kn}a_n^2 - (1-2\nu)(A_i + A_n)\right] \end{aligned} \tag{8.68}$$

For example,

$$4\pi(1-\nu)S_{1111} = 3B_{11}a_1^2 + (1-2\nu)A_1$$

$$
\begin{aligned}
4\pi(1-\nu)S_{1122} &= B_{12}a_2^2-(1-2\nu)A_1 \\
8\pi(1-\nu)S_{1212} &= B_{12}(a_1^2+a_2^2)+(1-2\nu)(A_1+A_2). \qquad (8.69)
\end{aligned}
$$

Equivalent relations appear in Eshelby (1957). Other elements of S_{injk} may be determined by cyclic permutation. Note that S_{injk} has some of the symmetry of a fourth rank tensor in an orthorhombic crystal, but $S_{iijj} \neq S_{jjii}$.
The stress tensor in the inclusion is $\sigma^I_{ij} = \sigma^C_{ij} - \sigma^T_{ij}$, where

$$
\begin{aligned}
\sigma^C_{ij} &= \frac{\mu}{4\pi(1-\nu)}\sum_{k,l}\left(\Psi_{,ijkl}-2\nu(\delta_{kl}\Phi_{,ij}+\delta_{ij}\Phi_{,kl})\right)e^T_{kl} \\
&\quad -\frac{\mu}{2\pi}\sum_k\left(\Phi_{,jk}e^T_{ik}+\Phi_{,ik}e^T_{jk}\right)+\delta_{ij}\frac{2\mu\nu}{1-2\nu}\frac{\nu}{1-\nu}\sum_k e^T_{kk} \qquad (8.70) \\
&= \frac{\mu}{2\pi(1-\nu)}\left(2\xi_{ij}e^T_{ij}+\delta_{ij}\sum_k(\xi_{ik}+2\nu(A_i+A_k))\,e^T_{kk}\right) \\
&\quad +\frac{\mu}{\pi}(A_i+A_j)e^T_{ij}+\delta_{ij}\frac{2\mu\nu}{1-2\nu}\frac{\nu}{1-\nu}\sum_k e^T_{kk}, \qquad (8.71)
\end{aligned}
$$

and

$$
\sigma^T_{ij} = 2\mu e^T_{ij}+\delta_{ij}\frac{2\mu\nu}{1-2\nu}\sum_k e^T_{kk}.
$$

8.6 The harmonic potential outside a homogeneous ellipsoid

8.6.1 Introduction

Continuing with the origin of the cartesian coordinate system at the centre of the ellipsoid, and its axes along the principal axes of the ellipsoid, let the position vector of a point outside the ellipsoid be $\mathbf{x} = (x_1, x_2, x_3)$. In spherical polar coordinates defined by eq 8.25 the ellipsoid with semi-axes a_1, a_2, a_3 is described by eq 8.26. The direct evaluation of the harmonic potential at $\mathbf{x}$ leads to the following triple integral:

$$
\Phi(\mathbf{x}) = \int_0^{2\pi} d\phi \int_0^{\pi} d\theta\, \sin\theta \int_0^{r(\theta,\phi)} d\rho\, \left[x^2+\rho^2-2\rho(x_1\sin\theta\cos\phi+x_2\sin\theta\sin\phi+x_3\cos\theta)\right]^{-\frac{1}{2}}, \qquad (8.72)
$$

where $x = |\mathbf{x}|$ and $r(\theta,\phi)$ is given by eq 8.26.

Mathematicians and astronomers of the 18th and 19th centuries were confronted with the intractability of this triple integral. But in 1809 an explicit formula for the harmonic potential at an external point emerged through a remarkable theorem due to Ivory[10]. In this section we shall outline Ivory's theorem. It is proved in detail in section 8.11.

Ivory showed[11] that the harmonic potential at a point external to a homogeneous ellipsoid is directly related to the potential at an interior point of a confocal ellipsoid[12], the surface of which passes through the external point. More precisely, for **x** outside the ellipsoid the harmonic potential is the same as eq 8.35 except the lower limits of the integrals in A_i and I_1 are equal to $\chi > 0$ rather than zero:

$$\Phi(\mathbf{x}) = \frac{I_1(\chi)}{2} - \sum_i A_i(\chi) x_i^2 \tag{8.73}$$

where

$$I_1(\chi) = 2\pi a_1 a_2 a_3 \int_\chi^\infty \frac{du}{\Delta(u)} \tag{8.74}$$

$$A_i(\chi) = \pi a_1 a_2 a_3 \int_\chi^\infty \frac{du}{\Delta(u)(a_i^2 + u)}. \tag{8.75}$$

and χ is the positive root of the cubic equation (the other two roots are negative):

$$\frac{x_1^2}{a_1^2+\chi} + \frac{x_2^2}{a_2^2+\chi} + \frac{x_3^2}{a_3^2+\chi} - 1 = 0. \tag{8.76}$$

A simple formula for χ is derived in section 8.12 in terms of a_1, a_2, a_3 and x_1, x_2, x_3.

The integrals $I_1(\chi)$, $A_1(\chi)$, $A_2(\chi)$ and $A_3(\chi)$ may be expressed in terms of elliptic integrals in exactly the same way as eqs 8.38 and 8.39 but with ϕ replaced by ϕ_χ where

$$\phi_\chi = \cos^{-1}\sqrt{(\chi + a_3^2)/(\chi + a_1^2)}. \tag{8.77}$$

Using this equation and eq 8.39 it is straightforward to derive the generalisations of eqs 8.40 and 8.41 for any value of $\chi \geq 0$:

$$A_1(\chi) + A_2(\chi) + A_3(\chi) = \frac{2\pi a_1 a_2 a_3}{\Delta(\chi)} \tag{8.78}$$

[10] James Ivory FRS, 1765-1847, British mathematician.

[11] Ivory, J, Phil. Trans R. Soc. **99**, 345-372 (1809).

[12] The ellipsoids $x_1^2/a_1^2 + x_2^2/a_2^2 + x_3^2/a_3^2 = 1$ and $x_1^2/(a_1^2+\chi) + x_2^2/(a_2^2+\chi) + x_3^2/(a_3^2+\chi) = 1$ are *confocal*, where χ is real positive number.

$$A_1(\chi)a_1^2 + A_2(\chi)a_2^2 + A_3(\chi)a_3^2 = \frac{I_1(\chi)}{2} - \frac{2\pi a_1 a_2 a_3 \chi}{\Delta(\chi)}, \tag{8.79}$$

which are consistent with eqs 8.40 and 8.41 when $\chi = 0$. Combining eqs 8.78 and 8.79 we obtain:

$$A_1(\chi)(a_1^2+\chi) + A_2(\chi)(a_2^2+\chi) + A_3(\chi)(a_3^2+\chi) = \frac{I_1(\chi)}{2}. \tag{8.80}$$

Similarly, the generalisation of eq 8.54 for any value of $\chi \geq 0$ is as follows:

$$\sum_j B_{ij}(\chi) = \frac{2\pi a_1 a_2 a_3}{\Delta(\chi)(a_i^2+\chi)} - 2B_{ii}(\chi), \tag{8.81}$$

where $B_{ij}(\chi)$ is the same as B_{ij} of eq 8.52 except the lower limit of the integral is equal to χ:

$$B_{ij}(\chi) = \pi a_1 a_2 a_3 \int_{\chi}^{\infty} \frac{\mathrm{d}u}{(a_i^2+u)(a_j^2+u)\Delta(u)}. \tag{8.82}$$

Let us outline how eq 8.73 is obtained. Consider the potential $\Phi(x_1, x_2, x_3)$ of an ellipsoid E with semi-axes a_1, a_2, a_3, where (x_1, x_2, x_3) is outside the ellipsoid. There is a unique confocal ellipsoid E' which passes through (x_1, x_2, x_3). It is described by eq 8.76. The three roots of this cubic equation are real and, as explained in section 8.12, and χ is equated to the positive root. The semi-axes of E' are $a_i' = (a_i^2+\chi)^{1/2}$. According to Laplace's theorem[13] the potential Φ' at (x_1, x_2, x_3) due to the ellipsoid E' is a factor $(a_1'a_2'a_3'/a_1a_2a_3)$ times greater than the potential Φ at the same point (x_1, x_2, x_3) due to the ellipsoid E. We know Φ' is given by eq 8.36:

$$\Phi'(x_1, x_2, x_3) = \pi a_1' a_2' a_3' \int_0^{\infty} \frac{\mathrm{d}u'}{(a_1'^2+u')^{\frac{1}{2}}(a_2'^2+u')^{\frac{1}{2}}(a_3'^2+u')^{\frac{1}{2}}} \left(1 - \sum_j \frac{x_j^2}{a_j'^2+u'}\right). \tag{8.83}$$

If we change the integration variable to $u = u' + \chi$ and use $a_j'^2 = a_j^2 + \chi$ we obtain :

$$\Phi(x_1, x_2, x_3) = \pi a_1 a_2 a_3 \int_{\chi}^{\infty} \frac{\mathrm{d}u}{(a_1^2+u)^{\frac{1}{2}}(a_2^2+u)^{\frac{1}{2}}(a_3^2+u)^{\frac{1}{2}}} \left(1 - \sum_j \frac{x_j^2}{a_j^2+u}\right), \tag{8.84}$$

which is identical to eq 8.73 with the integrals defined in eqs 8.74 and 8.75.

[13]As explained in section 8.11 a less general form of this theorem was developed by Maclaurin several decades before Laplace generalised it into its present form.

In eq 8.8 we saw that $\Phi(\mathbf{x})$ must satisfy Laplace's equation, $\nabla^2\Phi(\mathbf{x}) = 0$, if $\mathbf{x}$ is outside the ellipsoid. It is instructive to see how eq 8.73 satisfies Laplace's equation. Consider the first derivative $\partial\Phi/\partial x_k$:

$$\frac{\partial\Phi}{\partial x_k} = -2\pi a_1 a_2 a_3 x_k \int_{\chi}^{\infty} \frac{1}{\Delta(u)(a_k^2+u)} du - \pi a_1 a_2 a_3 \frac{1}{\Delta(\chi)} \left(1 - \sum_i \frac{x_i^2}{a_i^2+\chi}\right) \frac{\partial\chi}{\partial x_k} \quad (8.85)$$

The second term on the right arises from differentiating the lower limit, χ, of the integral. However, this term is zero because the term in large brackets is zero. This result may be generalised: the spatial first derivatives of the integrals involved in the harmonic potential do not involve spatial derivatives of χ. This leaves only the first term on the right of eq 8.85:

$$\frac{\partial\Phi}{\partial x_k} = -2A_k(\chi)x_k = -2\pi a_1 a_2 a_3 x_k \int_{\chi}^{\infty} \frac{1}{\Delta(u)(a_k^2+u)} du. \quad (8.86)$$

The dependence of χ on (x_1,x_2,x_3) through eq 8.76 provides the following partial derivatives:

$$\frac{\partial\chi}{\partial x_j} = \frac{2x_j/(a_j^2+\chi)}{\sum_i x_i^2/(a_i^2+\chi)^2}. \quad (8.87)$$

These partial derivatives make a crucial contribution to the Laplacian of $\Phi(\mathbf{x})$:

$$\begin{aligned} \nabla^2\Phi &= -2\pi a_1 a_2 a_3 \left\{ \int_{\chi}^{\infty} \frac{1}{\Delta(u)} \sum_j \frac{1}{(a_j^2+u)} du - \frac{1}{\Delta(\chi)} \sum_j \frac{x_j \partial\chi/\partial x_j}{(a_j^2+\chi)} \right\} \\ &= 4\pi a_1 a_2 a_3 \left\{ \int_{\chi}^{\infty} \frac{d}{du}\left(\frac{1}{\Delta(u)}\right) du + \frac{1}{\Delta(\chi)} \sum_j \frac{x_j^2/(a_j^2+\chi)^2}{\sum_i x_i^2/(a_i^2+\chi)^2} \right\} \\ &= 4\pi a_1 a_2 a_3 \left\{ -\frac{1}{\Delta(\chi)} + \frac{1}{\Delta(\chi)} \right\} = 0. \end{aligned} \quad (8.88)$$

In the second line we have used eqs 8.43 and 8.87.

8.7 The biharmonic potential outside a homogeneous ellipsoid

Inspired by Ivory's theorem for the harmonic potential outside the ellipsoid let us assume $\Psi_{,i}(\mathbf{x})$ is given by eq 8.61, but with the lower limit of the integrals A_i and B_{ij}

set equal to χ instead of zero as in eqs 8.75 and 8.82:

$$\begin{aligned}
\Psi_{,i}(\mathbf{x}) &= x_i\left(\Phi(\mathbf{x}) - a_i^2\Phi_i(\mathbf{x})\right)\\
\Phi(\mathbf{x}) &= \frac{I_1(\chi)}{2} - \sum_j A_j(\chi)x_j^2\\
&= \pi a_1a_2a_3\int_\chi^\infty \frac{1}{\Delta(u)}\left(1-\sum_j\frac{x_j^2}{(a_j^2+u)}\right)\mathrm{d}u\\
&= \pi a_1a_2a_3\int_\chi^\infty \frac{P(u)}{\Delta(u)}\mathrm{d}u \qquad (8.89)\\
\Phi_i(\mathbf{x}) &= A_i(\chi) - \sum_j B_{ij}(\chi)x_j^2\\
&= \pi a_1a_2a_3\int_\chi^\infty \frac{1}{\Delta(u)(a_i^2+u)}\left(1-\sum_j\frac{x_j^2}{(a_j^2+u)}\right)\mathrm{d}u\\
&= \pi a_1a_2a_3\int_\chi^\infty \frac{P(u)}{\Delta(u)(a_i^2+u)}\mathrm{d}u, \qquad (8.90)
\end{aligned}$$

where

$$P(u) = 1 - \sum_j \frac{x_j^2}{(a_j^2+u)} \qquad (8.91)$$

The second derivatives $\Psi_{i,j}(\mathbf{x})$ do not involve derivatives of χ because $P(\chi) = 0$ by virtue of eq 8.76. Therefore,

$$\begin{aligned}
\Psi_{,ij}(\mathbf{x}) &= \delta_{ij}(\Phi(\mathbf{x}) - a_i^2\Phi_i(\mathbf{x})) + x_i(\Phi_{,j}(\mathbf{x}) - a_i^2\Phi_{i,j}(\mathbf{x}))\\
&= \delta_{ij}\left(\Phi(\mathbf{x}) - a_i^2\Phi_i(\mathbf{x})\right) + 2x_ix_j\left(B_{ij}(\chi)a_i^2 - A_j(\chi)\right). \qquad (8.92)
\end{aligned}$$

Exercise 8.4 Using eqs 8.76, 8.78, 8.79, 8.81 and 8.92 show that $\nabla^2\Psi(\mathbf{x}) = 2\Phi(\mathbf{x})$. Through a different argument Eshelby (1959) deduced the following expression for $\Psi_{,12}$:

$$\Psi_{,12} = \frac{a_1^2}{a_1^2 - a_2^2} x_2 \Phi_{,1} - \frac{a_2^2}{a_1^2 - a_2^2} x_1 \Phi_{,2}$$

Show it is equivalent to $\Psi_{,12}$ obtained from eq 8.92. ■

Solution Using eq 8.92 we have:

$$\nabla^2\Psi = 3\Phi - \sum_i a_i^2\left(A_i(\chi) - \sum_j B_{ij}(\chi)x_j^2\right) + 2\sum_i \left(B_{ii}(\chi)a_i^2 - A_i(\chi)\right)x_i^2$$

Using eq 8.79 this may be rewritten as:

$$\begin{aligned}\nabla^2\Psi &= 3\Phi - \left(\frac{I_1(\chi)}{2} - \frac{2\pi a_1 a_2 a_3 \chi}{\Delta(\chi)}\right)\\ &\quad + \sum_i \sum_{j \neq i} a_i^2 B_{ij}(\chi)x_j^2 + \sum_i 3a_i^2 B_{ii}(\chi)x_i^2 - 2\sum_i A_i(\chi)x_i^2.\end{aligned}$$

It follows from eq 8.81 that:

$$3B_{ii}(\chi)a_i^2 = \frac{2\pi a_1 a_2 a_3 \chi}{(a_i^2 + \chi)\Delta(\chi)} a_i^2 - \sum_{j \neq i} B_{ij}(\chi)a_i^2.$$

Using this in the equation for $\nabla^2\Psi$ we get:

$$\begin{aligned}\nabla^2\Psi &= 3\Phi - \frac{I_1(\chi)}{2} + \frac{2\pi a_1 a_2 a_3 \chi}{\Delta(\chi)} + \sum_i \sum_{j \neq i} (a_i^2 - a_j^2) B_{ij}(\chi)x_j^2 - 2\sum_i A_i(\chi)x_i^2\\ &\quad + \frac{2\pi a_1 a_2 a_3}{\Delta(\chi)}\left(\sum_i x_i^2 - \chi \underbrace{\sum_i \frac{x_i^2}{a_i^2 + \chi}}_{=1}\right),\end{aligned}$$

where we have used eq 8.76. Using $(a_i^2 - a_j^2)B_{ij}(\chi) = A_j(\chi) - A_i(\chi)$ we obtain:

$$\nabla^2\Psi = 3\Phi - \frac{I_1(\chi)}{2} + \sum_i \left(\frac{2\pi a_1 a_2 a_3}{\Delta(\chi)} - 2A_i(\chi) + \sum_{j\neq i}(A_j(\chi) - A_i(\chi))\right)x_i^2$$

Using eq 8.78 this simplifies to:

$$\nabla^2\Psi = 3\Phi - \frac{I_1(\chi)}{2} + \sum_i A_i(\chi)x_i^2 = 2\Phi.$$

Equation 8.92 gives the following expression for $\Psi_{,12}$:

$$\begin{aligned}
\Psi_{,12} &= 2x_1x_2\left(B_{12}(\chi)a_1^2 - A_2(\chi)\right)\\
&= 2x_1x_2\left(\frac{A_1(\chi) - A_2(\chi)}{a_2^2 - a_1^2}a_1^2 - A_2(\chi)\right)\\
&= -\frac{2x_1x_2A_1(\chi)a_1^2}{a_1^2 - a_2^2} + \frac{2x_1x_2A_2(\chi)a_2^2}{a_1^2 - a_2^2}\\
&= \frac{a_1^2}{a_1^2 - a_2^2}x_2\Phi_{,1} - \frac{a_2^2}{a_1^2 - a_2^2}x_1\Phi_{,2}
\end{aligned}$$

The third derivatives of Ψ involve spatial derivatives of the integrals A_i and B_{ij}, which in turn involve derivatives of χ given by eq 8.87:

$$\begin{aligned}
\Psi_{,ijk}(\mathbf{x}) &= \delta_{ij}\left(\Phi_{,k}(\mathbf{x}) - a_i^2\Phi_{i,k}(\mathbf{x})\right) + \delta_{ik}\left(\Phi_{,j}(\mathbf{x}) - a_i^2\Phi_{i,j}(\mathbf{x})\right)\\
&\quad + x_i\left(\Phi_{,jk}(\mathbf{x}) - a_i^2\Phi_{i,jk}(\mathbf{x})\right)\\
&= 2\delta_{jk}\left(B_{ij}(\chi)a_j^2 - A_i(\chi)\right)x_i + 2\delta_{ki}\left(B_{jk}(\chi)a_k^2 - A_j(\chi)\right)x_j\\
&\quad + 2\delta_{ij}\left(B_{ki}(\chi)a_i^2 - A_k(\chi)\right)x_k + 4\chi\beta_{ijk}(\chi)x_ix_jx_k
\end{aligned} \tag{8.93}$$

where

$$\Phi_{,jk}(\mathbf{x}) = 4\alpha_{jk}(\chi)x_jx_k - 2A_j(\chi)\delta_{jk} \tag{8.94}$$

$$\Phi_{i,jk}(\mathbf{x}) = 4\beta_{ijk}(\chi)x_jx_k - 2B_{ij}(\chi)\delta_{jk} \tag{8.95}$$

and

$$\begin{aligned}
\frac{\partial A_j(\chi)}{\partial x_k} &= -2\alpha_{jk}(\chi)x_k && (8.96)\\
\alpha_{jk}(\chi) &= \frac{\pi a_1 a_2 a_3}{\Delta(\chi)(a_j^2+\chi)(a_k^2+\chi)\sum_m \frac{x_m^2}{(a_m^2+\chi)^2}} && (8.97)\\
\frac{\partial B_{ij}(\chi)}{\partial x_k} &= -2\beta_{ijk}(\chi)x_k && (8.98)\\
\beta_{ijk}(\chi) &= \frac{\alpha_{jk}(\chi)}{(a_i^2+\chi)}. && (8.99)
\end{aligned}$$

Note that eq 8.93 reproduces eq 8.64 when $\chi = 0$, as it should.

The fourth derivatives of $\Psi(\mathbf{x})$ are needed for the stress and strain fields outside the ellipsoid. They are as follows:

$$\begin{aligned}
\Psi_{,ijkl}(\mathbf{x}) &= \delta_{ij}\left(\Phi_{,kl}(\mathbf{x}) - a_i^2\Phi_{i,kl}(\mathbf{x})\right) + \delta_{ik}\left(\Phi_{,jl}(\mathbf{x}) - a_i^2\Phi_{i,jl}(\mathbf{x})\right)\\
&\quad +\delta_{il}\left(\Phi_{,jk}(\mathbf{x}) - a_i^2\Phi_{i,jk}(\mathbf{x})\right) + x_i\left(\Phi_{,jkl}(\mathbf{x}) - a_i^2\Phi_{i,jkl}(\mathbf{x})\right)\\
&= 2\left\{\delta_{jk}\delta_{il}\left(B_{ij}a_j^2 - A_i\right) + \delta_{ki}\delta_{jl}\left(B_{jk}a_k^2 - A_j\right) + \delta_{ij}\delta_{kl}\left(B_{ki}a_i^2 - A_k\right)\right\}\\
&\quad +4\chi\left[\beta_{lij}x_j(\delta_{lk}x_i + \delta_{ik}x_l) + \beta_{ljk}x_k(\delta_{il}x_j + \delta_{ij}x_l) + \beta_{lki}x_i(\delta_{jl}x_k + \delta_{jk}x_l)\right]\\
&\quad +4x_i x_j x_k\left(\beta_{ijk} + \chi\frac{\partial \beta_{ijk}}{\partial \chi}\right)\chi_{,l} && (8.100)
\end{aligned}$$

The derivatives $\partial\alpha_{jk}/\partial\chi$ and $\partial\beta_{ijk}/\partial\chi$ are as follows:

$$\begin{aligned}
\frac{\partial \alpha_{jk}}{\partial \chi} &= -\alpha_{jk}(\chi)\left(\frac{1}{a_j^2+\chi} + \frac{1}{a_k^2+\chi} + \frac{1}{2}\sum_m \frac{1}{(a_m^2+\chi)}\right.\\
&\qquad\left. -2\frac{\sum_m x_m^2/(a_m^2+\chi)^3}{\sum_m x_m^2/(a_m^2+\chi)^2}\right) && (8.101)\\
\frac{\partial \beta_{ijk}}{\partial \chi} &= \frac{1}{(a_i^2+\chi)}\left(\frac{\partial \alpha_{jk}}{\partial \chi} - \beta_{ijk}\right). && (8.102)
\end{aligned}$$

Exercise 8.5 Using eq 8.94 show that the second derivative $\Phi_{,ij}$ just outside the inclusion is equal to $\Phi_{,ij}$ inside the inclusion plus $4\pi n_i n_j$, where n_i is the i'th component of the local normal to the surface of the inclusion.

Using eq 8.100 show that the fourth derivative $\Psi_{,ijkl}$ just outside the inclusion is equal to $\Psi_{,ijkl}$ inside the inclusion plus $8\pi n_i n_j n_k n_l$. ■

Solution The normal to the ellipsoid at (x_1, x_2, x_3) is parallel to $[x_1/a_1^2, x_2/a_2^2, x_3/a_3^2]$. Therefore the unit normal is:

$$\hat{\mathbf{n}} = \frac{1}{\sqrt{x_1^2/a_1^4 + x_2^2/a_2^4 + x_3^2/a_3^4}} \left[\frac{x_1}{a_1^2}, \frac{x_2}{a_2^2}, \frac{x_3}{a_3^2} \right].$$

Inside the ellipsoid $\Phi_{,jk}$ is $-2A_j\delta_{jk}$.

Just outside the ellipsoid $\Phi_{,jk}$ is given by eq 8.94, with $\chi \to 0$:

$$\Phi_{,jk} \text{ outside } = 4\alpha_{jk}(0)x_j x_k - 2A_j(0)\delta_{jk},$$

where $A_j(0)$ equals A_j inside. Therefore $\Phi_{,jk}$ just outside the ellipsoid minus $\Phi_{,jk}$ inside the ellipsoid is:

$$4\alpha_{jk}(0)x_j x_k = 4\frac{\pi a_1 a_2 a_3 x_j x_k}{a_1 a_2 a_3 a_j^2 a_k^2 \sum_m \frac{x_m^2}{a_m^4}} = \frac{4\pi x_j x_k}{a_j^2 a_k^2 \sum_m \frac{x_m^2}{a_m^4}} = 4\pi n_j n_k.$$

Outside the inclusion $\Psi_{,ijkl}$ is given by 8.100 and inside the inclusion by eq 8.65. Taking the limit $\chi \to 0$, the outside value minus the inside value is:

$$4x_i x_j x_k \beta_{ijk}(0)\chi_{,l} = 8\pi \frac{\frac{x_i x_j x_k x_l}{a_i^2 a_j^2 a_k^2 a_l^2}}{\left(\sum_m \frac{x_m^2}{a_m^4} \right)^2} = 8\pi n_i n_j n_k n_l.$$

8.8 The elastic fields outside the inclusion

The elastic displacement field outside the inclusion is as follows:

$$
\begin{aligned}
4\pi(1-\nu)u_i^C(\mathbf{x}) = \sum_{j,k} \Big\{ & \delta_{jk}\left(B_{ij}(\chi)a_j^2 - A_i(\chi)\right)x_i + \delta_{ki}\left(B_{jk}(\chi)a_k^2 - A_j(\chi)\right)x_j \\
& + \delta_{ij}\left(B_{ki}(\chi)a_i^2 - A_k(\chi)\right)x_k + 2\chi\beta_{ijk}(\chi)x_ix_jx_k \\
& + 2\nu A_i(\chi)\delta_{jk}x_i + 4(1-\nu)A_k(\chi)\delta_{ij}x_k \Big\} e_{jk}^T. \qquad (8.103)
\end{aligned}
$$

Exercise 8.6 Verify that the elastic displacement field is continuous at the interface of the inclusion. ■

Solution The elastic displacement field inside the inclusion is given by eq 8.66. In the limit $\chi \to 0$, $A_i(\chi) \to A_i$ inside, and $B_{ij}(\chi) \to B_{ij}$ inside. Then the expression for the elastic displacement field outside the inclusion in eq 8.103 evaluated just outside the inclusion becomes the same as just inside. Hence the elastic displacement field is continuous at the interface.

The elastic strain field outside the inclusion is as follows:

$$
e_{ij}^C(\mathbf{x}) = \sum_{k,l} S_{ijkl}(\mathbf{x})\, e_{kl}^T \qquad (8.104)
$$

where

$$
\begin{aligned}
S_{ijkl}(\mathbf{x}) &= \frac{1}{8\pi(1-\nu)}\left(\Psi_{,ijkl}(\mathbf{x}) - 2\nu\delta_{kl}\Phi_{,ij}(\mathbf{x}) - 2(1-\nu)\left(\delta_{ik}\Phi_{,jl}(\mathbf{x}) + \delta_{jk}\Phi_{,il}(\mathbf{x})\right)\right) \\
&= \frac{1}{8\pi(1-\nu)}\Big\{\Psi_{,ijkl}(\mathbf{x}) + 4\nu\delta_{kl}\left(A_i(\chi)\delta_{ij} - 2\alpha_{ij}(\chi)x_ix_j\right) \\
&\quad + 4(1-\nu)\left(\delta_{ik}(A_j(\chi)\delta_{jl} - 2\alpha_{jl}(\chi)x_jx_l) + \delta_{jk}\left(A_i(\chi)\delta_{il} - 2\alpha_{il}(\chi)x_ix_l\right)\right)\Big\}
\end{aligned}
$$

(8.105)

and $\Psi_{,ijkl}(\mathbf{x})$ is given by eq 8.100. The dilation outside the inclusion is easily deduced:

$$
e^C(\mathbf{x}) = \sum_m e_{mm}^C(\mathbf{x}) = \sum_m S_{mmkl}(\mathbf{x})e_{kl}^T
$$

$$= -\frac{1-2\nu}{4\pi(1-\nu)} \sum_{k,l} \Phi_{,kl}(\mathbf{x}) e^T_{kl}, \tag{8.106}$$

where we have used $\nabla^2 \Psi(\mathbf{x}) = 2\Phi(\mathbf{x})$ and $\nabla^2 \Phi(\mathbf{x}) = 0$.

The stress field outside the inclusion is as follows:

$$\begin{aligned} \sigma^C_{ij}(\mathbf{x}) &= \frac{\mu}{4\pi(1-\nu)} \sum_{k,l} \left(\Psi_{,ijkl}(\mathbf{x}) - 2\nu \left(\delta_{kl}\Phi_{,ij}(\mathbf{x}) + \delta_{ij}\Phi_{,kl}(\mathbf{x})\right)\right) e^T_{kl} \\ &\quad -\frac{\mu}{2\pi} \sum_k \left(\Phi_{,jk}(\mathbf{x}) e^T_{ik} + \Phi_{,ik}(\mathbf{x}) e^T_{jk}\right). \end{aligned} \tag{8.107}$$

We note two differences between the stress tensors outside and inside (see eq 8.70) the inclusion. The first is that the second derivatives of Φ and fourth derivatives of Ψ are constant inside the inclusion, whereas they depend on position outside the inclusion. Secondly, the last term in eq 8.70 does not appear in eq 8.107. This term arises from the Laplacian of the harmonic potential, which is -4π inside the inclusion and zero outside.

Consider the stress tensors σ^{C+}_{ij} and σ^{C-}_{ij} just outside and just inside the ellipsoid respectively where the local normal to the surface of the ellipsoid is $\hat{\mathbf{n}}$. Since $\Phi_{,ij}$ and $\Psi_{,ijkl}$ just outside the inclusion are equal to $\Phi_{,ij}$ and $\Psi_{,ijkl}$ inside the inclusion plus $4\pi n_i n_j$ and $8\pi n_i n_j n_k n_l$ respectively we may write:

$$\begin{aligned} \sigma^{C+}_{ij} &= \sigma^{C-}_{ij} + \frac{2\mu}{1-\nu} \sum_{k,l} \left(n_i n_j n_k n_l - \nu(\delta_{kl} n_i n_j + \delta_{ij} n_k n_l)\right) e^T_{kl} \\ &\quad -2\mu \sum_k \left(n_j n_k e^T_{ik} + n_i n_k e^T_{jk}\right) - \delta_{ij} \frac{2\mu\nu}{1-2\nu} \frac{\nu}{1-\nu} \sum_k e^T_{kk}. \end{aligned} \tag{8.108}$$

Exercise 8.7 Using eq 8.108 show that:

$$\sum_j \sigma^{C+}_{ij} n_j = \sum_j \left(\sigma^{C-}_{ij} - \sigma^T_{ij}\right) n_j. \tag{8.109}$$

What is the physical meaning of this result? ■

Solution From eq 8.108 we have:

$$\sum_j \sigma^{C+}_{ij} n_j = \sum_j \sigma^{C-}_{ij} n_j + \frac{2\mu}{1-\nu} \sum_{j,k,l} \left(n_i n_j n_j n_k n_l - \nu(\delta_{kl} n_i n_j n_j + \delta_{ij} n_j n_k n_l)\right) e^T_{kl}$$

$$= -2\mu \sum_{k,j} \left(n_j n_j n_k e^T_{ik} + n_i n_j n_k e^T_{jk} \right) - \frac{2\mu\nu}{1-2\nu}\frac{\nu}{1-\nu}\sum_{j,k} n_j \delta_{ij} e^T_{kk}$$

$$= \sum_j \sigma^{C-}_{ij} n_j + \frac{2\mu}{1-\nu}\sum_{k,l}\left(\cancel{n_i n_k n_l} - \nu(\delta_{kl} n_i + \cancel{n_i n_k n_l})\right) e^T_{kl}$$

$$= -2\mu \sum_k n_k e^T_{ik} - \cancel{2\mu \sum_{k,l} n_i n_k n_l e^T_{kl}} - \frac{2\mu\nu}{1-2\nu}\frac{\nu}{1-\nu}\sum_k n_i e^T_{kk}$$

$$= \sum_j \sigma^{C-}_{ij} n_j - \frac{2\mu\nu}{1-\nu}\left(1 + \frac{\nu}{1-2\nu}\right) n_i \sum_k e^T_{kk} - 2\mu \sum_k e^T_{ik} n_k$$

$$= \sum_j \sigma^{C-}_{ij} n_j - \sum_j \left(2\mu e^T_{ij} + \frac{2\mu\nu}{1-2\nu}\delta_{ij}\sum_k e^T_{kk} \right) n_j$$

$$= \sum_j \left(\sigma^{C-}_{ij} - \sigma^T_{ij} \right) n_j$$

The meaning of this equation is that the tractions at the interface of the inclusion are continuous.

8.9 Closing remarks

The expressions derived for the displacement (eq 8.66), strain (eq 8.67) and stress (eq 8.71) fields inside the inclusion agree with those obtained by Eshelby (1957). The expression for the elastic displacement field outside the inclusion in eq 8.103 agrees with the equivalent displacement field in Eshelby (1959). I have not seen elsewhere the expressions for the strain and stress fields outside the inclusion in eqs 8.105 and 8.107. The approach taken here to derive the elastic fields inside and outside the inclusion differs from earlier treatments I have seen in that it is based on an evaluation of the harmonic and biharmonic potentials in both regions. Several pertinent checks were made on the derivations of the elastic fields in this chapter:

1. Inside the inclusion the potentials $\Phi(\mathbf{x})$ and $\Psi(\mathbf{x})$ satisfy the required relations $\nabla^2\Phi(\mathbf{x}) = -4\pi$ (eq 8.35 and eq 8.41) and $\nabla^2\Psi(\mathbf{x}) = 2\Phi(\mathbf{x})$ (Exercise 8.3). Outside the inclusion they satisfy the required relations $\nabla^2\Phi(\mathbf{x}) = 0$ (eq 8.88) and $\nabla^2\Psi(\mathbf{x}) = 2\Phi(\mathbf{x})$ (Exercise 8.4).
2. The second derivatives of $\Phi(\mathbf{x})$ and the fourth derivatives of $\Psi(\mathbf{x})$ display the same discontinuities on crossing the surface of the ellipsoid (Exercise 8.5) as noted by Eshelby (1957). These discontinuities are essential to satisfy the continuity of tractions at the surface of the ellipsoid (Exercise 8.7).

3. The elastic displacement field is continuous at the surface of the inclusion (Exercise 8.6).
4. The expression for the displacement field outside the inclusion, eq 8.103, agrees with eq 8.13 in the far field limit.
5. The expression for the strain field outside the inclusion, eq 8.105, agrees with eq 8.14 in the far field limit.
6. The expression for the stress field outside the inclusion, eq 8.107, agrees with eq 8.15 in the far field limit.

8.10 Appendix: Evaluation of the integrals I_2, I_{3i}, I_{4ij}

Using eqs 8.31, 8.33, 8.34, 8.40 and 8.41, the integral I_2 may be expressed as follows:

$$\begin{aligned} I_2 &= a_1^4 I_{21} + a_2^4 I_{22} + a_3^4 I_{23} \\ &= \frac{I_1}{2}\left(a_1^2 + a_2^2 + a_3^2\right) - \left(A_1 a_1^4 + A_2 a_2^4 + A_3 a_3^4\right) \\ &= (A_1 + A_2) a_1^2 a_2^2 + (A_2 + A_3) a_2^2 a_3^2 + (A_3 + A_1) a_3^2 a_1^2 \\ &= (2\pi - A_1) a_2^2 a_3^2 + (2\pi - A_2) a_3^2 a_1^2 + (2\pi - A_3) a_1^2 a_2^2. \end{aligned} \tag{8.110}$$

To find the integrals I_{31}, I_{32} and I_{33} of eq 8.48 we will need the partial derivatives $\partial A_i/\partial a_j$:

$$\frac{\partial A_i}{\partial a_j} = \frac{A_i}{a_j} - (1 + 2\delta_{ij}) B_{ij} a_j. \tag{8.111}$$

$\partial I_2/\partial a_1$ is evaluated as follows:

$$\begin{aligned} \frac{\partial I_2}{\partial a_1} &= -\frac{\partial A_1}{\partial a_1} a_2^2 a_3^2 - \frac{\partial A_2}{\partial a_1} a_3^2 a_1^2 - \frac{\partial A_3}{\partial a_1} a_1^2 a_2^2 + (2\pi - A_2) 2 a_3^2 a_1 + (2\pi - A_3) 2 a_2^2 a_1 \\ &= -\left(\frac{A_1}{a_1} - 3B_{11} a_1\right) a_2^2 a_3^2 - \left(\frac{A_2}{a_1} - B_{21} a_1\right) a_3^2 a_1^2 - \left(\frac{A_3}{a_1} - B_{31} a_1\right) a_1^2 a_2^2 \\ &\quad + (2\pi - A_2) 2 a_3^2 a_1 + (2\pi - A_3) 2 a_2^2 a_1 \\ &= -A_1 \frac{a_2^2 a_3^2}{a_1} - 3A_2 a_3^2 a_1 - 3A_3 a_2^2 a_1 + 3B_{11} a_1 a_2^2 a_3^2 + B_{21} a_1^3 a_3^2 + B_{31} a_1^3 a_2^2 \\ &\quad + 4\pi a_1 (a_2^2 + a_3^2). \end{aligned} \tag{8.112}$$

It follows that:

$$I_{31} = \frac{1}{a_1^4} \int_S r^6 l_1^2 \mathrm{d}\omega = \frac{1}{4a_1} \frac{\partial I_2}{\partial a_1}$$

$$= -A_1\frac{a_2^2a_3^2}{4a_1^2} - A_2\frac{3a_3^2}{4} - A_3\frac{3a_2^2}{4} + 3B_{11}\frac{a_2^2a_3^2}{4} + B_{21}\frac{a_1^2a_3^2}{4} + B_{31}\frac{a_1^2a_2^2}{4} + \pi(a_2^2+a_3^2)$$

$$= (2\pi - A_1)\frac{a_1^2a_2^2 + a_2^2a_3^2 + a_3^2a_1^2}{4a_1^2} + \frac{(\pi - A_2)a_3^2}{2} + \frac{(\pi - A_3)a_2^2}{2}. \tag{8.113}$$

Expressions for I_{32} and I_{33} may be obtained by cyclic permutation of the subscripts.

Adding $a_1^4I_{31} + a_2^4I_{32} + a_3^4I_{33}$ we obtain:

$$I_3 = \int_S r^6\,\mathrm{d}\omega = \pi\left\{(a_1^2+a_2^2+a_3^2)(a_1^2a_2^2+a_2^2a_3^2+a_3^2a_1^2) - \frac{1}{2}a_1^2a_2^2a_3^2\right\}$$

$$-\frac{I_1}{8}(a_1^2a_2^2+a_2^2a_3^2+a_3^2a_1^2) - \frac{1}{2}(a_1^2+a_2^2+a_3^2)(a_1^2a_2^2a_3^2)\left(\frac{A_1}{a_1^2}+\frac{A_2}{a_2^2}+\frac{A_3}{a_3^2}\right). \tag{8.114}$$

To evaluate I_{412} and I_{411} it is convenient to introduce the following integrals:

$$C_{ijk} = \pi a_1a_2a_3\int_0^\infty \frac{\mathrm{d}u}{(a_i^2+u)(a_j^2+u)(a_k^2+u)\Delta(u)}. \tag{8.115}$$

The order in which i, j and k appear is immaterial. There are three integrals with $i = j = k$, one integral with $i \neq j \neq k$ and six with $i = j \neq k$, making ten distinct integrals in all. They arise when the integrals B_{ij} of eq 8.52 are differentiated:

$$\frac{\partial B_{ij}}{\partial a_k} = \frac{B_{ij}}{a_k} - (2\delta_{ik} + 2\delta_{jk} + 1)\,a_kC_{ijk} \tag{8.116}$$

As in eqs 8.44 and 8.54, eq 8.43 may be used to derive relations between the integrals C_{ijk}. There are six such relations:

$$C_{111} + C_{112} + C_{113} = \frac{2\pi}{a_1^4} - 4C_{111} \tag{8.117}$$

$$C_{221} + C_{222} + C_{223} = \frac{2\pi}{a_2^4} - 4C_{222} \tag{8.118}$$

$$C_{331} + C_{332} + C_{333} = \frac{2\pi}{a_3^4} - 4C_{333} \tag{8.119}$$

$$C_{121} + C_{122} + C_{123} = \frac{2\pi}{a_1^2a_2^2} - C_{121} - C_{122} \tag{8.120}$$

$$C_{231}+C_{232}+C_{233} = \frac{2\pi}{a_2^2a_3^2}-C_{232}-C_{233} \tag{8.121}$$

$$C_{311}+C_{312}+C_{313} = \frac{2\pi}{a_3^2a_1^2}-C_{311}-C_{313} \tag{8.122}$$

It follows that there are only $10-6=4$ independent integrals C_{ijk}. They may be expressed in terms of the integrals A_i and I_1 using eqs 8.111, 8.53, 8.54 and 8.116.

The integral I_{412} is obtained from I_{31} by differentiation with respect to a_2:

$$\begin{aligned} I_{412} &= \frac{1}{a_1^4a_2^4}\int_S r^8 l_1^2 l_2^2\,\mathrm{d}\omega \\ &= \frac{1}{6a_2}\frac{\partial I_{31}}{\partial a_2} && (8.123)\\ &= -\frac{3a_3^2}{24}\left(\frac{A_1}{a_1^2}+\frac{A_2}{a_2^2}\right)-\frac{9A_3}{24}+\frac{9a_3^2}{24}(B_{11}+B_{22})+\frac{3}{24}\left(B_{31}a_1^2+B_{32}a_2^2\right) \\ &\quad +\frac{a_3^2}{24}\left(\frac{a_1^2}{a_2^2}+\frac{a_2^2}{a_1^2}\right)B_{21}-\frac{3a_3^2}{24}(C_{112}a_2^2+C_{221}a_1^2)-\frac{a_1^2a_2^2}{24}C_{312}+\frac{\pi}{3}. && (8.124)\end{aligned}$$

Note this expression is invariant with respect to interchanging 1 and 2, as required by the definition of I_{412}. The integrals I_{423} and I_{431} are found by cyclic permutation of the subscripts.

The integral I_{411} may also be obtained by differentiating I_{31}:

$$\begin{aligned} I_{411} &= \frac{1}{a_1^8}\int_S r^8 l_1^8\,\mathrm{d}\omega && (8.125)\\ &= \frac{1}{6a_1}\frac{\partial I_{31}}{\partial a_1}+\frac{2}{3a_1^2}I_{31} && (8.126)\end{aligned}$$

We obtain,

$$\begin{aligned} I_{411} &= -\frac{1}{8a_1^2}\left(A_1\frac{a_2^2a_3^2}{a_1^2}+5A_2a_3^2+5A_3a_2^2\right) \\ &\quad +\frac{1}{12a_1^2}\left(9B_{11}a_2^2a_3^2+5B_{21}a_1^2a_3^2+5B_{31}a_1^2a_2^2\right) \\ &\quad -\frac{1}{8}\left(5C_{111}a_2^2a_3^2+C_{211}a_1^2a_3^2+C_{311}a_1^2a_2^2\right)+\frac{2\pi}{3a_1^2}(a_2^2+a_3^2). && (8.127)\end{aligned}$$

The integrals I_{422} and I_{433} are found by cyclic permutation.

8.11 Appendix: Ivory's theorem

We begin with a geometrical property of confocal ellipsoids that plays a central role in Ivory's theorem.

Consider two confocal ellipsoids E and E' described by the equations:

$$E:\ \frac{x_1^2}{a_1^2}+\frac{x_2^2}{a_2^2}+\frac{x_3^2}{a_3^2}=1 \tag{8.128}$$

and

$$E':\ \frac{x_1'^2}{a_1'^2}+\frac{x_2'^2}{a_2'^2}+\frac{x_3'^2}{a_3'^2}=1 \tag{8.129}$$

where

$$a_i'^2=a_i^2+\chi. \tag{8.130}$$

The point $P=(x_1,x_2,x_3)$ in or on E is said to *correspond* to the point $P'=(x_1',x_2',x_3')$ in or on E' if:

$$\frac{x_i}{a_i}=\frac{x_i'}{a_i'}. \tag{8.131}$$

If P,P' and Q,Q' are pairs of corresponding points in or on the confocal ellipsoids we shall prove that $PQ'=P'Q$.

Let the coordinates of P and P' be (x_1,x_2,x_3) and (x_1',x_2',x_3'). Let the coordinates of Q and Q' be (z_1,z_2,z_3) and (z_1',z_2',z_3'). Then

$$\begin{aligned}(PQ')^2-(P'Q)^2 &= \sum_i (x_i-z_i')^2-(x_i'-z_i)^2\\ &= \sum_i \left(\frac{x_i'a_i}{a_i'}-z_i'\right)^2-\left(x_i'-\frac{z_i'a_i}{a_i'}\right)^2\\ &= \sum_i x_i'^2\left(\frac{a_i^2}{a_i'^2}-1\right)+z_i'^2\left(1-\frac{a_i^2}{a_i'^2}\right)\\ &= \chi\sum_i \frac{z_i'^2}{a_i'^2}-\frac{x_i'^2}{a_i'^2}=\chi(1-1)=0.\end{aligned} \tag{8.132}$$

Suppose we wish to evaluate the harmonic potential of the ellipsoid E at an external point $P'=(x_1',x_2',x_3')$. There is a unique confocal ellipsoid E' passing through P'. It is obtained by selecting the positive root of the cubic equation in χ that results when eq 8.130 is inserted in eq 8.129. A simple formula for χ is derived in section 8.12.

To derive Ivory's theorem we first focus on the force of attraction exerted on the point P' by the ellipsoid E. Like the astronomers of his time we may think of the force as arising from the gravitational attraction of the ellipsoid on a unit mass at $\mathbf{x}'$, but with the gravitational constant and the homogeneous density of E both set equal to unity. The force is minus the gradient of the potential. Consider the component of the force $\mathrm{d}f_1(\mathbf{x}')$ at P' in the x_1-direction due to a volume element $\mathrm{d}V_{\mathbf{x}}$ of E at $\mathbf{x}$. We have

$$\mathrm{d}f_1(\mathbf{x}') = -\frac{\partial}{\partial x_1}|\mathbf{x}' - \mathbf{x}|^{-1}\,\mathrm{d}V_{\mathbf{x}}. \tag{8.133}$$

Consider a column parallel to the x_1-axis of cross-sectional area $\mathrm{d}x_2\mathrm{d}x_3$ from the surface of the ellipsoid E at A to the surface at B, as illustrated schematically in Fig 8.2. There is a corresponding column in the ellipsoid E' between the points on its surface A' and B', which correspond to A and B on E. It follows from eq 8.131 that the cross-sectional area of the corresponding column in E' is:

$$\mathrm{d}x_2'\mathrm{d}x_3' = \frac{a_2'a_3'}{a_2a_3}\mathrm{d}x_2\mathrm{d}x_3. \tag{8.134}$$

Using eq 8.133, the force of attraction at P' due to the column between A and B in E is:

$$-\mathrm{d}x_2\,\mathrm{d}x_3\int_A^B \frac{\partial}{\partial x_1}|\mathbf{x}' - \mathbf{x}|^{-1}\,\mathrm{d}x_1 = \mathrm{d}x_2\,\mathrm{d}x_3\left(\frac{1}{(P'A)} - \frac{1}{(P'B)}\right). \tag{8.135}$$

Similarly, the force of attraction at P due to the column between A' and B' in E' is:

$$-\mathrm{d}x_2'\,\mathrm{d}x_3'\int_{A'}^{B'} \frac{\partial}{\partial x_1}|\mathbf{x}' - \mathbf{x}|^{-1}\,\mathrm{d}x_1 = \mathrm{d}x_2'\,\mathrm{d}x_3'\left(\frac{1}{(PA')} - \frac{1}{(PB')}\right). \tag{8.136}$$

By eq 8.132 we have $PA' = P'A$ and $PB' = P'B$. Therefore,

$$\frac{\text{the force of attraction of } E \text{ at } P' \text{ along } x_1 \text{ due to the column } \mathrm{d}x_2\mathrm{d}x_3}{\text{the force of attraction of } E' \text{ at } P \text{ along } x_1 \text{ due to the column } \mathrm{d}x_2'\mathrm{d}x_3'} = \frac{a_2a_3}{a_2'a_3'} \tag{8.137}$$

This result is independent of the choice of the column AB. Consequently, by adding the contributions of all such columns along x_1 we obtain the following result:

$$\frac{\text{the total force of attraction of } E \text{ at } P' \text{ along } x_1}{\text{the total force of attraction of } E' \text{ at } P \text{ along } x_1} = \frac{a_2a_3}{a_2'a_3'} = \frac{a_1a_2a_3a_1'}{a_1'a_2'a_3'a_1} \tag{8.138}$$

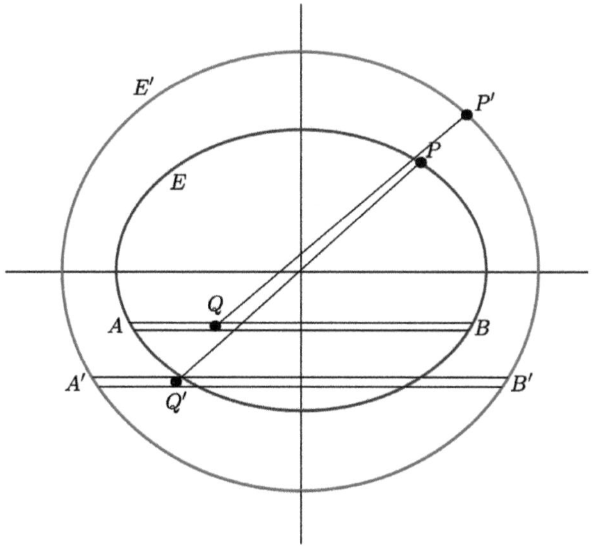

Figure 8.2: Schematic illustration of the construction used to derive eq 8.138 and Ivory's theorem. See text.

By repeating the argument along the x_2 and x_3-directions the ratios of the total forces of attraction of E at P' to the total forces of attraction of E' at P along x_2 and x_3 are $a_1a_3/(a'_1a'_3)$ and $a_1a_2/(a'_1a'_2)$ respectively. Therefore, we have:

$$\frac{\text{the total force of attraction of } E \text{ at } P' \text{ along } x_i}{\text{the total force of attraction of } E' \text{ at } P \text{ along } x_i} = \frac{a_1a_2a_3a'_i}{a'_1a'_2a'_3a_i} \tag{8.139}$$

Clearly the point P is internal to the ellipsoid E'. Therefore, we know the potential at P due to the ellipsoid E', and its gradient. The i'th component of its gradient is $-2A'_ix_i$, where A'_i follows from eq 8.34:

$$A'_i = \pi a'_1a'_2a'_3 \int_0^\infty \frac{du'}{(a'^2_i + u')\left[(a'^2_1 + u')(a'^2_2 + u')(a'^2_3 + u')\right]^{\frac{1}{2}}} \tag{8.140}$$

It follows from eq 8.139 that the i'th component of the force of attraction of the ellipsoid E at the external point P' is:

$$2A'_ix_i\frac{a_1a_2a_3a'_i}{a'_1a'_2a'_3a_i} = 2\pi a_1a_2a_3x'_i \int_0^\infty \frac{du'}{(a'^2_i + u')\left[(a'^2_1 + u')(a'^2_2 + u')(a'^2_3 + u')\right]^{\frac{1}{2}}} \tag{8.141}$$

where we have used eq 8.131. If we change the integration variable to $u = u' + \chi$ and use $a_i'^2 = a_i^2 + \chi$ the i'th component of the force of attraction of the ellipsoid E at the external point P' becomes:

$$2\pi a_1 a_2 a_3 x_i' \int_{\chi}^{\infty} \frac{\mathrm{d}u}{(a_i^2 + u)\left[(a_1^2 + u)(a_2^2 + u)(a_3^2 + u)\right]^{\frac{1}{2}}} = 2\pi a_1 a_2 a_3 x_i' \int_{\chi}^{\infty} \frac{\mathrm{d}u}{\Delta(u)(a_i^2 + u)}. \tag{8.142}$$

This result is very close to what we seek but it is expressed in terms of the force of attraction rather than the potential. We cannot integrate each component of the force to derive a potential because the lower limit of the integral in eq 8.142 depends implicitly on $\mathbf{x}'$ through eqs 8.129 and 8.130.

We are rescued by a theorem due to Maclaurin[14], and which follows directly from eq 8.142. The theorem states that the forces of attraction of two confocal homogeneous ellipsoids, at the same point outside *both*, are proportional to their volumes and in the same direction. (If the forces are due to gravity then the forces are proportional to the densities of the ellipsoids as well). If the confocal ellipsoids have semi-axes a_i and b_i it follows from eq 8.142 that each component of the forces of attraction at the same point outside both ellipsoids is in the ratio $a_1 a_2 a_3 : b_1 b_2 b_3$, which is the same as the ratio of their volumes. Since the components of the forces are in the same ratio the forces are parallel and Maclaurin's theorem follows. Since each component of the forces of attraction to the two ellipsoids is proportional to their volumes, the ratios of their potentials to their volumes can differ by no more than a constant. But since the potentials decay to zero at infinite separation that constant must be zero. *Therefore, the harmonic potentials of two homogeneous confocal ellipsoids, at a point external to both, are directly proportional to their volumes.*

The potential $\Phi'(\mathbf{x})$ due to the ellipsoid E' which passes through $\mathbf{x}$, and is confocal to an inner ellipsoid E, is $a_1' a_2' a_3' / (a_1 a_2 a_3)$ times the potential $\Phi(\mathbf{x})$ due to E. Since $\mathbf{x}$ is in E' we know $\Phi'(\mathbf{x})$. It is:

$$\Phi'(\mathbf{x}) = \pi a_1' a_2' a_3' \int_0^{\infty} \frac{\mathrm{d}u'}{\left[(a_1'^2 + u')(a_2'^2 + u')(a_3'^2 + u')\right]^{\frac{1}{2}}} \left(1 - \sum_i \frac{x_i^2}{a_i'^2 + u'}\right). \tag{8.143}$$

[14]Colin Maclaurin FRS, 1698-1746. British mathematician. Its most general form, which we give here, was proved by Pierre-Simon Laplace (1749-1827) in 1783 - see Todhunter, I *A History of the Mathematical Theories of Attraction and the Figure of the Earth from the time of Newton to that of Laplace*, **2**, paragraph 804, p31 (1873).

Hence, $\Phi(\mathbf{x})$ must be as follows:

$$\begin{aligned}\Phi(\mathbf{x}) &= \pi a_1 a_2 a_3 \int_0^{\infty} \frac{du'}{\left[(a_1'^2+u')(a_2'^2+u')(a_3'^2+u')\right]^{\frac{1}{2}}} \left(1-\sum_i \frac{x_i^2}{a_i'^2+u'}\right) \\ &= \pi a_1 a_2 a_3 \int_{\chi}^{\infty} \frac{du}{\Delta(u)} \left(1-\sum_i \frac{x_i^2}{a_i^2+u}\right),\end{aligned} \tag{8.144}$$

where $\Delta(u)$ is defined by eq 8.30 and the second line is obtained by changing the integration variable to $u = u' + \chi$ and by using $a_i'^2 = a_i^2 + \chi$. This is Ivory's theorem.[15] A simple formula for χ is derived in the next section.

8.12 Appendix: A simple formula for χ

χ is defined by the following equation for a confocal ellipsoid:

$$\frac{x_1^2}{a_1^2+\chi} + \frac{x_1^2}{a_2^2+\chi} + \frac{x_1^2}{a_3^2+\chi} = 1. \tag{8.145}$$

When $\chi = 0$ the point $\mathbf{x} = (x_1, x_2, x_3)$ is on the surface of the ellipsoid with semi-axes $a_1 > a_2 > a_3$, while points outside the ellipsoid require $\chi > 0$.

When eq 8.145 is rationalised it becomes the following cubic equation:

$$\chi^3 - A\chi^2 - B\chi - C = 0 \tag{8.146}$$

with

$$A = x_1^2 + x_2^2 + x_3^2 - a_1^2 - a_2^2 - a_3^2 \tag{8.147}$$

$$B = x_1^2(a_2^2+a_3^2) + x_2^2(a_3^2+a_1^2) + x_3^2(a_1^2+a_2^2) - a_1^2a_2^2 - a_2^2a_3^2 - a_3^2a_1^2 \tag{8.148}$$

$$C = x_1^2a_2^2a_3^2 + x_2^2a_3^2a_1^2 + x_3^2a_1^2a_2^2 - a_1^2a_2^2a_3^2 \tag{8.149}$$

Note that $C = 0$ for points $\mathbf{x}$ on the ellipsoid and $C > 0$ for points outside the ellipsoid.

---

[15] Perhaps because his theorem solved a problem that had exercised the minds of some of the greatest mathematicians for more than a century, Ivory was awarded the Copley Medal of the Royal Society in 1814, a year before he was elected FRS. The Copley Medal is the oldest and most prestigious of all the Royal Society awards.

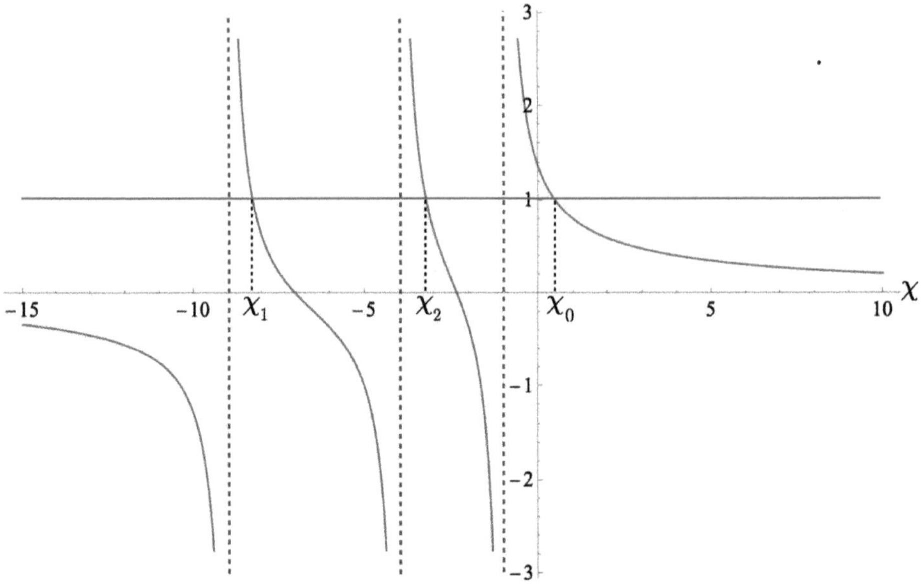

Figure 8.3: Graphical solution of eq 8.145. The left hand side of eq 8.145 is represented by the blue curves, and the right hand side by the red line. The broken majenta lines are asymptotes at $\chi = -a_1^2,\ -a_2^2$, and $-a_3^2$. The three solutions χ_0, χ_1 and χ_2 of the cubic equation, eq 8.146, are located where the red line crosses the blue curves. The solution we seek is χ_0.

Fig 8.3 is a graphical solution of eq 8.145 for the particular case $a_1 = 3,\ a_2 = 2,\ a_3 = 1$ and $x_1 = 1,\ x_2 = 1,\ x_3 = 1$. With these parameters we obtain $A = -11,\ B = -21,\ C = 13$. Fig 8.3 illustrates an important property of eq 8.146, namely that its three roots are real numbers. The root $\chi_0 > 0$ is the root we seek. It describes a confocal ellipsoid passing through **x**. The root $\chi_2 < 0$ describes a hyperboloid of one sheet passing through **x** and the root $\chi_1 < \chi_2$ describes a hyperboloid of two sheets passing through **x**.

The first step in solving eq 8.146 analytically is to eliminate the term in χ^2. This is achieved by introducing ζ through the equation:

$$\chi = \zeta + A/3 \tag{8.150}$$

Equation 8.146 becomes:

$$\zeta^3 - 3k\zeta + 2m = 0 \tag{8.151}$$

where

$$k = \frac{A^2}{9} + \frac{B}{3} \tag{8.152}$$

$$m = -\left(\frac{A^3}{27} + \frac{AB}{6} + \frac{C}{2}\right) \tag{8.153}$$

If we now make the substitution

$$\zeta = u + \frac{k}{u} \tag{8.154}$$

and multiply through by u^3, eq 8.151 becomes a quadratic equation in u^3:

$$\left(u^3\right)^2 + 2m\left(u^3\right) + k^3 = 0. \tag{8.155}$$

Therefore,

$$u^3 = -m \pm \sqrt{m^2 - k^3} \tag{8.156}$$

At first sight it appears there are not three but six roots. But if the root for u^3 obtained with the positive square root is $u_0^3 = \left(-m + \sqrt{m^2 - k^3}\right)$ then the root for u^3 obtained with the negative square root, i.e. $-m - \sqrt{m^2 - k^3}$, is k^3/u_0^3. That is because the product of the two roots is k^3. Therefore each root ζ of eq 8.151 involves both the positive and the negative square root solutions of eq 8.156, and therefore there are only three roots for ζ, as expected:

$$\zeta = u + \frac{k}{u} = \left(-m + \sqrt{m^2 - k^3}\right)^{1/3} + \left(-m - \sqrt{m^2 - k^3}\right)^{1/3}. \tag{8.157}$$

The three roots for ζ are real provided $k^3 > m^2$ because each root is then a sum of complex conjugates. Assuming $k^3 > m^2$ and writing $-m + i\sqrt{k^3 - m^2} = \rho e^{i\theta}$, where $\rho = k^{3/2}$ and $\cos\theta = -m/k^{3/2}$, the three roots of ζ are as follows:

$$\zeta = 2\sqrt{k}\cos\left(\frac{\theta}{3} + \frac{2n\pi}{3}\right), \text{ where } n = 0,\ 1,\ 2. \tag{8.158}$$

Therefore, the three roots for χ are as follows:

$$\chi_0 = 2\sqrt{k}\cos\left(\frac{\theta}{3}\right) + \frac{A}{3} \tag{8.159}$$

$$\chi_1 = 2\sqrt{k}\cos\left(\frac{\theta}{3}+\frac{2\pi}{3}\right)+\frac{A}{3} \tag{8.160}$$

$$\chi_2 = 2\sqrt{k}\cos\left(\frac{\theta}{3}+\frac{4\pi}{3}\right)+\frac{A}{3}. \tag{8.161}$$

For χ_0 to be the required root the following conditions must be satisfied:

1. When $C = 0$ the point $\mathbf{x}$ lies on the surface of the ellipsoid and χ is zero. When $C = 0$ we find $\chi_0 = 0$ and χ_1 and χ_2 are both negative.

2. $\chi_0 \to |\mathbf{x}|^2$ as $|\mathbf{x}| \to \infty$. This is the limiting behaviour for χ as the ellipsoid becomes infinitely large and almost spherical.

3. For points $\mathbf{x}$ outside the ellipsoid χ must be positive. Whereas χ_0 is always positive or zero, χ_1 and χ_2 are always negative.

Let us take each of these points in turn:

1. From eq 8.159 $\cos(\theta/3) = -A/(6k^{1/2})$ when $\chi_0 = 0$. Using the identity

   $$\cos\theta = 4\cos^3(\theta/3) - 3\cos(\theta/3)$$

   we deduce that $\cos\theta = \left(A^3/27 + AB/6\right)/k^{3/2}$. Comparing this result with $\cos\theta = -m/k^{3/2}$ and eq 8.153 for m we deduce that $C = 0$. When $\mathbf{x}$ is on the surface of the ellipsoid $-a_1^2 - a_2^2 \le A \le -a_2^2 - a_3^2$, and hence $0 < \theta/3 < \pi/2$. Equations 8.160 and 8.161 indicate that χ_1 and χ_2 are then both negative and $\chi_1 < \chi_2$.

2. As $|\mathbf{x}| \to \infty$ we find $\cos\theta = -m/k^{3/2} \to 1$. Therefore, $\theta \to 0$ and $\cos(\theta/3) \to 1$, and hence $\chi_0 \to 2\sqrt{k} + A/3 \to A \to |\mathbf{x}|^2$ as $|\mathbf{x}| \to \infty$.

3. χ_0 is zero for points $\mathbf{x}$ on the surface of the ellipsoid and increases for points on larger confocal ellipsoids. It is always positive or zero. Since the product of the roots $\chi_0\chi_1\chi_2 = C$, and $C > 0$ for points $\mathbf{x}$ outside the ellipsoid, the roots χ_1 and χ_2 are either both positive or both negative. Fig 8.3 indicates they are both negative because they occur between $-a_2^2$ and $-a_1^2$ and between $-a_3^2$ and $-a_2^2$ respectively.

For the example illustrated in Fig 8.3, eqs 8.159, 8.160 and 8.161 yield $\chi_0 = 0.4885$, $\chi_1 = -8.2710$ and $\chi_2 = -3.2175$. It may be verified that the point $\mathbf{x} = (1,1,1)$ lies on the confocal ellipsoid, eq 8.145, with $\chi = \chi_0$.

In conclusion, χ_0 is the value required for χ as the lower limit of the integrals of sections 8.6 and 8.7 for the harmonic and biharmonic potentials outside the ellipsoid. For convenience we list all the required formulae here:

$$\begin{aligned}
\chi_0 &= 2\sqrt{k}\cos(\theta/3) + A/3 \\
k &= \frac{A^2}{9} + \frac{B}{3} \\
\cos\theta &= -\frac{m}{k^{3/2}} \\
m &= -\left(\frac{A^3}{27} + \frac{AB}{6} + \frac{C}{2}\right) \\
A &= x_1^2 + x_2^2 + x_3^2 - a_1^2 - a_2^2 - a_3^2 \\
B &= x_1^2(a_2^2 + a_3^2) + x_2^2(a_3^2 + a_1^2) + x_3^2(a_1^2 + a_2^2) - a_1^2a_2^2 - a_2^2a_3^2 - a_3^2a_1^2 \\
C &= x_1^2a_2^2a_3^2 + x_2^2a_3^2a_1^2 + x_3^2a_1^2a_2^2 - a_1^2a_2^2a_3^2.
\end{aligned}$$

8.13 Problems

Problem 8.1 It was shown in section 8.6.1 that the stress in an inclusion, which would undergo a homogeneous strain e^T_{ij} if it were not constrained by the surrounding medium, is given by $\sigma^I_{ij} = \sigma^C_{ij} - \sigma^T_{ij}$, where σ^T_{ij} and σ^C_{ij} are the stresses related respectively to the strain e^T_{ij} and the constrained strain e^C_{ij} inside the inclusion by Hooke's law. The stress and strain fields in the medium surrounding the inclusion are the constrained fields σ^C_{ij} and e^C_{ij}. Let the strain related to σ^I_{ij} through Hooke's law be $e^I_{ij} = e^C_{ij} - e^T_{ij}$. Let the external surface of the body be denoted Σ and the surface of the inclusion be denoted $\mathcal{S}$. Let $\mathcal{R}$ be the region occupied by the inclusion and let $\mathcal{M}$ be the surrounding medium. Then the total elastic energy is:

$$E_{el} = \frac{1}{2}\int_{\mathcal{R}} \sum_{i,j} \sigma^I_{ij} e^I_{ij} \mathrm{d}V + \frac{1}{2}\int_{\mathcal{M}} \sum_{i,j} \sigma^C_{ij} e^C_{ij} \mathrm{d}V,$$

Using the continuity of surface tractions and displacements at all points on $\mathcal{S}$, and the divergence theorem, show that:

$$\frac{1}{2}\int_{\mathcal{M}} \sum_{i,j} \sigma_{ij}^{C} e_{ij}^{C} \mathrm{d}V = -\frac{1}{2}\int_{\mathcal{R}} \sum_{i,j} \sigma_{ij}^{I} e_{ij}^{C} \mathrm{d}V.$$

Hence show that the total elastic energy may be expressed as the following volume integral over the inclusion only, *whatever its shape*:

$$E_{el} = -\frac{1}{2}\int_{\mathcal{R}} \sum_{i,j} \sigma_{ij}^{I} e_{ij}^{T} \mathrm{d}V. \tag{8.162}$$

Solution The elastic energy in the medium may be converted into a surface integral using the divergence theorem:

$$\begin{aligned}
\frac{1}{2}\int_{\mathcal{M}} \sum_{i,j} \sigma_{ij}^{C} u_{i,j}^{C} \mathrm{d}V &= \frac{1}{2}\int_{\mathcal{M}} \sum_{i,j} \left(\sigma_{ij}^{C} u_{i}^{C}\right)_{,j} \mathrm{d}V \\
&= \frac{1}{2}\int_{\Sigma} \sum_{i,j} \sigma_{ij}^{C} u_{i}^{C} n_j \mathrm{d}S - \frac{1}{2}\int_{\mathcal{S}} \sum_{i,j} \sigma_{ij}^{C} u_{i}^{C} n_j \mathrm{d}S.
\end{aligned}$$

In the first line the absence of body forces has been used to equate $\sum_j \sigma_{ij,j}^{C}$ to zero. The minus sign in the second line is because the positive sense of the surface normal of the inclusion is into the medium, but the surface integral requires the opposite sense of the normal. Provided the external surface of the body is free of tractions then $\sum_j \sigma_{ij}^{C} n_j = 0$ on Σ, and the first integral is zero. Continuity of tractions on the surface of the inclusion requires $\sum_j \sigma_{ij}^{C} n_j = \sum_j \sigma_{ij}^{I} n_j$. Therefore,

$$\begin{aligned}
\frac{1}{2}\int_{\mathcal{M}} \sum_{i,j} \sigma_{ij}^{C} u_{i,j}^{C} \mathrm{d}V &= -\frac{1}{2}\int_{\mathcal{S}} \sum_{i,j} \sigma_{ij}^{I} u_{i}^{C} n_j \mathrm{d}S \\
&= -\frac{1}{2}\int_{\mathcal{R}} \sum_{i,j} \sigma_{ij}^{I} u_{i,j}^{C} \mathrm{d}V = -\frac{1}{2}\int_{\mathcal{R}} \sum_{i,j} \sigma_{ij}^{I} e_{ij}^{C} \mathrm{d}V.
\end{aligned}$$

In the second line the divergence theorem has been used to convert the integral over the surface of the inclusion into an integral over its volume. The total elastic energy is therefore:

$$E_{el} = \frac{1}{2}\int_{\mathcal{R}} \sum_{i,j} \sigma^{I}_{ij}\left(e^{I}_{ij} - e^{C}_{ij}\right) \mathrm{d}V = -\frac{1}{2}\int_{\mathcal{R}} \sum_{i,j} \sigma^{I}_{ij} e^{T}_{ij} \mathrm{d}V.$$

Comments

Consider the case where the surrounding medium is infinite in extent. If the inclusion is an ellipsoid the constrained stress and strain fields inside it are homogeneous. The total elastic energy of an ellipsoidal inclusion is then $-(V/2)\sigma^{I}_{ij}e^{T}_{ij}$, where $V = 4\pi a_1 a_2 a_3/3$ is the volume of the ellipsoid. If the inclusion is not an ellipsoid then the constrained stress and strain fields within it are inhomogeneous, and therefore σ^{I}_{ij} is a function of position.

When the surrounding medium is finite in extent, with no external tractions on Σ, the same expression for the total elastic energy applies, but the stress σ^{I}_{ij} is a function of position even for an ellipsoidal inclusion as a result of image interactions.

In all cases the total elastic energy is given by the integral over only the inclusion volume in eq 8.162.

Problem 8.2 The elastic interaction energy between a defect D and an applied field A is expressed as a surface integral in eq 6.6 where the integration surface encloses D only. In this question we apply this expression to an ellipsoidal inclusion to derive the interaction energy between an applied elastic field and an ellipsoidal inclusion. Take the surface S of the surface integral in eq 6.6 just outside the surface $\mathcal{S}$ of the inclusion. The interaction energy is then:

$$E_{int} = \int_{S} \sum_{i,j} \left(\sigma^{C}_{ij} u^{A}_{i} - \sigma^{A}_{ij} u^{C}_{i}\right) n_j \,\mathrm{d}S,$$

Show that the interaction energy may be expressed as the following volume integral over the interior of the inclusion $\mathcal{R}$:

$$E_{int} = \int_{\mathcal{R}} \sum_{i,j} \left(\sigma^{I}_{ij} e^{A}_{ij} - \sigma^{C}_{ij} e^{A}_{ij}\right) \mathrm{d}V.$$

Hence show that:

$$E_{int} = -\int_{\mathcal{R}} \sum_{i,j} \sigma^{A}_{ij} e^{T}_{ij} \,\mathrm{d}V. \tag{8.163}$$

Solution Using the continuity of tractions on the surface of the inclusion, $\sigma^C_{ij}n_j = \sigma^I_{ij}n_j$, and continuity of displacements u^C_i on either side of the surface of the inclusion, and $\sigma^I_{ij} = \sigma^C_{ij} - \sigma^T_{ij}$, and applying the divergence theorem we transform the surface integral into a volume integral over the inclusion:

$$
\begin{aligned}
E_{int} &= \int_S \sum_{i,j} \left(\sigma^I_{ij}u^A_i - \sigma^A_{ij}u^C_i\right) n_j \, dS \\
&= \int_{\mathcal{R}} \sum_{i,j} \left(\sigma^I_{ij}u^A_i - \sigma^A_{ij}u^C_i\right)_{,j} dV \\
&= \int_{\mathcal{R}} \sum_{i,j} \left(\sigma^I_{ij}u^A_{i,j} - \sigma^A_{ij}u^C_{i,j}\right) dV \\
&= \int_{\mathcal{R}} \sum_{i,j} \left(\sigma^I_{ij}e^A_{ij} - \sigma^A_{ij}e^C_{ij}\right) dV \\
&= \int_{\mathcal{R}} \sum_{i,j} \left(\sigma^I_{ij}e^A_{ij} - \sigma^C_{ij}e^A_{ij}\right) dV \\
&= -\int_{\mathcal{R}} \sum_{i,j} \sigma^T_{ij}e^A_{ij} dV \\
&= -\int_{\mathcal{R}} \sum_{i,j} \sigma^A_{ij}e^T_{ij} dV.
\end{aligned}
$$

In the third line we have equated the divergences of σ^I_{ij} and σ^A_{ij} to zero because there are no body forces in the inclusion.

Problem 8.3 Consider the *interaction energy between two inclusions* labelled A and B. With the origin of the coordinate system at the centre of one inclusion let the position vector of the centre of the other be $\mathbf{R}$. If $|\mathbf{R}|$ is much larger than either inclusion we may approximate the displacement field of one inclusion at the location of the other as in eq 8.2. In this approximation, and using eq 8.163 for the interaction energy, show that the interaction energy between the inclusions is as follows:

$$E_{int} \approx \sum_{i,j} \sum_{p,q} \rho^{(A)}_{ij} G_{ip,jq}(\mathbf{R}) \rho^{(B)}_{pq},$$

where $\rho_{ij}^{(A)} = \sigma_{ij}^{T(A)} V^{(A)}$ and $\sigma_{ij}^{T(A)}$ is the stress derived from the transformation strain $e_{ij}^{T(A)}$ using Hooke's law, $V^{(A)}$ is the volume of inclusion A, and similarly for inclusion B. The approximation amounts to treating the inclusions as point defects. $\rho_{ij}^{(A)}$ and $\rho_{ij}^{(B)}$ are dipole tensors (see section 4.5). This expressions for the interaction energy between two inclusions is also closely related to the interaction energy between two well separated dislocation loops in eq 6.17.

Solution Following eq 8.163, the interaction energy between the inclusions is given approximately by:

$$E_{int} \approx -V^{(A)} \sum_{i,l} \sigma_{il}^{T(A)} u_{i,l}^{(B)} = -V^{(B)} \sum_{i,l} \sigma_{il}^{T(B)} u_{i,l}^{(A)},$$

by the Maxwell reciprocity relation. Using eq 8.2 the displacement field at **R** from inclusion A is approximately:

$$u_i^{A)}(\mathbf{R}) \approx -V^{(A)} \sum_{j,k} G_{ij,k}(\mathbf{R}) \sigma_{jk}^{T(A)}. \tag{8.164}$$

Therefore, the interaction energy is as follows:

$$E_{int} \approx \sum_{i,j,k,l} V^{(B)} \sigma_{il}^{T(B)} G_{ij,kl}(\mathbf{R}) \sigma_{jk}^{T(A)} V^{(A)} = \sum_{i,j,k,l} \rho_{il}^{(B)} G_{ij,kl}(\mathbf{R}) \rho_{jk}^{(A)}.$$

Comment

This interaction energy has the same structure as that of two dislocation loops separated by a distance large compare to their sizes. See the discussion following eq 6.18.

Problem 8.4 Consider an inclusion in which the stress-free transformation strain is a dilation $e_{ij}^T = (e^T/3)\delta_{ij}$. Show that:

$$u_i^C(\mathbf{x}) = -\frac{(1+\nu)}{12\pi(1-\nu)} e^T \Phi_{,i}(\mathbf{x})$$

Hence prove that the dilation in the surrounding medium is zero, while the dilation in the inclusion is:

$$e^C = \frac{e^T}{3}\left(\frac{1+\nu}{1-\nu}\right).$$

Show that the hydrostatic stress in the inclusion is:

$$\frac{1}{3}\sum_i \sigma^I_{ii} = -\frac{4\mu e^T}{9}\left(\frac{1+\nu}{1-\nu}\right).$$

Solution From eq 8.9 the displacement field is as follows:

$$u^C_i(\mathbf{x}) = \frac{1}{8\pi(1-\nu)}\sum_{j,k}\left(\Psi_{,ijk}(\mathbf{x}) - 2\nu\Phi_{,i}(\mathbf{x})\delta_{jk} - 4(1-\nu)\delta_{ij}\Phi_{,k}(\mathbf{x})\right)e^T_{jk}.$$

For a pure dilation, $e^T_{jk} = \delta_{jk}e^T/3$, the displacement field becomes:

$$\begin{aligned} u^C_i(\mathbf{x}) &= \frac{e^T}{24\pi(1-\nu)}\sum_{j,k}\left(\Psi_{,ijk}(\mathbf{x})\delta_{jk} - 2\nu\Phi_{,i}(\mathbf{x})\delta_{jk}\delta_{jk} - 4(1-\nu)\delta_{ij}\delta_{jk}\Phi_{,k}(\mathbf{x})\right) \\ &= -\frac{e^T}{12\pi(1-\nu)}(1+\nu)\Phi_{,i}(\mathbf{x}) \end{aligned}$$

where we have used $\nabla^2\Psi(\mathbf{x}) = 2\Phi(\mathbf{x})$.

The dilation e^C is as follows:

$$e^C = \sum_i u^C_{i,i} = -\frac{(1+\nu)}{12\pi(1-\nu)}e^T\nabla^2\Phi(\mathbf{x}).$$

From eq 8.8 we have $\nabla^2\Phi(\mathbf{x}) = -4\pi$ inside the inclusion and $\nabla^2\Phi(\mathbf{x}) = 0$ outside the inclusion. Therefore there is no dilation outside the inclusion while the dilation inside the inclusion is:

$$e^C = \left(\frac{1+\nu}{1-\nu}\right)\frac{e^T}{3}.$$

The hydrostatic stress in the inclusion is evaluated as follows:

$$\frac{1}{3}\sum_i \sigma^I_{ii} = \frac{1}{3}\sum_i \sigma^C_{ii} - \sigma^T_{ii}$$

From Hooke's law we have:

$$\frac{1}{3}\sum_i \sigma_{ii}^C = \frac{1}{3}\left(2\mu e^C + \frac{6\mu\nu}{1-2\nu}e^C\right) = \frac{2\mu}{3}\frac{(1+\nu)^2}{(1-2\nu)(1-\nu)}\frac{e^T}{3}.$$

and

$$\frac{1}{3}\sum_i \sigma_{ii}^T = \frac{1}{3}\left(2\mu e^T + \frac{6\mu\nu}{1-2\nu}e^T\right) = \frac{2\mu}{3}\frac{(1+\nu)}{(1-2\nu)}e^T$$

Therefore,

$$\frac{1}{3}\sum_i \sigma_{ii}^I = \frac{1}{3}\sum_i \sigma_{ii}^C - \sigma_{ii}^T = -\frac{4\mu}{9}\left(\frac{1+\nu}{1-\nu}\right)e^T.$$

Comments

Remarkably, these results are independent of the shape of the inclusion, but they do assume the elastic constants inside the inclusion are the same as those outside, and that they are isotropic. For $\nu = \frac{1}{3}$ we find the surrounding medium reduces the unconstrained dilation of the inclusion by a third. Notice also that $e^C \to e^T$ when $\nu \to \frac{1}{2}$ because in that limit the material is rigid, and the bulk modulus becomes infinite. These results apply only when the inclusion is in an infinite medium. In a finite medium there is a dilation in the material outside the inclusion, as we found for point defects in section 5.2.

Problem 8.5 Show that eqs 8.103, 8.105 and 8.107 reproduce the displacement, strain and stress fields of the inclusion in the far-field limit, given by eqs 8.13, 8.14 and 8.15 respectively.

Hints: First show that $\chi \to |\mathbf{x}|^2 = x^2$ as $x \to \infty$. To find the limits of A_i and B_{ij} as $\chi \to \infty$ make the substitution $w = 1/u$ in the integrals

$$A_i = \pi a_1 a_2 a_3 \int_\chi^\infty \frac{\mathrm{d}u}{(a_i^2+u)\Delta(u)} \quad \text{and} \quad B_{ij} = \pi a_1 a_2 a_3 \int_\chi^\infty \frac{\mathrm{d}u}{(a_i^2+u)(a_j^2+u)\Delta(u)}.$$

Hence show that $A_i \to V/(2x^3)$, $B_{ij} \to 3V/(10x^5)$ and $\beta_{ijk} \to 3V/(4x^7)$ as $x \to \infty$.

Solution As discussed in section 8.12, χ is defined by the equation of the confocal ellipsoid:

$$\frac{x_1^2}{a_1^2+\chi}+\frac{x_2^2}{a_2^2+\chi}+\frac{x_3^2}{a_3^2+\chi}=1$$

As $x = |\mathbf{x}|$ increases the value of χ must also increase to balance this equation. The ellipsoid becomes almost spherical with radius $\chi \to x^2$.

Alternatively, χ is the positive root of the cubic equation:

$$\chi^3 - A\chi^2 - B\chi - C = 0,$$

When $x^2 \gg a_1^2, a_2^2, a_3^2$ the cubic equation reduces to $\chi^3 - A\chi^2 = 0$, so that $\chi \to A$. But $A = x_1^2 + x_2^2 + x_3^2 - a_1^2 - a_2^2 - a_3^2 \to x^2$. We conclude again that $\chi \to x^2$.

It follows that $\chi_{,k} \to 2x_k$.

Consider the limiting value of A_i when $x^2 \gg a_1^2, a_2^2, a_3^2$. A_i is defined by the following integral:

$$A_i = \pi a_1 a_2 a_3 \int_{\chi}^{\infty} \frac{du}{(a_i^2+u)\Delta(u)},$$

with $\Delta(u) = [(a_1^2+u)(a_2^2+u)(a_3^2+u)]^{\frac{1}{2}}$. Let $u = 1/w$. Then A_i becomes:

$$A_i = \pi a_1 a_2 a_3 \int_0^{1/\chi} dw \frac{w^{1/2}}{(1+a_i^2 w)(1+a_1^2 w)^{1/2}(1+a_2^2 w)^{1/2}(1+a_3^2 w)^{1/2}}.$$

This equation is still exact. But when we take the limit $\chi \gg 1$ we can expand the integrand to lowest order in w. The integrand becomes $w^{1/2}$, and

$$A_i \to \pi a_1 a_2 a_3 \int_0^{1/\chi} dw\, w^{1/2} = \frac{2}{3}\pi a_1 a_2 a_3 \frac{1}{\chi^{3/2}} = \frac{V}{2x^3},$$

where $V = (4\pi/3)a_1 a_2 a_3$ is the volume of the ellipsoid.

Consider the limiting value of B_{ij} when $x^2 \gg a_1^2, a_2^2, a_3^2$, where B_{ij} is defined by the following integral:

$$B_{ij} = \pi a_1 a_2 a_3 \int_{\chi}^{\infty} \frac{\mathrm{d}u}{(a_i^2 + u)(a_j^2 + u)\Delta(u)}.$$

Making the substitution $u = 1/w$ and retaining only the lowest order terms in w this integral becomes:

$$B_{ij} \to \pi a_1 a_2 a_3 \int_{0}^{1/\chi} \mathrm{d}w\, w^{3/2} = \frac{2}{5}\pi a_1 a_2 a_3 \frac{1}{\chi^{5/2}} = \frac{3V}{10x^5}.$$

Consider the limiting value of α_{jk}, which is defined by:

$$\alpha_{jk}(\chi) = \frac{\pi a_1 a_2 a_3}{\Delta(\chi)(a_j^2 + \chi)(a_k^2 + \chi)\sum_m \frac{x_m^2}{(a_m^2+\chi)^2}}.$$

In the limit $x^2 \gg a_1^2, a_2^2, a_3^2$ this becomes:

$$\alpha_{jk}(\chi) \to \frac{\pi a_1 a_2 a_3}{\chi^{3/2}\chi^2(x^2/\chi^2)} \to \frac{3V}{4\chi^{5/2}} \to \frac{3V}{4x^5}.$$

Consider the limiting value of β_{ijk}, which is defined by:

$$\beta_{ijk} = \frac{\alpha_{jk}}{(a_i^2 + \chi)}.$$

In the limit $x^2 \gg a_1^2, a_2^2, a_3^2$ this becomes:

$$\beta_{ijk} \to \frac{\alpha_{jk}}{\chi} \to \frac{3V}{4\chi^{7/2}} \to \frac{3V}{4x^7}.$$

and hence

$$\frac{\partial \beta_{ijk}}{\partial \chi} \to -\frac{21V}{8\chi^{9/2}} \to -\frac{21V}{8x^9}.$$

We now have what we need to derive the limiting forms of the exact displacement, strain and stress fields far from the inclusion.

The exact displacement field outside the inclusion is stated in eq 8.103. Far from the inclusion this exact expression becomes:

$$\begin{aligned}4\pi(1-\nu)u_i^C(\mathbf{x}) \quad \rightarrow \quad & \sum_{j,k} e^T_{jk}\left\{\delta_{jk}\left(\frac{3V}{10x^5}a_j^2 - \frac{V}{2x^3}\right)x_i + \delta_{ki}\left(\frac{3V}{10x^5}a_k^2 - \frac{V}{2x^3}\right)x_j\right. \\ & +\delta_{ij}\left(\frac{3V}{10x^5}a_i^2 - \frac{V}{2x^3}\right)x_k \\ & \left. +2x^2\frac{3V}{4x^7}x_i x_j x_k + 2\nu\frac{V}{2x^3}\delta_{jk}x_i + 4(1-\nu)\frac{V}{2x^3}\delta_{ij}x_k\right\}\end{aligned}$$

The dominant terms are of order $1/x^2$:

$$u_i^C(\mathbf{x}) \rightarrow \frac{V}{8\pi(1-\nu)}\sum_{j,k}\left\{2(1-2\nu)\frac{x_k}{x^3}\delta_{ij} - (1-2\nu)\frac{x_i}{x^3}\delta_{jk} + 3\frac{x_i x_j x_k}{x^5}\right\}e^T_{jk}.$$

which is identical to eq 8.13.

To derive the limiting forms of the strain and stress fields far from the inclusion from the exact expressions for the fields in eqs 8.105 and 8.107 we need the limiting forms of $\Phi_{,ij}$ and $\Psi_{,ijkl}$. They are:

$$\begin{aligned}\Phi_{,ij} \quad &\rightarrow \quad V\left(\frac{3x_i x_j}{x^5} - \frac{\delta_{ij}}{x^3}\right) \\ \Psi_{,ijkl} \quad &\rightarrow \quad -\frac{V}{x^3}\left(\delta_{ij}\delta_{kl} + \delta_{jk}\delta_{li} + \delta_{ik}\delta_{jl}\right) \\ & \quad +\frac{3V}{x^5}\left(x_i x_j\delta_{kl} + x_j x_k\delta_{li} + x_k x_l\delta_{ij} + x_l x_i\delta_{jk} + x_j x_l\delta_{ik} + x_i x_k\delta_{jl}\right) \\ & \quad -\frac{15V}{x^7}x_i x_j x_k x_l.\end{aligned}$$

Using these expressions in eqs 8.105 and 8.107 enables the limiting forms of the strain and stress fields to be derived far from the inclusion, after some tedious algebra. The limiting forms are identical to eqs 8.14 and 8.15 respectively.

Problem 8.6 Equation 8.2 suggests a systematic way of making the approximation to the far field of the ellipsoidal inclusion more accurate. If $|\mathbf{x}| \gg a_1, a_2, a_3$ then we may expand $G_{ij,k}(\mathbf{x}-\mathbf{x}')$ as a Taylor series:

$$G_{ij,k}(\mathbf{x}-\mathbf{x}') = G_{ij,k}(\mathbf{x}) - \sum_l x'_l G_{ij,kl}(\mathbf{x}) + \frac{1}{2}\sum_{l,m} x'_l x'_m G_{ij,klm}(\mathbf{x}) + \ldots \quad (8.165)$$

Inserting this expansion into eq 8.2 show that it becomes:

$$u_i^C(\mathbf{x}) = -V\sum_{j,k}\sigma^T_{jk}\left(G_{ij,k}(\mathbf{x}) + \frac{1}{10}\sum_m G_{ij,kmm}(\mathbf{x})a_m^2 + \ldots\right) \quad (8.166)$$

Solution Inserting the Taylor expansion, eq 8.165, into eq 8.2, the symmetry of the ellipsoid ensures that integrals that are odd in the coordinates x'_i are zero. The first term of the correction is therefore:

$$\frac{1}{2}\sum_{j,k}\sum_m G_{ij,kmm}(\mathbf{x})\int_{\mathcal{R}} \mathrm{d}V'\, x'^2_m.$$

Consider the integral for $m = 1$:

$$\int_{\mathcal{R}} \mathrm{d}V'\, x'^2_1 = 2\int_0^{a_1} \mathrm{d}x'_1\, \pi a_2 a_3\left(1 - \frac{x'^2_1}{a_1^2}\right)x'^2_1$$

$$= \frac{4\pi a_1^3 a_2 a_3}{15} = V\frac{a_1^2}{5}.$$

Therefore,

$$\int_{\mathcal{R}} \mathrm{d}V'\, x'^2_m = V\frac{a_m^2}{5}.$$

The first order correction to eq 8.2 is:

$$-V\sum_{j,k}\frac{\sigma^T_{jk}}{10}\sum_m G_{ij,kmm}(\mathbf{x})a_m^2.$$

Comment
Notice that the first order correction to the long-range displacement field of eq 8.13 involves third derivatives of the Green's functions, which vary as $1/|\mathbf{x}|^4$ while the associated stresses and strains vary as $1/|\mathbf{x}|^5$. A similar observation was made in connection with a finite centrosymmetric dislocation loop in eq 6.13. Further successive corrections involve the fifth, seventh, ninth, ... derivatives of the Green's function.

Problem 8.7 *A slip band.* As discussed in section 11.4.3, a slip band is a long, thin blade-like region that has undergone a simple shear through the collective motion of dislocations on parallel slip planes. It may be described as an ellipsoidal region with $a_1 \gg a_2 \gg a_3$ that has undergone a simple shear of 2ε on the $x_1 - x_2$ plane in the x_1-direction in response to a local shear stress σ^A_{13}. The local shear stress may arise from an externally applied load and from other defects in the body including other shear bands. As the shear band grows the local shear stress σ^A_{13} will change as a result of the interaction with other defects. In section 1.3.1 a simple shear was shown to comprise a pure shear and a rotation. The stress-free strain tensor has components $e^T_{13} = e^T_{31} = \varepsilon$. We will derive the elastic field inside and outside a slip band. These fields are useful because a slip band may be thought of as a defect in its own right characterised by a_1, a_2, a_3 and e^T_{13}.

For a slip band with $a_2/a_1 = 1/50$ and $a_3/a_1 = 1/500$ the expressions for A_1, A_2 and A_3 in eq 8.39 yield the following values:

$$\begin{aligned} A_1 &= 1.057 \times 10^{-3} \\ A_2 &= 0.571 \\ A_3 &= 5.711\,. \end{aligned}$$

For $\nu = 1/3$ the stress inside the slip band is as follows:

$$\begin{aligned} \sigma^C_{13} &= 1.818\mu e^T_{13} \\ \sigma^T_{13} &= 2\mu e^T_{13} \\ \sigma^I_{13} &= \sigma^C_{13} - \sigma^T_{13} = -0.182\mu e^T_{13}. \end{aligned} \tag{8.167}$$

There are no other stress components in the slip band apart from the applied stress σ^A_{13}. Notice that σ^I_{13} has the opposite sign to σ^T_{13}. The total shear stress acting on the dislocations that make up the shear band is $\sigma^A_{13} + \sigma^I_{13}$. Since σ^I_{13} is constant throughout the ellipsoid, all dislocations in the slip band experience the same shear stress if the variation of σ_{13}^A within the slip band is small. That is why they move and stop together. The shear band grows through the motion of dislocations until $\sigma_{13}^A + \sigma_{13}^I$ is equal to the friction stress acting on them. If the slip band grows in a self-similar manner, so that a_2/a_1 and a_3/a_1 remain constant, then σ_{13}^I also remains constant.

Using eq 8.103 show that the elastic displacement field outside the slip band is as follows:

$$2\pi(1-\nu)u_i^C(\mathbf{x}) = \left\{ (\delta_{i1}x_3 + \delta_{i3}x_1)\left(A_1(\chi)a_1^2 - A_3(\chi)a_3^2\right)/\left(a_3^2 - a_1^2\right) \right.$$
$$\left. + 2\chi\beta_{i13}(\chi)x_i x_1 x_3 + 2(1-\nu)\left(\delta_{i1}A_3(\chi)x_3 + \delta_{i3}A_1 x_1\right) \right\} e_{13}^T \quad (8.168)$$

where $A_1(\chi)$ and $A_3(\chi)$ are given by eq 8.39 and ϕ_χ by eq 8.77. The coefficients $\beta_{ijk}(\chi)$ are defined in eq 8.99 and a simple formula for χ is derived in section 8.12.

Show that the strain in the surrounding medium is given by:

$$e_{ij}^C(\mathbf{x}) = \frac{e_{13}^T}{4\pi(1-\nu)} \left(\Psi_{,ij13}(\mathbf{x}) \right.$$
$$\left. -(1-\nu)\left(\delta_{i1}\Phi_{,j3}(\mathbf{x}) + \delta_{j1}\Phi_{,i3}(\mathbf{x}) + \delta_{i3}\Phi_{,j1}(\mathbf{x}) + \delta_{j3}\Phi_{,i1}(\mathbf{x})\right) \right)$$

and that the dilation $e^C(\mathbf{x})$ is as follows:

$$e^C(\mathbf{x}) = -\frac{(1-2\nu)}{2\pi(1-\nu)}\Phi_{,13}(\mathbf{x})e_{13}^T = -\frac{2(1-2\nu)}{\pi(1-\nu)}\alpha_{13}(\mathbf{x})x_1 x_3 e_{13}^T \quad (8.169)$$

where $\alpha_{13}(\mathbf{x})$ is given by eq 8.97.

The stress field outside the slip band follows from Hooke's law (eq 8.107):

$$\sigma_{ij}^C(\mathbf{x}) = \frac{\mu e_{13}^T}{2\pi(1-\nu)} \left(\Psi_{,ij13}(\mathbf{x}) \right.$$
$$-(1-\nu)\left(\delta_{i1}\Phi_{,j3}(\mathbf{x}) + \delta_{j1}\Phi_{,i3}(\mathbf{x}) + \delta_{i3}\Phi_{,j1}(\mathbf{x}) + \delta_{j3}\Phi_{,i1}(\mathbf{x})\right)$$

$$-2\nu\delta_{ij}\Phi_{,13}(\mathbf{x})\Big). \tag{8.170}$$

Notice that the homogeneous shear strain $e^T_{13} = e^T_{31}$ inside the slip band generates finite values of all components of the stress tensor outside the slip band.

The five derivatives $\Phi_{,11}, \Phi_{,33}, \Phi_{,23}, \Phi_{,13}, \Phi_{,12}$ are readily evaluated using eq 8.94. We also need the 6 derivatives $\Psi_{,ij13}$ where $i, j = 1,2,3$. Show that:

$$\Psi_{,ij13} = 2\left(\delta_{j1}\delta_{i3}(B_{ij}(\chi)a_j^2 - A_i(\chi)) + \delta_{i1}\delta_{j3}(B_{j1}(\chi)a_1^2 - A_j(\chi))\right)$$
$$+4\chi\left(\delta_{i1}\beta_{ij3}x_jx_3 + \beta_{13j}x_1(\delta_{i3}x_j + \delta_{ij}x_3) + \beta_{i13}x_i(\delta_{j1}x_3 + \delta_{j3}x_1)\right)$$
$$+4x_ix_jx_1\left(\beta_{ij1} + \chi\frac{\partial\beta_{ij1}}{\partial\chi}\right)\chi_{,3}, \tag{8.171}$$

where $\partial\beta_{ijk}/\partial\chi$ and $\chi_{,3}$ are given by eqs 8.102, and 8.87 respectively.

Solution Using eq 8.103 the displacement field outside the slip band is evaluated as follows:

$$\begin{aligned}4\pi(1-\nu)u_i^C(\mathbf{x}) &= e^T_{13}\Big\{2\delta_{i1}\left(B_{31}(\chi)a_1^2 - A_3(\chi)\right)x_3 + 2\delta_{i3}\left(B_{13}(\chi)a_3^2 - A_1(\chi)\right)x_1\\ &\quad +4\chi\beta_{i13}(\chi)x_ix_1x_3 + 4(1-\nu)\left(\delta_{i1}A_3(\chi)x_3 + \delta_{i3}A_1(\chi)x_1\right)\Big\}\\ &= e^T_{13}\Big\{2\left(\frac{A_1(\chi)a_1^2 - A_3(\chi)a_3^2}{a_3^2 - a_1^2}\right)(\delta_{i1}x_3 + \delta_{i3}x_1)\\ &\quad +4\chi\beta_{i13}(\chi)x_ix_1x_3 + 4(1-\nu)\left(\delta_{i1}A_3(\chi)x_3 + \delta_{i3}A_1(\chi)x_1\right)\Big\}\end{aligned}$$

Using eq 8.105 the strain field outside the slip band is as follows:

$$e^C_{ij}(\mathbf{x}) = S_{ij13}(\mathbf{x})e^T_{13} + S_{ij31}(\mathbf{x})e^T_{31}$$

where

$$S_{ij13}(\mathbf{x}) = \frac{1}{8\pi(1-\nu)}\left(\Psi_{,ij13}(\mathbf{x}) - 2(1-\nu)(\delta_{i1}\Phi_{,j3}(\mathbf{x}) + \delta_{j1}\Phi_{,i3}(\mathbf{x}))\right)$$

$$S_{ij31}(\mathbf{x}) = \frac{1}{8\pi(1-\nu)}\left(\Psi_{,ij31}(\mathbf{x}) - 2(1-\nu)(\delta_{i3}\Phi_{,j1}(\mathbf{x}) + \delta_{j3}\Phi_{,i1}(\mathbf{x}))\right).$$

Since $\Psi_{,ij13}(\mathbf{x}) = \Psi_{,ij31}(\mathbf{x})$ it follows that:

$$e^C_{ij}(\mathbf{x}) = \frac{e^T_{jk}}{4\pi(1-\nu)}\Big\{\Psi_{,ij13}(\mathbf{x})$$
$$-(1-\nu)\left(\delta_{i1}\Phi_{,j3}(\mathbf{x}) + \delta_{i3}\Phi_{,j1}(\mathbf{x}) + \delta_{j1}\Phi_{,i3}(\mathbf{x}) + \delta_{j3}\Phi_{,i1}(\mathbf{x})\right)\Big\}$$

The dilation outside the slip band is the trace of the tensor $e^C_{ij}(\mathbf{x})$:

$$e^C = -\frac{(1-2\nu)e^T_{13}}{2\pi(1-\nu)}\Phi_{,13}(\mathbf{x}) = -\frac{(1-2\nu)e^T_{13}}{2\pi(1-\nu)}4\alpha_{13}(\mathbf{x})x_1x_3,$$

where

$$\alpha_{13}(\mathbf{x}) = \frac{\pi a_1a_2a_3}{\Delta(\chi)(a_1^2+\chi)(a_3^2+\chi)\sum_m \frac{x_m^2}{(a_m^2+\chi)^2}}.$$

To derive the stress field outside the slip band we use Hooke's law:

$$\sigma^C_{ij}(\mathbf{x}) = 2\mu e^C_{ij}(\mathbf{x}) + \frac{2\mu\nu}{1-2\nu}\delta_{ij}e^C$$
$$= \frac{\mu e^T_{13}}{2\pi(1-\nu)}\Big\{\Psi_{,ij13}(\mathbf{x})$$
$$-(1-\nu)\left(\delta_{i1}\Phi_{,j3}(\mathbf{x}) + \delta_{i3}\Phi_{,j1}(\mathbf{x}) + \delta_{j1}\Phi_{,i3}(\mathbf{x}) + \delta_{j3}\Phi_{,i1}(\mathbf{x})\right)$$
$$-2\nu\delta_{ij}\Phi_{,13}(\mathbf{x})\Big\}$$

From eq 8.100 we have:

$$\Psi_{,ij13}(\mathbf{x}) = 2\left\{\delta_{j1}\delta_{i3}\left(B_{ij}(\mathbf{x})a_j^2 - A_i(\mathbf{x})\right) + \delta_{i1}\delta_{j3}\left(B_{ij}(\mathbf{x})a_j^2 - A_i(\mathbf{x})\right)\right\}$$
$$+4\chi\Big(\delta_{i1}\beta_{3ij}(\mathbf{x})x_jx_3 + \beta_{13j}(\mathbf{x})x_1(\delta_{i3}x_j + \delta_{ij}x_3)$$
$$+\beta_{i13}(\mathbf{x})x_i(\delta_{j3}x_1 + \delta_{j1}x_3)\Big) + 4x_ix_jx_1\left(\beta_{ij1}(\mathbf{x}) + \chi\frac{\partial\beta_{ij1}(\mathbf{x})}{\partial\chi}\right)\chi_{,3}.$$

Comments

The stress field inside the slip band is finite. Putting numbers into eq 8.170 we find the stress field outside the slip band is finite everywhere[a]. However, the stresses change extremely rapidly just outside the slip band. For example, at $x_1 = a_1, x_2 = 0$ and $x_3 = 0$ we find $\sigma^C_{13} = -0.182\mu e^T_{13}$, and all other stress components are zero, which makes the tractions continuous across the interface. But slightly further along the x_1-axis at $x_1 = (1+10^{-5})a_1$ it increases to $\sigma^C_{13} = +0.842\mu e^T_{13}$, while at $x_1 = (1+10^{-2})a_1$ it is $+0.004\mu e^T_{13}$. If a_1 is less than 10 μm, then $10^{-5}a_1$ is the size of an atom. In reality these rapid stress variations occur within the cores of the dislocations that make up the slip band, and they are not described accurately by elasticity. At $x_1 = 5a_1$, $\sigma^C_{13} = +4.48\times10^{-7}\mu e^T_{13}$, which differs from the value ($4.27\times10^{-7}\mu e^T_{13}$) predicted by eq 8.15 by less than 5%. Not surprisingly the long range stress field of eq 8.15 fails to satisfy the condition of continuous tractions across the interface. But at distances from the centre of the slip band greater than about $5a_1$ it provides an acceptable approximation to the stress field.

[a]This is in contrast to the linear elastic field of a dislocation pileup, as will be discussed in Chapter 10.

9. The force on a defect

9.1 Introduction

We have already encountered the concept of the force on a defect with the Peach-Koehler force on a dislocation. In this chapter we will generalise the concept to other defects including point defects and interfaces. In Chapter 10 we will consider the force on a crack tip. The forces we are concerned with are often called configurational forces. They arise from changes in the total potential energy of the system when a defect is displaced. The total potential energy comprises the elastic energy of the body in which the defects reside and the potential energy of any external mechanism applying tractions to the surface of the body. The configurational force on the defect is then defined as minus the gradient of the total potential energy of the system with respect to the position of the defect. It follows that it is always a conservative force[1].

This chapter is based on a series of papers by Eshelby[2], who first introduced the concept of the force on an elastic singularity in a paper published in 1951[3]. He defined an elastic singularity in a homogeneous elastic medium in the following way. Draw a closed surface S inside the medium. There is an elastic singularity inside S if the stresses inside S could not be produced by body forces outside S or tractions on S. The stresses inside S are not necessarily singular in the mathematical sense, e.g. the stress field of a dislocation with a finite core width does not display a mathematical singularity. Nevertheless, a dislocation with a finite core width is an elastic singularity

[1]There are other forces on defects arising from an exchange of momentum with another field, such as an electric current flowing through the medium. Such current-induced forces may be regarded as body forces and it has been shown that they are generally non-conservative - see Todorov, T N, Dundas, D, Lü, J-T, Brandbyge M, and Hedegård, P, Eur. J. Phys. **35**, 065004 (2014).

[2]Eshelby, J D, Phil. Trans. R. Soc. London A **244**, 87-111 (1951); Solid State Physics **3**, 79-144 (1956); In *Inelastic behaviour of solids*, eds. M F Kanninen, W F Adler, A R Rosenfield and R I Jaffee, p. 77-115, McGraw-Hill: New York (1970); J. Elasticity **5**, 321-335 (1975); In *Continuum models of discrete systems*, eds. E Kröner and K-H Anthony, University of Waterloo Press: Waterloo, p.651-665, (1980). All 57 of Eshelby's papers are available in *Collected works of J. D. Eshelby*, eds. X Markenscoff and A Gupta, Springer: Dordrecht (2006). ISBN 9781402044168 hard back, ISBN 9789401776448 soft cover.

[3]published one year after his PhD in Physics at the University of Bristol.

in the sense defined by Eshelby.

9.2 The electrostatic force on a charge

To introduce the idea of the force on a defect we will first review the use of the Maxwell electrostatic stress tensor to calculate the force on a charge. Suppose there is some charge density $\rho(\mathbf{x})$ in a vacuum and an associated electric field $\mathbf{E}(\mathbf{x})$. The two are related by $\nabla \cdot \mathbf{E} = \rho(\mathbf{x})/\varepsilon_0$, where ε_0 is the permittivity of free space. The charge density may consist of a set of point charges or a continuous distribution. Consider a closed region $\mathcal{R}$ in the distribution of charges, and let $\mathcal{S}$ be the surface of $\mathcal{R}$. We can write down the force $\mathbf{F}$ acting on the charges inside $\mathcal{R}$ as the following volume integral over $\mathcal{R}$:

$$\mathbf{F} = \int_{\mathcal{R}} \rho(\mathbf{x})\, \mathbf{E}(\mathbf{x})\, \mathrm{d}V. \tag{9.1}$$

In this expression $\mathbf{E}(\mathbf{x})$ includes contributions to the electric field at $\mathbf{x}$ from both inside and outside $\mathcal{R}$. If the integrand could be expressed as the divergence of a second rank tensor $M_{ij}(\mathbf{x})$, that is to say if $\rho(\mathbf{x})E_i(\mathbf{x}) = M_{ij,j}$, it would be possible to use the divergence theorem to express the force as a surface integral over $\mathcal{S}$:

$$F_i = \int_{\mathcal{R}} \rho(\mathbf{x})\, E_i(\mathbf{x})\, \mathrm{d}V = \int_{\mathcal{R}} M_{ij,j}\, \mathrm{d}V = \int_{\mathcal{S}} M_{ij}\, n_j\, \mathrm{d}S. \tag{9.2}$$

As shown in texts on electromagnetism[4] the tensor M_{ij} does exist and it is the Maxwell electrostatic stress tensor:

$$M_{ij} = \varepsilon_0 \left\{ E_i E_j - \frac{E^2}{2}\delta_{ij} \right\} \tag{9.3}$$

Exercise 9.1 Verify that $M_{ij,j} = \rho E_i$. ■

Solution The divergence of M_{ij} is as follows:

$$M_{ij,j} = \varepsilon_0 \left(E_{i,j}E_j + E_i E_{j,j} - E_k E_{k,j}\delta_{ij} \right),$$

where we have used $\left(E^2\right)_{,j} = 2E_k E_{k,j}$. According to Poisson's equation $E_{j,j} = \rho/\varepsilon_0$.

[4] e.g. Griffiths, D J, *Introduction to electrodynamics*, International edn., Prentice-Hall (1999), p.351. ISBN 978-9332550445.

Therefore,

$$\begin{aligned} M_{ij,j} &= \varepsilon_0 \left(E_{i,j} E_j + E_i \frac{\rho}{\varepsilon_0} - E_k E_{k,i} \right) \\ &= \rho E_i + \varepsilon_0 \left(E_{i,j} E_j - E_k E_{k,i} \right) \\ &= \rho E_i + \varepsilon_0 \left(E_{i,k} E_k - E_k E_{k,i} \right) \end{aligned}$$

Since $E_i = -V_{,i}$, where V is the electrostatic potential, then $E_{i,k} = -V_{,ik} = -V_{,ki} = E_{k,i}$. Therefore $E_{i,k}E_k - E_k E_{k,i} = 0$, and $M_{ij,j} = \rho E_i$.

Imagine the charge distribution is a set of point charges and we wish to calculate the force on one of them. Suppose the surface $\mathcal{S}$ surrounding our chosen point charge is deformed into another surface $\mathcal{S}'$ in such a way that no other point charges are included within $\mathcal{S}'$. Since there are no charges between $\mathcal{S}$ and $\mathcal{S}'$ the divergence $M_{ij,j}$ is zero everywhere between the surfaces, and therefore the surface integrals $\int_S M_{ij} n_j \mathrm{d}S$ and $\int_{S'} M_{ij} n_j \mathrm{d}S$ are equal.

In the following sections we will derive a tensor whose divergence can replace the integrand in a volume integral representing the force on a defect. Then, using the divergence theorem this volume integral can be converted into a surface integral enclosing the defect, which can be deformed arbitrarily so long as it does not include any other defects, to give an expression for the configurational force on the defect.

9.3 The pressure on an interface

In this section we derive the force per unit area on an interface separating two regions A and B of a body. The interface may enclose a region or it may extend through the body and terminate at its surface. The forces we are concerned with arise from differences in the elastic energy densities W^A and W^B, for example stemming from different elastic constants in A and B when A and B are elastically deformed, or differences in the densities of dislocations and point defects which drive recrystallisation. The interface may also be an agent of deformation, as in mechanical twinning, martensitic transformations and a slip band, where movement of the interface is accompanied by deformation of the region through which it sweeps. There are other sources of forces on interfaces, which we will not be concerned with, arising from curvature of the interface, and chemical free energy differences on either side of the interface during a phase change. In the following we assume the interface is coherent in the sense that the elastic displacement field and tractions are continuous across the interface.

Let the local normal vector to the interface be directed from B to A, with its tail on the interface. At each point on the interface $\mathcal{I}$ we imagine it migrates by the infinitesimal vector $\delta\xi n_i$ to a new position I'. With $\delta\xi > 0$ this results in the consumption of A and the growth of B. Let δE_{tot} be the change in the potential energy of the system resulting from this migration.

Cut out and remove the sliver of A between I and I' before the interface has migrated, applying suitable tractions to the freshly created surfaces to prevent relaxation. The change in the energy of A accompanying this first step is

$$\delta E^{(1)} = -\int_I W^A \delta\xi \mathrm{d}S \tag{9.4}$$

The elastic displacements on the newly created surface of A are $u_i^A + \delta\xi n_k u_{i,k}^A$, and the tractions are $\sigma_{ij}^A n_j + O(\delta\xi)$. The elastic displacements on the surface of A will relax to final values u_i^{FA}, where the difference $u_i^{FA} - \left(u_i^A + \delta\xi n_k u_{i,k}^A\right)$ is of order $\delta\xi$. Therefore, the surface tractions on the surface of A do work, which gives a second contribution to the change of the energy:

$$\delta E^{(2)} = -\int_I \sigma_{ij}^A n_j \left[u_i^{FA} - \left(u_i^A + \delta\xi n_k u_{i,k}^A\right)\right] \mathrm{d}S \quad + \; O(\delta\xi^2) \tag{9.5}$$

Now consider changes on the B-side. First there is the elastic energy arising from the creation of the sliver B between I and I':

$$\delta E^{(3)} = \int_I W^B \delta\xi \mathrm{d}S. \tag{9.6}$$

The elastic displacements on the newly created surface of B are $u_i^B + \delta\xi n_k u_{i,k}^B$ and they relax in the final state to u_i^{FB}. The tractions on the surface of B are $\sigma_{ij}^B n_j + O(\delta\xi)$. The work done by them during the relaxation to the final state is:

$$\delta E^{(4)} = +\int_I \sigma_{ij}^B n_j \left(u_i^{FB} - (u_i^B + \delta\xi n_k u_{i,k}^B)\right) \mathrm{d}S \quad + \; O(\delta\xi^2) \tag{9.7}$$

The minus sign in front of the integral in eq 9.5 is a plus sign in eq 9.7 because the sense of the normal is reversed.

In the final configuration the tractions and elastic displacements are continuous across the interface. Therefore, $\sigma_{ij}^A n_j = \sigma_{ij}^B n_j$ and $u_i^{FA} - u_i^A = u_i^{FB} - u_i^B$. The total change in the energy is then:

$$\delta E_{tot} \quad = \quad \delta E^{(1)} + \delta E^{(2)} + \delta E^{(3)} + \delta E^{(4)}$$

$$= -\int_I n_j \left\{ (W^A \delta_{jk} - \sigma^A_{ij} u^A_{i,k}) - (W^B \delta_{jk} - \sigma^B_{ij} u^B_{i,k}) \right\} \delta\xi n_k \mathrm{d}S$$

$$= -\int_I n_j \left\{ P^A_{jk} - P^B_{jk} \right\} \delta\xi n_k \mathrm{d}S \tag{9.8}$$

where

$$P_{jk} = W\delta_{jk} - \sigma_{ij} u_{i,k} \tag{9.9}$$

is called the static energy-momentum tensor of the elastic field. A formal derivation of the static energy-momentum tensor is given in section 9.5. This is the first time we meet it and it plays a central role in the configurational force on any defect, not only an interface. The pressure acting on an element of area of the interface is then:

$$p = -\frac{\delta E_{tot}}{\delta\xi \, \mathrm{d}S} = n_j \left(P^A_{jk} - P^B_{jk} \right) n_k. \tag{9.10}$$

Exercise 9.2 Consider a flat interface with unit normal $\hat{\mathbf{n}}$ between two semi-infinite media with elastic constants c^A_{ijkl} and c^B_{ijkl} with reference to a common coordinate system. Suppose a uniform displacement gradient $u_{i,j}$ is applied to both media, such that continuity of tractions and displacements at the interface is maintained. Show that the pressure generated on the interface is equal to the difference in the elastic energy densities. ■

Solution The common displacement gradient generates different stress fields: $\sigma^A_{ij} = c^A_{ijkl} u_{k,l}$ and $\sigma^B_{ij} = c^B_{ijkl} u_{k,l}$, such that $\sigma^A_{ij} n_j = \sigma^B_{ij} n_j$ to maintain continuity of tractions.

The pressure on the interface is:

$$\begin{aligned} p &= n_j \left(P^A_{jk} - P^B_{jk} \right) n_k \\ &= n_j \left(W^A - W^B \right) \delta_{jk} n_k - \underbrace{n_j \left(\sigma^A_{ji} - \sigma^B_{ji} \right)}_{=0} u_{i,k} n_k \\ &= W^A - A^B \end{aligned}$$

9.4 The force on a static defect

Consider a body occupying a region $\mathcal{R}$ containing a defect D at (ξ_1,ξ_2,ξ_3). For simplicity we will assume the elastic constants are the same throughout the body. The external surface S of the body may be loaded with tractions by some external mechanism. If F_k is the force on the defect and if it undergoes a small displacement $\delta\xi_m$ then the change in the potential energy of the body and the external loading mechanism, is

$$\delta E_{tot} = -F_m \delta\xi_m \tag{9.11}$$

where

$$F_m = \int_S P_{mj} n_j \mathrm{d}S \tag{9.12}$$

and P_{jk} is the static energy-momentum tensor given by eq 9.9.

To prove eq 9.11 we proceed by first rigidly displacing the entire displacement field of the defect, thus $u_i^D(\mathbf{x}) \to u_i^D(\mathbf{x}-\delta\boldsymbol{\xi})$. This will then need to be adjusted at the surface of the body to satisfy the boundary conditions on S. To calculate the total change in the potential energy we consider changes in both the elastic energy of the body and the potential energy of the loading mechansim.

In the first step the change in elastic energy of the body when the displacement field of D is rigidly translated is given by:

$$\begin{aligned} \delta E_{el}^{(1)} &= \int_{\mathcal{R}} W(\mathbf{x}-\delta\boldsymbol{\xi}) - W(\mathbf{x})\mathrm{d}V \\ &= -\delta\xi_m \int_{\mathcal{R}} \frac{\partial W}{\partial x_m}\mathrm{d}V \quad + O(\delta\xi^2) \\ &= -\delta\xi_m \int_S W n_m \mathrm{d}S \quad + O(\delta\xi^2) \end{aligned} \tag{9.13}$$

The derivative of the elastic energy in the integrand of the volume integral may diverge, and this raises the question of whether it is integrable. But by converting the volume integral into a surface integral in the last line only a thin sliver around the surface of the body contributes to the integral, which is far from the elastic singularity at D and therefore the surface integral always converges.

In the second step we make a correction to satisfy the boundary conditions on the surface S of the body. Following the rigid displacement of the elastic field of D in the first step the surface tractions on $\mathcal{S}$ are $\{\sigma_{ij}(\mathbf{x}) - \delta\xi_m\sigma_{ij,m}(\mathbf{x})\}n_j + O(\delta\xi^2)$. After correcting these surface tractions they become to $\sigma_{ij}(\mathbf{x})n_j$ plus a term of order $\delta\xi$ depending on the *hardness* of the mechanism applying the surface tractions. If the mechanism is soft the loading will be *dead* and the traction will remain $\sigma_{ij}(\mathbf{x})n_j$. If the loading is rigid (infinitely hard) the change in the surface traction from $\sigma_{ij}(\mathbf{x})n_j$ will still be of order $\delta\xi$. The surface displacements after the rigid translation of the field of D are $u_i(\mathbf{x}) - \delta\xi_m u_{i,m}(\mathbf{x})$, where $\mathbf{x}$ is on $\mathcal{S}$. After the adjustment they become $u_i^F(\mathbf{x})$, which can be determined only by detailed calculation. Nevertheless, $u_i^F(\mathbf{x}) - u_i(\mathbf{x})$ will be first order in $\delta\xi$. Therefore the change in the elastic energy of the body resulting from the restoration of the original surface tractions on the surface $\mathcal{S}$ is:

$$\delta E_{el}^{(2)} = \int_{S} \sigma_{ij}\left\{u_i^F(\mathbf{x}) - u_i(\mathbf{x}) + \delta\xi_m u_{i,m}(\mathbf{x})\right\} n_j \, \mathrm{d}S \qquad + \; O(\delta\xi^2). \tag{9.14}$$

Since the term in the brackets $\{\ldots\}$ is first order in $\delta\xi$ this expression is independent of the hardness of the loading mechanism because the contribution to $\delta E_{el}^{(2)}$ arising from any change in the surface tractions is second order in $\delta\xi$. The work done by the external loading mechanism at the end of both steps is also independent of any change in the surface tractions to first order in $\delta\xi$:

$$\delta w_{ext} = \int_{S} \sigma_{ij}\left\{u_i^F(\mathbf{x}) - u_i(\mathbf{x})\right\} n_j \mathrm{d}S \qquad + \; O(\delta\xi^2) \tag{9.15}$$

The change in the potential energy of the loading mechanism is $-\delta w_{ext}$. Adding $\delta E_{el}^{(1)}$, $\delta E_{el}^{(2)}$ and $-\delta w_{ext}$ we obtain the change in the total potential energy of the body and loading mechanism:

$$\delta E_{tot} = -\delta\xi_m \int_{S} (W\delta_{mj} - \sigma_{ij}u_{i,m})n_j \mathrm{d}S + O(\delta\xi^2), \tag{9.16}$$

from which eq 9.12 follows.

So far we have considered the somewhat unrealistic case of a body containing just one defect D. It is straightforward to extend the treatment to the more realistic case of the force on D arising from the elastic fields of other defects within the body, in addition to any loads applied to the surface $\mathcal{S}$ of the body. Construct an internal surface S surrounding D and no other defects. The above steps may then be repeated with the region within S loaded by surface tractions on S that arise from other defects in the body and from external loads applied to $\mathcal{S}$. Thus, we arrive at the more general result for the force on D:

$$F_m = \int_S (W\delta_{mj} - \sigma_{ij}u_{i,m})n_j \mathrm{d}S, \tag{9.17}$$

where S is any surface enclosing only D inside the body.

The invariance of the force in eq 9.17 if the surface S is changed to S' follows because the divergence of the integrand is zero provided no defects exist between S and S'. We will now prove this for a general elastic energy density W that depends on the elastic displacement field and the displacement gradients[5]: $W = W(u_i, u_{i,j})$.

The partial derivative of W with respect to x_m is as follows:

$$\begin{aligned}\frac{\partial W}{\partial x_m} &= \frac{\partial W}{\partial u_p}\frac{\partial u_p}{\partial x_m} + \frac{\partial W}{\partial u_{p,q}}\frac{\partial u_{p,q}}{\partial x_m} \\ &= -f_p\, u_{p,m} + \sigma_{pq}\, u_{p,qm},\end{aligned} \tag{9.18}$$

where $-\partial W/\partial u_p = f_p$ is a body force and $\partial W/\partial u_{p,q} = \sigma_{pq}$ is a component of the stress tensor. Therefore the divergence of the integrand in eq 9.17 is as follows:

$$\begin{aligned}P_{mj,j} &= (W\delta_{mj} - \sigma_{ij}u_{i,m})_{,j} \\ &= -f_p\, u_{p,m} + \sigma_{pq}\, u_{p,qm} - \sigma_{ij,j}u_{i,m} - \sigma_{ij}u_{i,mj}\end{aligned} \tag{9.19}$$

We see that $P_{mj,j} = 0$ because the equilibrium condition ensures $\sigma_{ij,j} + f_i = 0$. It follows that S may be deformed into S' and the integral in eq 9.17 is invariant.

The divergence $P_{mj,j}$ is non-zero when there is a defect because the elastic energy density then has an explicit dependence on position: $W = W(u_i, u_{i,j}, x_i)$. For then the gradient of W with respect to position becomes:

$$\frac{\partial W}{\partial x_m} = -f_p\, u_{p,m} + \sigma_{pq}\, u_{p,qm} + \left(\frac{\partial W}{\partial x_m}\right)_{exp}. \tag{9.20}$$

The explicit partial derivative $(\partial W/\partial x_m)_{exp}$ is evaluated by displacing the point x_m by a virtual amount δx_m while keeping fixed all other points in the body and keeping the elastic displacements and displacement gradients fixed. If there is a defect at $\mathbf{X}$ then when $\mathbf{x} = \mathbf{X}$ the partial derivative $(\partial W/\partial x_m)_{exp}$ is nonzero because the defect undergoes a virtual displacement. The divergence $P_{mj,j}$ is then $(\partial W/\partial x_m)_{exp}$ and the force F_m becomes:

$$F_m = \int_R \left(\frac{\partial W}{\partial x_m}\right)_{exp} \mathrm{d}V. \tag{9.21}$$

With just one defect in the region R at $\mathbf{X}$ then $(\partial W/\partial x_m)_{exp} = (\partial W/\partial X_m)_{exp}\,\delta(\mathbf{x}-\mathbf{X})$.

[5]Note that the relation between stress and strain may be non-linear

Exercise 9.3 For the elastic energy density $W = W(u_i, u_{i,j})$ show that the elastic energy of a body is stationary when:

$$\frac{\partial W}{\partial u_i} - \frac{\partial}{\partial x_j}\left(\frac{\partial W}{\partial u_{i,j}}\right) = 0. \tag{9.22}$$

What is the physical meaning of this equation? ■

Solution The elastic energy of the body is:

$$E_{el} = \int W(u_i, u_{i,j})\mathrm{d}V.$$

Consider a variation δu such that $\delta u_i = 0$ on the surface S of the body. The change in the elastic energy is:

$$\begin{aligned} \delta E_{el} &= \int \frac{\partial W}{\partial u_i}\delta u_i + \frac{\partial W}{\partial u_{i,j}}\delta u_{i,j}\,\mathrm{d}V \\ &= \int \frac{\partial W}{\partial u_i}\delta u_i + \left(\frac{\partial W}{\partial u_{i,j}}\delta u_i\right)_{,j} - \left(\frac{\partial W}{\partial u_{i,j}}\right)_{,j}\delta u_i\,\mathrm{d}V \end{aligned}$$

Applying the divergence theorem to the second term we get:

$$\int \left(\frac{\partial W}{\partial u_{i,j}}\delta u_i\right)_{,j}\mathrm{d}V = \int_S \frac{\partial W}{\partial u_{i,j}}\delta u_i n_j\,\mathrm{d}S.$$

The integral over the surface Σ is zero because $\delta u_i = 0$ on Σ. Therefore the change in the elastic energy is:

$$\delta E_{el} = \int \left\{\frac{\partial W}{\partial u_i} - \left(\frac{\partial W}{\partial u_{i,j}}\right)_{,j}\right\}\delta u_i\,\mathrm{d}V.$$

Therefore the elastic energy is stationary when

$$\frac{\partial W}{\partial u_i} - \left(\frac{\partial W}{\partial u_{i,j}}\right)_{,j} = 0.$$

From eq 2.16 it follows that $\partial W/\partial u_{i,j}(\mathbf{x})$ is the stress tensor component $\sigma_{ij}(\mathbf{x})$. The derivative $-\partial W/\partial u_i(\mathbf{x})$ is the body force component $f_i(\mathbf{x})$. Therefore, this equation is the equation of mechanical equilibrium, eq 2.15.

Comment

Notice that in this derivation of the equation of mechanical equilibrium no assumption is made about the relation between elastic stress and strain: it applies in linear and nonlinear elasticity.

9.5 Relationship to the static energy-momentum tensor

In this section we will use the formal methods of classical field theory to derive the static energy-momentum tensor. We will show that it is equal to the integrand of the expression in eq 9.17 for the force on a defect.

We begin with the time-independent Lagrangian density, which we write as $L = L(u_i, u_{i,j}, x_m)$. As usual u_i and $u_{i,j}$ are the displacements and displacement gradients. The presence of inhomogeneities such as defects introduces an explicit dependence of the Lagrangian density on position, as in eq 9.20. Since we are ignoring kinetic energy the Lagrangian density is just $-W$, where W is the elastic energy density.

The integral of the Lagrangian density over the volume of the body is minimised with respect to variations δu_i of the displacement field, where $\delta u_i = 0$ on the surface of the body, when the following Euler-Lagrange equation is satisfied:

$$\frac{\partial}{\partial x_j}\left(\frac{\partial L}{\partial u_{i,j}}\right) - \frac{\partial L}{\partial u_i} = 0. \tag{9.23}$$

With $L = -W$ this is the equation of mechanical equilibrium.

Consider the partial derivative of the Lagrangian density with respect to x_l. To be clear, it is defined as:

$$\frac{\partial L}{\partial x_l} = \lim_{\delta x_l \to 0} \frac{L(x_l + \delta x_l) - L(x_l)}{\delta x_l}. \tag{9.24}$$

It has both implicit contributions and an explicit contribution:

$$\begin{aligned}
\frac{\partial L}{\partial x_l} &= \frac{\partial L}{\partial u_i} u_{i,l} + \frac{\partial L}{\partial u_{i,j}} u_{i,jl} + \left(\frac{\partial L}{\partial x_l}\right)_{exp} \\
&= \left[\frac{\partial L}{\partial u_i} u_{i,l} - \left(\frac{\partial}{\partial x_j}\left(\frac{\partial L}{\partial u_{i,j}}\right)\right) u_{i,l}\right] + \left(\frac{\partial}{\partial x_j}\left(\frac{\partial L}{\partial u_{i,j}}\right)\right) u_{i,l} + \frac{\partial L}{\partial u_{i,j}} u_{i,jl} + \left(\frac{\partial L}{\partial x_l}\right)_{exp} \\
&= \frac{\partial}{\partial x_j}\left\{\left(\frac{\partial L}{\partial u_{i,j}}\right) u_{i,l}\right\} + \left(\frac{\partial L}{\partial x_l}\right)_{exp}
\end{aligned} \tag{9.25}$$

The term in square brackets is zero, owing to the Euler-Lagrange equation eq 9.23. Equation 9.25 can be rewritten as follows:

$$\left(\frac{\partial L}{\partial x_l}\right)_{exp} = -\frac{\partial}{\partial x_j} P_{jl} \tag{9.26}$$

where

$$P_{jl} = \frac{\partial L}{\partial u_{i,j}} u_{i,l} - L\delta_{jl} \tag{9.27}$$

is the static energy-momentum tensor. Setting $L = -W$, and using $\partial W/\partial u_{i,j} = \sigma_{ij}$, we see that eq 9.27 is the same as eq 9.9. Note that in eq 9.26 the explicit partial derivative of the Lagrangian is equated to the divergence of the static energy-momentum tensor, which is analogous to the divergence of the Maxwell electrostatic stress tensor in eq 9.3 as an expression of the electrostatic force on a point charge in Exercise 9.1.

9.6 The force due to an applied stress and image forces

The elastic energy density W, the stress tensor σ_{ij} and the displacement gradient $u_{i,m}$ in eq 9.17 include contributions from all elastic fields present in the body. In addition to the elastic fields $\sigma_{ij}^{D\infty}, u_i^{D\infty}$ of the defect in an infinite medium there are contributions from other defects and from forces applied to the surface of the body, which we collectively label σ_{ij}^A, u_i^A, and from satisfying the boundary conditions on the surface of the body σ_{ij}^I, u_i^I which are often called image fields. In linear elasticity we have:

$$\begin{aligned} \sigma_{ij} &= \sigma_{ij}^{D\infty} + \sigma_{ij}^A + \sigma_{ij}^I \\ u_i &= u_i^{D\infty} + u_i^A + u_i^I \end{aligned}$$

When these sums are inserted in eq 9.17 the force becomes:

$$F_m = \int_\Sigma (D\infty, D\infty) + (A,A) + (I,I) + (D\infty, A) + (A,I) + (I, D\infty)\, \mathrm{d}S. \tag{9.28}$$

The diagonal terms are given by

$$(X,X) = \left(\frac{1}{2}\sigma_{pq}^X u_{p,q}^X \delta_{mj} - \sigma_{ij}^X u_{i,m}^X\right) n_j, \tag{9.29}$$

and the off-diagonal terms are given by

$$(X,Y) = \left(\frac{1}{2}\left(\sigma_{pq}^X u_{p,q}^Y + \sigma_{pq}^Y u_{p,q}^X\right)\delta_{mj} - \left(\sigma_{ij}^X u_{i,m}^Y + \sigma_{ij}^Y u_{i,m}^X\right)\right) n_j. \tag{9.30}$$

To proceed we need an unfamiliar relation between integrals over a closed surface Σ involving a differentiable function w:

$$\int_{\Sigma} w_{q,q} n_m \, \mathrm{d}S = \int_{\Sigma} w_{j,m} n_j \, \mathrm{d}S. \tag{9.31}$$

where each component of the vector function $\mathbf{w}$ is differentiable. One way to derive eq 9.31 is to consider the change in the integral of $\mathbf{w} \cdot \hat{\mathbf{n}} = w_j n_j$, where $\hat{\mathbf{n}}$ is the local unit outward normal, taken over a closed surface Σ when the surface is displaced by an infinitesimal amount ε along the x_m-axis. The change in the surface integral is:

$$\varepsilon \int_{\Sigma} \frac{\partial}{\partial x_m} \mathbf{w} \cdot \hat{\mathbf{n}} \, \mathrm{d}S = \int_{\Sigma} w_{j,m} n_j \mathrm{d}S.$$

But this is also equal to:

$$\int_{\Sigma'} w_j n_j \mathrm{d}S - \int_{\Sigma} w_j n_j \mathrm{d}S,$$

where Σ' is the surface following the displacement of Σ by ε along x_m. By the divergence theorem this difference in surface integrals is equal to the volume integral of the divergence of $\mathbf{w}$ taken over the thin sliver between Σ and Σ'. A volume element of this sliver is $\varepsilon \hat{\mathbf{e}}_m \cdot \hat{\mathbf{n}} \mathrm{d}S = \varepsilon n_m \, \mathrm{d}S$, where $\hat{\mathbf{e}}_m$ is a unit vector along the x_m-axis. Therefore,

$$\varepsilon \int_{\Sigma} \nabla \cdot \mathbf{w} \, (\hat{\mathbf{e}}_l \cdot \hat{\mathbf{n}}) \, \mathrm{d}S = \varepsilon \int_{\Sigma} \frac{\partial}{\partial x_l} \mathbf{w} \cdot \hat{\mathbf{n}} \, \mathrm{d}S.$$

Equation 9.31 then follows.

Using eq 9.31 with $w_q = \sigma^X_{pq} u^Y_p + \sigma^Y_{pq} u^X_p$ we obtain:

$$\int_{\Sigma} \left(\sigma^X_{pq} u^Y_{p,q} + \sigma^Y_{pq} u^X_{p,q} \right) n_m \mathrm{d}S =$$

$$\int_{\Sigma} \left(\sigma^X_{pj} u^Y_p + \sigma^Y_{pj} u^X_p \right)_{,m} n_j \, \mathrm{d}S = \int_{\Sigma} \left(\sigma^X_{pj,m} u^Y_p + \sigma^X_{pj} u^Y_{p,m} + \sigma^Y_{pj,m} u^X_p + \sigma^Y_{pj} u^X_{p,m} \right) n_j \, \mathrm{d}S, \tag{9.32}$$

where the equilibrium condition $\sigma_{pq,q} = 0$ has been used. The contribution $F^{(XY)}_m$ to the force F_m on the defect arising from the (XY) term in eq 9.30 is then:

$$F^{(XY)}_m = \frac{1}{2} \int_{\Sigma} \left(\sigma^X_{pj,m} u^Y_p + \sigma^Y_{pj,m} u^X_p - \sigma^X_{pj} u^Y_{p,m} - \sigma^Y_{pj} u^X_{p,m} \right) n_j \, \mathrm{d}S \tag{9.33}$$

Using the divergence theorem we may convert this to a volume integral over the volume V enclosed by Σ:

$$F_m^{(XY)} = \frac{1}{2}\int_V \left(\sigma^X_{pj,m}u^Y_{p,j} + \sigma^Y_{pj,m}u^X_{p,j} - \sigma^X_{pj}u^Y_{p,mj} - \sigma^Y_{pj}u^X_{p,mj}\right) \mathrm{d}V, \tag{9.34}$$

where we have again used the equilibrium condition $\sigma_{ij,j} = 0$. If $\sigma^X_{pj} = c_{pjkl}u^X_{k,l}$ and $\sigma^Y_{pj} = c_{pjkl}u^Y_{k,l}$ then it is easy to show that this volume integral is zero remembering that $c_{ijkl} = c_{klij}$. But if X and/or Y is $D\infty$ this is no longer true because $e^{D\infty}_{pj} \neq (1/2)(u^{D\infty}_{p,j} + u^{D\infty}_{j,p})$ throughout V. Thus all the off-diagonal terms in eq 9.30 are zero except those involving $D\infty$. The diagonal terms (A, A) and (I, I) are zero by the same argument. The diagonal term $(D\infty, D\infty)$ is zero because Σ may be taken at infinity where the integral tends to zero provided σ_{ij} decays at least as rapidly as $1/r^2$ in three dimensions and at least as rapidly as $1/r$ in two dimensions.

The conclusion is that only the two off-diagonal terms involving $D\infty$ are non-zero: $(D\infty, A)$ and $(D\infty, I)$. These are the forces arising from the applied stress and the image field. The force arising from an applied stress is:

$$F_m^A = \int_\Sigma \left(\sigma^{D\infty}_{pj,m}u^A_p - \sigma^A_{pj}u^{D\infty}_{p,m}\right) n_j \mathrm{d}S, \tag{9.35}$$

and the force arising from the image field is:

$$F_m^I = \int_\Sigma \left(\sigma^{D\infty}_{pj,m}u^I_p - \sigma^I_{pj}u^{D\infty}_{p,m}\right) n_j \mathrm{d}S. \tag{9.36}$$

In eq 9.35 we see that the force on a defect in a finite body due to other defects or tractions applied to the external surface of the body involves the field of the defect in an *infinite* medium, not those in the *finite* body. This is a very significant simplification.

9.7 The Peach-Koehler force revisited

In this section we apply the general expression for the force on a defect in eq 9.17 to rederive the Peach-Koehler formula for the force per unit length on a dislocation due to an applied stress σ^A_{ij}, which was first obtained in section 6.6.

To apply eq 9.35 to a dislocation consider a straight Volterra dislocation with its positive line sense along the positive x_3-axis of a cartesian coordinate system with an arbitrary Burgers vector b_i. The surface S of the integral in eq 9.35 has to enclose only

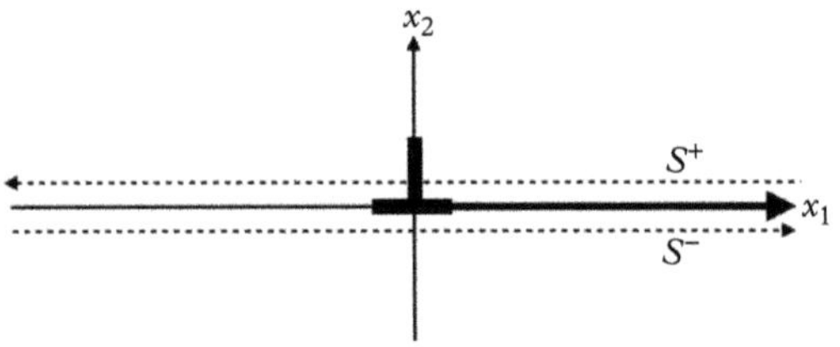

Figure 9.1: To illustrate the evaluation of the Peach-Koehler force. A straight dislocation lies along the x_3 axis, with its positive line sense along the positive x_3 direction. The cut is in the plane $x_2 = 0$, along $x_1 \geq 0$. The surface integral in eq 9.38 is taken in the clockwise sense along the positive direction of the dislocation line sense, which appears anti-clockwise in the figure because the positive direction of the dislocation line is out of the page. S^+ and S^- are surfaces infinitesimally above and below the x_1-axis.

the dislocation, but is otherwise arbitrary. The Burgers vector is defined by the line integral (see eq 6.2):

$$b_i = \oint_C \frac{\partial u_i^{D\infty}}{\partial x_k} \mathrm{d}x_k, \tag{9.37}$$

where the circuit C is any clockwise circuit taken around the positive sense of the dislocation line. The absence of a resultant body force associated with the dislocation ensures that the first term in the integrand in eq 9.35 integrates to zero. We are left with:

$$F_m^A = -\int_S \sigma_{ij}^A u_{i,m}^{D\infty} n_j \mathrm{d}S. \tag{9.38}$$

Define axes x_1 and x_2 in the plane normal to the dislocation line. Let the cut plane be the half-plane $x_2 = 0, x_1 > 0$. We choose S to consist of the surfaces S^+ and S^- extending from $x_1 = -\infty$ to $x_1 = +\infty$, infinitesimally above and below the x_1-axis, as shown in Fig 9.1, and closed by infinitesimal segments along x_2 at $x_1 = \pm\infty$. On S^+ and S^- we have $\sigma_{ij}^A(x_1, 0^+, x_3)n_j = \sigma_{ij}^A(x_1, 0^-, x_3)n_j$. We also have $u_i^{D\infty}(x_1, 0^+, x_3) - u_i^{D\infty}(x_1, 0^-, x_3) = -b_i$ for all $x_1 > 0$ and for all x_3, and $u_i^{D\infty}(x_1, 0^+, x_3) - u_i^{D\infty}(x_1, 0^-, x_3) = 0$ for all $x_1 < 0$ and for all x_3. Therefore, $u_{i,1}^{D\infty}(x_1, 0^+, x_3) - u_{i,1}^{D\infty}(x_1, 0^-, x_3) = -b_i\delta(x_1)$, which is consistent with eq 9.37. Following section 6.6 the positive normal to the cut half-plane is along $-x_2$, that is $n_j = -\delta_{j2}$. Carrying out the integration around S, in the sense indicated in Fig 9.1, we obtain the force per unit length of dislocation line, $F_1^A = \sigma_{i2}^A b_i$, where σ_{i2}^A is evaluated at the origin. This agrees with eq 6.10 with $t_p = \delta_{p3}$.

To find F_2^A we choose the cut plane to be the half-plane $x_1 = 0, x_2 > 0$ and we choose S^+ and S^- to extend from $x_2 = -\infty$ to $x_2 = +\infty$ with $x_1 = \pm\varepsilon$, where ε is a positive infinitesimal number. Repeating the argument with $n_j = \delta_{j1}$ and $u_{i,2}^{D\infty}(0^+, x_2, x_3) - u_{i,2}^{D\infty}(0^-, x_2, x_3) = b_i \delta(x_2)$, which is consistent with eq 9.37, we obtain $F_2^A = -\sigma_{i1} b_i$. This also agrees with eq 6.10.

9.8 Problems

Problem 9.1 In an invariant plane strain deformation an interface separates deformed and undeformed regions of a medium. The interface is an invariant plane of the deformation, so that it is neither deformed nor rotated. Consider a flat interface between semi-infinite regions A and B. The positive sense of the normal $\hat{\mathbf{n}}$ to the interface points from region B to region A. Region A has not undergone the invariant plane strain, but region B has. If $\hat{\mathbf{e}}$ is a unit vector perpendicular to the unit normal $\hat{\mathbf{n}}$ show that the displacement gradient of a general invariant plane strain is as follows:

$$u_{i,k}^B = s\hat{e}_i \hat{n}_k + \lambda \hat{n}_i \hat{n}_k,$$

and state the meaning of s and λ. Verify that any vector lying in the interface is unchanged by this deformation, and hence the interface is invariant. In deformation twinning $\lambda = 0$, and the twin crystal and the parent crystal are related by a simple shear.

If only *linear* elastic forces were at play then any movement of the interface induced by an applied stress would be reversed when the stress is removed. In both deformation twinning and martensitic phase changes the elastic energy density W becomes a *nonlinear* function of the displacement gradient. That is because after the crystal has been twinned or transformed martensitically it corresponds to a local minimum of the elastic energy density W in the space of displacement gradients. The elastic energy density W then describes the change in the energy of the crystal along a path between the initial and final states of the crystal in this space. The variation of the elastic energy density along such a path can be calculated using modern methods based on density functional theory (DFT). When the displacement gradient from a local minimum in this space is sufficiently small the change in the energy is described adequately by linear elasticity. But linear elasticity fails to capture the existence of other local minima.

In the absence of an applied stress we write $W^A = W^{A0}$ where W^{A0} is the energy density of region A at the local minimum. Similarly, $W^B = W^{B0}$ is the energy density of region B at its local minimum. Show that the pressure on the interface is then

$W^{A0} - W^{B0}$. This difference in the energy densities may depend on other variables such as temperature and applied magnetic or electric fields. The pressure $W^{A0} - W^{B0}$ may the regarded as a *chemical* pressure because it arises from atomic interactions in undistorted crystals A and B. In mechanical twinning $W^{A0} = W^{B0}$ because the crystal structures of states A and B are related by a rotation.

When there is an applied stress the displacement field in region A is just the elastic displacement u_i^A from the atomic configuration at the local energy minimum. The displacement field in region B is $u_i^{B0} + u_i^B$ where u_i^{B0} is the displacement due to the invariant plane strain and u_{iB} is the additional elastic displacement caused by the applied stress. Show that the pressure on the interface in the presence of an applied stress becomes:

$$p = n_j \left[\left(W^{A0} + W_{el}^A - W^{B0} - W_{el}^B \right) \delta_{jk} + \sigma_{ij} \left(u_{i,k}^{B0} + u_{i,k}^B - u_{i,k}^A \right) \right] n_k$$

In general the components of the displacement gradient $u_{i,k}^{B0}$ due to the invariant plane strain are much larger than those due to the applied stress. In that case we may ignore the elastic displacement gradients $u_{i,k}^A$ and $u_{i,k}^B$ in comparison to $u_{i,k}^{B0}$. Show that the pressure on the interface then becomes:

$$p = \left(W^{A0} - W^{B0} + W_{el}^{A0} - W_{el}^{B0} \right) + \left[\{ n_i \sigma_{ij} e_j \} s + \{ n_i \sigma_{ij} n_j \} \lambda \right].$$

Solution Let the origin of the Cartesian coordinate system coincide with a point in the interface at some initial time. Let $\mathbf{x}$ be the position of a point in region A with respect to this origin. At some later time the interface moves into region A and reaches the point that was at $\mathbf{x}$. The displacement undergone by the point that was at $\mathbf{x}$ is:

$$u_i^B(\mathbf{x}) = u_{i,k}^B x_k = s e_i n_k x_k + \lambda n_i n_k n_k$$

or in vector notation:

$$\mathbf{u}^B(\mathbf{x}) = s(\hat{\mathbf{n}} \cdot \mathbf{x})\hat{\mathbf{e}} + \lambda(\hat{\mathbf{n}} \cdot \mathbf{x})\hat{\mathbf{n}}.$$

If $\mathbf{x}$ lies in the interface then $\hat{\mathbf{n}} \cdot \mathbf{x} = 0$ and hence $\mathbf{u}^B(\mathbf{x}) = 0$. Since this holds true for all points in the interface the interface is invariant. The meaning of s is that it is the magnitude of a simple shear strain in the direction of $\hat{\mathbf{e}}$. The meaning of λ is that it is the magnitude of a strain normal to the interface.

In the absence of an applied stress the pressure on the interface is:

$$p = n_j\left(P^A_{jk} - P^B_{jk}\right)n_k = n_j\left(W^{A0} - W^{B0}\right)n_k\delta_{jk} = W^{A0} - W^{B0}.$$

If $W^{A0} > W^{B0}$ then $p > 0$, so that A is consumed and B grows.

In the presence of an applied stress,

$$P^A_{jk} = \left(W^{A0} + W^A_{el}\right)\delta_{jk} - \sigma_{ij}u^A_{i,k}$$

and

$$P^B_{jk} = \left(W^{B0} + W^B_{el}\right)\delta_{jk} - \sigma_{ij}\left(u^{B0}_{i,k} + u^B_{i,k}\right),$$

so that

$$p = n_j\left[W^{A0} - W^{B0} + W^A_{el} - W^B_{el} + \sigma_{ij}\left(u^{B0}_{i,k} + u^B_{i,k} - u^A_{i,k}\right)\right]n_k.$$

If $u^A_{i,k}, u^B_{i,k} \ll u^{B0}_{i,k}$ then the pressure on the interface is:

$$\begin{aligned} p &= n_j\left[\left(W^{A0} - W^{B0} + W^A_{el} - W^B_{el}\right)\delta_{jk} + \sigma_{ij}\left(se_in_k + \lambda n_in_k\right)\right]n_k \\ &= \left(W^{A0} - W^{B0} + W^A_{el} - W^B_{el}\right) + \left[\left(e_i\sigma_{ij}n_j\right)s + \left(n_i\sigma_{ij}n_j\right)\lambda\right]. \end{aligned}$$

Comment

In a martensitic transformation the term in round brackets may be influenced by other fields such as thermal, magnetic or electric. The term in square brackets shows that resolved components of the applied stress may drive the transformation, converting region A into region B, by promoting the simple shear and the tensile strain normal to the interface. In shape memory alloys these two terms are played off against each other. Even though W^{B0} may be slightly greater than W^{A0} a suitably applied stress may drive the transformation and convert region A into the martensite phase B: this is *stress-induced martensite*. Then by adjusting the temperature or some other field $W^{B0} - W^{A0}$ may become sufficiently negative to drive the transformation in the reverse direction. This reversibility of the martensitic phase change is the basis of the *shape memory effect.*

Problem 9.2 In the derivation of the pressure on an interface in eq 9.10 it was assumed there is continuity of displacements and tractions across the interface. These assumptions apply when the interface is *coherent*. At high temperatures, or when there is a large misfit strain between the crystal lattices on either side of the interface, the interface may become incoherent. In that case only the normal displacements and normal tractions are continuous across the interface. The tractions parallel to the interface are assumed to be relaxed to zero through sliding of one crystal with respect to the other as if the interface were greased. In general there are then discontinuous changes in the displacement vector parallel to the interface. Show that the pressure on an incoherent interface is:

$$p = n_i \left(Q_{ij}^A - Q_{ij}^B \right) n_j$$

where

$$Q_{ij} = W\delta_{ij} - \sigma_{ij} \left(n_k u_{m,k} n_m \right).$$

Solution Repeating the four steps of section 9.3 we have:

$$\delta E^{(1)} = -\int_I W^A \, \delta\xi \, \mathrm{d}S$$

which is the same as in section 9.3. Also, as in section 9.3 the elastic displacements on the newly created surface of A are $u_i^A + \delta\xi n_k u_{i,k}^A$. The surface tractions on the newly created surface of A are the same as those before the interface was displaced plus a correction of order $\delta\xi$. Before the interface was displaced the traction was only normal to the interface, i.e. $(n_p \sigma_{pq}^A n_q) n_i$, because the tractions parallel to the interface were relaxed to zero. Let the final relaxed displacements on the A-side be u_i^{FA}. Then the difference $u_i^{FA} - (u_i^A + \delta\xi n_k u_{i,k}^A)$ is of order $\delta\xi$. The tractions on the A-side do work:

$$\delta E^{(2)} = -\int_I \left(n_p \sigma_{pq}^A n_q \right) n_i \left[u_i^{FA} - \left(u_i^A + \delta\xi n_k u_{i,k}^A \right) \right] \mathrm{d}S + O(\delta\xi^2).$$

On the B-side we have,

$$\delta E^{(3)} = +\int_I W^B \, \delta\xi \, \mathrm{d}S$$

and

$$\delta E^{(4)} = +\int_I \left(n_p \sigma^B_{pq} n_q\right) n_i \left[u_i^{FB} - \left(u_i^B + \delta\xi n_k u^B_{i,k}\right)\right] \mathrm{d}S + O(\delta\xi^2).$$

The change in the total energy to first order in $\delta\xi$ is:

$$\begin{aligned}
\delta E_{tot} &= \delta E^{(1)} + \delta E^{(2)} + \delta E^{(3)} + \delta E^{(4)} \\
&= -\int_I \Bigg\{ \left(W^A - W^B\right)\delta\xi - \left(n_p \sigma^A_{pq} n_q\right) n_i \left[u_i^{FA} - \left(u_i^A + \delta\xi n_k u^A_{i,k}\right)\right] \\
&\quad + \left(n_p \sigma^B_{pq} n_q\right) n_i \left[u_i^{FB} - \left(u_i^B + \delta\xi n_k u^B_{i,k}\right)\right] \Bigg\} \mathrm{d}S
\end{aligned}$$

Continuity of normal tractions requires $n_p \sigma^A_{pq} n_q = n_p \sigma^B_{pq} n_q$. Continuity of normal displacements requires $n_i\,(u_i^{FA} - u_i^A) = n_i\,(u_i^{FB} - u_i^B)$. Therefore,

$$n_p \sigma^A_{pq} n_q \left(u_i^{FA} - u_i^A\right) = n_p \sigma^B_{pq} n_q \left(u_i^{FB} - u_i^B\right).$$

Hence:

$$\delta E_{tot} = -\int_I \left[W^A - \left(n_p \sigma^A_{pq} n_q\right)\left(n_i u^A_{i,k} n_k\right) - W^B + \left(n_p \sigma^B_{pq} n_q\right)\left(n_i u^B_{i,k} n_k\right)\right] \delta\xi \, \mathrm{d}S$$

and the pressure on the interface is:

$$p = n_p \left(Q^A_{pq} - Q^B_{pq}\right) n_q$$

where

$$Q_{pq} = W\delta_{pq} - \sigma_{pq}\left(n_i u_{i,k} n_k\right).$$

10. Cracks

10.1 Introduction

The nucleation and propagation of cracks leads to fracture. For this reason cracks have been studied for about a century. There are many shapes and sizes of cracks, and a range of theoretical and computational methods has been developed to solve their elastic fields. In this chapter we will focus on just one theoretical technique in which cracks are represented by continuous distributions of dislocations. There are three reasons for choosing this technique. As first pointed out by Zener[1], the nucleation of cracks in the interior of metals and alloys is always associated with plastic deformation. Secondly, cracks in all but the most brittle crystalline materials are associated with some degree of plasticity in their vicinity, called plastic zones. Plastic zones contain crystal dislocations which interact with the elastic field of the crack. Thirdly, the representation of the crack itself in terms of dislocations enables these interactions to be described conveniently using dislocation theory. The elastic fields of cracks are intimately related to the elastic fields of dislocation pileups, where dislocations pile up against barriers such as grain boundaries and precipitates during plastic deformation. Dislocation pileups may nucleate cracks[2]. Thus, there are close physical and mathematical relationships between cracks and dislocations.

Various criteria for crack propagation have been proposed in the literature. The most fundamental is the Griffith criterion, which states that for a crack to grow the total potential energy of the system must decrease, where the 'system' includes the external loading mechanism. The Griffith criterion is discussed in section 10.5. A recurring feature of the theory of fracture is its incompleteness - the lack of knowledge about atomic mechanisms at the root of a crack and how they are influenced by temperature, strain rate, local stress, local chemical composition and microstructure. In 1921 Griffith wrote in the seminal paper containing his criterion for fracture that no criterion for

[1]Zener, C M, in *Fracturing of metals*, American Society for Metals: Cleveland, Ohio, p.3 (1948). Clarence Melvin Zener 1905-1993, US physicist.

[2]Problem 10.6 considers a model for such nucleation.

strength or fracture could be considered complete without taking into consideration interatomic forces. This incompleteness will emerge several times in this chapter. We will see in section 10.6 that dislocations ahead of a crack may either reduce or increase the stress acting on the crack tip. Forces acting on atoms at the crack tip are thus strongly influenced by groups of dislocations, and other defects, that may be further away. The multi-scale nature of fracture is also one of its most characteristic features. Screening, or 'shielding', of applied stresses acting at crack tips by dislocations is also a central theme of this chapter.

10.2 Mathematical preliminaries

In this chapter we will encounter singular integral equations of the Cauchy type. They have the following form:

$$P\int_{-1}^{1} \frac{F(u)}{v-u}\mathrm{d}u = H(v), \tag{10.1}$$

where v lies between -1 and 1. The P in front of the integral signifies it is a principal value type. The standard way to deal with these equations is to use complex analysis following the work[3] in this area by Muskhelishvili[4]. We shall follow a simpler approach using Chebyshev[5] polynomials[6]. This will limit our use of complex analysis to evaluating contour integrals, of which there will be many in this chapter.

10.2.1 Chebyshev polynomials

Chebyshev polynomials of the first kind are defined as follows:

$$T_n(\cos\theta) = \cos n\theta, \quad \text{where } n \geq 0. \tag{10.2}$$

Using the addition formula for cosines it is easy to show:

$$T_{n+1}(\cos\theta) = 2\cos\theta\, T_n(\cos\theta) - T_{n-1}(\cos\theta),$$

[3] *Singular integral equations*, Muskhelishvili, N I, Dover Publications: New York (2008), ISBN 9780486462424.

[4] Nikolay Ivanovich Muskhelishvili 1891-1976, Georgian mathematician, physicist and engineer.

[5] Pafnuty Lvovich Chebyshev 1821-1894, Russian mathematician.

[6] This is the approach taken by Bilby, B A, and Eshelby, J D, in *Dislocations and the Theory of Fracture* forming Chapter 2 of *Fracture*, ed. H Liebowitz, volume 1, Academic Press: New York (1968). Although these authors did not explicitly mention Chebyshev polynomials the mathematics they used is the mathematics of Chebyshev polynomials.

where $n \geq 1$. Replacing $\cos\theta$ by x we have the recurrence relation:

$$T_{n+1}(x) = 2xT_n(x) - T_{n-1}(x), \tag{10.3}$$

where $T_n(x) = \cos\left(n\cos^{-1}x\right)$, and $T_0(x) = 1$, $T_1(x) = x$.

Chebyshev polynomials of the second kind are defined as follows:

$$U_n(\cos\theta) = \frac{\sin(n+1)\theta}{\sin\theta}, \quad \text{where } n \geq 0. \tag{10.4}$$

Using the addition formula for sines it is easy to show:

$$U_{n+1}(\cos\theta) = 2\cos\theta U_n(\cos\theta) - U_{n-1}(\cos\theta),$$

where $n \geq 1$. Replacing $\cos\theta$ by x we have the recurrence relation:

$$U_{n+1}(x) = 2xU_n(x) - U_{n-1}(x), \tag{10.5}$$

where $U_n(x) = \sin\left((n+1)\cos^{-1}x\right)/\sin\left(\cos^{-1}x\right)$, and $U_0(x) = 1$, $U_1(x) = 2x$.

Exercise 10.1 Using the recurrence relations determine the Chebyshev polynomials of the first and second kind up to $n = 6$.

Prove the following orthogonality relations on the interval $-1 \leq x \leq 1$:

$$\int_{-1}^{1} \frac{T_m(x)T_n(x)}{\sqrt{1-x^2}}\,\mathrm{d}x = \begin{cases} 0, & n \neq m \\ \pi, & n = m = 0 \\ \pi/2 & n = m \neq 0, \end{cases} \tag{10.6}$$

and

$$\int_{-1}^{1} U_m(x)U_n(x)\sqrt{1-x^2}\,\mathrm{d}x = \begin{cases} 0, & n \neq m \\ \pi/2 & n = m. \end{cases} \tag{10.7}$$

Hint Make the substitution $x = \cos\theta$. ∎

Solution With $T_0(x) = 1$ and $T_1(x) = x$ the recurrence relation $T_{n+1}(x) = 2xT_n(x) - T_{n-1}(x)$ yields:

$$\begin{aligned} T_2(x) &= 2x^2 - 1 \\ T_3(x) &= 4x^3 - 3x \end{aligned}$$

$$
\begin{aligned}
T_4(x) &= 8x^4 - 8x^2 + 1 \\
T_5(x) &= 16x^5 - 20x^3 + 5x \\
T_6(x) &= 32x^6 - 48x^4 + 18x^2 - 1
\end{aligned}
$$

With $U_0(x) = 1$ and $U_1(x) = 2x$ the recurrence relation $U_{n+1}(x) = 2xU_n(x) - U_{n-1}(x)$ yields:

$$
\begin{aligned}
U_2(x) &= 4x^2 - 1 \\
U_3(x) &= 8x^3 - 4x \\
U_4(x) &= 16x^4 - 12x^2 + 1 \\
U_5(x) &= 32x^5 - 32x^3 + 6x \\
U_6(x) &= 64x^6 - 80x^4 + 24x^2 - 1
\end{aligned}
$$

Let

$$
I_1 = \int_{-1}^{1} \frac{T_m(x)T_n(x)}{\sqrt{1-x^2}}\,\mathrm{d}x,
$$

and let $x = \cos\theta$. Then $T_p(x) = \cos p\theta$. I_1 is then:

$$
I_1 = \int_{0}^{\pi} \cos m\theta \cos n\theta \,\mathrm{d}\theta = \begin{cases} 0, & n \neq m \\ \pi, & n = m = 0 \\ \pi/2 & n = m \neq 0. \end{cases}
$$

Let

$$
I_2 = \int_{-1}^{1} U_m(x)U_n(x)\sqrt{1-x^2}\,\mathrm{d}x,
$$

and let $x = \cos\theta$. Then $U_p(x) = \sin[(p+1)\theta]/\sin\theta$. I_2 is then:

$$
I_2 = \int_{0}^{\pi} \sin(m+1)\theta \sin(n+1)\theta \,\mathrm{d}\theta = \begin{cases} 0, & n \neq m \\ \pi/2 & n = m. \end{cases}
$$

As a result of the orthogonality relations, eq 10.6, any integrable function $F(x)$ on the interval $-1 \leq x \leq 1$ may be expanded in terms of Chebyshev polynomials of the first kind:

$$F(x) = \sum_{n=0}^{\infty} a_n T_n(x) \tag{10.8}$$

where

$$\begin{aligned} a_0 &= \frac{1}{\pi}\int_{-1}^{1} \frac{F(x)}{\sqrt{1-x^2}}\,\mathrm{d}x \\ a_m &= \frac{2}{\pi}\int_{-1}^{1} \frac{F(x)T_m(x)}{\sqrt{1-x^2}}\,\mathrm{d}x \end{aligned} \tag{10.9}$$

Chebyshev polynomials are related by the following principal value integrals:

$$P\int_{-1}^{1} \frac{T_n(y)}{(y-x)\sqrt{1-y^2}}\,\mathrm{d}y = \pi U_{n-1}(x), \ |x| \leq 1 \tag{10.10}$$

$$P\int_{-1}^{1} \frac{\sqrt{1-y^2}\,U_{n-1}(y)}{(y-x)}\,\mathrm{d}y = -\pi T_n(x), \ |x| \leq 1. \tag{10.11}$$

Exercise 10.2 Prove eqs 10.10 and 10.11.
Hint Making the substitutions $y = \cos\theta$ and $x = \cos\phi$ show that eq 10.10 becomes:

$$P\int_{0}^{\pi} \frac{\cos n\theta}{\cos\theta - \cos\phi}\,\mathrm{d}\theta = \frac{\pi \sin n\phi}{\sin\phi},$$

and eq 10.11 becomes:

$$-P\int_{0}^{\pi} \frac{\sin\theta \sin n\theta}{\cos\theta - \cos\phi}\,\mathrm{d}\theta = \pi\cos n\phi.$$

These integrals may be expressed as contour integrals in the complex plane by making the substitution $z = e^{i\theta}$. ■

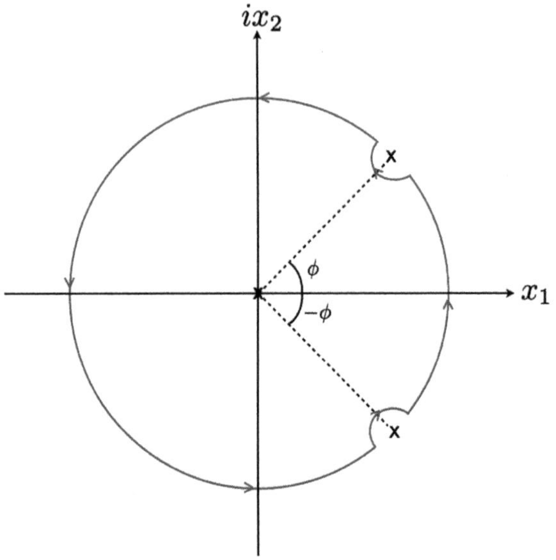

Figure 10.1: Contour to evaluate the integral in eq 10.12. There are two simple poles on the unit circle and a pole of order n at the origin, shown by crosses.

Solution To prove

$$I_1 = P\int_{-1}^{1} \frac{T_n(y)}{(y-x)\sqrt{1-y^2}}\,\mathrm{d}y = \pi U_{n-1}(x),\ |x| \le 1$$

we may let $y = \cos\theta$ and $x = \cos\phi$ since $|y|$ and $|x| \le 1$. Then I_1 becomes

$$I_1 = \frac{P}{2}\int_{-\pi}^{\pi} \frac{\cos n\theta}{\cos\theta - \cos\phi}\,\mathrm{d}\theta.$$

This integral may be evaluated using contour integration. Let $z = e^{i\theta}$. Then I_1 is transformed into:

$$I_1 = \frac{P}{2}\oint_C \frac{z^n + z^{-n}}{(z + z^{-1} - 2\cos\phi)}\frac{\mathrm{d}z}{iz} = \frac{P}{2i}\oint_C \frac{z^n + z^{-n}}{(z - e^{i\phi})(z - e^{-i\phi})}\mathrm{d}z, \tag{10.12}$$

where the contour C is shown in red in Fig 10.1. The residue at $z = e^{i\phi}$ is $-i\cos(n\phi)/\sin\phi$. The residue at $z = e^{-i\phi}$ is $i\cos(n\phi)/\sin\phi$. Therefore the contributions to I_1 from these two poles cancel. The residue at z=0 is:

$$\frac{1}{(n-1)!}\frac{1}{2i\sin\phi}\left(\frac{\mathrm{d}^{n-1}}{\mathrm{d}z^{n-1}}\left[\frac{1}{z-e^{i\phi}}-\frac{1}{z-e^{-i\phi}}\right]\right)_{z=0}$$

Since

$$\left(\frac{\mathrm{d}^{n-1}}{\mathrm{d}z^{n-1}}\frac{1}{z-e^{i\phi}}\right)_{z=0} = (-1)^{n-1}(n-1)!\left((z-e^{i\phi})^{-n}\right)_{z=0} = -(n-1)!\,e^{-in\phi}$$

$$\left(\frac{\mathrm{d}^{n-1}}{\mathrm{d}z^{n-1}}\frac{1}{z-e^{-i\phi}}\right)_{z=0} = (-1)^{n-1}(n-1)!\left((z-e^{-i\phi})^{-n}\right)_{z=0} = -(n-1)!\,e^{in\phi}$$

then

$$I_1 = \frac{2\pi i}{2i}\frac{1}{2i\sin\phi}\left(e^{in\phi}-e^{-in\phi}\right) = \pi\frac{\sin n\phi}{\sin\phi} = \pi U_{n-1}(\cos\phi)$$

which proves eq 10.10.

To prove

$$I_2 = P\int_{-1}^{1}\frac{\sqrt{1-y^2}\,U_{n-1}(y)}{(y-x)}\mathrm{d}y = -\pi T_n(x),\ |x|\leq 1$$

again we make the substitutions $y = \cos\theta$ and $x = \cos\phi$, whereupon I_2 becomes:

$$I_2 = \frac{P}{2}\int_{-\pi}^{\pi}\frac{\sin\theta\sin n\theta}{\cos\theta-\cos\phi}\mathrm{d}\theta = \frac{P}{4}\int_{-\pi}^{\pi}\frac{\cos(n-1)\theta-\cos(n+1)\theta}{\cos\theta-\cos\phi}\mathrm{d}\theta.$$

We showed above that:

$$\frac{P}{2}\int_{-\pi}^{\pi}\frac{\cos n\theta}{\cos\theta-\cos\phi}\mathrm{d}\theta = \pi\frac{\sin n\phi}{\sin\phi}$$

Using this result we obtain:

$$I_2 = -\pi \cos n\phi = -\pi T_n(\cos\phi),$$

which proves eq 10.11.

Comment

The integrals of eqs 10.10 and 10.11 are stated without proof in Appendix C of the paper by Bilby and Eshelby (1968).

10.2.2 Solution of the Cauchy type integral equation

We now have the tools we need to solve the integral equation, eq 10.1. Let $F(u) = L(u)/\sqrt{1-u^2}$, where $L(u)$ is an integrable function on $[-1,1]$ to be determined by solving the integral equation. We may expand $L(u)$ in terms of Chebyshev polynomials of the first kind:

$$L(u) = a_0 + \sum_{n=1}^{\infty} a_n T_n(u). \tag{10.13}$$

Inserting this expansion into eq 10.1 we obtain:

$$P\int_{-1}^{1} \left(a_0 + \sum_{n=1}^{\infty} a_n T_n(u) \right) \frac{1}{\sqrt{(1-u^2)}} \frac{1}{(v-u)} \, du = H(v). \tag{10.14}$$

Using eq 10.10 to evaluate the integral this equation becomes:

$$\sum_{n=1}^{\infty} a_n \pi U_{n-1}(v) = -H(v). \tag{10.15}$$

Note that a_0 is undetermined because

$$P\int_{-1}^{1} \frac{1}{\sqrt{1-u^2}} \frac{1}{(v-u)} \, du = 0 \tag{10.16}$$

as may be verified by contour integration. The constant a_0 is the solution of the homogeneous equation, where $H(v)$ is zero, and we shall see it is determined by the boundary conditions at $u = \pm 1$.

If we now multiply both sides of eq 10.15 by $\sqrt{1-v^2}/(v-u)$ and integrate over v from $v = -1$ to $v = 1$, making use of eq 10.11, we obtain:

$$\pi^2 \sum_{n=1}^{\infty} a_n T_n(u) = P\int_{-1}^{1} \frac{H(v)\sqrt{1-v^2}}{v-u} \, dv. \tag{10.17}$$

Using eq 10.13 and the definition $L(u) = \sqrt{1-u^2}\,F(u)$ we reach the solution of the integral equation:

$$F(u) = \frac{a_0}{\sqrt{1-u^2}} + \frac{1}{\pi^2} P \int_{-1}^{1} \frac{H(v)}{(v-u)} \sqrt{\frac{1-v^2}{1-u^2}}\, \mathrm{d}v. \tag{10.18}$$

Quite often we are confronted with a different range of integration for the integral equation, such as:

$$P \int_a^b \frac{D(y)}{x-y}\, \mathrm{d}y = Q(x), \tag{10.19}$$

where $a \le x \le b$. But if we let $u = (2y - a - b)/(b - a)$ and $v = (2x - a - b)/(b - a)$ this integral equation becomes identical to eq 10.1, and the corresponding solution is readily obtained from eq 10.18:

$$D(y) = \frac{1}{\pi^2 \sqrt{(y-a)(b-y)}} \left[C + P \int_a^b \frac{Q(x)}{x-y} \sqrt{(x-a)(b-x)}\, \mathrm{d}x \right], \tag{10.20}$$

where C is again an arbitrary constant determined by the boundary conditions on $D(y)$ at $y = a, b$. The solution in eq 10.20 diverges at $y = a, b$. If a solution is sought where $D(y)$ is bounded at $y = a$ then the term in square brackets in eq 10.20 is set to zero at $y = a$, which yields:

$$C = -P \int_a^b \frac{Q(x)}{x-a} \sqrt{(x-a)(b-x)}\, \mathrm{d}x.$$

When this is substituted back into eq 10.20 we obtain:

$$D(y) = \frac{1}{\pi^2} \sqrt{\frac{(y-a)}{(b-y)}}\, P \int_a^b Q(x) \sqrt{\frac{(b-x)}{(x-a)}} \frac{1}{(x-y)}\, \mathrm{d}x. \tag{10.21}$$

Similarly, a solution bounded at $y = b$ is as follows:

$$D(y) = \frac{1}{\pi^2} \sqrt{\frac{(b-y)}{(y-a)}}\, P \int_a^b Q(x) \sqrt{\frac{(x-a)}{(b-x)}} \frac{1}{(x-y)}\, \mathrm{d}x. \tag{10.22}$$

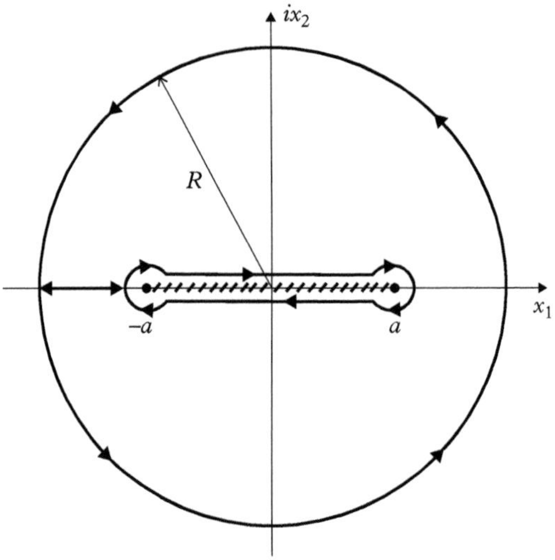

Figure 10.2: Contour to evaluate the integral in eq 10.25. The limit $R \to \infty$ is taken.

If this solution is also required to be bounded at $y = a$ then the integral in eq 10.22 has to be zero at $y = a$, or equivalently the integral in eq 10.21 has to be zero at $y = b$. Both are satisfied if the following relationship holds:

$$\int_a^b \frac{Q(x)}{\sqrt{(x-a)(b-x)}}\,\mathrm{d}x = 0. \tag{10.23}$$

The solution of the integral equation then becomes:

$$\begin{aligned} D(y) &= \frac{1}{\pi^2}\sqrt{\frac{(b-y)}{(y-a)}}\left\{ P\int_a^b Q(x)\sqrt{\frac{(x-a)}{(b-x)}}\frac{1}{(x-y)}\,\mathrm{d}x - \int_a^b \frac{Q(x)}{\sqrt{(x-a)(b-x)}}\,\mathrm{d}x \right\} \\ &= \frac{1}{\pi^2}\sqrt{(y-a)(b-y)}\,P\int_a^b \frac{Q(x)}{\sqrt{(x-a)(b-x)}}\frac{1}{(x-y)}\,\mathrm{d}x. \end{aligned} \tag{10.24}$$

10.2.3 Some contour integrals

In this chapter we use contour integration extensively. There are some contour integrals that involve a careful analysis of the contribution from the closure of the contour at infinity. For example, consider the following familiar integral:

$$I = \int_{-a}^{a} \sqrt{a^2 - x^2}\,\mathrm{d}x.$$

It is trivial to evaluate this integral by making the substitution $x = a\sin\theta$ or $x = a\cos\theta$ to obtain $I = \pi a^2/2$. It may also be evaluated in the complex plane by considering the following contour integral:

$$J_1 = \oint_C \sqrt{z^2 - a^2}\,\mathrm{d}z, \tag{10.25}$$

where the contour C is illustrated in Fig 10.2. On the upper surface of the branch cut between $-a$ and $+a$ we find $\sqrt{z^2 - a^2} = i\sqrt{a^2 - x^2}$, and on the lower surface $\sqrt{z^2 - a^2} = -i\sqrt{a^2 - x^2}$. Therefore the contribution to the contour integral from the loop around the branch cut is $2iI$. The contributions from the small circles at $z = \pm a$ are zero. The contour is closed at infinity by a large circle $z = Re^{i\theta}$, where $0 \le \theta \le 2\pi$ and we take the limit[7] $R \to \infty$. The contribution from this large circle is evaluated as follows:

$$\begin{aligned}
\lim_{R\to\infty} \int_0^{2\pi} \left(R^2 e^{2i\theta} - a^2\right)^{\frac{1}{2}} iRe^{i\theta}\,\mathrm{d}\theta &= \lim_{R\to\infty} \int_0^{2\pi} Re^{i\theta}\left(1 - \frac{a^2}{R^2 e^{2i\theta}}\right)^{\frac{1}{2}} iRe^{i\theta}\,\mathrm{d}\theta \\
&= \lim_{R\to\infty} \int_0^{2\pi} Re^{i\theta}\left(1 - \frac{a^2}{2R^2 e^{2i\theta}} + \ldots\right) iRe^{i\theta}\,\mathrm{d}\theta \\
&= -\pi i a^2.
\end{aligned}$$

Since there are no poles inside the contour Cauchy's theorem tells us that $2iI - \pi i a^2 = 0$, and therefore $I = \pi a^2/2$, as before.

We will also encounter principal value integrals of the form:

$$I = P\int_{-a}^{a} \frac{\sqrt{a^2 - x^2}}{x - b}\,\mathrm{d}x, \qquad -a \le b \le a,$$

[7]The process of taking limits may be illustrated by this example attributed to Konrad Zacharias Lorenz 1903-1989, Nobel prize winning, Austrian zoologist. Philosophers are people who know less and less about more and more, until they know nothing about everything. Scientists are people who know more and more about less and less, until they know everything about nothing.

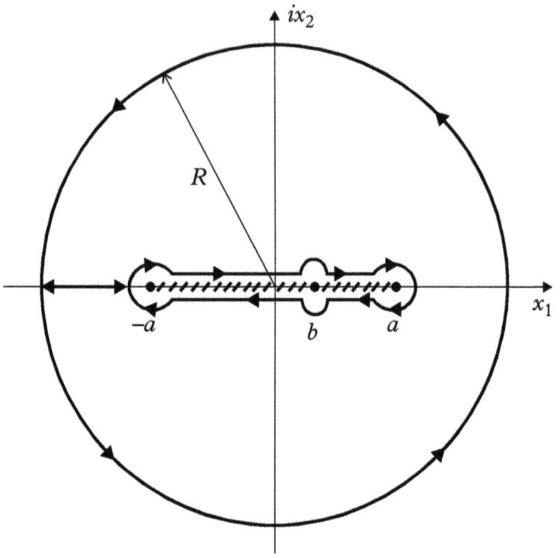

Figure 10.3: Contour to evaluate the integral in eq 10.26. The limit $R \to \infty$ is taken.

It may be evaluated by considering the contour integral:

$$J_2 = \oint_C \frac{\sqrt{z^2 - a^2}}{z - b} \mathrm{d}z, \tag{10.26}$$

where C is the contour shown in Fig 10.3. There is a simple pole on the branch cut at $z = b$. The contributions from the semi-circles above and below the pole are equal and opposite because the square root changes sign on either side of the branch cut. Therefore the contribution from the loop around the branch cut is $2iI$, including the semi-circles around the pole. The contour is again closed by a large circle of radius R, giving the following contribution:

$$\begin{aligned}
\lim_{R\to\infty} \int_0^{2\pi} \frac{(R^2 e^{2i\theta} - a^2)^{\frac{1}{2}}}{(Re^{i\theta} - b)} iRe^{i\theta} \mathrm{d}\theta &= \lim_{R\to\infty} \int_0^{2\pi} \frac{Re^{i\theta}\left(1 - \frac{a^2}{R^2 e^{2i\theta}}\right)^{\frac{1}{2}}}{Re^{i\theta}\left(1 - \frac{b}{Re^{i\theta}}\right)} iRe^{i\theta} \mathrm{d}\theta \\
&= \lim_{R\to\infty} \int_0^{2\pi} \left(1 - \frac{a^2}{2R^2 e^{2i\theta}}\right)\left(1 + \frac{b}{Re^{i\theta}}\right) iRe^{i\theta} \mathrm{d}\theta \\
&= 2\pi i b.
\end{aligned}$$

Since there are no poles inside the contour C Cauchy's theorem tells us that $2iI+2\pi ib=0$, or $I=-\pi b$.

Exercise 10.3 Prove the following integrals using contour integration:

$$\int_{-a}^{a}\frac{1}{\sqrt{a^2-x^2}}\,\mathrm{d}x = \pi$$

$$P\int_{-a}^{a}\frac{1}{\sqrt{a^2-x^2}}\frac{1}{x-b}\,\mathrm{d}x = 0, \qquad -a\le b\le a$$

$$\int_{-a}^{a}\frac{1}{\sqrt{a^2-x^2}}\frac{1}{x-b}\,\mathrm{d}x = -\frac{\pi}{\sqrt{b^2-a^2}}, \qquad b>a$$

■

Solution To evaluate

$$\int_{-a}^{a}\frac{1}{\sqrt{a^2-x^2}}\,\mathrm{d}x$$

as a contour integral, consider

$$J=\oint_C\frac{1}{\sqrt{z^2-a^2}}\,\mathrm{d}z,$$

where C is the contour shown in Fig 10.2. On the upper surface of the branch cut $1/\sqrt{z^2-a^2}=-i/\sqrt{a^2-x^2}$, while on the lower surface it is $i/\sqrt{a^2-x^2}$. The contributions from the small circles at $z=\pm a$ are zero. Therefore the contribution from the loop around the branch cut is:

$$-2i\int_{-a}^{a}\frac{1}{\sqrt{a^2-x^2}}\,\mathrm{d}x.$$

The contribution from the circle of radius $R \to \infty$, where $z = Re^{i\theta}$, is:

$$\lim_{R\to\infty}\int_0^{2\pi}\frac{iRe^{i\theta}}{\sqrt{R^2e^{2i\theta}-a^2}}\,\mathrm{d}\theta = \lim_{R\to\infty}\int_0^{2\pi}\frac{iRe^{i\theta}}{Re^{i\theta}\left(1-\frac{a^2}{R^2e^{2i\theta}}\right)^{\frac{1}{2}}}\,\mathrm{d}\theta = 2\pi i$$

Since there no poles inside the contour $J = 0$. Therefore,

$$-2i\int_{-a}^{a}\frac{1}{\sqrt{a^2-x^2}}\,\mathrm{d}x + 2\pi i = 0,$$

and hence

$$\int_{-a}^{a}\frac{1}{\sqrt{a^2-x^2}}\,\mathrm{d}x = \pi.$$

It is trivial to verify this result by making the substitution $x = a\sin\theta$.

To evaluate

$$P\int_{-a}^{a}\frac{1}{\sqrt{a^2-x^2}}\frac{1}{x-b}\,\mathrm{d}x, \qquad -a \le b \le a,$$

as a contour integral, consider:

$$J = \oint_C \frac{1}{\sqrt{z^2-a^2}}\frac{1}{z-b}\,\mathrm{d}z,$$

where C is the contour illustrated in Fig 10.3. The contribution from the semi-circles above and below the pole on the branch cut is again zero because the square root changes sign on either side of the branch cut. The contributions from the infinitesimal circles at $z = \pm a$ may also be shown to be zero. Hence, the contribution from the loop around the branch cut is:

$$-2iP\int_{-a}^{a}\frac{1}{\sqrt{a^2-x^2}}\frac{1}{x-b}\,\mathrm{d}x.$$

The contribution from the circle of radius $R \to \infty$ is:

$$\lim_{R\to\infty}\int_0^{2\pi}\frac{iRe^{i\theta}}{\sqrt{R^2e^{2i\theta}-a^2}}\frac{1}{(Re^{i\theta}-b)}d\theta = \lim_{R\to\infty}\int_0^{2\pi}\frac{iRe^{i\theta}}{Re^{i\theta}\sqrt{1-\frac{a^2}{R^2e^{2i\theta}}}}\frac{1}{(Re^{i\theta}-b)}d\theta = 0.$$

Since there are no poles inside the contour $J = 0$. Hence:

$$P\int_{-a}^{a}\frac{1}{\sqrt{a^2-x^2}}\frac{1}{x-b}dx = 0.$$

To evaluate

$$\int_{-a}^{a}\frac{1}{\sqrt{a^2-x^2}}\frac{1}{x-b}dx \qquad \text{where } b > a$$

by contour integration consider

$$J = \oint_C \frac{1}{\sqrt{z^2-a^2}}\frac{1}{z-b}dz,$$

where the contour C is the same as Fig 10.3 except the pole at $z = b$ is no longer on the branch cut but at a point on the real axis beyond $z = a$. The contribution from the loop around the branch cut is again:

$$-2iP\int_{-a}^{a}\frac{1}{\sqrt{a^2-x^2}}\frac{1}{x-b}dx.$$

The contribution from the circle at $R \to \infty$ is again zero. The residue at the simple pole at $z = b$ is $1/\sqrt{b^2-a^2}$. Therefore,

$$-2iP\int_{-a}^{a}\frac{1}{\sqrt{a^2-x^2}}\frac{1}{x-b}dx = 2\pi i\frac{1}{\sqrt{b^2-a^2}},$$

and hence

$$P\int_{-a}^{a}\frac{1}{\sqrt{a^2-x^2}}\frac{1}{x-b}\mathrm{d}x=-\frac{\pi}{\sqrt{b^2-a^2}}.$$

10.3 Representation of loaded cracks as distributions of dislocations

Consider a slit crack in the plane $x_2 = 0$ between $x_1 = a$ and $x_1 = b$, with $a < b$, and infinitely long along the x_3-axis. A slit crack is a rectangular cut in an infinite elastic continuum, which is infinitely long in one direction and finite in the perpendicular direction. The faces of the cut undergo a relative displacement in response to an applied load. Suppose a uniform tensile stress $\sigma_{22} = \sigma$ is applied to the medium at $x_2 = \pm\infty$. The faces of the crack separate slightly under the action of the applied stress. Let $s(x_1)$ denote the x_2-coordinate of the upper crack face at $a \leq x_1 \leq b$ minus the x_2-coordinate of the lower crack face at x_1, where $s(a) = s(b) = 0$. The function $s(x_1)$ is called the crack opening displacement.

The variation of $s(x_1)$ with x_1 may be modelled mathematically as the result of a continuous distribution of edge dislocations in $a \leq x_1 \leq b$, with their lines parallel to the x_3-axis and their Burgers vectors along the x_2-axis. Let the positive sense of the dislocation lines be along the positive direction of the x_3-axis. Let $f(x_1)\mathrm{d}x_1$ be the Burgers vector of dislocations between x_1 and $x_1 + \mathrm{d}x_1$. Then the relationship between the crack opening displacement and $f(x_1)$ is as follows:

$$s(x_1) = -\int_{a}^{x_1} f(x_1')\,\mathrm{d}x_1' \tag{10.27}$$

where the negative sign in front of the integral is consistent with the FS/RH convention and the definition of $s(x_1)$. The distribution of dislocations in $a \leq x_1 \leq b$ representing the crack opening is illustrated in Fig 10.4(a)[8]. In Fig 10.4(b) it is shown that the

[8]There is a drawing similar to Fig 10.4 on p.215 of the book *Les dislocations* by J Friedel, Paris: Gauthier-Villars (1956), where he called the dislocations representing the crack *les dislocations de clivage* - cleavage dislocations.
Jacques Friedel, ForMemRS 1921-2014, French materials physicist and President of the French Academy of Sciences 1992-94.
The first suggestion that cracks and slip bands may both be represented by arrays of dislocations appears to be due to Zener(1948). The close relationship between a crack and a pileup of dislocations was also discussed in Eshelby, J D, Frank, F C, and Nabarro, F R N, Phil. Mag. **42**, 351-364 (1951). The mathematics for the replacement of arrays of discrete dislocations in pileups by continuous distributions of

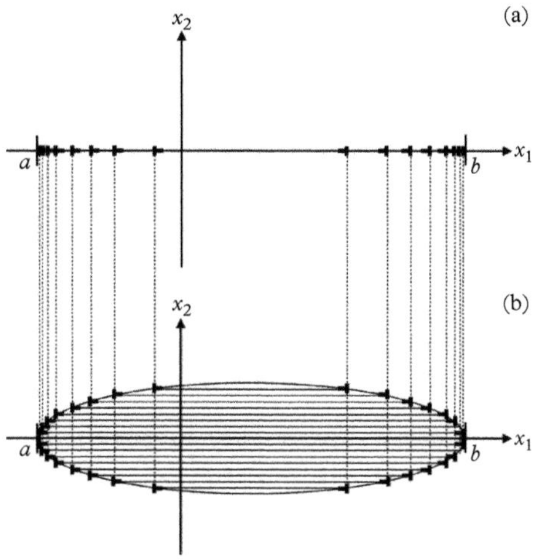

Figure 10.4: To illustrate the representation of a slit crack under a normal tensile load by a distribution of dislocations. (a) a slit crack occupies $a \leq x_1 \leq b$ and is subjected to a tensile stress along x_2. The crack faces separate in (b) forming an ellipse in cross-section. The elliptical crack faces may also be thought of as the interface surrounding an inclusion comprising a stack of interstitial loops shown in (b). When the locations of the edge dislocations bounding these interstitial loops are projected by the vertical broken lines onto the slit crack in (a) we obtain the representation of the loaded slit crack as a distribution of edge dislocations. As the spacing between the horizontal planes on which the interstitial loops lie becomes infinitesimal the distribution of dislocations representing the crack becomes continuous, and their Burgers vectors become infinitesimal. The normal tractions created on the crack faces by the dislocations cancel the tractions arising from the applied normal stress so that the crack faces are free of tractions.

distribution in (a) may be viewed as a distribution of interstitial loops collapsed onto the plane $x_2 = 0$. The loops transform the slit crack into the ellipse shown in (b). Since these loops are interstitial in character they create a compressive stress field inside the ellipse which generates tractions on the crack faces that exactly cancel the tractions arising from the applied load. *The requirement that there are no net tractions on the faces of the loaded slit crack is a boundary condition that must be satisfied by* $f(x_1)$.

So far we have discussed the loading of a slit crack with normal along x_2 by a

infinitesimal dislocations was developed by Leibfried (Leibfried, G, Z. Phys. **130**, 244 (1951), in German) and by Head and Louat (Head, A K, and Louat, N, Australian J. Phys., **8**, 1 (1955)). Alan Kenneth Head FRS 1925-2010, Australian physicist. Norman P Louat 1920-2010, Australian and US physicist.

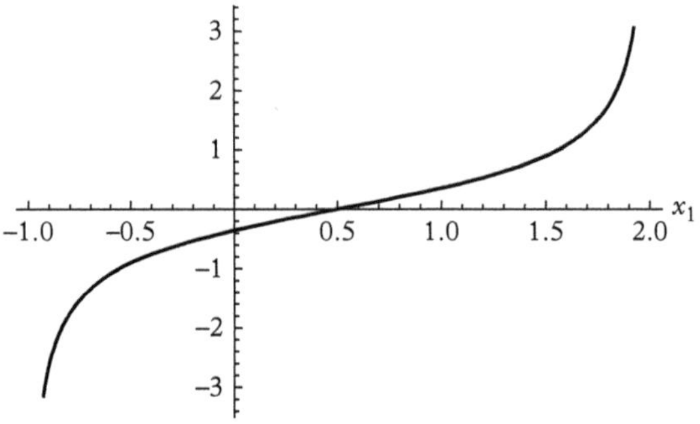

Figure 10.5: Plot of the Burgers vector density for an elastic mode I crack, given by eq 10.32. In this example the crack tips are at $x_1 = -1$ and $x_1 = 2$, where the Burgers vector density diverges. The vertical axis is in units of $2(1-\nu)\sigma/\mu$.

normal stress σ_{22}. This is called mode I loading. The crack may also be loaded by a shear stress σ_{12} and this is called mode II loading. A slit crack loaded in mode II may be represented by a distribution of edge dislocations with Burgers vectors parallel to x_1. If the slit crack is loaded by a shear stress σ_{23} this is called mode III loading. A slit crack loaded in mode III may be represented by a distribution of screw dislocations along the x_3 axis. More complex loadings on a slit crack may be modelled by combining these dislocation representations.

10.4 Elastic field of a mode I slit crack and the stress intensity factor

In this section we will set up and solve the integral equation for the Burgers vector density $f(x_1)$ representing the mode I slit crack loaded by an applied normal tensile stress σ of the previous section. Far from the crack the stress field is just the applied normal stress. But this field is perturbed very significantly near the crack, and we shall see it becomes singular at the crack tips at $x_1 = a, b$. In this section we assume the response of the medium is purely elastic so that no plasticity takes place.

The applied normal stress σ creates tractions $-\sigma\hat{\mathbf{e}}_2$ on the crack faces. These tractions have to be cancelled by tractions created by the edge dislocations representing the crack. An edge dislocation at $(x_1',0)$ with Burgers vector $\mathbf{b} = [0,b,0]$ creates a stress field at (x_1,x_2) with the following components:

$$\sigma_{11}(x_1,x_2) = \frac{\mu b}{2\pi(1-\nu)} \frac{(x_1-x_1')\left((x_1-x_1')^2-x_2^2\right)}{\left((x_1-x_1')^2+x_2^2\right)^2}$$

$$\sigma_{22}(x_1,x_2) = \frac{\mu b}{2\pi(1-\nu)} \frac{(x_1-x_1')\left(3x_2^2+(x_1-x_1')^2\right)}{\left((x_1-x_1')^2+x_2^2\right)^2}$$

$$\sigma_{12}(x_1,x_2) = \sigma_{21}(x_1,x_2) = \frac{\mu b}{2\pi(1-\nu)} \frac{x_2\left((x_1-x_1')^2-x_2^2\right)}{\left((x_1-x_1')^2+x_2^2\right)^2}$$

$$\sigma_{33}(x_1,x_2) = \nu\left(\sigma_{11}(x_1,x_2)+\sigma_{22}(x_1,x_2)\right)$$

$$\sigma_{13}(x_1,x_2) = \sigma_{31}(x_1,x_2) = \sigma_{23}(x_1,x_2) = \sigma_{32}(x_1,x_2) = 0 \tag{10.28}$$

In the plane of the slit crack $x_2 = 0$ and the only component of the stress field of a dislocation at $(x_1',0)$ contributing to tractions on the crack faces is $\sigma_{22}(x_1,0)$:

$$\sigma_{22}(x_1,0) = \frac{\mu b}{2\pi(1-\nu)} \frac{1}{(x_1-x_1')} \tag{10.29}$$

The total normal stress at $(x_1,0)$ arising from the distribution of edge dislocations in the crack is:

$$\Sigma_{22}(x_1,0) = \frac{\mu}{2\pi(1-\nu)} P\int_a^b \frac{f(x_1')}{(x_1-x_1')}\,\mathrm{d}x_1'. \tag{10.30}$$

The integral is a principal value integral because the edge dislocation at x_1 does not exert a stress upon itself. The requirement of zero tractions on the crack faces leads to the following integral equation for the Burgers vector density:

$$P\int_a^b \frac{f(x_1')}{(x_1-x_1')}\,\mathrm{d}x_1' + \frac{2\pi(1-\nu)}{\mu}\sigma = 0 \tag{10.31}$$

The solution to this equation follows from eq 10.20. The arbitrary constant C is determined by the condition that the total Burgers vector content of the distribution is zero:

$$f(x_1) = \frac{2(1-\nu)\sigma}{\mu} \frac{1}{\sqrt{(x_1-a)(b-x_1)}} \left(x_1 - \left(\frac{a+b}{2}\right)\right) \tag{10.32}$$

This solution is plotted in Fig 10.5.

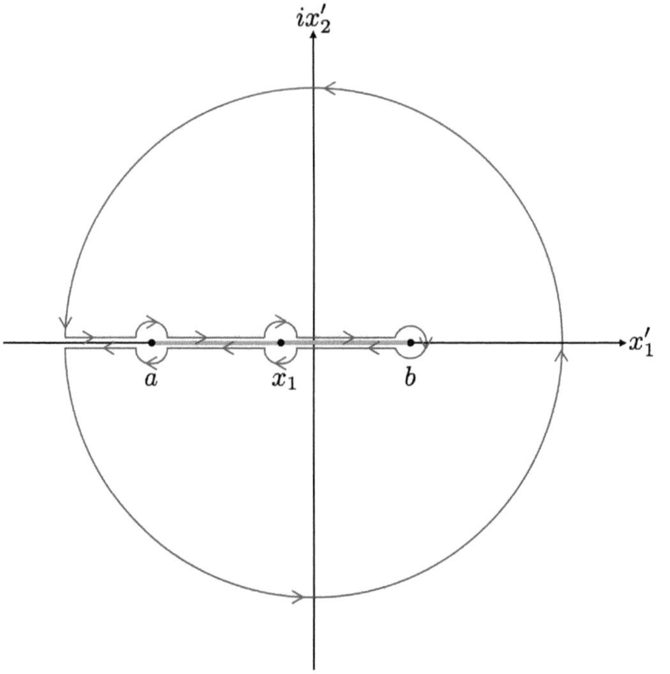

Figure 10.6: The contour used to evaluate the integral in eq 10.34. The thick green line is the branch cut on $a \le x'_1 \le b$. There is a simple pole on the branch cut at $x'_1 = x_1$. The radius of the large circle is taken to infinity.

Exercise 10.4 Derive eq 10.32 from eq 10.20.

Compare the continuous distribution $f(x_1)$ with the distribution of discrete dislocations sketched in Fig 10.4.

Show that under the action of the applied stress the crack opening displacement $s(x)$ attains a maximum value of $(1-\nu)(\sigma/\mu)(b-a)$ in the middle of the crack, and that the crack shape is an ellipse with the following equation:

$$\left(x_1 - \left(\frac{a+b}{2}\right)\right)^2 + \left(\frac{\mu x_2}{(1-\nu)\sigma}\right)^2 = \left(\frac{b-a}{2}\right)^2 \tag{10.33}$$

■

Solution It follows from eq 10.20 that:

$$f(x_1) = \frac{1}{\pi^2\sqrt{(x_1-a)(b-x_1)}}\left[C - \frac{2\pi(1-\nu)\sigma}{\mu} P\int_a^b \frac{\sqrt{(x_1'-a)(b-x_1')}}{x_1'-x_1}\mathrm{d}x_1'\right].$$

Here C is an arbitrary constant determined by the condition that there is no net Burgers vector associated with the crack, so that $\int_a^b f(x_1)\mathrm{d}x_1 = 0$.

To evaluate the principal value integral, consider the contour integral:

$$J = \oint_C \frac{\sqrt{(z-a)(z-b)}}{z-x_1}\mathrm{d}z, \qquad a \le x_1 \le b. \tag{10.34}$$

The contour C is shown in Fig 10.6. On the upper surface of the branch cut $\sqrt{(z-a)(z-b)} = i\sqrt{(x_1'-a)(b-x_1')}$. On the lower surface of the branch cut $\sqrt{(z-a)(z-b)} = -i\sqrt{(x_1'-a)(b-x_1')}$. Therefore the contributions from the semi-circles around the pole at $z = x_1$ cancel. The contributions from the circles around $z = a$ and $z = b$ are both zero. Therefore the contribution from the loop around the branch cut is:

$$2iP\int_a^b \frac{\sqrt{(x_1'-a)(b-x_1')}}{x_1'-x_1}\mathrm{d}x_1'.$$

Let the radius of the large circle be R. The contribution from this circle, in the limit $R \to \infty$, is found as follows:

$$\lim_{R\to\infty}\int_0^{2\pi} \frac{\sqrt{(Re^{i\theta}-a)(Re^{i\theta}-b)}}{Re^{i\theta}-x_1} iRe^{i\theta}\,\mathrm{d}\theta =$$

$$= \lim_{R\to\infty}\int_0^{2\pi} \frac{\sqrt{\left(1-\frac{a}{Re^{i\theta}}\right)\left(1-\frac{b}{Re^{i\theta}}\right)}}{1-\frac{x_1}{Re^{i\theta}}} iRe^{i\theta}\,\mathrm{d}\theta$$

$$= \lim_{R\to\infty} \int_0^{2\pi} \left(1 + \frac{x_1}{Re^{i\theta}} + \ldots\right)\left(1 - \frac{a}{2Re^{i\theta}} + \ldots\right)\left(1 - \frac{b}{2Re^{i\theta}} + \ldots\right) iRe^{i\theta}\mathrm{d}\theta$$

$$= 2\pi i\left(x_1 - \frac{a+b}{2}\right).$$

Since there are no poles inside the contour C the contour integral $J = 0$. Hence

$$2iP\int_a^b \frac{\sqrt{(x_1' - a)(b - x_1')}}{x_1' - x_1}\mathrm{d}x_1' + 2\pi i\left(x_1 - \frac{a+b}{2}\right) = 0$$

and therefore,

$$P\int_a^b \frac{\sqrt{(x_1' - a)(b - x_1')}}{x_1' - x_1}\mathrm{d}x_1' = -\pi\left(x_1 - \frac{a+b}{2}\right)$$

It is straightforward to show that the condition $\int_a^b f(x_1)\mathrm{d}x_1 = 0$ is satisfied only if the arbitrary constant C is zero. In that case eq 10.32 follows.

The continuously distributed Burgers vector density of eq 10.32 is similar to the distribution of discrete dislocations seen in Fig 10.4. They are both symmetrical about the middle of the crack, where the Burgers density changes sign. They both increase rapidly towards the crack tips.

The crack opening displacement $s(x_1)$ is given by:

$$s(x_1) = -\int_a^{x_1} f(x_1')\mathrm{d}x_1' = -\frac{2(1-\nu)\sigma}{\mu}\int_a^{x_1} \frac{x_1' - (a+b)/2}{\sqrt{(x_1' - a)(b - x_1')}}\mathrm{d}x_1', \qquad a \le x_1 \le b.$$

If we make the substitution $y = x_1' - (a+b)/2$ this integral is evaluated easily and we obtain:

$$s(x_1) = \frac{2(1-\nu)\sigma}{\mu}\sqrt{\left(\frac{b-a}{2}\right)^2 - \left(x_1 - \frac{a+b}{2}\right)^2} = \frac{2(1-\nu)\sigma}{\mu}\sqrt{(x_1 - a)(b - x_1)}.$$

At $x_1 = (a+b)/2$ the crack opening displacement is a maximum:

$$s((a+b)/2) = \frac{(1-\nu)\sigma}{\mu}(b-a).$$

Setting $x_2 = s(x_2)/2$ then

$$x_2 = \frac{(1-\nu)\sigma}{\mu}\sqrt{\left(\frac{b-a}{2}\right)^2 - \left(x_1 - \frac{a+b}{2}\right)^2}$$

Therefore,

$$\left(x_1 - \frac{a+b}{2}\right)^2 + \frac{\mu^2}{(1-\nu)^2\sigma^2}x_2^2 = \left(\frac{b-a}{2}\right)^2.$$

which is the equation of an ellipse centred at $[(a+b)/2, 0]$ and eccentricity $\sqrt{1-\{(1-\nu)\sigma/\mu\}^2}$.

Along the x_1-axis at $x_1 > b$ the total stress $\sigma^T_{22}(x_1, 0)$ is given by:

$$\begin{aligned}\sigma^T_{22}(x_1,0) &= \sigma + \frac{\sigma}{\pi}\int_a^b \frac{1}{(x_1 - x_1')}\frac{x_1' - (a+b)/2}{\sqrt{(x_1'-a)(b-x_1')}}\,\mathrm{d}x_1' \\ &= \sigma\left(\frac{x_1-(a+b)/2}{\sqrt{(x_1-a)(x_1-b)}}\right). \qquad (10.35)\end{aligned}$$

The crack generates stress singularities at its tips. In eq 10.35 let $x_1 = b + \Delta$ where $\Delta \ll (b-a)$. We obtain:

$$\sigma^T_{22}(b+\Delta,0) = \frac{\sigma}{2}\sqrt{\frac{(b-a)}{\Delta}} + O\left(\sqrt{\frac{\Delta}{(b-a)}}\right) \qquad (10.36)$$

Thus, the applied stress field is intensified near the crack tips, displaying an inverse square root singularity . The *stress intensity factor* is defined by

$$K_I = \lim_{\Delta\to 0}\sqrt{\Delta}\,\sigma^T_{22}(b+\Delta,0) = \sigma\frac{\sqrt{(b-a)}}{2} = \sigma\sqrt{c/2}, \qquad (10.37)$$

where $2c = (b-a)$ is the crack length. The units of the stress intensity factor are $\mathrm{Pa}\cdot\mathrm{m}^{1/2}$ or equivalently $\mathrm{N}\cdot\mathrm{m}^{-3/2}$.

Expressions for the stress field of the loaded crack throughout the medium may be obtained using the Burgers vector density in eq 10.32 and the stress components of an individual edge dislocation, eq 10.28. For example,

$$\sigma_{12}^{T}(x_1,x_2) = \frac{\sigma x_2}{\pi}\int_a^b \frac{((x_1-x_1')^2-x_2^2)}{\left((x_1-x_1')^2+x_2^2\right)^2}\frac{(x_1'-(a+b)/2)}{\sqrt{(x_1'-a)(b-x_1')}}\,\mathrm{d}x_1', \tag{10.38}$$

which can be evaluated by contour integration. See problem 10.1.

10.5 Energy considerations and Griffith's fracture criterion

It is sometimes said that when a crack grows the elastic energy stored in the body is reduced. Thus, one sees in the literature the expression *the elastic energy release rate*, which is supposed to drive the growth of a crack. Following Bilby and Eshelby (1968) we will show that the elastic energy of the body *increases* as the crack grows at constant applied load[9]. However, the *sum* of the elastic energy and the potential energy of the external loading mechanism decreases when the crack grows. It is this sum which drives crack growth and leads to Griffith's thermodynamic criterion for fracture. Since the elastic energy stored in the body is a form of potential energy we may call the rate of decrease of the sum of the elastic energy stored in the body and the potential energy of the external loading mechanism the *potential energy release rate* rather than the elastic energy release rate.

Consider a body subjected to constant forces on its external surface. The body contains a crack which can grow. Let the stress and displacement fields in the body when the crack has a length L be σ_{ij} and u_i. The elastic energy stored in the body is the usual expression:

$$E_{el} = \frac{1}{2}\int_V \sigma_{ij}u_{i,j}\,\mathrm{d}V = \frac{1}{2}\int_{S_0} \sigma_{ij}u_i n_j\,\mathrm{d}S.$$

The integral over the volume of the body V is transformed into a surface integral using the divergence theorem. The surface comprises the external surface S_0 of the body and the faces of the crack. But the tractions, $\sigma_{ij}n_j$, on the surface of the crack are zero, and therefore the crack faces do not contribute to the surface integral, which may therefore be taken over the external surface only. The equilibrium condition $\sigma_{ij,j} = 0$ has been used in the second integral.

[9]If you are unconvinced consider a catapult. For a given applied force the elastic energy stored in the catapult bands increases as the compliance of the bands increases. As a crack grows the body it is in becomes more deformable, i.e. more compliant. Therefore, for the same forces applied to the surface of the body the elastic energy increases as the crack grows.

When the crack grows by δL let the stress and displacement fields in the body become σ'_{ij} and u'_i. The elastic energy of the body changes by:

$$\begin{aligned}\delta E_{el} &= \frac{1}{2}\int_{S_0} \left(\sigma'_{ij}u'_i - \sigma_{ij}u_i\right) n_j \mathrm{d}S \\ &= \frac{1}{2}\int_{S_0} \sigma_{ij}\left(u'_i - u_i\right) n_j \mathrm{d}S, \end{aligned} \tag{10.39}$$

where use has been made of $\sigma_{ij}n_j = \sigma'_{ij}n_j$ because the loading is constant. At the same time the potential energy of the external loading mechanism is reduced because it does work on the body:

$$\delta U_{ext} = -\int_{S_0} \sigma_{ij}\left(u'_i - u_i\right) n_j \mathrm{d}S. \tag{10.40}$$

Therefore the sum δU of the changes in the elastic energy of the body and the potential energy of the external loading mechanism is as follows:

$$\delta U = \delta E_{el} + \delta U_{ext} = -\frac{1}{2}\int_{S_0} \sigma_{ij}\left(u'_i - u_i\right) n_j \mathrm{d}S = -\delta E_{el} = \frac{1}{2}\delta U_{ext}. \tag{10.41}$$

Thus, the potential energy release rate is equal in magnitude to the change of the elastic energy but they have opposite signs. As anticipated at the beginning of this section the elastic energy of the body increases as the crack grows under a constant load.

To calculate the change in the total potential energy of the body and the loading mechanism when the crack is introduced we may extend the same argument. Let σ^A_{ij} and u^A_i be the stress and displacement fields in the loaded body before the crack is introduced. Let σ'_{ij} and u'_i be the stress and displacement fields in the body after the crack has been introduced, and where the surface of the body is subjected to the same surface tractions. The change in elastic energy of the body associated with the introduction of the crack is then:

$$\begin{aligned}\Delta E_{el} &= \frac{1}{2}\int_{V} \sigma'_{ij}u'_{i,j} - \sigma^A_{ij}u^A_{i,j}\,\mathrm{d}V \\ &= \frac{1}{2}\int_{S_0} \left(\sigma'_{ij}u'_i - \sigma^A_{ij}u^A_i\right) n_j \mathrm{d}S\end{aligned}$$

$$= \frac{1}{2}\int_{S_0} \sigma_{ij}^{A}\left(u_i' - u_i^{A}\right) n_j \mathrm{d}S$$

$$= \frac{1}{2}\int_{S_0} \sigma_{ij}'\left(u_i' - u_i^{A}\right) n_j \mathrm{d}S. \tag{10.42}$$

The surface S_0 includes the faces of the crack. But in the last line $\sigma_{ij}' n_j = 0$ on the crack faces, so the contributions from the crack faces are zero. Therefore, S_0 may be taken as the external surface of the body.

The change in the potential energy of the external loading mechanism following the introduction of the crack is:

$$\Delta U_{ext} = -\int_{S_0} \sigma_{ij}^{A}\left(u_i' - u_i^{A}\right) n_j \,\mathrm{d}S \tag{10.43}$$

The change in the total potential energy following the introduction of the crack is:

$$\Delta U = \Delta E_{el} + \Delta U_{ext} = -\frac{1}{2}\int_{S_0} \sigma_{ij}^{A}\left(u_i' - u_i^{A}\right) n_j \,\mathrm{d}S = -\frac{1}{2}\int_{S_0} \left(\sigma_{ij}^{A} u_i' - \sigma_{ij}' u_i^{A}\right) n_j \,\mathrm{d}S. \tag{10.44}$$

If there are no other defects present in the body the divergence of $\left(\sigma_{ij}^{A} u_i' - \sigma_{ij}' u_i^{A}\right)$ is zero throughout the body except at the crack. The surface S_0 may therefore be deformed to a closed surface S just outside and infinitesimally close to the crack faces. Equation 10.44 then becomes:

$$\Delta U = -\frac{1}{2}\int_{S} \left(\sigma_{ij}^{A} u_i' - \sigma_{ij}' u_i^{A}\right) n_j \,\mathrm{d}S = -\frac{1}{2}\int_{S} \sigma_{ij}^{A} u_i' n_j \,\mathrm{d}S, \tag{10.45}$$

where use has been made of $\sigma_{ij}' n_j = 0$ on the crack faces. Since u_i^{A} and σ_{ij}^{A} are continuous at the crack we obtain:

$$\Delta U = -\frac{1}{2}\int_{S} \sigma_{ij}^{A} u_i^{C} n_j \,\mathrm{d}S, \tag{10.46}$$

where u_i^{C} is the displacement of the crack faces due to the applied load.

When this equation is applied to the slit crack loaded in mode I of the previous section it becomes:

$$\Delta U = -\frac{1}{2}\int_{a}^{b} \sigma \, s(x_1)\,\mathrm{d}x_1$$

$$= -\frac{1}{2}\frac{(1-\nu)\sigma^2}{\mu}\int_a^b \sqrt{(b-a)^2-(2x_1-(a+b))^2}\,\mathrm{d}x_1$$

$$= -\frac{\pi(1-\nu)\sigma^2 c^2}{2\mu}, \tag{10.47}$$

where ΔU is the change of potential energy per unit length of the slit crack, and σ is the applied normal stress far from the crack.

The total energy of the body per unit length of the slit crack is ΔU plus the energy of the crack surfaces. If γ is the energy per unit area of the crack surfaces then since there are two surfaces, each of length $2c$, the crack surface energy is $4\gamma c$. Griffith[10] argued[11] that for the crack to grow the total energy of the system must decrease with increasing crack length. The critical crack length c^*, above which the total energy of the system decreases with increasing crack length (and below which it increases with increasing crack length), is thus:

$$c^* = \frac{4\gamma\mu}{\pi(1-\nu)\sigma^2}. \tag{10.48}$$

This equation is equivalent to saying that the stress intensity factor, $K_I = \sigma\sqrt{c/2}$ has to reach a critical value K_{Ic} for the crack to grow:

$$K_{Ic} = \sqrt{\frac{2\gamma\mu}{\pi(1-\nu)}} = \sqrt{\frac{\gamma Y}{\pi(1-\nu^2)}}, \tag{10.49}$$

where Y is the Young's modulus.

This is a very important result because K_{Ic} is a property of the material. For $\gamma \approx 1\ \mathrm{J\cdot m^{-2}}$, $\mathrm{Y} \approx 10^{11}$ Pa, $\nu \approx 1/3$, we find $K_{Ic} \approx 0.2\ \mathrm{MPa\cdot m^{1/2}}$. The presence of the surface energy in K_{Ic} reflects the work that has to be done to break bonds if the crack is to grow. The presence of the elastic constants reflects the increase in the elastic energy of the entire system if the crack were to grow under the influence of a constant external load. It is evident that the energetics of crack growth involves a wide range of length scales from atomic bonds to the size of the component.

The analysis of this section has excluded plasticity, and other forms of irreversibility such as the excitation of crystal lattice vibrations. Griffith's analysis applies to a crack in a perfectly elastic medium where the crack may grow or shrink quasistatically and hence reversibly. In practice there are extremely few cases where this applies. The stress

[10]Alan Arnold Griffith FRS 1893-1963, British engineer.

[11]Griffith, A A, Phil. Trans. R. Soc. A, **221**, 163-198 (1921).

ahead of a crack in metallic systems and many other crystalline materials invariably leads to some dislocation activity even in so-called *brittle* crystals (see section 11.6). The work of fracture in such materials is typically at least an order of magnitude larger than twice the surface energy, and in more ductile materials it may be several orders of magnitude larger. It is therefore essential to include some degree of plasticity in a more realistic model of a crack in a crystal. Whenever the spatial extent of the crack tip plasticity at fracture is small compared to the crack length, Griffith's criterion can be modified to include plasticity by reinterpreting the γ in eq 10.48 as the sum of the plastic work of fracture and the surface energy of the crack[12]. It is important to recognise that the surface energy of the crack remains significant even when it is only a small fraction of the plastic work of fracture. This becomes obvious if the surface energy approaches zero, for then there is no plasticity because no stress can be sustained at the crack tip. There is a nonlinear, monotonically increasing relationship between the surface energy and the plastic work of transgranular fracture[13].

The stress intensity factor for any particular crack in a specimen depends on the size and geometry of both the crack and the sample in which it is located. Calculating the stress intensity factors in different geometries is the field of linear elastic fracture mechanics. It has enabled standardised sample geometries to be introduced to measure critical stress intensity factors. Although the critical stress intensity factor is a material property it does depend on temperature and the strain rate of the loading mechanism. It also depends sensitively on the microstructure of the material, which is determined by the history of its thermal and mechanical treatment and its impurity content. As a material parameter the critical stress intensity factor is similar in this respect to the yield stress.

10.6 The interaction between a dislocation and a slit crack

The stresses created by defects such as dislocations may act as loads on cracks. The principal difference compared to the previous sections is that the loading on the crack faces is no longer uniform. But the boundary condition that there can be no tractions on the crack faces remains the same. Such interactions between other defects and cracks may decrease or increase the local loading on a crack by externally applied forces on the body. In this section we consider the case of an edge dislocation with Burgers vector $\mathbf{b} = [0, b, 0]$ with its line parallel to the x_3-axis at $(D, 0)$. It is placed ahead of a slit crack with $a = -2c$ and $b = 0$, as shown in Fig 10.7. No external forces are applied

[12] Orowan, E, Trans. Inst. Eng. Shipbuilders, Scotland **89**, 165 (1945). Unavailable online. See also Orowan, E, Rep. Prog. Phys., **12**, 185-232 (1949).

[13] This insight was due to Jokl M L, Vitek V and McMahon Jr., C J, Acta Metall., **28**, 1479-1488 (1980). Charles J McMahon Jr. 1933-2022, US metallurgist.

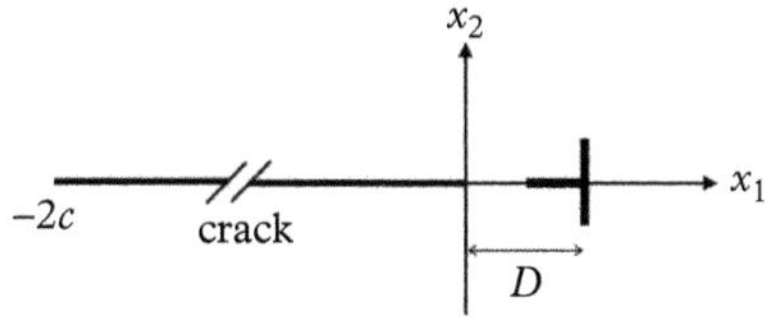

Figure 10.7: To illustrate the geometry for the calculation of the Burgers vector density induced by an edge dislocation at a distance D ahead of a long crack of length $2c$, where $D \ll 2c$. The crack is not subjected to any loading apart from that created by the edge dislocation at D.

to the body, so that the only tractions on the crack faces are those arising from the dislocation at $(D,0)$.

The traction created on the crack faces at x_1 by the dislocation at $(D,0)$ arises from its stress component $\sigma_{22}^D(x_1,0)$:

$$\sigma_{22}^D(x_1,0) = \frac{\mu b}{2\pi(1-\nu)} \frac{1}{x_1 - D}. \tag{10.50}$$

The crack surfaces deform in such a way as to annihilate these tractions. The induced deformation may be modelled by a continuous distribution of dislocations in the crack with Burgers vector density $f(x_1')$. As before, $f(x_1')\mathrm{d}x_1'$ is the infinitesimal Burgers vector of induced edge dislocations between x_1' and $x_1' + \mathrm{d}x_1'$ with their lines parallel to the x_3 axis. The requirement that there are no resultant tractions on the crack faces then leads to the following Cauchy principal value integral equation:

$$\frac{\mu}{2\pi(1-\nu)} P\int_{-2c}^{0} \frac{f(x_1')}{x_1 - x_1'}\mathrm{d}x_1' = -\frac{\mu}{2\pi(1-\nu)} \frac{b}{x_1 - D}, \tag{10.51}$$

where $-2c \le x_1 \le 0$. Using eq 10.20 we can write down the solution:

$$f(x_1) = \frac{b}{\pi^2\sqrt{(x_1+2c)(-x_1)}}\left[C - P\int_{-2c}^{0} \frac{1}{x_1' - D} \frac{\sqrt{(x_1'+2c)(-x_1')}}{x_1' - x_1}\mathrm{d}x_1'\right] \tag{10.52}$$

Evaluating the integral by contour integration we obtain:

$$f(x_1) = \frac{b}{\pi\sqrt{(x_1+2c)(-x_1)}}\left[C' - \frac{\sqrt{D(D+2c)}}{D - x_1}\right], \tag{10.53}$$

where the constant C' is determined by the condition that the integral of $f(x_1)$ over the length of the crack must be zero. This condition follows from the conservation of

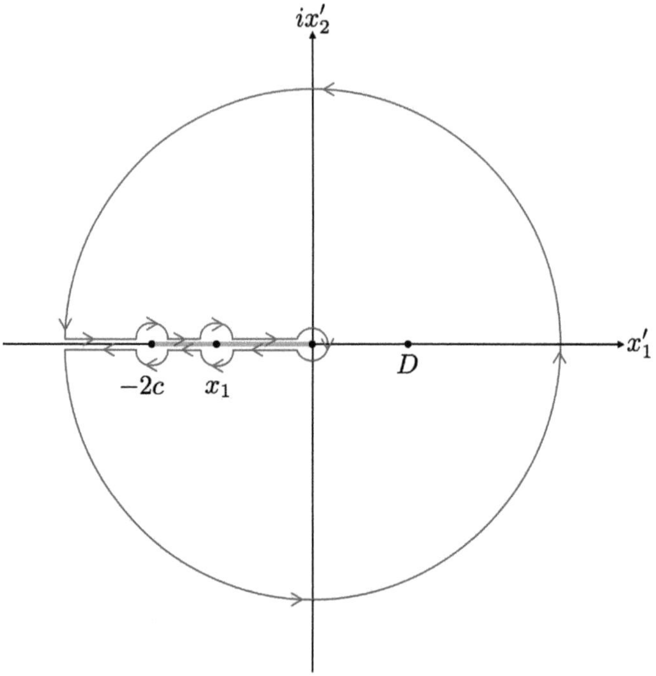

Figure 10.8: The contour used to evaluate the integral in eq 10.55. The thick green line is the branch cut on $-2c \le x_1' \le 0$. There is a simple pole at $x_1' = D > 0$. The radius of the large circle is taken to infinity.

the total Burgers vector, and it yields $C' = 1$. The induced Burgers vector density is therefore as follows:

$$f(x_1) = \frac{b}{\pi\sqrt{(x_1+2c)(-x_1)}}\left[1 - \frac{\sqrt{D(D+2c)}}{D-x_1}\right]. \tag{10.54}$$

Exercise 10.5 Verify that the Burgers vector density of eq 10.54 satisfies eq 10.51. ■

Solution The task is to show that:

$$\frac{b}{\pi} P \int_{-2c}^{0} \frac{1}{\sqrt{(x_1'+2c)(-x_1')}}\left(1 + \frac{\sqrt{D(D+2c)}}{x_1'-D}\right)\frac{1}{x_1'-x_1}\,\mathrm{d}x_1' = \frac{b}{x_1 - D},$$

where $-2c \leq x_1 \leq 0$ and $D > 0$.

Consider the following contour integral

$$J = \oint_C \frac{1}{\sqrt{(z+2c)z}} \left(1 + \frac{\sqrt{D(D+2c)}}{z-D}\right) \frac{1}{z-x_1} \mathrm{d}z. \tag{10.55}$$

The contour C is illustrated in Fig 10.8. Again we find there are cancelling contributions from the semicircles above and below the pole at $x_1' = x_1$. The contributions from the semicircles at $x_1' = -2c$ and the circle at $x_1' = 0$ are zero. Therefore, the contribution to the contour integral from the loop around the branch cut is:

$$-2iP \int_{-2c}^{0} \frac{1}{\sqrt{(x_1'+2c)(-x_1')}} \left(1 + \frac{\sqrt{D(D+2c)}}{x_1'-D}\right) \frac{1}{x_1'-x_1} \mathrm{d}x_1'$$

There is no contribution to the contour integral from the large circle as its radius tends to infinity. The residue at the pole at $z = D$ inside the contour is $1/(D-x_1)$. Applying Cauchy's theorem we obtain:

$$-2iP \int_{-2c}^{0} \frac{1}{\sqrt{(x_1'+2c)(-x_1')}} \left(1 + \frac{\sqrt{D(D+2c)}}{x_1'-D}\right) \frac{1}{x_1'-x_1} \mathrm{d}x_1' = \frac{2\pi i}{D-x_1}$$

from which the required result follows.

10.6.1 Shielding and anti-shielding of cracks by dislocations

It is interesting to see how the stress field $\sigma_{22}(x_1,0)$ of the dislocation is modified by the presence of the crack. For this purpose we evaluate $\sigma_{22}(x_1,0)$ with $x_1 > 0$:

$$\sigma_{22}(x_1,0) = \frac{\mu b}{2\pi(1-\nu)} \frac{1}{(x_1-D)} + \frac{\mu}{2\pi(1-\nu)} \int_{-2c}^{0} \frac{f(x_1')}{(x_1-x_1')} \mathrm{d}x_1'. \tag{10.56}$$

Inserting the Burgers vector density in eq 10.54 into the integral we obtain:

$$\sigma_{22}(x_1,0) = \frac{\mu b}{2\pi(1-\nu)} \frac{1}{\sqrt{x_1(x_1+2c)}} \left(\frac{\sqrt{D(2c+D)}}{(x_1-D)} + 1\right) \tag{10.57}$$

This is plotted in Fig 10.9.

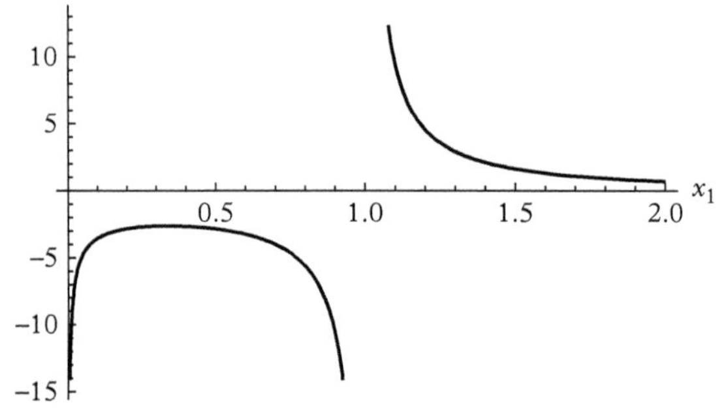

Figure 10.9: Plot of the normal stress component $\sigma_{22}(x_1, 0)$, as given by eq 10.57, for an edge dislocation with Burgers vector $[0, b, 0]$ located at $x_1 = 1, x_2 = 0$ in front of an otherwise unloaded slit crack between $x_1 = -100$ and $x_1 = 0$ on $x_2 = 0$. The vertical axis is in units of $\mu b/[2\pi(1-\nu)]$. In addition to the normal $1/x_1$ divergence at the dislocation core there is a $1/\sqrt{x_1}$ divergence in the field near the crack tip.

The presence of the crack introduces the square root pre-factor into the stress field of the dislocation. This can be made clearer by considering the case where the crack length $2c$ is much larger than D and x_1. In that case:

$$\sigma_{22}(x_1, 0) \approx \frac{\mu b}{2\pi(1-\nu)} \sqrt{\frac{D}{x_1}} \frac{1}{(x_1 - D)}. \tag{10.58}$$

For x_1 close to D the stress field is very close to that of the isolated dislocation. But as x_1 approaches the crack tip at $x_1 = 0$ the stress field has an inverse square root singularity. The stress intensity factor K_I^D at the crack tip due to the dislocation is as follows:

$$K_I^D \approx \lim_{x_1 \to 0} \sqrt{x_1} \frac{\mu b}{2\pi(1-\nu)} \sqrt{\frac{D}{x_1}} \frac{1}{(x_1 - D)} \approx -\frac{\mu b}{2\pi(1-\nu)\sqrt{D}}. \tag{10.59}$$

If the crack tip is subjected to an applied stress intensity factor K_I, dislocations with positive Burgers vectors will reduce K_I. Such dislocations are described as *shielding* because they screen the crack tip from the stress field that is creating K_I. Dislocations with negative Burgers vectors are described as *anti-shielding* because they increase the stress intensity factor K_I. The resultant stress intensity factor comprising the applied stress intensity factor and the intensity factors contributed by dislocations outside the crack is called the local stress intensity factor. The local stress intensity factor is the resultant stress intensity factor acting on the crack tip. At the other side of the crack, at $x_1 = -2c$, the Burgers vectors of shielding and anti-shielding dislocations are negative and positive respectively.

The induced Burgers vector density in eq 10.54 generates a Peach-Koehler force along the x_1-axis per unit length on the dislocation at $(D > 0, 0)$ given by:

$$\begin{aligned} F_1^{ind}(D,0) &= \frac{\mu b}{2\pi(1-\nu)} \int_{-2c}^{0} \frac{f(x_1)}{(D-x_1)} \mathrm{d}x_1 \\ &= \frac{\mu b^2}{2\pi^2(1-\nu)} \int_{-2c}^{0} \frac{\mathrm{d}x_1}{(D-x_1)} \frac{1}{\sqrt{(-x_1)(x_1+2c)}} \left[1 - \frac{\sqrt{D(D+2c)}}{(D-x_1)} \right] \\ &= -\frac{\mu b^2}{2\pi(1-\nu)} \left[\frac{(D+c)}{D(D+2c)} - \frac{1}{\sqrt{D(D+2c)}} \right]. \end{aligned} \quad (10.60)$$

This is often called an image force and it always attracts the dislocation towards the crack tip. If $c \gg D$ then the image force is the force of attraction between two edge dislocations with Burgers vectors $[0, b, 0]$ and $[0, -b, 0]$ spaced $2D$ apart along the x_1-axis. As with image charges in a metal surface the *image dislocation* has the opposite sign Burgers vector and it is located at D on the other side of the crack tip from the real dislocation.

Using eq 10.35 we can write down the stress $\sigma_{22}(D, 0)$ arising from the applied normal stress σ and the distribution of dislocations in the crack required to annihilate the surface tractions on the crack faces:

$$\sigma_{22}(D,0) = \sigma \frac{(D+c)}{\sqrt{D(D+2c)}}. \quad (10.61)$$

Combining the last two equations, the total force per unit length acting on the edge dislocation at $(D > 0, 0)$ when there is an applied normal stress σ is therefore as follows:

$$F_1^T(D,0) = \left[\sigma \frac{(D+c)}{\sqrt{D(D+2c)}} - \frac{\mu b}{2\pi(1-\nu)} \left(\frac{(D+c)}{D(D+2c)} - \frac{1}{\sqrt{D(D+2c)}} \right) \right] b. \quad (10.62)$$

Thus, anti-shielding dislocations at $(D > 0, 0)$, with Burgers vectors $[0, -b, 0]$, are attracted to the crack tip, whereas shielding dislocations, with Burgers vectors $[0, +b, 0]$, move away from the crack tip under a sufficiently high applied stress σ to overcome the image force. It follows that dislocations emitted by the crack are shielding dislocations[14]. The square brackets contain the resultant normal stress acting on the edge dislocation

[14]This is analogous to the reduction in stress ahead of a pileup at a grain boundary by transmission of slip into the adjacent grain.

at $(D,0)$. The dislocation will move only if this stress is greater in magnitude than the friction stress opposing dislocation motion.

Some comments are in order concerning the interpretation of the dislocation friction stress. Since we have focussed on mode I cracks the dislocations that formally make up the crack, and the real dislocation ahead of the crack, are edge dislocations with their Burgers vectors along x_2. Therefore if they are to move along x_1 they will do so by climb. But if the edge dislocation ahead of the crack had had a Burgers vector parallel to x_1 both the induced formal dislocations in the crack and the real dislocation would have been glide edge dislocations. Furthermore the integral equation defining the Burgers vector density of the induced glide dislocation density in the mode II crack would have been identical to the induced climb dislocation density in eq 10.51. The only difference would have beeen the involvement in the tractions on the crack faces of the stress component $\sigma_{12}(x_1 - x_1')$ for $\mathbf{b} = [b, 0, 0]$ rather than $\sigma_{22}(x_1 - x_1')$ for $\mathbf{b} = [0, b, 0]$, but these stress components are mathematically identical. Therefore, from a mathematical point of view we can map the mode I loading of the slit crack involving climb dislocations onto a mode II loading of the slit crack involving glide dislocations. From a physical point of view we should acknowledge that the geometry of these one dimensional models is a gross simplification of the reality of three dimensional cracks interacting with dislocation loops. Therefore, it would not be appropriate to interpret the friction stress as anything other than the resistance to dislocation glide since that is how dislocations move in the vicinity of a real crack, except possibly during creep rupture at elevated temperatures when climb may be involved.

Exercise 10.6 When $c \gg D$ in eq 10.62 show that the force $F_1^T(D,0)$ becomes:

$$F_1^T(D,0) = \left(\frac{K_I}{\sqrt{D}} - \frac{\mu b}{4\pi(1-\nu)D} \right) b, \tag{10.63}$$

where $K_I = \sigma\sqrt{c/2}$ is the stress intensity factor of the bare, elastic crack. For a shielding dislocation b is positive, and for an anti-shielding dislocation b is negative.

■

Solution When $c \gg D$ eq 10.62 becomes:

$$F_1^T(D,0) = \left\{ \sigma\sqrt{\frac{c}{2D}} - \frac{\mu b}{4\pi(1-\nu)D} \right\} b$$

$$= \left(\frac{K_I}{\sqrt{D}} - \frac{\mu b}{4\pi(1-\nu)D} \right) b,$$

which is eq 10.63.

10.7 Dugdale-Bilby-Cottrell-Swinden (DBCS) model

This model[15] extends the representation of an elastic slit crack as an array of dislocations to include regions of plasticity adjacent to the crack tips. These regions are the plastic zones and they are also represented by dislocation arrays. In contrast to the formal representation of a loaded crack as a continuous distribution of dislocations, the dislocations in the plastic zone are not formal but real objects, although they are treated mathematically in the same way as the dislocations representing the loaded crack. The idea is that the singularities in the elastic solution at the crack tips generate stresses that exceed the elastic limit and lead to plasticity in the plastic zones. In the absence of work hardening the stress in the plastic zones is assumed to be a constant, which is a friction stress that is equated to the yield stress, σ_1. As a result of the plastic zones the stress singularities at the crack tips are eliminated and the local stress intensity factors are zero. This section is based on the treatment by Lardner[16].

Consider a slit crack loaded in mode I, as shown in Fig 10.10. The crack is between $x_1 = -c$ and $x_1 = c$ on the $x_2 = 0$ plane and extends from $x_3 = -\infty$ to $x_3 = +\infty$. There are plastic zones between $x_1 = -a$ and $x_1 = -c$ and between $x_1 = c$ and $x_1 = a$. We shall find that the size of each plastic zone, $a - c$, is determined by the crack length $2c$, the yield stress σ_1 and the applied load normal stress σ at $x_2 = \pm\infty$.

The crack and the plastic zones are represented by a continuous distribution of edge dislocations along the x_1-axis with Burgers vectors along $\pm x_2$ and their positive line directions along the positive x_3-axis. As before, $f(x_1)\mathrm{d}x_1$ is the Burgers vector of dislocations between x_1 and $x_1 + \mathrm{d}x_1$. There are two boundary conditions to be satisfied:

$$\frac{\mu}{2\pi(1-\nu)} P \int_{-a}^{a} \frac{f(x_1')}{(x_1 - x_1')} \mathrm{d}x_1' + \sigma = \begin{cases} 0 & \text{in } |x_1| < c, \\ \sigma_1 & \text{in } c < |x_1| < a. \end{cases} \tag{10.64}$$

At the ends of the plastic zones where $x_1 = \pm a$ the distribution $f(x_1) \to 0$ because there are no barriers for the dislocations to pile up against and eqs 10.23 and 10.24

[15] Dugdale, D S, J. Mech. Phys. Solids **8**, 100-104 (1960), David S. Dugdale, British engineer. Bilby, B A, Cottrell A H, and Swinden K H, Proc. R. Soc. A **272**, 304-314 (1963). Sir Alan Howard Cottrell FRS 1919-2012, British metallurgist.

[16] Lardner, R W, *Mathematical theory of dislocations and fracture*, University of Toronto Press, Chapter 5 (1974), ISBN 0-8020-5277-0. Robin W Lardner, Canadian applied mathematician.

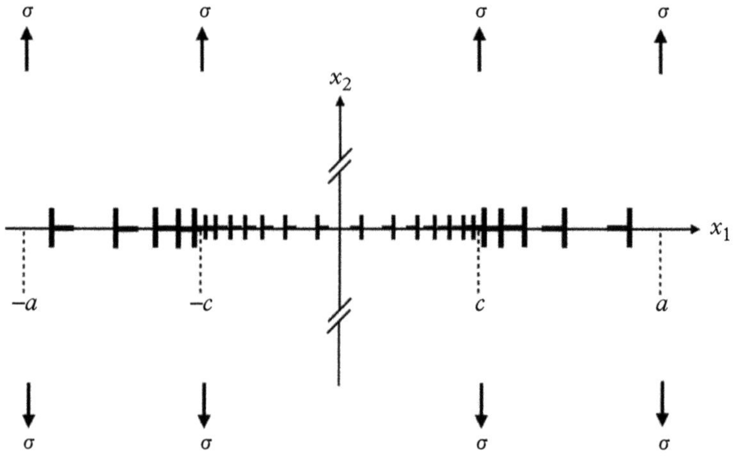

Figure 10.10: To illustrate the geometry of the Dugdale-Bilby-Cottrell-Swinden (DBCS) model.

apply. The solution to the integral equation is then as follows:

$$f(x_1) = -\frac{1}{\pi^2}\frac{2\pi(1-\nu)}{\mu}\sqrt{a^2-x_1^2}\left[\sigma P\int_{-c}^{c}\frac{1}{\sqrt{a^2-x_1'^2}}\frac{\mathrm{d}x_1'}{x_1'-x_1}\right.$$
$$\left.+(\sigma-\sigma_1)P\int_{-a}^{-c}\frac{1}{\sqrt{a^2-x_1'^2}}\frac{\mathrm{d}x_1'}{x_1'-x_1}+(\sigma-\sigma_1)P\int_{c}^{a}\frac{1}{\sqrt{a^2-x_1'^2}}\frac{\mathrm{d}x_1'}{x_1'-x_1}\right]. \tag{10.65}$$

Since

$$P\int_{-a}^{a}\frac{1}{\sqrt{a^2-x_1'^2}}\frac{\mathrm{d}x_1'}{x_1'-x_1} = 0, \tag{10.66}$$

eq 10.65 may be rewritten as follows:

$$f(x_1) = -\frac{\sigma_1}{\pi^2}\frac{2\pi(1-\nu)}{\mu}\sqrt{a^2-x_1^2}\,P\int_{-c}^{c}\frac{1}{\sqrt{a^2-x_1'^2}}\frac{\mathrm{d}x_1'}{x_1'-x_1}. \tag{10.67}$$

The integral may be evaluated using the following indefinite integral, valid for $|x_1| < a$:

$$\int\frac{1}{\sqrt{a^2-x_1'^2}}\frac{\mathrm{d}x_1'}{x_1'-x_1} = -\frac{1}{\sqrt{a^2-x_1^2}}\ln\left\{\frac{\sqrt{(a-x_1)(a+x_1')}+\sqrt{(a+x_1)(a-x_1')}}{\sqrt{(a-x_1)(a+x_1')}-\sqrt{(a+x_1)(a-x_1')}}\right\}. \tag{10.68}$$

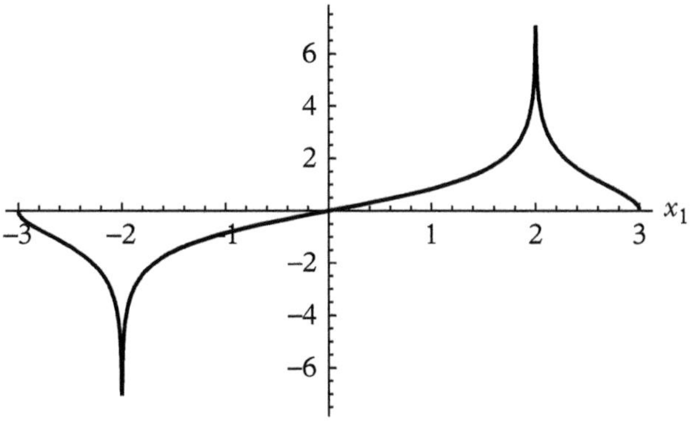

Figure 10.11: Plot of the Burgers vector density $f(x_1)$ for the DBCS model, given by eq 10.69. In this example the half-length of the crack $c = 2$, and the plastic zones have length 1, so that $a = 3$. The vertical axis is in units of $2(1-\nu)\sigma_1/(\pi\mu)$. Note the logarithmic divergences at the crack tips at $x_1 = \pm 2$.

Thus, we obtain the solution:

$$f(x_1) = \frac{2(1-\nu)}{\pi}\frac{\sigma_1}{\mu}\ln\left|\frac{x_1\sqrt{a^2-c^2}+c\sqrt{a^2-x_1^2}}{x_1\sqrt{a^2-c^2}-c\sqrt{a^2-x_1^2}}\right|. \tag{10.69}$$

This solution is plotted in Fig 10.11. The discontinuity in the stresses at the crack tips gives rise to logarithmic singularities at $|x_1| = c$.

The condition, eq 10.23, for this solution in eq 10.69 to exist is as follows:

$$\int_{-a}^{-c}\frac{(\sigma-\sigma_1)}{\sqrt{a^2-x_1'^2}}\mathrm{d}x_1' + \int_{-c}^{c}\frac{\sigma}{\sqrt{a^2-x_1'^2}}\mathrm{d}x_1' + \int_{c}^{a}\frac{(\sigma-\sigma_1)}{\sqrt{a^2-x_1'^2}}\mathrm{d}x_1' = 0. \tag{10.70}$$

These integrals lead to the following expression for the size, $a-c$, of each plastic zone:

$$a - c = c\left(\sec\left(\frac{\pi\sigma}{2\sigma_1}\right) - 1\right), \text{ or } c/a = \cos(\pi\sigma/2\sigma_1). \tag{10.71}$$

Thus, as the yield stress σ_1 increases the size of the plastic zone decreases; when the plastic zone is small compared to the crack this condition is known as small scale yielding. As $\sigma \to \sigma_1$ the plastic zone becomes much larger than the crack, a condition known as general yielding.

As a result of the plastic zones the crack opening displacements are increased. This may be understood physically as a result of the dislocations in each plastic zone having been emitted from the adjoining crack tip, thereby opening the crack. From eq 10.27 the crack tip opening displacement at $x_1 = c$ is:

$$s(c) = -\int_a^c f(x_1)\mathrm{d}x_1. \tag{10.72}$$

The Burgers vector density given by eq 10.69 may be integrated by making the change of variable $x_1 = a\cos\theta$. The integral in eq 10.72 then becomes:

$$s(c) = -\frac{2(1-\nu)}{\pi}\frac{\sigma_1}{\mu} a \int_0^{\phi} \ln\left|\frac{\sin(\phi+\theta)}{\sin(\phi-\theta)}\right| \sin\theta\,\mathrm{d}\theta, \tag{10.73}$$

where $\cos\phi = c/a$, and $\phi = \pi\sigma/(2\sigma_1)$. After integrating by parts the following crack opening displacement is obtained[17]:

$$\begin{aligned} s(c) &= -\frac{2(1-\nu)}{\pi}\frac{\sigma_1}{\mu} a \left[\cos\theta \ln\left|\frac{\sin(\phi+\theta)}{\sin(\phi-\theta)}\right| + \cos\phi \ln\left|\frac{\sin\theta - \sin\phi}{\sin\theta + \sin\phi}\right|\right]_{\theta=0}^{\theta=\phi} \\ &= \frac{2(1-\nu)}{\pi}\frac{\sigma_1}{\mu} c \ln(\sec\phi) = \frac{2(1-\nu)}{\pi}\frac{\sigma_1}{\mu} c \ln(a/c). \end{aligned} \tag{10.74}$$

The physical significance of the crack tip opening displacement is that it is often observed that fracture in a material where plastic deformation can occur is preceded by a critical amount of plastic deformation at the crack tip. The crack tip opening displacement then has to reach a critical value for fracture to occur. This introduces a dependence of whether fracture will occur on the size of the sample: if it is less than the size of the plastic zone then fracture will occur only when yielding has occurred throughout the sample, which is the general yielding condition. But if the sample is much larger than the size of the plastic zone, fracture can occur at a smaller stress once the plastic zone has reached its critical size, which amounts to a critical value of the crack tip opening displacement. Understanding this size dependence of the fracture criterion in an elastic-plastic material was the initial motivation for the seminal paper by Bilby, Cottrell and Swinden (1963). They argued σ_1 depends on the size of the plastic zone. If the plastic zone is contained within one grain it may be identified with the friction stress opposing dislocation motion within that grain. That might consist of the Peierls stress and the friction stress created by impurities in solution or as precipitates. But when the plastic zone is larger than a single grain plasticity has to propagate across grain boundaries, and that increases the value of σ_1 because it involves renucleating plasticity in each grain.

[17]We note this is a factor of 2 less than the value given by Lardner (1974) in his equation 5.55, p.165.

10.8 The dislocation free zone model

The assumption of the DBCS model that the plastic zone extends from the crack tip eliminates the elastic stress singularity at the crack tip. As a result the local stress intensity factor is zero. Furthermore, the stress acting at the crack tip in their model is σ_1, which is typically of the order of 10 – 100 MPa. A stress of 100 MPa creates forces between atoms of order 0.01 eV $\cdot$ Å^{-1}. As Thomson[18] points out on p.88 of his review article[19] this is far too small to break bonds at the crack tip, which requires forces of order 0.1 – 1 eV $\cdot$ Å^{-1}. The shielding of the crack tip by the plastic zone in the DBCS model is too complete to enable the crack to propagate by breaking bonds at the crack tip. Instead the plastic zone increases in size as the applied stress σ increases, and the crack tip opening displacement increases indefinitely. The DBCS model always predicts ductile fracture.

The plastic zone is made up of shielding dislocations which reduce the local stress intensity factor. In section 10.6.1 we found that, at a sufficiently high applied stress to overcome the attraction of the image force, shielding dislocations are repelled from the crack tip. Anti-shielding dislocations are always attracted towards the crack where they can be absorbed creating steps on the crack surfaces. If shielding dislocations were emitted from the crack tip at a sufficiently high applied stress to escape the attraction to the crack they will move from the crack tip until they are stopped by the friction stress σ_1. Alternatively, both shielding and anti-shielding dislocations may be generated from sources near the crack tip: anti-shielding dislocations run into the crack, while shielding dislocations move away from the crack and establish the plastic zone. These scenarios raise the possibility of a dislocation free zone between the crack tip and the plastic zone. Inside the dislocation free zone shielding dislocations are repelled into the plastic zone and anti-shielding dislocations are attracted into the crack.

If the plastic zone is located further from the crack tip, shielding of the crack tip decreases and the local stress intensity factor rises. Also, as the size of the dislocation free zone increases the size of the plastic zone decreases because the stress field of the crack decreases with distance from its tip. Eventually shielding by the plastic zone diminishes to such an extent that we recover the elastic limit and the local stress intensity factor is that of the bare elastic crack, $\sigma\sqrt{c/2}$ (see eq 10.37). This describes the completely brittle limit. In the DBCS limit there is no dislocation free zone and the local stress intensity factor is zero. This describes the completely ductile limit. As the size of the dislocation free zone varies between these limiting cases we span the range of behaviour from purely brittle to purely ductile.

Experimental evidence for the existence of dislocation free zones at cracks has been

[18]Robb Milton Thomson 1925-, US materials physicist
[19]Thomson, R, *Solid State Physics*, **39**, 2-129 (1986).

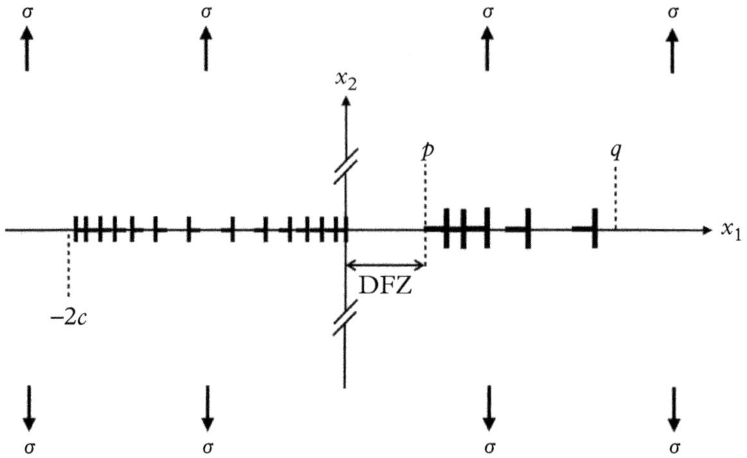

Figure 10.12: To illustrate the geometry of the dislocation free zone model.

published[20]. Chang and Ohr[21] extended the DBCS model to include a dislocation free zone. Here we consider the force balance on dislocations in a plastic zone separated from the crack by a dislocation free zone. This section is based on the paper[22] by Majumdar[23] and Burns[24] (1983), but for a mode I crack rather than a mode III crack.

As before we consider a slit crack in the $x_2 = 0$ plane between $x_1 = -2c$ and $x_1 = 0$. It is loaded by a tensile stress $\sigma_{22} = \sigma$ at $x_2 = \pm\infty$. There is a plastic zone between $x_1 = p$ and $x_1 = q$, where $q > p > 0$. There is a dislocation free zone between $x_1 = 0$ and $x_1 = p$. The crack length $2c$ is assumed to be much larger than p and q, which amounts to the assumption of small scale yielding. This setup is illustrated in Fig 10.12. The friction stress, which we again equate to the yield stress, is σ_1.

Let $f(x_1)$ be the continuous distribution of Burgers vector density in the plastic zone. The infinitesimal Burgers vector between x_1 and $x_1 + \mathrm{d}x_1$, where $p \leq x_1 < q$, is $\mathrm{d}b(x_1) = f(x_1)\mathrm{d}x_1$. Equation 10.63 gives the force on the dislocation at x_1 arising from the loaded crack and from the image interaction. This force is:

$$\mathrm{d}F(x_1) = \frac{K_I \,\mathrm{d}b(x_1)}{\sqrt{x_1}} - \frac{\mu \,\mathrm{d}b(x_1)\,\mathrm{d}b(x_1)}{4\pi(1-\nu)x_1}. \tag{10.75}$$

[20]Horton J A and Ohr S M, J. Mater. Sci. **17**, 3140-3148 (1982). Chia, K Y, and Burns, S J, Scripta Metall., **18**, 467-472 (1984).

[21]Chang, S-J, and Ohr, S M, J. Appl. Phys. **52**, 7174-7181 (1981). Chang, S-J, and Ohr, S M, Int. J. Fracture **23**, R3-R6 (1983).

[22]Majumdar, B S, and Burns, S J, Int. J. Fracture **21**, 229-240 (1983).

[23]Bhaskar S. Majumdar, Indian and US engineer

[24]Stephen J Burns, US engineer

If this dislocation were the only dislocation in the plastic zone this would be the total force acting on it in the presence of the crack, apart from the friction force. But it also experiences a force arising from the stresses σ_{22} generated in the presence of the crack by other dislocations in the plastic zone. These stresses are given by eq 10.58. Thus the force balance on the dislocation at x_1 becomes the following:

$$\frac{K_I\,\mathrm{d}b(x_1)}{\sqrt{x_1}} - \frac{\mu\,\mathrm{d}b(x_1)\mathrm{d}b(x_1)}{4\pi(1-\nu)x_1} + \mathrm{d}b(x_1)\,P\int_{x_1'=p}^{q} \frac{\mu}{2\pi(1-\nu)}\frac{f(x_1')}{x_1-x_1'}\sqrt{\frac{x_1'}{x_1}}\mathrm{d}x_1' = \sigma_1\mathrm{d}b(x_1). \tag{10.76}$$

Since the image force involves the product of two infinitesimal quantities it may be neglected. Thus we arrive at the following integral equation:

$$\int_p^q \frac{g(x_1')}{x_1-x_1'}\mathrm{d}x_1' = \frac{2\pi(1-\nu)}{\mu}\left(\sigma_1\sqrt{x_1}-K_I\right), \tag{10.77}$$

where $g(x_1) = f(x_1)\sqrt{x_1}$. Since the distribution must be zero at $x_1 = p, q$ the solution of this integral equation is given by eq 10.24, provided the condition embodied in eq 10.23 is satisfied. Thus,

$$f(x_1) = \frac{2(1-\nu)}{\pi\mu}\sqrt{\frac{(x_1-p)(q-x_1)}{x_1}}\,P\int_p^q \frac{(\sigma_1\sqrt{x_1'}-K_I)}{\sqrt{(x_1'-p)(q-x_1')}}\frac{1}{(x_1'-x_1)}\mathrm{d}x_1', \tag{10.78}$$

provided the following condition is satisfied:

$$\int_p^q \frac{(K_I-\sigma_1\sqrt{x_1})}{\sqrt{(x_1-p)(q-x_1)}}\mathrm{d}x_1 = 0. \tag{10.79}$$

This condition leads to the following relationship:

$$\frac{\pi K_I}{2\sigma_1\sqrt{q}} = E\left(\frac{q-p}{q}\right), \tag{10.80}$$

where $E(z)$ is the complete elliptic integral of the second kind:

$$E(z) = \int_0^{\pi/2}\sqrt{\left(1-z\sin^2\theta\right)}\,\mathrm{d}\theta. \tag{10.81}$$

For given values of K_I and σ_1 we can use eq 10.80 to show that as the plastic zone is located further from the crack, the width of the plastic zone decreases. The argument is as follows. When $p = 0$ the plastic zone begins at the crack tip, which is the DBCS limit. Then $(q-p)/q = 1$ and $E(1) = 1$. When $p = q$ the plastic zone vanishes and we have the elastic limit. In that case $(q-p)/q = 0$ and $E(0) = \pi/2$. Let $q = q_{BCS}$ in the DBCS limit, and $q = q_{el}$ in the elastic limit. Using eq 10.80 we find $q_{el} = (4/\pi^2)q_{BCS}$. Thus, as p increases from zero to q_{el}, q decreases monotonically from q_{BCS} to $(4/\pi^2)q_{BCS}$, and the width of the plastic zone decreases monotonically from q_{BCS} to zero. From eq 10.80 the width, q_{BCS}, of the plastic zone in the DBCS limit is $\pi^2\sigma^2 c/(8\sigma_1^2)$. This agrees with eq 10.71 in the limit $\sigma \ll \sigma_1$, which is the appropriate limit when the crack is much longer than the plastic zone.

The integral in eq 10.78 may be expressed in terms of elliptic integrals:

$$\begin{aligned} f(x_1) &= \frac{4(1-\nu)}{\pi}\frac{\sigma_1}{\mu}\sqrt{\frac{(x_1-p)}{(q-x_1)}}\sqrt{\frac{q}{x_1}}\,P\int_0^{\pi/2}\frac{1-\frac{q-p}{q}\sin^2\theta}{\left(1-\frac{q-p}{q-x_1}\sin^2\theta\right)\sqrt{1-\frac{q-p}{q}\sin^2\theta}}\,\mathrm{d}\theta \\ &= \frac{4(1-\nu)}{\pi}\frac{\sigma_1}{\mu}\sqrt{\frac{(x_1-p)}{(q-x_1)}}\sqrt{\frac{q}{x_1}}\left\{\frac{x_1}{q}\Pi\left(\frac{q-p}{q-x_1},\frac{q-p}{q}\right)+\left(1-\frac{x_1}{q}\right)K\left(\frac{q-p}{q}\right)\right\} \end{aligned} \tag{10.82}$$

where

$$\Pi(n,z) = \int_0^{\pi/2}\frac{\mathrm{d}\theta}{\left(1-n\sin^2\theta\right)\sqrt{1-z\sin^2\theta}} \tag{10.83}$$

is the complete elliptic integral of the third kind, and

$$F(z) = \int_0^{\pi/2}\frac{\mathrm{d}\theta}{\sqrt{1-z\sin^2\theta}} \tag{10.84}$$

is the complete elliptic integral of the first kind.

Exercise 10.7 Derive eq 10.82 from eq 10.78. ■

Solution First, we note that for $p \le x_1 \le q$:

$$P\int_p^q \frac{1}{\sqrt{(x_1'-p)(q-x_1')}}\frac{1}{x_1'-x_1}\,\mathrm{d}x_1' = 0.$$

Therefore, eq 10.78 becomes:

$$f(x_1) = \frac{2(1-\nu)\sigma_1}{\pi\mu}\sqrt{\frac{(x_1-p)(q-x_1)}{x_1}}\,P\int_p^q \frac{\sigma_1\sqrt{x_1'}}{\sqrt{(x_1'-p)(q-x_1')}}\frac{1}{(x_1'-x_1)}\,\mathrm{d}x_1'$$

Let

$$I = P\int_p^q \sqrt{\frac{x_1'}{(x_1'-p)(q-x_1')}}\frac{1}{x_1'-x_1}\,\mathrm{d}x_1', \qquad p \le x_1 \le q.$$

Let

$$\begin{aligned} x_1' &= \frac{q+p}{2}+\frac{q-p}{2}\cos\theta \\ x_1 &= \frac{q+p}{2}+\frac{q-p}{2}\cos\phi \end{aligned}$$

Then,

$$\begin{aligned} I &= P\int_0^\pi \frac{\sqrt{\frac{q+p}{2}+\frac{q-p}{2}\cos\theta}}{\cancel{\frac{q-p}{2}\sin\theta}}\,\frac{\cancel{\frac{q-p}{2}\sin\theta}}{\frac{q-p}{2}(\cos\theta-\cos\phi)}\,\mathrm{d}\theta \\ &= P\int_0^\pi \frac{\sqrt{(\frac{q-p}{2})(2-2\sin^2\theta/2)+p}}{(q-p)(\sin^2\phi/2-\sin^2\theta/2)}\,\mathrm{d}\theta \\ &= \frac{\sqrt{q}}{q-p}P\int_0^\pi \frac{\sqrt{1-k\sin^2\theta/2}}{\sin^2\phi/2-\sin^2\theta/2}\,\mathrm{d}\theta, \end{aligned}$$

where $k = 1 - p/q$. Let $\alpha = \theta/2$ and $\beta = (q-p)/(q-x_1)$. Continuing,

$$
\begin{aligned}
I &= \frac{2\sqrt{q}}{q-p} P\int_0^{\pi/2} \frac{\sqrt{1-k\sin^2\alpha}}{\left(\frac{q-x_1}{q-p}\right)\left(1-\left(\frac{q-p}{q-x_1}\right)\sin^2\alpha\right)} \\
&= \frac{2\sqrt{q}}{q-x_1} P\int_0^{\pi/2} \left\{ \frac{1-k/\beta}{\left(1-\beta\sin^2\alpha\right)\sqrt{1-k\sin^2\alpha}} + \frac{k/\beta}{\sqrt{1-k\sin^2\alpha}} \mathrm{d}\alpha \right\} \\
&= \frac{2\sqrt{q}}{q-x_1} \left\{ \left(1-\frac{k}{\beta}\right)\Pi(\beta,k) + \frac{k}{\beta}K(k) \right\}
\end{aligned}
$$

Therefore,

$$
\begin{aligned}
f(x_1) &= \frac{2(1-\nu)\sigma_1}{\pi\mu}\sqrt{\frac{(x_1-p)(q-x_1)}{x_1}}\, I \\
&= \frac{4(1-\nu)}{\pi}\frac{\sigma_1}{\mu}\sqrt{\frac{x_1-p}{q-x_1}}\sqrt{\frac{q}{x_1}}\left\{\frac{x_1}{q}\Pi\left(\frac{q-p}{q-x_1},\frac{q-p}{q}\right) + \left(1-\frac{x_1}{q}\right)K\left(\frac{q-p}{q}\right)\right\}
\end{aligned}
$$

In the previous section we found that in the DBCS model the local stress intensity factor at the crack tip is zero - the crack tip is completely screened by the plastic zone. We may use eq 10.59 and eq 10.82 to calculate the local stress intensity factor in the dislocation free zone model:

$$
\begin{aligned}
k_I^{DFZ} &= K_I - \frac{\mu}{2\pi(1-\nu)}\int_p^q \frac{f(x_1)}{\sqrt{x_1}}\mathrm{d}x_1 \\
&= K_I - \frac{2\sigma_1}{\pi^2}\int_0^{\pi/2} \mathrm{d}\theta\sqrt{q-(q-p)\sin^2\theta}\; P\int_p^q \mathrm{d}x_1 \frac{\sqrt{(x_1-p)(q-x_1)}}{x_1(q-(q-p)\sin^2\theta - x_1)} \\
&= K_I - \frac{2\sigma_1}{\pi}\int_0^{\pi/2} \mathrm{d}\theta\sqrt{q-(q-p)\sin^2\theta}\left(1-\frac{\sqrt{pq}}{q-(q-p)\sin^2\theta}\right) \\
&= K_I - \frac{2\sigma_1\sqrt{q}}{\pi}\left(E\left(\frac{q-p}{q}\right) - \sqrt{p/q}\,K\left(\frac{q-p}{q}\right)\right) \qquad (10.85)
\end{aligned}
$$

If we now use eq 10.80 the local stress intensity factor may be rewritten more succinctly

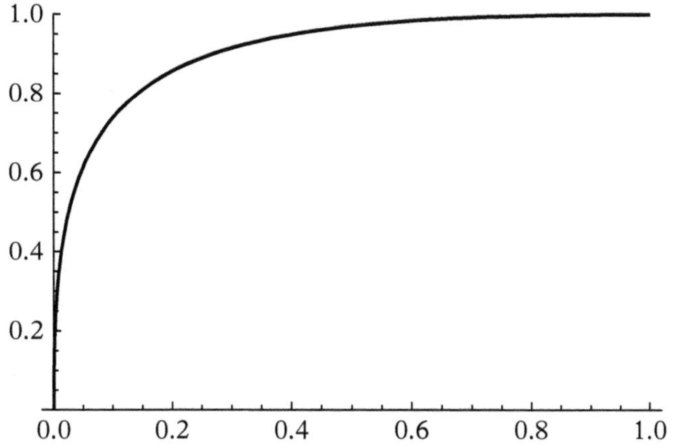

Figure 10.13: A plot of eq 10.86. The vertical axis is k_I^{DFZ}/K_I. The horizontal axis is p/q.

as follows:

$$k_I^{DFZ} = \frac{\sqrt{p/q}\,K(1-(p/q))}{E(1-(p/q))}\,K_I \tag{10.86}$$

When $p = 0$ we recover the DBCS limit and eq 10.86 confirms the local stress intensity factor on the crack tip is zero. When $p = q$ the plastic zone vanishes and we recover the elastic limit where eq 10.86 confirms the local stress intensity factor is the applied stress intensity factor, K_I.

Fig 10.13 shows a plot of k_I^{DFZ}/K_I against p/q, obtained using eq 10.86. It is seen that k_I^{DFZ}/K_I increases very rapidly for small values of p/q and approaches unity asymptotically. The asymptotic expansion for the elliptic function of the first kind near $z = 1$ is as follows[25]:

$$F(1-\varepsilon) \to \frac{1}{2}\ln\left(\frac{16}{\varepsilon}\right) + O(\varepsilon\ln\varepsilon), \tag{10.87}$$

where $0 < \varepsilon \ll 1$. Using this expansion, eqs 10.80 and 10.86, and $E(z) \to 1$ as $z \to 1$ we obtain the following relations for small values of p/q:

$$K_I = \frac{2\sigma_1\sqrt{q}}{\pi} \tag{10.88}$$

[25] https://functions.wolfram.com/EllipticIntegrals/EllipticK/introductions/CompleteEllipticIntegrals/05/

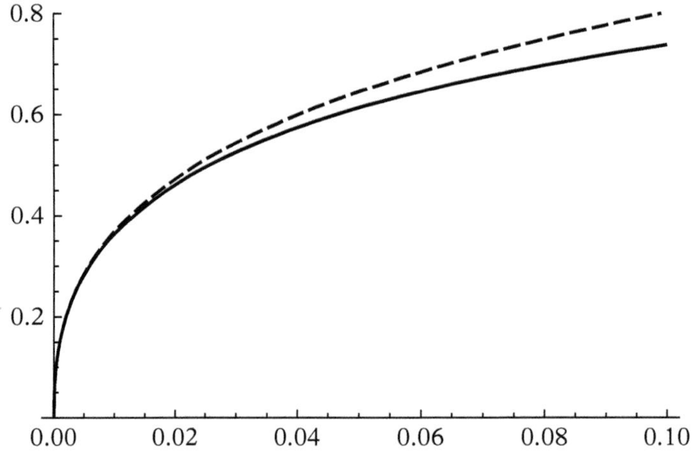

Figure 10.14: Comparison of k_I^{DFZ}/K_I (vertical axis) as a function of p/q for $0 \leq p/q \leq 0.1$ as given exactly (solid line) by eq 10.86 and approximately (dashed line) by eq 10.90.

$$k_I^{DFZ} = \frac{2\sigma_1}{\pi}\sqrt{p}\left(\ln 4 - \frac{1}{2}\ln\frac{p}{q}\right) \tag{10.89}$$

$$\frac{k_I^{DFZ}}{K_I} = \sqrt{\frac{p}{q}}\left(\ln 4 - \frac{1}{2}\ln\frac{p}{q}\right) \tag{10.90}$$

eq 10.88 is an equation for q (which is approximately the width of the plastic zone when $p/q \ll 1$) in terms of the applied stress intensity factor K_I and the yield stress σ_1. Equation 10.89 expresses the local stress intensity factor of the crack tip in terms of the width p of the dislocation free zone, the approximate width of the plastic zone q, and the yield stress σ_1. It is seen in this equation, and graphically in Fig 10.13 at small values of p/q, that small changes in the width p of the dislocation free zone have a more marked influence on the local stress intensity factor than changes in the width of the plastic zone. Equations similar to eqs 10.88 to 10.90 were published by Weertman et al. in 1983[26]. The accuracy of the approximation in eq 10.90 may be gauged by plotting it and the exact result, eq 10.86, on the same graph at small values of p/q. This is done in Fig 10.14 where it is seen that the approximation underestimates the screening by the plastic zone. Nevertheless it is reasonably accurate up to $p/q \approx 0.1$, which spans the range $0 \leq k_I^{DFZ}/K_I \lesssim 0.7$.

[26]Weertman, J, Lin, I-H, and Thomson, R, Acta Metallurgica **31**, 473-482 (1983). Johannes Weertman 1925-2018, US materials scientist and geophysicist.

Both the DBCS and dislocation free zone models assume dislocations are available and mobile to populate the plastic zone. One possibility is that shielding dislocations are emitted from the crack tip, under the influence of the high stresses present there. Another is that sources operate near the crack, also driven by the stress field near the tip, sending anti-shielding dislocations into the crack and shielding dislocations away from it to form the plastic zone.

10.9 The influence of interatomic forces on slit cracks

The stress singularities at the crack tips of section 10.4 are a consequence of the use of linear elasticity to describe the elastic field. These singularities lead to infinite stresses at the crack tips, which are unrealistic. Even the elliptical shape of the crack is doubtful, particularly at the crack tips where it is rounded. In reality interatomic forces at the tip will pull the faces of the crack towards each other, so that far from being rounded the profile is more like a cusp. In other words, the Burgers vector density $f(x_1)$ should tend to zero at the crack tips, not the infinite value predicted by linear elasticity in eq 10.32.

The Barenblatt[27] model[28] of slit cracks addresses these shortcomings of the theory of the elastic crack presented in section 10.4 by including cohesive forces acting between the crack faces. In a metal these forces are significant only when the crack faces are separated by no more than a few Ångstroms. The regions where these cohesive forces are significant are limited to *cohesive zones* of the order of 10Å in length from the crack tips. Where the cohesive zones begin the attraction between the crack faces is considered just about negligible. It quickly rises towards the cracks tips, becoming much larger than the tractions on the crack faces arising from the applied normal loads. It is reasonable to assume that the cohesive forces in the cohesive zones are independent of the surface tractions arising from the applied normal loads.

As shown[29] by Willis[30] and more generally by Rice[31] the Barenblatt model does not alter the Griffith criterion for fracture. This is because, regardless of the existence of the cohesive zones, if the crack length increases by δc the surface area of the two crack faces increases by $2\delta c$ per unit length, and thus the energy cost remains $2\gamma\delta c$.

The Griffith and Barenblatt models are continuum models. For a given applied load they predict a unique crack length where the crack is in equilibrium: if it is smaller than the equilibrium length the crack will tend to close completely, if it is larger it will

[27]Grigory Isaakovich Barenblatt ForMemRS 1927-2018, Russian mathematician and physicist

[28]Barenblatt, G I, J. Appl. Math. Mechs. **23**, 622-636 (1959); Barenblatt, G I, J. Appl. Math. Mechs. **23**, 1009-1029 (1959); Barenblatt, G I, Adv. Appl. Mech. **7**, 55-129 (1962).

[29]Willis, J R, J. Mech. Phys. Solids **15**, 151-162 (1967); Rice, J R, J. Appl. Mech. **35**, 379-386 (1968).

[30]John Raymond Willis FRS, British mathematician

[31]James Robert Rice ForMemRS 1940- , US engineer and geophysicist

tend to continue to grow. This picture changes when we take into account the discrete atomic structure of a crack. Recall that in the treatment of a single dislocation in a continuum the energy of the dislocation is independent of its position. But the energy of a single dislocation in a crystal is a periodic function of its position in the slip plane. The peaks in the energy oscillations are the Peierls barriers discussed in section 7.5. Similar barriers may exist for cracks in a crystal lattice, and if they become sufficiently large cracks may become trapped in the energy minima. This phenomenon is known as *lattice trapping*[32]. The Peierls barriers for dislocations decrease as the core width increases. We may expect a similar trend in the case of cracks: the longer the cohesive zone the smaller the barrier to crack growth. The Frenkel-Kontorova model would suggest that the length of the cohesive zone increases with the stiffness of the bonds between the crack faces and decreases with the amplitude of the periodic variations of the energy of the crack as a function of its position.

10.10 Problems

Problem 10.1 Consider a slit crack between $x_1 = -2c$ and $x_1 = 0$ loaded in mode I by a tensile stress $\sigma_{22} = \sigma$ far from the crack. Using eqs 10.28 and 10.32 show that at $(x_1, x_2) = (\rho\cos\alpha, \rho\sin\alpha)$, where $\rho \ll c$, the stress field of the crack is as follows:

$$\begin{aligned} \sigma_{11}(\rho\cos\alpha,\rho\sin\alpha) &= \frac{K_I}{\sqrt{\rho}}\left(\cos(\alpha/2) - \frac{1}{2}\sin(3\alpha/2)\sin\alpha\right) \\ \sigma_{22}(\rho\cos\alpha,\rho\sin\alpha) &= \frac{K_I}{\sqrt{\rho}}\left(\cos(\alpha/2) + \frac{1}{2}\sin(3\alpha/2)\sin\alpha\right) \\ \sigma_{12}(\rho\cos\alpha,\rho\sin\alpha) &= \frac{K_I}{2\sqrt{\rho}}\cos(3\alpha/2)\sin\alpha, \end{aligned} \tag{10.91}$$

where $K_I = \sigma\sqrt{c/2}$.

Hint Write down an integral for each stress component like eq 10.38. Evaluate these integrals using contour integration, noting the second order poles at $\rho e^{\pm i\alpha}$. Then identify the dominant terms when $c \gg \rho$.

Hence show that the elastic energy density close to the crack tip is given by:

$$W = \frac{K_I^2}{4\mu\rho}\left(2(1-2\nu)\cos^2(\alpha/2) + \frac{1}{2}\sin^2\alpha\right). \tag{10.92}$$

[32] Thomson, R, Hsieh, C, and Rana, V, J. Appl. Phys. **42**, 3154-3160 (1971).

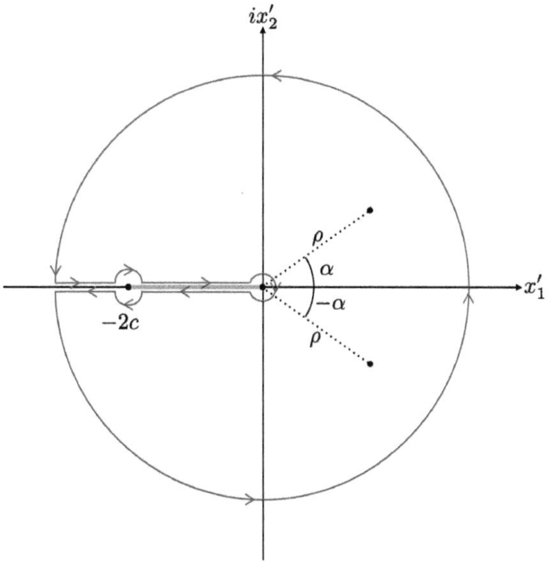

Figure 10.15: Contour to evaluate the integral in eq 10.93. There is a branch cut, shown in green, on $-2c \le x_1 \le 0$ and there are second order poles at $z = \rho e^{i\alpha}$ and $z = \rho e^{-i\alpha}$. The radius of the large circle is taken to infinity.

Solution The stress components $\sigma_{11}(x_1,x_2), \sigma_{22}(x_1,x_2)$ and $\sigma_{12}(x_1,x_2)$ have the following form:

$$\sigma_{ij}(x_1,x_2) = \frac{\sigma}{\pi} \int_{-2c}^{0} \frac{x_1'+c}{\sqrt{(x_1'+2c)(-x_1')}} \frac{g_{ij}(x_1',x_1,x_2)}{\left((x_1-x_1')^2+x_2^2\right)^2} \mathrm{d}x_1',$$

where

$$\begin{aligned} g_{11}(x_1',x_1,x_2) &= (x_1-x_1')\left((x_1-x_1')^2-x_2^2\right) \\ g_{22}(x_1',x_1,x_2) &= (x_1-x_1')\left((x_1-x_1')^2+3x_2^2\right) \\ g_{12}(x_1',x_1,x_2) &= x_2\left((x_1-x_1')^2-x_2^2\right) \end{aligned}$$

Note there are no singularities in the functions $g_{ij}(x_1', x_1, x_2)$.

To evaluate $\sigma_{ij}(x_1, x_2)$ consider the contour integral:

$$J = \frac{\sigma}{\pi} \oint_C \frac{z+c}{\sqrt{(z+2c)z}} \frac{g_{ij}(z, x_1, x_2)}{\left((x_1 - z)^2 + x_2^2\right)^2} \mathrm{d}z.$$

Writing $x_1 = \rho\cos\alpha$ and $x_2 = \rho\sin\alpha$, this may be expressed as fol]lows:

$$J = \frac{\sigma}{\pi} \oint_C \frac{z+c}{\sqrt{(z+2c)z}} \frac{g_{ij}(z, \rho\cos\alpha, \rho\sin\alpha)}{(z - \rho e^{i\alpha})^2 (z - \rho e^{-i\alpha})^2} \mathrm{d}z. \tag{10.93}$$

Fig 10.15 shows the contour C. There is a branch cut on the real axis between $x_1 = -2c$ and $x_1 = 0$. There are second order poles at $z = \rho e^{i\alpha}$ and $z = \rho e^{-i\alpha}$. In all three cases of g_{ij} there are no contributions to the contour integral from the small circles at $z = -2c$ and $z = 0$. The contribution from the large circle is $-2i\sigma$ for g_{11} and g_{22} and zero for g_{12}. The contribution to J from the loop around the branch cut $-2i\sigma_{ij}$. Therefore, by Cauchy's theorem:

$$J = -2i\sigma_{ij} - 2i\sigma\delta_{ij} = 2\pi i\left(\text{residue at } z = \rho e^{i\alpha} + \text{residue at } z = \rho e^{-i\alpha}\right)$$

where i and $j = 1$ or 2. Since the residues are $z = \rho e^{\pm i\alpha}$ are complex conjugates we may write:

$$\sigma_{ij} = -\sigma\delta_{ij} - 2\sigma\,\mathrm{Re}[\text{residue at } z = \rho e^{i\alpha}], \tag{10.94}$$

where Re[z] is the real part of the complex number z. The residue at $z = \rho e^{i\alpha}$ is as follows:

$$\frac{\sigma}{\pi}\left\{\left(\frac{(\rho e^{i\alpha}+c)^2}{d^3 e^{3i\theta}} - \frac{1}{de^{i\theta}}\right)\frac{g_{ij}}{4\rho^2\sin^2\alpha} - \left(\frac{\rho e^{i\alpha}+c}{de^{i\theta}}\right)\frac{\left(\rho\sin\alpha\frac{\mathrm{d}g_{ij}}{\mathrm{d}z} + ig_{ij}\right)}{4\rho^3\sin^3\alpha}\right\}$$

where g_{ij} and $\mathrm{d}g_{ij}/\mathrm{d}z$ are evaluated at $z = \rho e^{i\alpha}$, $x_1 = \rho\cos\alpha$ and $x_2 = \rho\sin\alpha$. We have introduced $de^{i\theta}$, which is defined as follows:

$$de^{i\theta} = \sqrt{(\rho e^{i\alpha} + 2c)\rho e^{i\alpha}} \to \sqrt{2\rho c}\, e^{i\alpha/2} \text{ when } \rho/c \ll 1.$$

In this limit the residue at $z = \rho e^{i\alpha}$ becomes:

$$\text{residue at } z = \rho e^{i\alpha}: \sqrt{\frac{c}{2\rho}}\left\{\frac{e^{-3i\alpha/2} g_{ij}}{8\rho^3 \sin^2\alpha} - e^{-i\alpha/2}\left(\frac{\rho \sin\alpha \frac{dg_{ij}}{dz} + i g_{ij}}{4\rho^3 \sin^3\alpha}\right)\right\}$$

For $z = \rho e^{i\alpha}$ we have

$$\begin{aligned}
g_{11} &= 2i\rho^3 \sin^3\alpha \\
dg_{11}/dz &= 4\rho^2 \sin^2\alpha \\
g_{22} &= -2i\rho^3 \sin^3\alpha \\
dg_{22}/dz &= 0 \\
g_{12} &= -2\rho^3 \sin^3\alpha \\
dg_{12}/dz &= 2i\rho^2 \sin^2\alpha
\end{aligned}$$

In eq 10.94 we may drop the term $\sigma\delta_{ij}$ because it is negligible comparison to $\sigma\sqrt{c/(2\rho)}$. The stress components listed in the question then follow from eq 10.94.

In plane strain the elastic energy density, W, is:

$$\begin{aligned}
W &= \frac{1}{4\mu}\left(\sigma_{11}^2 + \sigma_{22}^2 + 2\sigma_{12}^2 - \nu(\sigma_{11} + \sigma_{22})^2\right) \\
&= \frac{1}{4\mu}\frac{K_I^2}{\rho}\left\{\left(\cos\frac{\alpha}{2} - \frac{1}{2}\sin\alpha\sin\frac{3\alpha}{2}\right)^2 + \left(\cos\frac{\alpha}{2} + \frac{1}{2}\sin\alpha\sin\frac{3\alpha}{2}\right)^2\right. \\
&\quad \left. + \frac{1}{2}\sin^2\alpha\cos^2\frac{3\alpha}{2} - 4\nu\cos^2\frac{\alpha}{2}\right\} \\
&= \frac{K_I^2}{4\mu\rho}\left(\frac{1}{2}\sin^2\alpha + 2(1-\nu)\cos^2\frac{\alpha}{2}\right).
\end{aligned}$$

Problem 10.2 Show that the stress field components in eq 10.91 are related to the following displacement field components through Hooke's law:

$$u_1(\rho\cos\alpha, \rho\sin\alpha) = \frac{K_I\sqrt{\rho}}{\mu}\cos(\alpha/2)\left(1 - 2\nu + \sin^2(\alpha/2)\right)$$

$$u_2(\rho\cos\alpha, \rho\sin\alpha) = \frac{K_I\sqrt{\rho}}{\mu}\sin(\alpha/2)\Big(2(1-\nu)-\cos^2(\alpha/2)\Big). \qquad (10.95)$$

Solution Using the chain rule we obtain:

$$\frac{\partial}{\partial x_1} = \cos\alpha\frac{\partial}{\partial\rho} - \frac{\sin\alpha}{\rho}\frac{\partial}{\partial\alpha}$$
$$\frac{\partial}{\partial x_2} = \sin\alpha\frac{\partial}{\partial\rho} + \frac{\cos\alpha}{\rho}\frac{\partial}{\partial\alpha}$$

Hence:

$$u_{1,1} = \frac{K_I}{2\mu\sqrt{\rho}}\left((1-2\nu)\cos\left(\frac{\alpha}{2}\right) - \frac{1}{2}\sin\alpha\sin\left(\frac{3\alpha}{2}\right)\right)$$
$$u_{2,2} = \frac{K_I}{2\mu\sqrt{\rho}}\left((1-2\nu)\cos\left(\frac{\alpha}{2}\right) + \frac{1}{2}\sin\alpha\sin\left(\frac{3\alpha}{2}\right)\right)$$
$$u_{1,2} = \frac{K_I}{2\mu\sqrt{\rho}}\left(2(1-\nu)\sin\left(\frac{\alpha}{2}\right) + \frac{1}{2}\sin\alpha\cos\left(\frac{3\alpha}{2}\right)\right)$$
$$u_{2,1} = \frac{K_I}{2\mu\sqrt{\rho}}\left(-2(1-\nu)\sin\left(\frac{\alpha}{2}\right) + \frac{1}{2}\sin\alpha\cos\left(\frac{3\alpha}{2}\right)\right)$$

Putting these into Hooke's law:

$$\sigma_{ij} = \mu(u_{i,j} + u_{j,i}) + \frac{2\mu\nu}{1-2\nu}\delta_{ij}(u_{m,m})$$

we obtain the stresses of eq 10.91.

Comment

The force per unit length on the mode I crack tip at $x_1 = 0$, tending to make it advance along the x_1-axis, may be obtained by evaluating the integral in eq 9.17, where S is any surface enclosing the tip. In the context of fracture this integral is called the J-integral, where it was discovered independently by Rice (1968). To use the stress components of eq 10.91 we should choose S to be close to the crack tip. Since the crack geometry is invariant along x_3 the surface integral becomes a line integral around the crack tip. The terms required to evaluate the line integral are contained in eqs 10.91, 10.92 and the four displacement gradients above. The following result is obtained:

$$F_1 = \frac{\pi(1-\nu)K_I^2}{\mu} \equiv J_I, \tag{10.96}$$

where $K_I = \sigma\sqrt{c/2}$ as before. When K_I reaches the critical value for the crack to grow according to Griffith's criterion, eq 10.49, the force on the crack tip, J_I, reaches 2γ. At this critical condition the force J_I, which arises from the change in the sum of the potential energy of the external loading mechanism and the elastic energy of the crack, exactly balances the force 2γ required to increase the area of the two crack faces. Thus, for an elastic crack, criteria for crack growth based on critical values of K_I and J_I are equivalent. But this equivalence breaks down when there is plasticity because J_I becomes dependent on whether the contour S includes the plastic zone. However, when there is a dislocation free zone the elastic field at the crack tip again becomes singular. The singularity is characterised by a local stress intensity factor k_I (see eq 10.85) which reflects the screening of the applied stress by the plastic zone. S may then be chosen to enclose the crack tip only, in which case eq 10.96 is recovered with K_I replaced by the local stress intensity factor k_I.

Problem 10.3 Consider a slit crack between $x_1 = -2c$ and $x_1 = 0$ on $x_2 = 0$ loaded in mode III by a shear stress $\sigma_{23} = \sigma^A$ at $x_2 = \pm\infty$. The tractions created on the crack faces by the applied stress must be eliminated. This is achieved by introducing a distribution of screw dislocations in $-2c \leq x_1 \leq 0$, with lines parallel to x_3, and a Burgers vector density $f(x_1)$. Using the stress field of a screw dislocation in eq 6.41 show that the integral equation governing the distribution $f(x_1)$ is as follows:

$$P\int_{-2c}^{0} \frac{f(x_1')}{x_1 - x_1'} \mathrm{d}x_1' = -\frac{2\pi\sigma^A}{\mu}. \tag{10.97}$$

Show that solution of this equation is:

$$f(x_1) = \frac{2\sigma^A}{\mu} \frac{(x_1 + c)}{\sqrt{(x_1 + 2c)(-x_1)}}. \tag{10.98}$$

Show that non-zero stress field components of the crack at (x_1, x_2) are as follows:

$$\sigma_{13}(x_1, x_2) = -\frac{\sigma^A x_2}{\pi} \int_{-2c}^{0} \frac{(x_1' + c)}{\sqrt{(x_1' + 2c)(-x_1')}} \frac{1}{(x_1 - x_1')^2 + x_2^2} \mathrm{d}x_1'$$

$$\sigma_{23}(x_1,x_2) \;=\; \frac{\sigma^A}{\pi}\int_{-2c}^{0}\frac{(x_1'+c)}{\sqrt{(x_1'+2c)(-x_1')}}\frac{(x_1-x_1')}{(x_1-x_1')^2+x_2^2}\,\mathrm{d}x_1' \qquad (10.99)$$

Evaluate these integrals by contour integration to obtain the following expressions:

$$\sigma_{13}(x_1,x_2) \;=\; \sigma^A \operatorname{Im}\left[\frac{x_1+ix_2+c}{\sqrt{(x_1+ix_2+2c)(x_1+ix_2)}}\right]$$

$$\sigma_{23}(x_1,x_2) \;=\; \sigma^A \operatorname{Re}\left[\frac{x_1+ix_2+c}{\sqrt{(x_1+ix_2+2c)(x_1+ix_2)}}\right] \qquad (10.100)$$

For $(x_1,x_2) = (\rho\cos\alpha, \rho\sin\alpha)$, where $\rho/c \ll 1$, show that these stress components become the following close to the crack tip at $(0,0)$:

$$\sigma_{13} \;=\; -\sigma^A\sqrt{\frac{c}{2\rho}}\,\sin(\alpha/2)$$

$$\sigma_{23} \;=\; \sigma^A\sqrt{\frac{c}{2\rho}}\,\cos(\alpha/2). \qquad (10.101)$$

It follows that the stress intensity factor for the mode III slit crack is also $\sigma^A\sqrt{c/2}$. Show that the crack opening displacement $s(x_1)$ is as follows:

$$s(x_1) = \frac{2\sigma^A}{\mu}\sqrt{c^2-(c+x_1)^2}. \qquad (10.102)$$

Show that the change in the total potential energy following the introduction of the crack is:

$$\Delta U = -\frac{\pi\left(\sigma^A\right)^2 c^2}{2\mu} \qquad (10.103)$$

Hence show that the Griffith criterion for the growth of a mode III slit crack is as follows:

$$\sigma^A\sqrt{c/2} = \sqrt{\frac{2\gamma\mu}{\pi}}. \qquad (10.104)$$

Solution The stress field of a screw dislocation was derived in problem 6.3. For a screw dislocation along the x_3-axis the stress field components are:

$$\sigma_{13} = -\frac{\mu b}{2\pi}\frac{x_2}{x_1^2+x_2^2}$$

$$\sigma_{23} = \frac{\mu b}{2\pi}\frac{x_1}{x_1^2+x_2^2}$$

The relevant component is σ_{23} because the applied stress is $\sigma_{23}=\sigma^A$ at $x_2=\pm\infty$. On the interval $-2c \le x_1 \le 0$ of the plane $x_2=0$ there has to be a continuous distribution of screw dislocations, with their lines parallel to the x_3-axis, to ensure that the tractions on the crack faces are zero. Let $f(x_1')\mathrm{d}x_1'$ be the Burgers vector of screw dislocations between x_1' and $x_1'+\mathrm{d}x_1'$. For a screw dislocation located at $(x_1',0)$ with Burgers vector $f(x_1')\mathrm{d}x_1'$ the stress σ_{23} at $(x_1,0)$ is:

$$\sigma_{23}(x_1,0) = \frac{\mu}{2\pi}\frac{f(x_1')\mathrm{d}x_1'}{x_1-x_1'}.$$

The condition for there to be no resultant stress σ_{23}, and hence no traction at $(x_1,0)$, is obtained by integrating the contributions from the continuous distribution of dislocations to the stress σ_{23} at $(x_1,0)$:

$$\frac{\mu}{2\pi}P\int_{-2c}^{0}\frac{f(x_1')\mathrm{d}x_1'}{x_1-x_1'}\mathrm{d}x_1' + \sigma_A = 0,$$

or

$$P\int_{-2c}^{0}\frac{f(x_1')\mathrm{d}x_1'}{x_1-x_1'}\mathrm{d}x_1' = -\frac{2\pi\sigma^A}{\mu},$$

where the integral is a principal value because the screw dislocation at x_1 does not exert a stress on itself.

The solution follows from eq 10.20, with $a=-2c$ and $b=0$:

$$f(x_1) = \frac{1}{\pi^2\sqrt{(x_1+2c)(-x_1)}}\left\{C - \frac{2\pi\sigma^A}{\mu}P\int_{-2c}^{0}\frac{\sqrt{(x_1'+2c)(-x_1')}}{x_1'-x_1}\mathrm{d}x_1'\right\},$$

where C is determined by the condition that there is no net Burgers vector associated with the crack: $\int_{-2c}^{0} f(x_1)\mathrm{d}x_1 = 0$. As shown in the solution to exercise 10.4, the distribution $f(x_1)$ is as follows:

$$f(x_1) = \frac{2\sigma^A}{\mu} \frac{x_1 + c}{\sqrt{(x_1 + 2c)(-x_1)}}.$$

The stresses at (x_1, x_2) due to the applied stress and the crack are then as follows:

$$\sigma_{13}(x_1, x_2) = -\frac{\sigma^A}{\pi} \int_{-2c}^{0} \frac{x_1' + c}{\sqrt{(x_1' + 2c)(-x_1')}} \frac{x_2}{(x_1 - x_1')^2 + x_2^2} \mathrm{d}x_1'$$

$$\sigma_{23}(x_1, x_2) = \frac{\sigma^A}{\pi} \int_{-2c}^{0} \frac{x_1' + c}{\sqrt{(x_1' + 2c)(-x_1')}} \frac{x_1 - x_1'}{(x_1 - x_1')^2 + x_2^2} \mathrm{d}x_1' + \sigma^A$$

Let the integral in the expression for $\sigma_{13}(x_1, x_2)$ be I_1:

$$I_1 = \int_{-2c}^{0} \frac{x_1' + c}{\sqrt{(x_1' + 2c)(-x_1')}} \frac{x_2}{(x_1' - (x_1 + ix_2))(x_1' - (x_1 - ix_2))} \mathrm{d}x_1'.$$

To evaluate this integral consider the following contour integral:

$$\oint_C \frac{z + c}{\sqrt{(z + 2c)z}} \frac{x_2}{(z - (x_1 + ix_2))(z - (x_1 - ix_2))} \mathrm{d}z.$$

The contour C is shown in Fig 10.15, but the poles at $z = \rho e^{\pm i\alpha} = x_1 \pm ix_2$ are now first order. There are no contributions to the contour integral from the small circles at $z = -2c$ and $z = 0$ or from the large circle when its radius is taken to infinity. The contribution from the loop around the branch cut is $-2iI_1$. The residue at $z = z_1 + ix + 2$ is:

$$\frac{1}{2i} \frac{x_1 + ix_2 + c}{\sqrt{(x_1 + ix_2 + 2c)(x_1 + ix_2)}}.$$

The residue at $z = x_1 - ix_2$ is:

$$\frac{-1}{2i} \frac{x_1 - ix_2 + c}{\sqrt{(x_1 - ix_2 + 2c)(x_1 - ix_2)}}.$$

The sum of the residues is:

$$\mathrm{Im}\left[\frac{x_1+ix_2+c}{\sqrt{(x_1+ix_2+2c)(x_1+ix_2)}}\right]$$

where Im[z] means the imaginary part of the complex number z. By Cauchy's theorem, $-2iI_1 = 2\pi i$ times the sum of the residues. Therefore,

$$\sigma_{13}(x_1,x_2) = \sigma^A \mathrm{Im}\left[\frac{x_1+ix_2+c}{\sqrt{(x_1+ix_2+2c)(x_1+ix_2)}}\right]$$

For $\sigma_{23}(x_1,x_2)$ we evaluate the integral I_2:

$$I_2 = \int_{-2c}^{0} \frac{x_1'+c}{\sqrt{(x_1'+2c)(-x_1')}} \frac{x_1-x_1'}{(x_1'-(x_1+ix_2))\,(x_1'-(x_1-ix_2))}\mathrm{d}x_1'.$$

Consider the contour integral:

$$\oint_C \frac{z+c}{\sqrt{(z+2c)z}} \frac{x_1-z}{(z-(x_1+ix_2))\,(z-(x_1-ix_2))}\mathrm{d}z.$$

where the contour C is the same as for I_1 in this problem. Again, there are no contributions to the contour integral from the small circles at $z=-2c$ and at $z=0$. But there is a contribution of $-2\pi i$ from the large circle as its radius tends to infinity. The contribution from the loop around the branch cut is $-2iI_2$. The residue at $z = x_1+ix_2$ is:

$$-\frac{1}{2}\frac{x_1+ix_2+c}{\sqrt{(x_1+ix_2+2c)(x_1+ix_2)}}.$$

The residue at $z = x_1-ix_2$ is:

$$-\frac{1}{2}\frac{x_1-ix_2+c}{\sqrt{(x_1-ix_2+2c)(x_1-ix_2)}}.$$

The sum of the residues is:

$$-\mathrm{Re}\left[\frac{x_1+ix_2+c}{\sqrt{(x_1+ix_2+2c)(x_1+ix_2)}}\right].$$

By Cauchy's theorem $-2iI_2 - 2\pi i = 2\pi i$ times the sum of the residues. Therefore,

$$\sigma_{23}(x_1,x_2) = \sigma^A \mathrm{Re}\left[\frac{x_1 + ix_2 + c}{\sqrt{(x_1 + ix_2 + 2c)(x_1 + ix_2)}}\right].$$

Note this includes the far-field stress $\sigma_{23}(x_1, x_2 \to \pm\infty) \to \sigma^A$.

Writing $(x_1, x_2) = (\rho\cos\alpha, \rho\sin\alpha)$ and in the limit $\rho/c \ll 1$ we obtain:

$$\sigma_{13} = \sigma^A \mathrm{Im}\left[\frac{\rho e^{i\alpha} + c}{(\rho e^{i\alpha} + 2c)\rho e^{i\alpha}}\right] \to \sigma^A \mathrm{Im}\frac{c}{2c\rho e^{i\alpha}} = -\sigma^A\sqrt{\frac{c}{2\rho}}\sin\left(\frac{\alpha}{2}\right)$$

$$\sigma_{23} = \sigma^A \mathrm{Re}\left[\frac{\rho e^{i\alpha} + c}{(\rho e^{i\alpha} + 2c)\rho e^{i\alpha}}\right] \to \sigma^A \mathrm{Re}\frac{c}{2c\rho e^{i\alpha}} = \sigma^A\sqrt{\frac{c}{2\rho}}\cos\left(\frac{\alpha}{2}\right)$$

The crack opening displacement is:

$$s(x_1) = -\int_{-2c}^{x_1} f(x_1')\mathrm{d}x_1' = -\frac{2\sigma^A}{\mu}\int_{-2c}^{x_1}\frac{x_1' + c}{\sqrt{(x_1' + 2c)(-x_1')}}\mathrm{d}x_1' = \frac{2\sigma^A}{\mu}\sqrt{c^2 - (c + x_1)^2}.$$

The change in the total potential energy associated with the formation of the crack is:

$$\Delta U = -\frac{1}{2}\int_{-2c}^{0}\sigma^A s(x_1)\mathrm{d}x_1 = -\frac{(\sigma^A)^2}{\mu}\int_{-2c}^{0}\sqrt{c^2 - (c + x_1)^2}\mathrm{d}x_1 = -\frac{\pi(\sigma^A)^2c^2}{2\mu}.$$

The Griffith criterion is:

$$\frac{\mathrm{d}}{\mathrm{d}c}\left(-\frac{\pi(\sigma^A)^2c^2}{2\mu} + 4\gamma c\right) = 0, \text{ or } (\sigma^A)^2c/2 = 2\gamma\mu/\pi.$$

Therefore, the critical stress intensity factor is

$$K_{IIIc} = \sigma^A\sqrt{\frac{c_{crit}}{2}} = \sqrt{\frac{2\gamma\mu}{\pi}}.$$

Problem 10.4 Show that the force along the x_1-axis on unit length of the tip of the mode III slit crack of the previous question is $J_{III} = \pi\left(\sigma^A\right)^2 c/(2\mu)$. When Griffith's criterion is satisfied show that $J_{III} = 2\gamma$.

Solution From eq 9.17, the force in the x_1 direction, acting on unit length of the crack tip at $x_1 = 0, x_2 = 0$, is as follows:

$$F_1 = \int_S (W\delta_{1j} - \sigma_{ij}u_{i,1})\, n_j \mathrm{d}S.$$

For the surface S we may choose a cylinder of unit length and radius ρ surrounding the crack tip. By making $\rho \ll c$ we may use eqs 10.101 for the stress field components σ_{13} and σ_{23}. We have:

$$\sigma_{ij}u_{i,1}n_j = (\sigma_{31}n_1 + \sigma_{32}n_2)u_{3,1}$$

where $u_{3,1} = \sigma_{13}/\mu$ because $u_1 = 0$. We will also need $u_{3,2}$ for the elastic energy density, and it is σ_{23}/μ because $u_2 = 0$. Since $n_1 = \cos\alpha$ and $n_2 = \sin\alpha$ then using eqs 10.101 we obtain:

$$\sigma_{ij}u_{i,1}n_j = -\frac{\left(\sigma^A\right)^2}{\mu}\frac{c}{2\rho}\sin^2\left(\frac{\alpha}{2}\right).$$

The elastic energy density is:

$$W = \frac{1}{2}\sigma_{kl}u_{k,l} = \frac{1}{2\mu}\left(\sigma_{31}^2 + \sigma_{32}^2\right) = \frac{\left(\sigma^A\right)^2}{2\mu}\frac{c}{2\rho}.$$

Substituting these terms into the integral for the force F_1:

$$F_1 = \int_{-\pi}^{\pi}\left\{\frac{\left(\sigma^A\right)^2}{2\mu}\frac{c}{2\rho}\cos\alpha + \frac{\left(\sigma^A\right)^2}{\mu}\frac{c}{2\rho}\sin^2\left(\frac{\alpha}{2}\right)\right\}\rho\mathrm{d}\alpha = \pi\frac{\left(\sigma^A\right)^2 c}{2\mu} = J_{III}.$$

When the Griffith criterion is satisfied:

$$\sigma^A\sqrt{\frac{c}{2}} = \sqrt{\frac{2\gamma\mu}{\pi}}$$

In that case $J_{III} = 2\gamma$.

Comment

When the Griffith criterion is satisfied the reduction in the total potential energy is just sufficient to supply the increase in the energy of the two crack surfaces. Per unit length of crack along x_3, the increase in the energy of the two crack surfaces when the crack advances by δx_1 is $2\gamma\delta x_1$. This is the energy supplied by the work done by the force $J_{III} = 2\gamma$ on the crack tip per unit length along x_3. Thus when K_{III} satisfies the Griffith criterion, so does J_{III}. This equivalence assumes no plasticity is involved in the crack growth.

Problem 10.5 A Frank-Read source in the centre of a rectangular grain emits dislocation loops under the action of a shear stress resolved on the slip plane in the direction of the Burgers vector. If the boundaries surrounding the grain are impenetrable obstacles the loops pile up at them, as illustrated in Fig 10.16. If one side of the grain is parallel to the Burgers vector the dislocations parallel to that side have pure screw character, and on the perpendicular sides they have pure edge character. If there are many dislocations in each pileup we may make the continuum approximation to derive the stress field created by the pileups.

Consider the two pileups of edge dislocations and approximate the edge dislocations lines as having infinite length. Let the normal to the slip plane be along the positive x_2-axis, and the Burgers vector of each dislocation be parallel to x_1 (see Fig 10.16). The edge dislocation line directions are then along the $\pm x_3$-axis. The edge dislocations on either side of the source have opposite signs because their line directions are reversed (since they form loops). We may treat them equivalently as having the same line sense but opposite sign Burgers vectors. Let σ^A be the effective resolved component of the applied stress acting on the slip plane in the direction of the Burgers vector[33]. Let $f(x_1)$ be the Burgers vector density of dislocations in the two pileups. The source continues to emit dislocations until σ^A is counteracted by the shear stress created by the pileups. The system is then in mechanical equilibrium.

Using the stress field of an edge dislocation given in eq 6.33 show that the condition for mechanical equilibrium of the two pileups is as follows:

$$\frac{\mu}{2\pi(1-\nu)} P \int_{-d}^{d} \frac{f(x_1')}{x_1 - x_1'} \, dx_1' = -\sigma^A, \tag{10.105}$$

where d is the length of each pileup. By now this should look very familiar. The

[33]by 'effective' we mean after the shear stress required to operate the Frank-Read source has been subtracted from the resolved component of the applied stress.

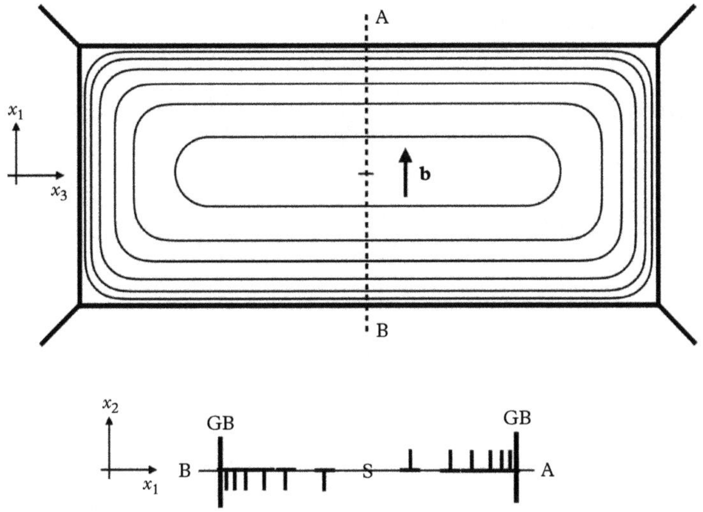

Figure 10.16: Upper: Plan view of a rectangular grain in a polycrystal containing a Frank-Read source at its centre which is emitting dislocation loops that pile up at the grain boundaries, shown as thicker lines. The Burgers vector **b** is along x_1, and the grain sides are parallel to $x_1 - x_2$ and $x_3 - x_2$ planes of a cartesian coordinate system. The normal to the page is along x_2. The broken line AB passes through edge dislocations piled up at the upper and lower grain boundaries. Lower: Side view of the edge dislocation pileups along AB looking along the x_3-axis. S signifies the Frank-Read source, GB signifies a grain boundary. The same coordinate system is used in both figures.

solution follows immediately from eq 10.32:

$$f(x_1) = \frac{2(1-\nu)\sigma^A}{\mu}\frac{x_1}{\sqrt{d^2 - x_1^2}} \tag{10.106}$$

If b is the magnitude of the Burgers vector of the discrete dislocations in the pileups show that the number n of dislocations in each pileup is:

$$n = \frac{2(1-\nu)\sigma^A}{\mu}\frac{d}{b}. \tag{10.107}$$

Following the same procedure that led to the stress components in eq 10.91, it may be shown that at $(x_1, x_2) = (d + \rho\cos\alpha, \rho\sin\alpha)$, where $\rho/d \ll 1$ the stress field close to the tip of the pileup has the following components:

$$\begin{aligned}
\sigma_{11}(d+\rho\cos\alpha,\rho\sin\alpha) &= -\sigma^A\sqrt{\frac{d}{2\rho}}\left(2\sin\left(\frac{\alpha}{2}\right)+\frac{1}{2}\sin\alpha\cos\left(\frac{3\alpha}{2}\right)\right)\\
\sigma_{22}(d+\rho\cos\alpha,\rho\sin\alpha) &= \sigma^A\sqrt{\frac{d}{2\rho}}\left(\frac{1}{2}\sin\alpha\cos\left(\frac{3\alpha}{2}\right)\right)\\
\sigma_{12}(d+\rho\cos\alpha,\rho\sin\alpha) &= \sigma^A\sqrt{\frac{d}{2\rho}}\left(\cos\left(\frac{\alpha}{2}\right)-\frac{1}{2}\sin\alpha\sin\left(\frac{3\alpha}{2}\right)\right) \qquad (10.108)
\end{aligned}$$

These stress components $\sigma_{ij}(\rho,\alpha)$ are expressed in the coordinate system of the grain containing the pileup[34]. In the adjacent grain let the normal to a slip plane and the corresponding slip direction be $\hat{\mathbf{n}}(\beta) = [-\sin\beta,\cos\beta,0]$ and $\hat{\mathbf{s}}(\beta) = [\cos\beta,\sin\beta,0]$ respectively, expressed in the coordinate system of the grain containing the pileup. Let there be a source in the adjacent grain at a distance $\rho = \rho_s$ from the end of the pileup, where $\rho_s/d \ll 1$. If the resolved shear on the adjacent slip system required to activate the source is τ_s show that the following condition must be satisfied if the source is to be activated:

$$\frac{\sigma^A}{4}\sqrt{\frac{d}{2\rho_s}}\left[\cos\left(\frac{\beta}{2}\right)+3\cos\left(\frac{3\beta}{2}\right)\right] = \tau_s \qquad (10.109)$$

Solution The resolved shear stress at $(x_1,0)$ created by an edge dislocation with Burgers vector components $b_i = b\delta_{i1}$ at $(x_1',0)$ is:

$$\sigma_{12}(x_1,0) = \frac{\mu b}{2\pi(1-\nu)}\frac{1}{x_1-x_1'}.$$

Therefore, the resolved shear stress at $(x_1,0)$ due to the edge dislocations with Burgers vector density $f(x_1')$ distributed between $x_1' = -d$ and $x_1' = d$ is:

$$\frac{\mu}{2\pi(1-\nu)}P\int_{d}^{d}\frac{f(x_1')}{x_1-x_1'}\,\mathrm{d}x_1'.$$

[34]They also describe the stress field close to the tip of a mode II slit crack.

When this is added to σ^A the resultant stress must be zero for mechanical equilibrium (ignoring the friction stress). Therefore,

$$\frac{\mu}{2\pi(1-\nu)} P \int_{d}^{d} \frac{f(x_1')}{x_1 - x_1'} \mathrm{d}x_1' = -\sigma^A,$$

for which the solution is given in eq 10.106. The total Burgers vector content of each pile up is:

$$\frac{2(1-\nu)\sigma^A}{\mu} \int_0^d \frac{x_1'}{\sqrt{d^2 - x_1'^2}} \mathrm{d}x_1' = \frac{2(1-\nu)\sigma^A}{\mu} d.$$

For discrete dislocations, with Burgers vector b, an estimate of their number is:

$$\frac{2(1-\nu)\sigma^A}{\mu} \frac{d}{b}.$$

The condition to activate the source in the adjacent grain is that the resolved shear stress on the adjacent slip system at $\rho = \rho_s$, $\alpha = \beta$ is equal to τ_s. The resolved shear shear stress ahead of the pile up on the plane with normal $\hat{\mathbf{n}}(\beta)$ in the direction of $\hat{\mathbf{s}}(\beta)$ is:

$$\begin{aligned}
s_i(\beta)\sigma_{ij}(\rho_s,\beta)n_j(\beta) &= \left(\frac{\sigma_{22}(\rho_s,\beta) - \sigma_{11}(\rho_s,\beta)}{2}\right)\sin(2\beta) + \sigma_{12}(\rho_s,\beta)\cos(2\beta) \\
&= \frac{\sigma^A}{2}\sqrt{\frac{d}{2\rho_s}}\Big(\sin\beta\cos(3\beta/2)\sin(2\beta) + 2\sin(\beta/2)\sin(2\beta) \\
&\quad + 2\cos(\beta/2)\cos(2\beta) - \sin\beta\sin(3\beta/2)\cos(2\beta)\Big) \\
&= \frac{\sigma^A}{2}\sqrt{\frac{d}{2\rho_s}}\Big(\sin\beta\sin(\beta/2) + 2\cos(3\beta/2)\Big) \\
&= \frac{\sigma^A}{4}\sqrt{\frac{d}{2\rho_s}}\Big(\cos(\beta/2) + \cos(3\beta/2)\Big).
\end{aligned}$$

Comment

The average of value of $[\cos(\beta/2) + 3\cos(3\beta/2)]$ between $\beta = -\pi/2$ and $\beta = +\pi/2$ is $4\sqrt{2}/\pi$. Writing $\sigma^A = \sigma_y - \sigma_f$, where σ_y is the yield stress at which plastic deformation is propagated from one grain to another, and σ_f is the friction stress opposing dislocation motion, eq 10.109 may be written in the following form for a polycrystal:

$$\sigma_y = \sigma_f + k_y d^{-1/2}, \tag{10.110}$$

where $k_y = \pi\tau_s\sqrt{\rho_s}$ is a factor that determines how easily slip is transmitted from one grain to the next. Equation 10.110 is called the Hall-Petch relation after Hall[a] and Petch[b] who found it experimentally[c]. From a technological point of view it is useful because it shows the yield stress may be raised by decreasing the grain size. A smaller grain size may be achieved by a combination of plastic deformation and carefully controlled annealing to induce and arrest recrystallisation to freeze in a small grain size.

[a]Eric O Hall, New Zealand physicist

[b]Norman James Petch FRS 1917-1992, British metallurgist

[c]Hall, E O, Proc. Phys. Soc. **B64**, 747-753 (1951); Petch, N J, J. Iron and Steel Inst. **174**, 25 (1953).

Problem 10.6 Griffith's criterion for crack growth assumes a pre-existing crack, which raises the question of how cracks form in a metal. Following an earlier suggestion[35] by Mott[36], Stroh[37] showed[38] during his doctorate how the stress concentration ahead of a dislocation pileup may nucleate a crack. Consider the nucleation of a mode I crack ahead of a pileup. We assume this happens when the tensile stress across some plane just ahead of the pileup exceeds a critical value. For this to happen the total energy of the system must decrease as a result of the nucleation of the crack. Using eq 10.47 for the potential energy released when a mode I slit crack of length $2c$ forms, and taking into account the energy $4\gamma c$ of creating the two crack surfaces, show that the total

[35]Mott, N F, Proc. R. Soc. A **220**, 1 (1953).

[36]Sir Nevill Francis Mott FRS 1905-1996, British Nobel prize winning theoretical physicist, who established solid-state physics in the UK.

[37]Alan N Stroh 1926-1962, South African theoretical physicist, PhD student of Eshelby and Mott at Bristol University, creator of the elegant and widely used sextic formalism of anisotropic elasticity, his career was cut short at age 36 by a fatal car accident in Colorado. A more extensive biography is available on p.159-161 of *Anisotropic Elasticity* by Ting, T C T, Oxford University Press: Oxford and New York (1996) ISBN: 0195074475.

[38]Stroh, A N, Proc. R. Soc. **223**, 404-414 (1954).

energy decreases provided:

$$\sigma \geq \sqrt{\frac{8\gamma\mu}{\pi(1-\nu)c}} \tag{10.111}$$

The stress field ahead of the pileup is given by eq 10.108. The next step is to identify the angle α in eq 10.108 associated with the largest tensile stress along the normal $\hat{\mathbf{n}} = [-\sin\alpha, \cos\alpha]$. This is given by $\sigma(\alpha) = n_i\sigma_{ij}n_j$. Show that:

$$\sigma(\alpha) = -3\sigma^A\sqrt{\frac{d}{2\rho}}\sin(\alpha/2)\cos^2(\alpha/2) \tag{10.112}$$

$\sigma(\alpha)$ is a maximum when $\alpha = -\cos^{-1}(1/3)$, which is $-70.5°$. Show that the corresponding maximum tensile stress is

$$\sigma_{max} = \frac{2}{\sqrt{3}}\sqrt{\frac{d}{2\rho}}\sigma^A. \tag{10.113}$$

If the crack is nucleated its length $2c$ will equal ρ. Hence show that the condition for the pileup to nucleate a mode I crack is as follows:

$$\sigma^A\sqrt{d} \geq 2\sqrt{\frac{6\gamma\mu}{\pi(1-\nu)}} \tag{10.114}$$

Solution The change in the total energy when a mode I crack of length $2c$ is nucleated is:

$$\Delta E_{tot} = -\frac{\pi(1-\nu)\sigma^2c^2}{2\mu} + 4\gamma c,$$

where σ is the local applied tensile stress normal to the crack plane. The change in the total energy has to be negative for the crack to be nucleated. Therefore, the condition for crack nucleation is:

$$\sigma \geq \sqrt{\frac{8\gamma\mu}{\pi(1-\nu)c}}.$$

Using the stress field components of an edge dislocation pileup in eq 10.108, the normal stress acting on a plane with normal $\hat{\mathbf{n}} = [-\sin\alpha, \cos\alpha]$ ahead of the pileup is:

$$
\begin{aligned}
\sigma(\alpha) &= \sigma_{11}\sin^2\alpha + \sigma_{22}\cos^2\alpha - 2\sigma_{12}\sin\alpha\cos\alpha \\
&= -\sigma^A\sqrt{\frac{d}{2\rho}}\left\{2\sin(\alpha/2)\sin^2\alpha + \frac{1}{2}\sin^3\alpha\cos(3\alpha/2)\right. \\
&\quad -\frac{1}{2}\sin\alpha\cos(3\alpha/2)\cos^2\alpha \\
&\quad \left. + \ \sin(2\alpha)\cos(\alpha/2) - \frac{1}{2}\sin(2\alpha)\sin\alpha\sin(3\alpha/2)\right\} \\
&= -3\sigma^A\sqrt{\frac{d}{2\rho}}\sin(\alpha/2)\cos^2(\alpha/2).
\end{aligned}
$$

The tensile stress is a maximum of $(2/\sqrt{3})\sigma^A\sqrt{d/(2\rho)}$ when $\alpha = \cos^{-1}(1/3) = -70.53°$. Equating this maximum tensile stress to σ in eq 10.111 and equating ρ to $2c$ the condition for a mode I crack to be nucleated by the pileup is:

$$
\frac{2}{\sqrt{3}}\sigma^A\sqrt{\frac{d}{4c}} \geq \sqrt{\frac{8\gamma\mu}{\pi(1-\nu)c}}
$$

which simplifies to

$$
\sigma^A\sqrt{d} \geq 2\sqrt{\frac{6\gamma\mu}{\pi(1-\nu)}}.
$$

Comment

Since the total energy of the system continues to decrease once a crack has been nucleated in Stroh's model, the crack will continue to grow. This implies that according to Stroh's model brittle fracture is controlled by the nucleation of a crack, not its growth. However, just as grain boundaries act as obstacles to slip, they may also impede the growth of cracks whenever fracture occurs preferentially on particular crystallographic planes. A reduction of the grain size has the dual benefit of strengthening the material by raising the yield stress while also making it less susceptible to the nucleation of cracks and possibly their subsequent growth.

11. Open questions

1.1 Introduction

In this chapter five areas of current research in the physics of crystal defects are introduced. The presentation differs from earlier chapters in being more like research seminars than detailed expositions. References to the literature are given where further information may be found. There are other large areas of current research in the physics of defects such as very high strain rate deformation and irradiation damage. The five topics discussed here have been chosen because they can be introduced relatively briefly, and questions to frame further research can be formulated.

1.2 Electroplasticity

.2.1 Introduction and experimental observations

In 1963 in the former Soviet Union it was discovered that the flow of a direct electric current through a metal can reduce its flow stress and enhance its overall ductility. During irradiation of zinc single crystals with 1MeV electrons, Troitskii and Likhtman[1] observed a significant decrease in the flow stress as the crystals were deformed in uniaxial tension at liquid nitrogen temperatures. The elongation to fracture was greater when the electron beam was in the basal plane than normal to the basal plane. The predominant slip systems in zinc are in the basal plane. These observations suggested the possibility that the passage of a directed electron current through a metal may increase the mobility of dislocations. Troitskii and other Soviet scientists carried out further experiments revealing the effects of ≈ 100 μs *pulses* of direct current densities between 10^3 and 10^5 $\text{A}\cdot\text{cm}^{-2}$, on the flow stress, stress relaxation, creep, brittle fracture, fatigue and metal-working. The pulses were short to minimise Joule heating and the current densities were high to maximise their effect. The influence of the current pulses on plasticity was called the electroplastic effect.

[1]Troitskii, O A and Likhtman, V I, Dokl. Akad. Nauk SSSR **148**, 332-334 (1963). In Russian.

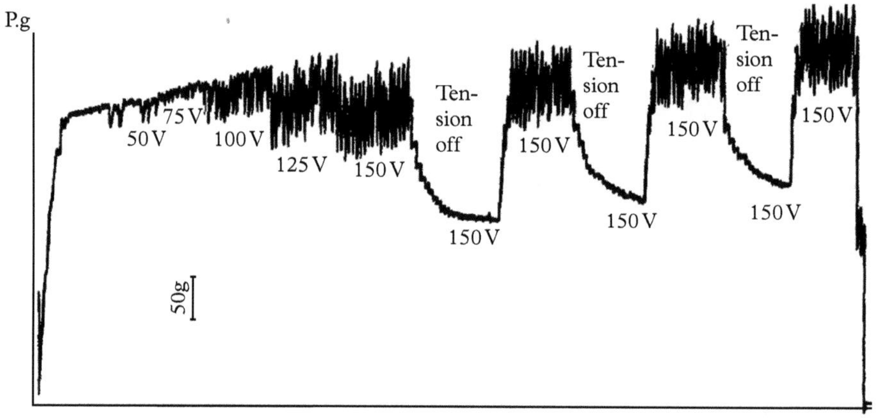

Figure 11.1: Load-displacement curve for a single crystal of zinc, subjected to DC current pulses while it is deformed in uniaxial tension at 78 K at a strain rate of 1.1×10^{-5} s^{-1}. The horizontal axis is time. The voltage is increased in steps of 25 V from 50 V to 150 V, where 100 V corresponds to a current of approximately 1.5×10^{5} A $\cdot$ cm^{-2}. The load falls with each current pulse, and rises again when the pulse ends. The size of the load drops increases with the voltage (and current) of the current pulses. When the test machine is shut off there is stress relaxation and the current pulses then have a much smaller influence. Reproduced with permission from Troitskii, O A, JETP Letters **10**, 11-14 (1969).

Fig 11.1 shows the load against displacement for a wire single crystal of zinc[2], diameter 1 mm and length 15 mm, stretched at a constant speed of 0.01 cm $\cdot$ min^{-1} equivalent to a strain rate of 1.1×10^{-5} s^{-1}. Pulses of direct current of duration $\approx 100\ \mu$s and between 600 A and 1800 A were passed through the wire. The current between the pulses did not exceed 0.3 A. As soon as each pulse was applied the load dropped. Troitskii and coworkers found the load drop ΔP varied between 10% and 40% of the applied load, it increased with the current density J, it varied with the direction of **J** relative to crystal axes, it increased with decreasing temperature, it tended to decrease with increasing strain rate, and it occurs in compression as well as tension.

Stimulated by these results experimentalists in the US extended these studies to polycrystalline samples of Al, Cu, Pb, Ni, Fe, Nb, W, Sn and Ti. For details see

[2]Troitskii, O A, JETP Letters **10**, 11-14 (1969).

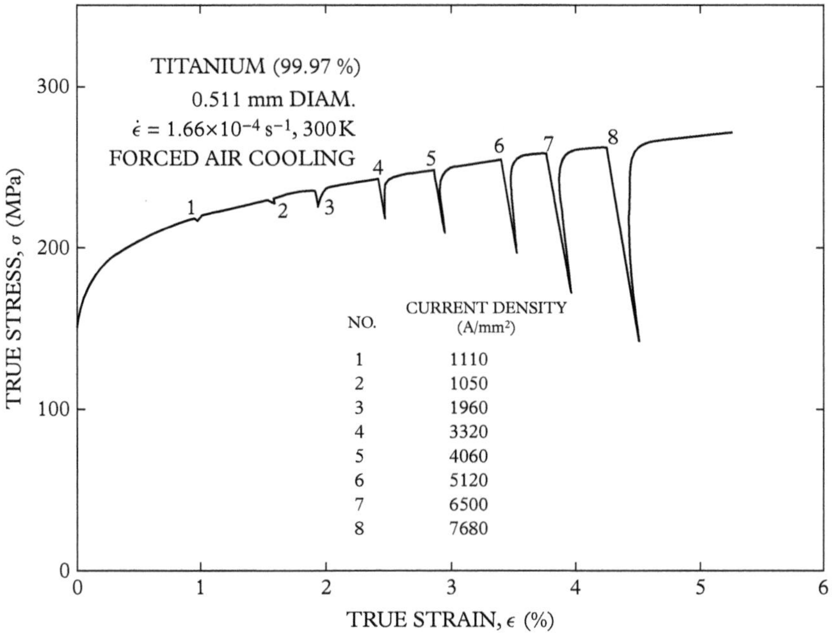

Figure 11.2: True stress-true strain relation for a Ti wire deformed in tension and subjected to current pulses of increasing current density. Reprinted from Okazaki, K, Kagawa, M and Conrad, H, Scr. Metall. **12**, 1063-1068 Copyright 1978, with permission from Elsevier.

reviews[3] by Conrad[4] and coworkers. For a more recent review see Dimitrov et al.[5].

Fig 11.2 from Okazaki et al.[6] shows the relation between true stress and true strain for a polycrystalline wire of Ti (99.97% purity), ~ 50 mm gauge length and 0.51 mm diameter, subjected to a uniaxial tensile loading and pulses of current of duration $\lesssim$ 100 μs at a nominal temperature of 300 K with forced air cooling. Note the stress drop when the pulse is applied and the return to a continuation of the upper envelope of the stress-strain curve after the pulse. They also found the upper envelope of the stress-strain curve with current pulses was the same as that for a separate specimen

[3]Sprecher, A F, Mannan, S L and Conrad, H Acta Metallurgica **34**, 1145-1162 (1986); Conrad, H and Sprecher, A F in *Dislocations in Solids* **8** Chapter 43, 497-541 (1989). North-Holland: Amsterdam. ISBN 978-0444705150; Conrad, H, Mat. Sci. Eng. A **287** (2000); Antolovich, S D and Conrad, H, Materials and Manufacturing Processes, **19**, 587-610 (2004). Stephen D Antolovich, US materials engineer.

[4]Hans Conrad 1922-2022. US materials scientist.

[5]Dimitrov, N K, Liu, Y and Horstemeyer, M F, Mechanics of Advanced Materials and Structures **29**, 705 (2022).

[6]Okazaki, K, Kagawa, M and Conrad, H, Scr. Metall. **12**, 1063-1068 (1978).

which had been deformed without current pulses. These observations indicate that whatever changes occur in the wire, as a result of the current pulses, they appear to be reversible. It is difficult to reconcile such reversibility with changes in dislocation configurations, which are normally irreversible.

Using pulses of much smaller current densities (5×10^2 $\text{A} \cdot \text{cm}^{-2}$) and longer duration (0.1 s) on titanium 7 at.% aluminium samples, Zhao et al.[7], showed the pulses alter dislocation configurations and ductility of the alloy. Significantly, these changes were *not* found when the current was continuous. Zhao et al. concluded the pulses were essential for the effect. In another study Shibkov et al.[8], showed that pulses of increasing current density ($10 - 50 \times 10^2$ $\text{A} \cdot \text{cm}^{-2}$) applied to certain aluminium – magnesium alloys delayed the onset of jerky flow to higher plastic strains and eventually suppressed it completely. Although these studies on alloys are very interesting the presence of solute atoms complicates the interpretation of the results significantly. For this reason we shall confine our attention to pure metals.

Electroplasticity has started to be exploited in engineering applications, particularly as a means of facilitating bulk deformation and sheet-metal forming[9] Some of these applications use continuous currents where Joule heating could be significant. There are some instances of pulsed currents being used, which have the advantage of reduced energy costs.

Although the literature on electroplasticity is extensive there are relatively few systematic explorations of the parameters affecting it, such as the current density, the shape and duration of the current pulses and their frequency, the width of the sample in relation to the skin depth, the influence of the magnetic permeability of the sample, the orientation of the crystal axes in single crystal samples to the wire axis, the level of interstitial and substitutional point defects, grain size and the degree of prior work hardening. There have also been only a few observations by transmission electron microscopy of dislocation configurations in samples before and after electropulsing treatments. The paucity of systematic experimentation is one of the principal reasons why there is still no widely accepted understanding of the mechanism(s) of electroplasticity.

Another unfortunate aspect of the literature of electroplasticity is that a good deal of the more fundamental experimental and theoretical research has been published in

---

[7]Zhao, S, Zhang, R, Chong, Y, Li, X, Abu-Odeh, A, Rothchild, E, Chrzan, D C, Asta, M, Morris, J W and Minor, A M, Nature Materials **20**, 468-472 (2021).

[8]Shibkov, A A, Denisov, A A, Zheltov, M A, Zolotov, A E, and Gasanov, M F, Materials Science and Engineering A **610**, 338-343,(2014).

[9]Guan, L, Tang, G and Chu, P K, J. Mat. Res. **25**,1215-1224 (2010).
Huu-Duc Nguyen-Tran, Hyun-Seok Oh, Sung-Tae Hong, Heung Nam Han, Jian Cao, Sung-Hoon Ahn and Doo-Man Chun, International Journal of Precision Engineering and Manufacturing Green Technology **2**, 365-376 (2015).

Russian in journals that are not taken by most institutions in the West either in print or electronically.

11.2.2 Mechanisms

An obvious question is whether the current pulses inject pulses of heat leading to thermal expansion and hence the observed load drops. The question has been carefully considered by Troitskii[10] and in the reviews already cited by Conrad and coworkers. They showed that in some experiments, especially those involving pulsing of the current, the load drops could not be explained by Joule heating.

In the absence of a current it has been established experimentally that conduction electrons in a metal exert a drag force on moving dislocations. One source of electron drag is the atomic restructuring in the core as a dislocation glides, which generates electron-hole pairs in the vicinity of the Fermi energy, thereby dissipating energy. Electron-hole pairs are also created by the motion of the scattering potential the dislocation represents for conduction electrons.

The relation between the drag force F per unit length on a dislocation and its speed v_d, at low speeds where inertial effects are negligible, is $F = Bv_d$. The drag coefficient B is often expressed as $B_e + B_p$ where B_e is the electron contribution and B_p is the phonon contribution. The electron contribution dominates only at cryogenic temperatures. Experiments by Hikata et al.[11] found, at temperatures below 40 K, the drag coefficient of dislocations in aluminium is approximately constant at 1.4×10^{-6} Pa · s. They also found that above 40 K the drag coefficient increases with temperature due to phonon drag.

Viewed in the frame of a dislocation moving through the metal, there is a current of conduction electrons moving past it in the opposite direction. In this reference frame the drag force is seen as a 'wind force' acting on a stationary dislocation in the direction of an electron current flowing past it. Antolovich and Conrad (2004) identified this wind force as a current-induced force that leads to electromigration of a stationary dislocation immersed in an electron current. Following earlier work these authors wrote the force per unit length f_c on a dislocation due to interactions with electrons as follows:

$$f_c = B_e(v_e - v_d), \tag{11.1}$$

where v_e is the electron drift speed. If the dislocation is stationary f_c is the current-induced wind force $B_e v_e$. If it is moving at the same speed as the drift speed of the

[10] Troitskii, O A, Strength of Materials **7** 804 (1975).
[11] Hikata, A, Johnson, R A and Elbaum, C, Phys. Rev. B **2**, 4856 (1970).

electrons there is no current-induced or electron drag force acting on it. If it is moving faster than the electron drift speed it experiences a drag force.

The electron drift speed is related to the current density, J, by $v_e = J/ne$. Therefore, using the above measured value of the electron drag coefficient for aluminium, $B_e = 1.4 \times 10^{-6}$ Pa·s, and a current density of $J = 10^6$ A·cm^{-2}, we obtain the current-induced force per unit length on a stationary dislocation $F_e = 4.8 \times 10^{-7}$ N·m^{-1}. Equating this force per unit length to an equivalent shear stress acting on the dislocation, multiplied by its Burgers vector, we find it is equivalent to a shear stress of 1.7 kPa. Even if B_e were two orders of magnitude larger the equivalent stress would be only a little more than atmospheric pressure. These stresses are much too small to explain the stress drops observed during electropulsing experiments.

The modern theory of current-induced forces on an isolated point defect in a large metal crystal may be summarised as follows. Assume the crystal is in a fully relaxed configuration before a constant current is passed through the crystal. Current-carrying electrons are scattered by the point defect and atoms close to it. The scattering redistributes electronic charge around the point defect. In the steady state, the charge redistribution is self-consistent with the total potential acting on the electrons. The current-induced force on the point defect is then equal to the charge of its nucleus multiplied by the self-consistent electric field at its nucleus.

A point defect presents a scattering potential V to the electron current, measured from the bottom of the conduction band. If $V > 0$ the point defect obstructs the current and electrons impinging on the point defect are reflected back creating an excess of electrons on the electron source side of the defect and a deficit on the electron drain side. The resulting dipole of charge establishes an electric field in the opposite direction to the current flow. If $V < 0$ the defect locally facilitates the current flow and the current-induced dipole electric field is in the same direction as the electron current flow. The current-induced electric field exerts a body force on the defect.

For an electron current to exert a direct force on a dislocation[12], such that the dislocation is induced to glide, the current has to generate a resolved shear stress on the slip plane in the direction of the Burgers vector on atoms in the dislocation core. This requires atoms in the core on one side of the slip plane to experience forces parallel to the Burgers vector **b**, and anti-parallel to **b** on the other side of the slip plane. Since an electrostatic dipole creates a net force on a point defect the current has to induce an electrostatic *quadrupole* along the dislocation line to create a shear stress. The quadrupole generates a *dipole* of forces, with the directions of the forces parallel and anti-parallel to **b**, as sketched in Fig 11.3 for an edge dislocation. At the

[12]By a 'direct force' on the dislocation we mean a force arising from the local scattering of current-carrying electrons caused by the dislocation.

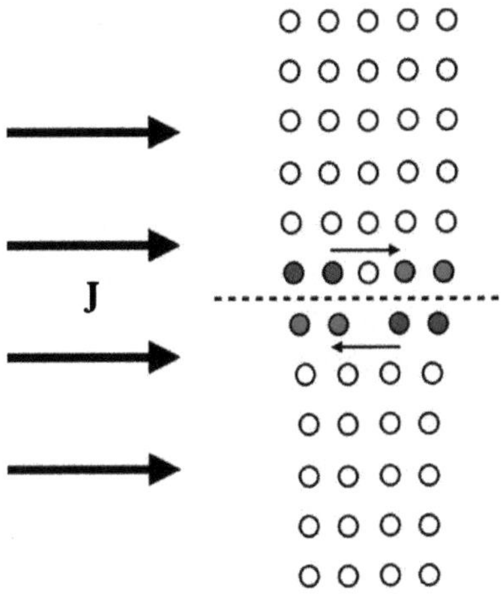

Figure 11.3: Sketch to illustrate a current-induced electrostatic quadrupole at an edge dislocation. The large arrows show an incident electron current of density **J** from the left. The red atoms signify an excess of electronic charge and the blue a deficit. The quadrupole generates forces, shown by small arrows, on atoms either side of the slip plane (broken line) in the dislocation core. These body forces create a shear stress on the dislocation core.

termination of the extra half plane atoms are forced closer together which facilitates current flow leading to a dipole as shown. Just below the termination of the extra half-plane atoms are forced apart, obstructing current flow which creates a dipole of the opposite sense. Thus, the current-induced charge distribution is a quadrupole[13]. Scattering of current-carrying electrons by the dislocation may also create a body force monopole. However, a monopolar body force does not induce a dislocation to glide (or climb). It merely strains the crystal lattice in the vicinity of the dislocation.

In Fig 11.2 the flow stress in Ti reduces by $\sim$ 100 MPa when the current density is of order $10^6\ \mathrm{A\cdot cm^{-2}}$. For atoms ~ 2 Å apart a shear stress of order $\sim$ 100 MPa requires current-induced forces on atoms in the core of the order of $2\times10^{-3}\ \mathrm{eV}\cdot\text{Å}^{-1}$. These forces are some three orders of magnitude larger than might be expected[14] to be produced by a current density of order $10^6\ \mathrm{A\cdot cm^{-2}}$. This order of magnitude estimate leads to the conclusion that it is very unlikely the load drops seen in Fig 11.2 are caused by current-induced shear stresses on dislocation cores.

[13] It may be thought that since a kink on a dislocation is effectively a point defect only a current-induced dipole is needed to make it drift. But the movement of a kink also requires a local shear stress and hence a quadrupole of current-induced charge.

[14] Hoekstra, J, Sutton, A P, Todorov, T N and Horsfield, A P, Phys. Rev. B **62**, 8568-8571 (2000).

Even if the above analysis is rejected there are further problems with the notion that current-enhanced dislocation motion can account for the observed load drops. Consider again the edge dislocation depicted in Fig 11.3. If we rotate the crystal by 180° about the dashed line the extra half plane is below the slip plane and the signs of the dipoles above and below the slip plane are reversed. This reverses the stress acting on the dislocation. The line direction of the dislocation is also reversed by this operation, but the Burgers vector of the dislocation remains the same. Therefore, the Peach-Koehler force, eq 6.10, acting on the dislocation does not change after this rotation by 180°.

The Burgers vector is reversed if the crystal is rotated by 180° about the dislocation line. The current-induced shear stress acting on the dislocation is reversed because the extra half plane is again beneath the slip plane. But the direction of the dislocation line does not change. The result is that the current-induced Peach-Koehler force on the dislocation is again unchanged.

It follows that for an edge dislocation the current-induced force per unit length is independent of the signs of its Burgers vector and line direction. If there are current-induced forces on screw dislocations they are likely to display the same independence of the signs of the Burgers vector and line direction. Therefore, electromigration of equal numbers of positive and negative dislocations does not increase plastic strain because they move in the same direction under the influence of the current. Similarly dislocation loops will be translated by the current but not expanded. We conclude that current-induced forces on dislocations, even when they are significant, do not lead to changes in the overall plastic strain.

Another mechanism proposed for the load drops is the pinch effect. There are two versions: static and dynamic. The static pinch effect applies when there is a steady current. The dynamic pinch effect applies when the current changes in time, as during a pulse. Both effects stem from Lorentz body forces acting on the conduction electrons of the metallic conductor. The analysis below follows Sutton and Todorov (2021)[15].

Consider a homogeneous, cylindrical, metallic wire carrying an electric current. The metal comprises conduction electrons and ions. Each ion consists of the atomic nucleus and core electrons which remain bound to the nucleus. Conduction electrons drift in response to the applied electric field according to Ohm's law. Their drift speed v_e is many orders of magnitude less than the speed of light c. The relaxation time of the electrons τ_e is the time taken by conduction electrons in a metal to screen a charge. For most metals it is of order $10^{-14} - 10^{-15}$ s. Since the Thomas-Fermi screening length in a metal is comparable to the interatomic distance and the charge relaxation time is so small, local charge neutrality is maintained at all times except at plasmon frequencies when electrons and ions are induced to move in opposite directions.

[15] Sutton, A P and Todorov, T N, Phys. Rev. Materials **5**, 113605 (2021).

Since local charge neutrality is maintained at all times, even when the current is pulsed with a pulse duration of microseconds, $\nabla \cdot \mathbf{J} = 0$ at all times. The total momentum of electrons entering and leaving each transverse slice of the wire per unit time does not change. This remains true even when there are defects present. Therefore, there is no net force on the conduction electrons. There are two forces on conduction electrons: the force exerted by the applied electric field and the electrostatic force exerted by the ions. The force exerted by conduction electrons on ions is equal and opposite to the force exerted by ions on conduction electrons. The force exerted by the applied electric field on ions is equal and opposite to the force it exerts on conduction electrons. Therefore the net force on ions is equal and opposite to the net force on conduction electrons. Since the latter is zero there is no net force on ions due to an electric field. This is the zero sum rule[16]. It follows that there is no net force on a metallic conductor due to an electric field even if it induces a current to flow[17].

However, the zero sum rule is modified by the presence of magnetic fields inside a metallic conductor. Electrons driven along a wire by an applied electric field create a magnetic field inside (and outside) the conductor. As a result there is a Lorentz force on the conduction electrons. The Lorentz force on the electrons is directed towards the axis of the wire. To maintain local charge neutrality it is balanced by a force $-\rho_0 \mathcal{E}(r)$ due to a radial electric field $\mathcal{E}(r)$, sometimes called a Hall field. Here $-\rho_0 = -ne$ is the conduction electron charge density, where n is the number of conduction electrons per unit volume and e is the electronic charge. The ionic charge density is ρ_0. Conduction electrons also experience a force exerted by ions, as in the previous paragraph. The net force on the electrons remains zero because in any transverse slice the momentum of electrons entering and leaving the slice per unit time is the same, since local charge neutrality is maintained.

Let $\mathbf{F}_e(\mathbf{r})$ be the net force per unit volume acting on conduction electrons at $\mathbf{r}$. Let $\mathbf{F}_e^i(\mathbf{r})$ be the electrostatic force per unit volume exerted by the ions in the metal on conduction electrons at $\mathbf{r}$. Let $\mathbf{F}_e^E(\mathbf{r})$ be the force per unit volume the applied and Hall electric fields exert on conduction electrons at $\mathbf{r}$. Let $\mathbf{F}_e^B(\mathbf{r})$ be the force per unit volume the magnetic field exerts on conduction electrons at $\mathbf{r}$. Then

$$\mathbf{F}_e(\mathbf{r}) = \mathbf{F}_e^i(\mathbf{r}) + \mathbf{F}_e^E(\mathbf{r}) + \mathbf{F}_e^B(\mathbf{r}) = \mathbf{0}. \tag{11.2}$$

The net force on the ions arises from the electron-ion electrostatic interaction and

[16]Hoekstra, J, Sutton, A P and Todorov, T N, J. Phys.: Condens. Matter **14**, L137 (2002).

[17]This statement assumes there are no other forces acting on the conductor, such as may be applied through its attachment to a substrate. In a free-standing metallic conductor electromigration of atoms during the passage of a current does not result in a displacement of its centre of mass. However, if the conductor is attached to a substrate there is a net force acting on the conductor and its centre of mass is displaced during electromigration.

the force due the applied and Hall electric fields. These forces are $-\mathbf{F}_e^i(\mathbf{r})$ and $-\mathbf{F}_e^E(\mathbf{r})$ respectively. We see from eq 11.2 that their sum is the Lorentz force on the electrons $\mathbf{F}_e^B(\mathbf{r})$. Therefore the net force on the ions due to the current is identical to the Lorentz force acting on the conduction electrons. This is a counter-intuitive result because the ions are not moving, and therefore the magnetic field does not exert a Lorentz force on them directly. Instead, the Lorentz force is transmitted to the ions by the conduction electrons through their electrostatic interaction. Forces acting on conduction electrons of a magnetic origin have a particular significance for electroplasticity because they create body forces on ions which are the sources of current-induced stresses throughout the conductor. These body forces act radially inwards in a homogeneous cylindrical conductor: they are the origin of the pinch effect.

If the wire is under tension at constant length the tensile stress in the wire is reduced by the pinch effect. That is because the radial compression would increase the wire length through the Poisson effect if its length were not constrained. But since the wire length is constrained the Poisson effect results in a compressive stress along the wire, which reduces the applied tensile stress.

Consider a long straight cylindrical wire of radius r_m. Introduce a cylindrical coordinate system (r, ϕ, z) with the z-axis along the axis of the wire. We assume the wire is homogeneous and elastically isotropic. The wire is carrying a constant current of density J_0 along the positive z-axis, which is uniformly distributed within the wire. The magnetic field inside the wire at a radius r is:

$$\mathbf{B}(r) = \frac{\mu_r \mu_0 J_0 r}{2} \widehat{\boldsymbol{\phi}} \tag{11.3}$$

where $\widehat{\boldsymbol{\phi}}$ is a unit vector along the positive azimuthal direction, μ_r is the relative magnetic permeability of the wire and μ_0 is the magnetic permeability of free space ($\mu_0 = 4\pi \times 10^{-7}$ NA^{-2}). The Lorentz force on the conduction electrons per unit volume, $\mathbf{f}_L = \mathbf{J}_0 \times \mathbf{B}$, is then

$$\mathbf{f}_L = -\frac{\mu_r \mu_0 j_0^2 r}{2} \hat{\mathbf{r}} \tag{11.4}$$

Inserting this body force into the equations of mechanical equilibrium in an elastic solid the following stress fields are obtained:

$$\sigma_{rr}(r) = -\frac{(3-2\nu)}{16(1-\nu)} \left(r_m^2 - r^2 \right) \mu_r \mu_0 J_0^2 \tag{11.5}$$

$$\sigma_{\phi\phi}(r) = -\frac{((3-2\nu) r_m^2 - (1+2\nu) r^2)}{16(1-\nu)} \mu_r \mu_0 J_0^2 \tag{11.6}$$

$$\sigma_{zz}(r) \;=\; \nu\left(\sigma_{rr}(r)+\sigma_{\phi\phi}(r)\right) = -\frac{\nu}{8(1-\nu)}\left((3-2\nu)r_m^2 - 2r^2\right)\mu_r\mu_0 J_0^2 \quad (11.7)$$

where as usual ν is Poisson's ratio. Since $\nu \le \frac{1}{2}$ we see that $\sigma_{zz}(r)$ is compressive at all radii[18]. The mean value of $\sigma_{zz}(r)$ is:

$$\langle\sigma_{zz}\rangle = -\nu\mu_r\mu_0 J_0^2 r_m^2/4. \quad (11.8)$$

For $\nu = \frac{1}{3}$, $\mu_r = 1$, $J_0 = 10^4$ A·mm^{-2} and $r_m = 0.5$ mm, we obtain $\langle\sigma_{zz}\rangle \approx 2.5$ MPa, and the maximum value of σ_{zz} is approximately 5 MPa. These stresses are too small by an order of magnitude to account for the stress drops observed in Fig 11.2.

The mechanisms considered so far have ignored the time dependence of the current pulses. Indeed they apply equally well when the current is constant in time. In the dynamic pinch effect of Sutton and Todorov (2021) the time dependence of the current pulses leads to additional currents through electromagnetic induction. Thus, the rate at which the current rises and falls during a pulse features as prominently in this mechanism as the maximum of the current density from the external current source.

The significance of induced currents in a cylindrical wire of radius r_m is determined by the parameter $R_m = r_m/L_D$. Here, L_D is the characteristic length in the vector diffusion equation governing the evolution in space and time of the induced current density in an infinitely long, homogeneous, cylindrical wire. For a Lorentzian pulse, in which the full width at half maximum is $2p$, the parameter L_D is $\sqrt{2p/(\mu_r\mu_0\sigma_c)}$. More generally, p is a characteristic time taken by the applied electric field to rise to its maximum value from zero, or fall to zero from its maximum value. The electrical conductivity is σ_c. The parameter L_D is the same as the skin depth of a conductor carrying an alternating current of angular frequency $1/p$. When R_m is less than 1 the effects of electromagnetic induction of the current density are small. The current density is then adequately described by the static pinch effect, and it is almost uniformly distributed in the wire. The condition $R_m > 1$ indicates when the effects of electromagnetic induction are significant. The value of R_m increases when the wire radius r_m increases, the pulse width $2p$ decreases, or the wire has a larger electrical conductivity σ_c or larger relative magnetic permeability μ_r.

Okazaki et al. (1978) measured stress drops as a function of current density in 99.9% iron wires of 0.35mm diameter and in 99.97% titanium wires of diameters 1.31 and 1.36 mm diameter. They found the stress drops in the iron wire were about an order of magnitude greater than in the titanium wires at current densities between 500 and 1000 A·mm^{-2}. This observation is consistent with the smaller value of L_D in iron, due to its large magnetic permeability, compared with titanium, raising the value of R_m in the iron wire despite the smaller value of r_m. Such a large difference between the

[18] or zero if $\nu = \frac{1}{2}$.

stress drops in iron and titanium cannot be explained by other mechanisms proposed for the stress drops.

For a given metal the static pinch effect predicts, at constant current density, that the average stress $\langle\sigma_{zz}\rangle$is proportional to r_m^2. The dynamic pinch effect predicts a much stronger dependence of the peak stress on r_m as R_m increases above one, while maintaining the current density and all other parameters constant. This can be tested experimentally.

The Lorentz force on conduction electrons is conveyed as a body force to the ions of the crystal lattice by the electron-ion interaction. This body force generates mechanical stresses through the equation of mechanical equilibrium. Pulsing the current creates body force pulses, which in turn generate elastic waves. Mechanical vibrations have been detected experimentally by applying piezoelectric sensors to the surface of solid cylindrical or tubular conductors subjected to current pulses[19].

11.2.3 Suggestions for further research

The mechanism for the stress drops caused by current pulses observed experimentally in Figs 11.1 and 11.2 remains uncertain. The mechanism based on electromagnetic induction is yet to be tested experimentally. This could involve a *systematic* study of the stress drops using current pulses with the same peak applied current but different rates of the current rise and fall. For example a pulse approximating an isosceles triangular wave form with the same peak current should show larger stress drops the shorter the pulse duration. A pulse approximating a square wave form should show spikes in the stress drops when the current rises *and* when it falls.

The model of Sutton and Todorov (2021) treated an infinitely long, homogeneous wire. It would be interesting to include microstructure such as second phase particles with different electrical conductivity and/or magnetic permeability, although this would probably require a numerical calculation. A possibly illuminating simple two-phase geometry would be an infinite cylindrical wire comprising two coaxial conductors, one occupying $0 \leq r \leq r_1$ and the other occupying $r_1 < r \leq r_m$. By attributing different electrical conductivities and/or magnetic permeabilities to the two conductors it would be interesting to calculate the stresses on the interface between them for different current pulse profiles.

11.3 The mobility of dislocations in pure single crystals

In this section we consider the factors limiting the mobility of a single dislocation in a pure single crystal where there are no other defects or microstructure to limit its

[19]Troitskii, O A, Skvortsov, O B and Stashenko, V I, Russian electrical engineering **89**,143-146 (2018).

motion[20]. We may think of this mobility as an intrinsic dislocation mobility, limited only by properties of the dislocation itself and the pure crystal it occupies.

3.1 Static vs. dynamic friction

In section 7.5 the Peierls-Nabarro model was used to estimate the stress required to enable a dislocation to glide in a crystal lattice. According to this model, once the Peierls stress has been overcome a dislocation continues to move indefinitely until it meets an obstacle, or it exits the crystal. As the dislocation glides over the landscape of the periodic Peierls potential the rises and falls of its potential energy are exactly compensated by the falls and rises of the kinetic energy. The Peierls-Nabarro model treats the motion of a dislocation in the absence of any energy dissipation.

Experimental reality could not be more different. Observations indicate that even in the purest single crystals dislocation glide is overdamped. As a dislocation glides in a metal the breaking and making of bonds between atoms in the core excites crystal lattice vibrations and electronic transitions to states above the Fermi energy. The Peierls stress may be thought of as the *static* friction stress that has to be overcome to set the dislocation in motion. But once it is moving the physics of the *dynamic* friction stress is the coupling to the elementary excitations of the crystal. The dissipative nature of dislocation glide is consistent with the observation in section 11.4.2 that more than 90% of the work done on a metal during plastic deformation is converted into heat. It also gives rise to energy absorption in internal friction experiments.

In the elastodynamic theory of moving dislocations the self-energy of a dislocation comprises an elastic potential energy and a kinetic energy associated with the motion of volume elements throughout the continuum. The self-energy diverges at the shear wave speed c_t suggesting that the shear wave speed is an upper limit on the speed of a dislocation. However, the dependence of the self-energy on the speed of the dislocation is a separate matter from dissipation of its energy due to friction. In elastodynamics an isolated dislocation in an infinite continuum moving at constant velocity experiences no energy dissipation: it continues to move at the same velocity indefinitely. That is because in the inertial frame of the dislocation core the system is time-invariant. No elastic waves are radiated from a dislocation moving at a constant velocity in elastodynamics, and therefore the self-energy of the dislocation is constant. Elastic waves are radiated from a dislocation in elastodynamics only when it is accelerated or created or annihilated in the continuum.

In an attempt to include the physics that is missing in elasticity theory of a dislocation moving at a constant speed v it is usual to introduce an *ad hoc* viscous drag force f_{drag}

[20]For a recent review of many of the topics covered in this section and much more see Gurrutxaga-Lerma, B, Verschueren, J, Sutton, A P, and Dini, D International Materials Reviews **66**, 215-255 (2021).

per unit length through an empirical relation of the form:

$$f_{drag} = B \cdot v, \tag{11.9}$$

where B is the drag coefficient, with units $\mathrm{N \cdot m^{-2} \cdot s = Pa \cdot s}$. The drag coefficient is thus an inverse mobility. When $v \ll c_t$ a dislocation moves at a constant speed of $\tau b/B$ where τb is the usual Peach-Koehler glide force on the dislocation. At such low dislocation speeds it is reasonable to assume the drag coefficient is independent of v and dependent only on whether the dislocation is edge or screw or mixed, and dependent on the vibrational and electronic properties of the crystal. In the following we estimate contributions to the drag coefficient from physical principles.

11.3.2 Drag forces due to interactions with lattice vibrations

The scattering of phonons by stationary dislocations reduces the phonon contribution to the thermal conductivity of the crystal. This is particularly significant for insulators, where phonons are the dominant mechanism of thermal transport, and for the efficiency of thermoelectric materials. There are two mechanisms. The distortion of the crystal lattice created by the dislocation alters bond stiffnesses, because interatomic forces are anharmonic, which leads to scattering of phonons. This is the *anharmonic mechanism.* In effect this mechanism treats the time-average atomic structure of the dislocation as a region of inhomogeneity in the bond stiffnesses.

The second mechanism is called *fluttering*[21]. A dislocation segment pinned at two ends may be induced to vibrate in the slip plane, like a string under tension, by an oscillatory Peach-Koehler force through its interaction with transverse phonons in the slip plane. The vibrating dislocation segment becomes a source of new phonons that are radiated into the crystal lattice. The net result is that those phonons that excite the vibration of the dislocation are scattered and the thermal conductivity is reduced. The fluttering mechanism has been treated both classically and quantum mechanically by Lund[22] and coworkers[23].

These two mechanisms of the reduction of the thermal conductivity of a crystal by stationary dislocations also reduce the mobility of a moving dislocation through a crystal in thermal equilibrium. However, they cannot be the only mechanisms by which phonons exert a drag force on a moving dislocation because their effect diminishes to zero as the temperature approaches absolute zero. As discussed by Swinburne and

[21]Alshits, V I in *Elastic strain fields and dislocation mobility*, eds. V L Indenbom and J Lothe, Chapter 11, p 625-697, North-Holland: Amsterdam (1992). ISBN 0444887733.
Vladimir Iosifovich Alshits 1941-. Russian materials physicist.

[22]Fernando Lund, Chilean materials physicist.

[23]see Lund, F and Scheihing H., B, Phys. Rev. B **99** 214102 (2019) and references therein.

Dudarev[24] there is a significant contribution to phonon drag on a moving dislocation that is temperature independent. These authors discuss the origin of the temperature independent contribution for defects such as nanoscale dislocation loops. In the next section we describe another temperature independent contribution to phonon drag.

.3.3 Radiation of atomic vibrations from a moving dislocation

In section 6.9 we introduced the concept of a force dipole to generate the displacement by the Burgers vector of a dislocation. As a dislocation glides, atoms on either side of its slip plane experience equal and opposite impulses to raise them over the Peierls barrier and bring about the relative displacement across the slip plane by the Burgers vector. Once the Peierls barrier has been surmounted the dislocation falls into the next Peierls valley where the energy of surmounting the barrier is radiated into the surrounding crystal by phonons. This is the origin of acoustic emission of dislocations when they glide. It also contributes directly to the heating of the crystal by dislocation motion. The drag force is independent of temperature because phonons are created by the movement of the dislocation, not by scattering of pre-existing crystal lattice vibrations.

Recent research[25] has suggested that the crystal lattice vibrations created by a dislocation gliding at a sufficiently high speed can spawn the creation of further dislocations. This is a result of a resonance between the vibrations created by the dislocation and the natural vibrations of the crystal lattice. Although these resonances were identified in earlier papers[26], it was not recognised that they may lead to mechanical instabilities resulting in the formation of new dislocations. A resonance arises when the group and phase velocities of the lattice waves travelling in the same direction as the dislocation equal the dislocation velocity. When this condition is satisfied the energy dissipated in the dislocation core cannot escape and its accumulation eventually leads to the creation of two further dislocations with equal and opposite Burgers vectors, thus conserving the total Burgers vector. This mechanism of generating dislocations kinematically was confirmed by molecular dynamics simulations in the NVE-ensemble of screw dislocations in tungsten by Verscheuren et al. (2018). A resonance was found at $0.24c_t$, where c_t is the transverse wave speed, accompanied by an avalanche of dislocation loops. The kinematic generation mechanism takes a small but finite time to act. A dislocation accelerated from rest will behave in one of two ways. If there is

[24]Swinburne, T D and Dudarev, S L, Phys. Rev. B **92**, 134302, (2015).

[25]Verschueren, J, Gurrutxaga-Lerma, B, Balint, D S, Sutton, A P and Dini D, Phys. Rev. Lett. **121**, 145502 (2018).

[26]Atkinson, W and Cabrera, N Phys. Rev. **138**, A763-A766 (1965); Celli, V and Flytzanis, N, J. Appl. Phys. **41**, 4443-4447 (1970); Ishioka, S, J. Phys. Soc. Jpn. **30**, 323-327 (1971); Caro, J A and Glass N, J. Phys. (Paris), Lett. bf 45, 1337-1345 (1984).

sufficient time for the generation mechanism to act when resonances occur, it will act repeatedly and avalanches of dislocations will be created, limiting the speeds of the dislocations to those at which the resonances occur[27]. Alternatively, if the dislocations are accelerated so rapidly there is insufficient time for the generation mechanism to act their speeds will not be limited by the resonance speeds. If dislocations are injected at speeds above those of the resonances they are avoided altogether.

The kinematic generation of dislocations at resonant speeds may be relevant to adiabatic shear. Failure of metallic components subjected to very high rate deformation often occurs through the formation of narrow shear bands, $5 - 500\ \mu$m wide, in which the shear strain is extremely high. These shear bands are *adiabatic* because they are formed so rapidly there is insufficient time for most of the heat generated within them to escape. It would be interesting and useful to test this idea by simulating an adiabatic shear failure allowing for the possibility of kinematic generation of dislocations at resonant dislocation speeds.

11.3.4 Drag forces due to electronic excitations

When a dislocation glides bonds are broken and remade. These events excite electrons which provides another source of drag on a dislocation. The modern approach to electron-phonon interactions is based on first principles DFT methods and has predictive capabilities[28]. As far as I know these methods have not been applied to a gliding dislocation.

A direct approach would be to use time-dependent DFT to simulate the excitation of electrons by a gliding dislocation with Ehrenfest dynamics. In Ehrenfest dynamics the atomic nuclei satisfy Newtonian equations of motion with forces coupled explicitly to the evolving electronic structure, while the electrons are treated fully quantum mechanically and self-consistently by solving the time-dependent Schrödinger equation[29]. I am not aware this has been done either.

Existing approaches in the literature have been superseded by the potential capabilities of these modern methods. For example, consider the analysis by Holstein[30], which appeared in the appendix to a paper[31] by Tittmann and Bömmel. His analysis is based on the concept of a deformation potential to provide the link between the

[27]These limiting speeds may be much less than the shear wave speed and they have nothing to do with the divergence of the self-energy of the dislocation at the shear wave speed in elastodynamics.

[28]see Giustino, F, Rev. Mod. Phys. **89** 015003, 63 pages (2017); Erratum Rev. Mod. Phys. **91**, 019901 (2019).

[29]see, for example, Mason, D R, Le Page, J, Race, C P, Foulkes, W M C, Finnis, M W and Sutton, A P, J. Phys. Condens. Matter **19**, 436209 (2007).

[30]Theodore David Holstein 1915-1985, US theoretical physicist.

[31]Tittmann, B R and Bömmel, H E, Phys. Rev. **151**, 178-189 (1966).

moving strain field of a dislocation and a potential that excites electrons in first order, time-dependent perturbation theory. For an approach based on deformation potentials to be valid the strain field has to be slowly varying relative to the atomic spacing, which breaks down in the dislocation core. This is particularly serious in metals because the rapidly changing potentials in the dislocation core are screened in reality, which cannot be treated adequately by deformation potentials. Modern methods do not have these deficiencies. But perhaps the most significant concern about the use of deformation potentials is that in a cubic crystal they have no shear components to first order in the strain. Since shear is the key result of the glide of a dislocation this is indeed a serious weakness. For example, the application of Holstein's analysis to a screw dislocation in isotropic elasticity would predict no electronic excitations.

To calculate the electronic excitations in simulations of moving dislocations using modern DFT methods would be a major undertaking. It would provide, for the first time, reliable information about the contributions of electronic excitations to the drag on dislocations as a function of their speed. It would also be useful to investigate electronic excitations in a metal at moving kinks on a dislocation, which might be more tractable.

.4 Work hardening

.1 The nature of the problem

Work hardening, or strain hardening as it is sometimes called, is the increase in the flow stress of a metal as a result of plastic deformation. As Cottrell noted[32] in 1953 it is a spectacular effect enabling the yield strengths of pure copper and aluminium to be raised a hundred fold. The goal of a theory of work hardening is to explain and predict the stress-strain relation of a material as a function of temperature, strain rate, grain size, alloy composition, and microstructural features such as second phase particles and the distribution of grain orientations known as texture. Since dislocations are the agents of plastic deformation it is their collective behaviour that leads to the observed stress-strain relation. Their collective behaviour involves short and long-range interactions with each other, and with other microstructural features. In 1953 Cottrell pointed out that work hardening was the first problem to be tackled by dislocation theory, and may well prove to be the last to be solved. Almost fifty years later he expressed the magnitude of the enduring challenge of developing a successful theory of work hardening as follows:

It is sometimes said that the turbulent flow of liquids is the most difficult remaining problem in classical physics. Not so. Work hardening is worse. ... Whereas fluid

[32]Cottrell, A H, *Dislocations and plastic flow in crystals*, Chapter 10, Clarendon Press: Oxford (1953).

dynamics can be treated by continuum methods, so that everything can be reduced to the purely mathematical problem of solving standard differential equations, there is no similar escape in work hardening, for the discrete structures of dislocations render the theory intrinsically atomistic, even though in their lengthwise dimensions dislocations are macroscopic objects, governed mainly by a classical physics which is unusual in not being reducible to continuum theory. ... Another unusual and extremely complicating feature is of course that dislocations are lines, not the familiar point particles of mainstream physics - and flexible lines at that - so that the standard methods of particle theory are inapplicable. ... Furthermore, neither of the two main strategies of theoretical many-body physics - the statistical mechanical approach, and the reduction of the many-body problem to that of the behaviour of a single element of the assembly - is available to work hardening. The first fails because the behaviour of the whole system is governed by that of weakest links, not the average, and is thermodynamically irreversible. The second fails because dislocations are flexible lines, interlinked and entangled, so that the entire system behaves more like a single object of extreme structural complexity and deformability.... Of course, the properties of single dislocations have long been well-established: glide, climb, cross slip and sessile kinematic modes; reactions, combinations and dissociations; Frank-Read sources, etc. These properties provide the alphabet in which the story of work hardening must be written. But only the alphabet, no more than that.[33]

Cottrell's observations in this passage are particularly salutary for anyone teaching a course on dislocations and plasticity. In section 11.4.3 a different way to think about work-hardening will be introduced. It is based on recent ideas related to the theory of self-organised criticality. But first we review briefly some experimental work on work-hardening.

11.4.2 Work hardening of fcc single crystals

Early experiments on tensile stress-strain relations of single crystals of ductile fcc metals such as copper and aluminium revealed at least three stages of work hardening. If the crystal is oriented so that only one slip system is activated initially the crystal enters stage I where the slope $d\sigma/de$ is very small at about $10^{-4}\mu$. This small hardening, often called *easy glide*, arises from having to overcome attractive multipolar interactions between trains of dislocations on parallel slip planes. The trains are moving in opposite directions because the Burgers vectors in one train have the opposite sign to those in

[33]Reprinted from *Dislocations in Solids* **11**, eds. F R N Nabarro and M S Duesbery, *Commentary: A brief view of work hardening*, by A H Cottrell, pages vii - xvii, Copyright 2002, with permission from Elsevier. Elsevier: Amsterdam, ISBN 0-444-50966-6.

the other train. As plastic strain increases the crystal axes rotate and the resolved shear stress due to the applied load on a second slip system increases. When the resolved shear stress on the second slip system becomes comparable to that on the first, the two slip systems are equally active, and the stress-strain curve has then fully entered stage II. However, secondary slip is activated by local internal stresses before the resolved stress due to the applied load is sufficient to activate it. Consequently the transition from stage I to stage II occurs gradually over a range of plastic strains. Stage II is the dominant part of the stress-strain curve, where the slope is remarkably constant at about $\mu/300$ and independent of temperature. Slip takes place on inclined slip planes and dislocations on one slip plane have to cut through those on inclined planes forming steps or *jogs* on the dislocation lines. If the jogs are sessile additional stress leads to the formation of dislocation dipoles and point defects. Dislocations in inclined slip planes may also combine and form sessile dislocations which may block the passage of further dislocations on either slip plane. These interactions between dislocations on inclined slip planes are collectively called *forest* interactions and they are responsible for the large increase in the hardening rate in stage II compared to stage I. The Orowan flow stress predicts that in Stage II the increase in the flow stress due to these forest interactions is $\alpha\mu b\sqrt{\rho_f}$ where ρ_f is the density of forest interactions[34] and $\alpha \approx 0.3$. Stage II ends and stage III begins when the slope starts to decrease and ductile fracture eventually occurs at a stress which depends on temperature. If the crystal is oriented so that two or more slip systems are activated by the external load as soon as plasticity begins there is no stage I and stage II is entered straight away. For the same reason, in a fcc polycrystal stage II is entered as soon as plasticity begins. Conversely, in some hexagonal crystals where slip occurs only on one slip plane, stage II is not entered before fracture occurs.

It is easy to show that during stage II most of the work of plastic deformation is not stored in the deformed body as potential energy but is dissipated as heat[35]. During stage II the stress-strain relation is $\sigma \approx (\mu/300)e$. The work done per unit volume during an increment $\mathrm{d}e$ of strain is $\mathrm{d}W = \sigma \mathrm{d}e \approx 300(\sigma/\mu)\mathrm{d}\sigma$. The increment in the density of stored potential energy is $\mathrm{d}E \approx \frac{1}{2}\mu b^2 \mathrm{d}\rho$, where ρ is the average dislocation density. Assuming $\sigma = 0.3\mu b\sqrt{\rho_f}$, where ρ_f is the density of forest interactions, and $\rho_f = \beta\rho$ where[36] $\beta \approx \frac{1}{2}$, we obtain $\mathrm{d}E/\mathrm{d}W \approx 1/14$. Plastic deformation is a dissipative process.

[34]It is a sobering fact that the dependence of the flow stress on the square root of dislocation density first appeared in G I Taylor's paper of 1934.

[35]Nabarro, F R N, Basinski, Z S and Holt, D B, Adv. Phys. **13**, 193-323(1964).
Zbigniew Stanislaw Basinski FRS 1928-1999. Canadian and British materials physicist, born in Poland.

[36]Basinski, Z S and Basinski,, S J, Phil. Mag. **9**, 51-80 (1964).
Sylvia J Basinski 1929-2020, Canadian and British materials physicist.

11.4.3 Work hardening and self-organised criticality

The density of dislocations in a heavily deformed ductile metal, such as copper, may reach 10^{12} cm^{-2}. It is remarkable that this is equivalent to ten million kilometres of dislocation line in each cubic centimetre, or roughly one in every thousand atoms is in a dislocation core. However, one of the most significant discoveries by transmission electron microscopy of deformed samples is that dislocations are not distributed uniformly during plastic deformation. A cellular structure is formed, with the cell walls made of dense bundles or *braids* of dislocations separated by regions where the dislocation density is much smaller. This is an example of self-organisation of dislocations. During work-hardening dislocations tend to organise themselves into low energy configurations in which their long-range elastic fields are mutually screened[37]. For example dislocations of opposite sign tend to form dipoles, while edge dislocations of the same sign tend to form small angle grain boundaries, often called subgrain boundaries. Screw dislocations of opposite sign will also tend to form dipoles and they may also annihilate each other by cross slip. If a group of dislocations within a region has a net Burgers vector its elastic field will stimulate the migration of neighbouring dislocations into the region to reduce the net Burgers vector to zero, possibly through the activation of sources. Observations have also shown that most dislocations are at rest most of the time in metals subjected to heavy deformation at moderate strain rates, indicating that dislocations are in temporary states of mechanical equilibrium. When dislocation motion occurs it is in bursts or avalanches involving groups of dislocations forming slip bands. This has led Brown[38] to describe plasticity as *constantly intermittent*[39], and it contrasts with the assumption made in continuum plasticity of continuous deformation. These characteristic features of plasticity are also common to systems displaying self-organised criticality.

Self-organised criticality (SOC) combines the two concepts of self-organisation and criticality[40]. Self-organisation is the ability of a non-equilibrium system to develop structures and patterns without any external interference. Criticality is precisely defined in statistical mechanics of second order phase transitions where, at the transition temperature, a localised disturbance can propagate across the entire system even though interactions between constituent particles have much shorter range. In the words of Bak[41], who developed the concept of SOC:

---

[37]Kuhlmann-Wilsdorf, D, Mat. Sci. Eng. **86**, 53-66 (1987). Doris Kuhlmann-Wilsdorf 1922-2010, US metallurgist born in Germany.

[38]Lawrence Michael Brown FRS, 1936 -. British materials physicist born in Canada.

[39]Brown, L M, Materials Science and Technology **28**, 1209-1232 (2012).

[40]Jensen, H J, *Self-organized criticality*, Cambridge University Press: Cambridge (1998). ISBN: 0521483719.

[41]Per Bak 1948-2002. Danish theoretical physicist, who worked in Denmark, the US and the UK.

... complex behaviour in nature reflects the tendency of large systems with many components to evolve into a poised, critical state, way out of balance, where minor disturbances may lead to events, called avalanches, of all sizes. Most of the changes take place through catastrophic events rather than by following a smooth gradual path. The evolution to this very delicate state occurs without design from any outside agent. The state is established solely because of the dynamical interactions among individual elements of the system: the critical state is self-organized.[42]

In the context of work-hardening SOC describes self-organisation of the dislocation structure resulting in every part of it having an equal probability over time of being a site where further slip is initiated. Brown describes[43] this self-organised, critical state as one of maximum *suppleness*. Before the material reaches this state plastic deformation is initiated at particular susceptible sites. As deformation proceeds all regions are invaded by slip bands. Eventually the material enters a critical state where all regions are on the verge of an avalanche of plastic deformation. In the language of dynamical systems the self-organised critical state is an *attractor*[44] towards which the dislocation structure of the metal evolves. When it is slightly disturbed by the creation of a new slip band it returns to the self-organised critical state. Brown (2016) argued that ductile crystals provide model systems to study SOC free from uncontrollable variables. One of features of SOC is that it leads to power laws between various observable quantities. Brown (2016) discusses a number of such power laws in work hardening.

11.4.4 Slip lines and slip bands

Slip lines are traces of a slip plane where dislocation glide has introduced a relative displacement on either side of the slip plane. The slip *line* is created where the slip plane exits at a free surface and it is made visible by the formation of a step on the surface. Slip *bands* are long blade-like regions with an aspect ratio of about 50:1 that have been sheared by the collective motion of dislocation loops on parallel slip planes. Thus slip bands comprise groups of slip lines. Brown (2012) has compiled the following list of experimental observations, all referenced in his paper, about slip lines and slip bands during work hardening of pure metals not subjected to cyclic loading:

[42]Reproduced with permission by Oxford University Press from Bak, P, *How nature works*, p.1-2, Oxford University Press: Oxford (1997). ISBN 019850164. Copyright ©1997.

[43]Brown, L M, Philosophical Magazine **96**, 2696-2713 (2016).

[44]Prigogine, I, *From being to becoming*, pages 7-8, W.H. Freeman and Co.: San Francisco. ISBN 0-7167-1108-7. Ilya Romanovich Prigogine 1917-2003. Russian born, Nobel prize winning physical chemist, who worked in Belgium and the US.

- The average plastic displacement associated with a slip line, as determined by the height of steps where they emerge at a surface, varies by at most a factor of three, typically between 3 and 10 nm, as the stress level changes by a factor of fifty. This behaviour is found in Cu, Zn and NaCl. Although there is much greater variability in the plastic displacement associated with individual slip lines the *average* plastic displacement is observed to be roughly constant and independent of the material, throughout the deformation.
- Over the same range of stress and strain the average length of slip lines decreases inversely with stress by a factor of sixty, from about 20 μm at the beginning of stage II.
- Acoustic emissions from plastically deformed copper single crystals indicate that the bursts of slip last a few microseconds at the onset of stage II. The intensity of an acoustic emission is proportional to the slipped area multiplied by the plastic displacement. Since the average plastic displacement is approximately constant, and the slipped area is inversely proportional to the stress squared, the intensity of the acoustic emission is expected to vary with the inverse square of the stress, which is observed in acoustic emission experiments.
- To produce the observed plastic displacements and the observed intensity of acoustic emissions some tens of dislocations move during a slip burst, which implies they move in a cooperative manner. The speed of the dislocations during a burst is of order metres per second.
- Once a slip line has emerged at a surface the height of the surface step is stable and does not increase with time. The steps formed when slip bands intersect inside the crystal also do not change after they have been formed. These observations indicate that once a slip band has been formed it becomes inactive and does not continue to grow. Further slip occurs through the formation of new slip bands. The crystal becomes a palimpsest[45] where each region of the crystal is sheared repeatedly through participation in slip bands. Brown (2012) estimated that after 10% plastic strain each atom has participated in a slip band about 100 times.
- There are observations of slip bands inclined by $1-5°$ to the slip direction. A slip band is nearly planar. By analogy with aerodynamics the *pitch* is the angle between the slip band and the slip direction. The angle of *roll* is between the perpendicular to the slip direction and the slip band. Thus, slip bands have two more degrees of freedom than dislocation pileups on single slip planes.

[45]A palimpsest is a document that has been cleaned and written on again, such as a slate written on with chalk and wiped clean to be written on again.

11.4.5 Slip bands as the agents of plastic deformation

The observations listed in the previous section suggest a coarse-grained model in which slip bands are the agents of plastic deformation, an idea apparently first suggested[46] by Jackson[47] in 1985 who treated slip bands as ellipsoidal regions undergoing simple shear. This was one of the principal motivations for deriving the elastic field inside and outside an ellipsoidal slip band in problem 8.7. Brown modelled slip bands as long blade-like ellipsoids. During the few microseconds it takes for a slip band to form it lengthens and it can change the orientation of its principal axes so long as it remains ellipsoidal. *An as yet unanswered question is whether the intermittency of slip and self-organised criticality observed experimentally can be captured by a model in which slip bands, treated as ellipsoids, interact with each other through their elastic fields.*

Consider first a shear band represented by an ellipsoid with no pitch or roll undergoing a simple shear. In problem 8.7 the stress fields inside and outside a shear band were discussed. When there is a small angle of pitch, θ, the strain tensor e_{ij}^{T} inside the ellipsoid develops some normal components, the magnitudes of which are proportional to θ. They lead to normal stresses inside the shear band which are proportional to the shear strain and θ. Resolved components of these normal stresses can activate slip in directions and on planes inclined to the primary slip plane within the shear band. When loops of the primary and secondary slip systems meet on the periphery of the ellipsoid they interact and form obstacles - these are the *forest interactions*. Thus the loops of the primary slip system block further slip, or contraction, of loops on the secondary slip systems and *vice versa.* The ellipsoid is thus stabilised by these forest interactions on its periphery. The walls of the slip band consist of arrays of primary and secondary dislocations. They are almost parallel to the primary slip plane because of the long thin shape of the slip band. The interior of the slip band is virtually free of dislocations. In cross section this produces the cellular structure observed in transmission electron microscopy. Brown[48] has estimated the linear hardening rate on the primary slip system in this ellipsoidal slip band model. On the primary slip plane it involves the spacing of obstacles arising from forest interactions with secondary dislocations. These obstacles act as pinning points between which the primary dislocations have to bow out to escape. The stress required for them to bow out is inversely proportional to the spacing of the secondary dislocations, which is directly proportional to the simple shear ε on the primary slip system, and hence linear hardening results.

Because the thickness (a_3) of the ellipsoid is so much less than its width (a_2) and

[46] Jackson, P J, Acta Metallurgica **33**, 449-454 (1985).

[47] Paul J Jackson, South African materials physicist

[48] Brown, L M, Metallurgical Transactions A, **22**, 1693-1708 (1991).

its length (a_1) the extent of secondary slip is much less than primary slip. This is why the rotation of the crystal is due almost entirely to slip on the primary system. As the crystal rotates the resolved shear stress on one or more secondary slip systems increases, while the resolved shear stress on the primary slip system decreases. Eventually the sum of the applied stress and the normal stresses resolved onto one or more secondary slip systems enables secondary dislocations to bow out between the obstacles provided by the primary dislocations and create shear bands on secondary systems.

We have here the outline of a coarse-grained model for work hardening. Elastic interactions between slip bands can be modelled using the explicit expressions for the stress and strain fields inside and outside transformed ellipsoids developed in Chapter 8. A protocol would need to be developed for a simulation in which slip bands are created in response to local stresses and overwrite each other. It would be interesting, not only as a model of work-hardening but as a possible paradigm for SOC, to see whether the model leads to a self-organised critical state and whether it captures the 'constantly intermittent nature of slip'.

11.5 Hydrogen assisted cracking in metals

11.5.1 Introduction

Hydrogen assisted cracking is one of the most common forms of embrittlement of metals and alloys. In this section we review briefly the mechanisms that have been proposed in the light of experimental observations. It is fair to say that after nearly a century and a half of research there is no single widely accepted mechanism for hydrogen assisted cracking, except when brittle hydride particles are formed. But hydrogen assisted cracking remains a major concern for many industries including energy production, automobiles, aerospace, shipping, construction, defence, oil extraction and the burgeoning hydrogen economy.

In 1875 Johnson[49] presented a remarkable paper[50] to the Royal Society in which he described a series of elegant experiments that established a good deal of what is known today about the embrittlement of iron and steel by hydrogen. He observed that when iron is immersed for a few minutes in acids that result in the generation of hydrogen the toughness and ductility of the metal are reduced significantly. When a fractured surface of the embrittled iron was moistened it frothed as bubbles of hydrogen gas were released from the metal. The embrittlement was reversible because when the iron was left for sufficiently long for the hydrogen to escape, its ductility returned. The time taken for the ductility to return decreased if the iron was warmed. These

[49]William Henry Johnson 1850-1914, British metallurgist

[50]Johnson, W H, Proc. R. Soc. Lond. **23**, 168-179 (1875).

observations showed it was hydrogen that diffused in the metal and was responsible for the embrittlement. If there were any remaining hydrogen trapped in the metal it was harmless, although, as discussed below, this is not true if the hydrogen formed brittle hydride phases. He observed the toughness and ductility of unstressed iron exposed to hydrogen gas were not reduced, proving that hydrogen in its molecular form was not responsible for the embrittlement[51]. Johnson also proved that the embrittlement was due to the ingress of atomic hydrogen, and not to the acid, by connecting iron wires to the electrodes of an electrochemical cell in which the electrolyte was Manchester town water containing no acid. He compared the embrittlement of iron by hydrogen with that of zinc by liquid mercury and suggested that the hydrogen impedes the "motion of one molecule over another" (shear) as the mechanism of the embrittlement. As discussed in section 11.5.2, the analogy with liquid metal embrittlement has been developed significantly over the past fifty years. Johnson's conjecture about hydrogen influencing shear processes has also played a central role in discussions of mechanisms of hydrogen embrittlement. Johnson's paper was prescient and seminal.

11.5.2 Possible mechanisms

Hydrogen enhanced decohesion (HEDE) was proposed by Pfeil[52] in his 1926 paper[53] regarding tensile tests during "pickling" of ferritic steels. He observed intergranular and transgranular failures, and he suggested they were caused by a reduction in strength of metal-metal bonds at grain boundaries and cleavage planes respectively. He did not offer an explanation for the reduction, and even today there does not seem to be any direct experimental evidence for weakening of metallic bonds caused by hydrogen[54]. Nevertheless hydrogen induced decohesion remains a plausible mechanism, particularly for some intergranular fractures as a result of segregation of hydrogen to grain boundaries.

In 1969 Westlake[55] noticed that whenever hydrides form they are almost always

[51]We now know that hydrogen molecules dissociate on clean surfaces of some metals, especially when they are stressed.

[52]Leonard Bessemer Pfeil FRS 1898-1969, British metallurgist

[53]Pfeil, L B, Proc. R. Soc. A **112**, 182-195 (1926).

[54]Some theories of hydrogen enhanced decohesion are based on the assumption that hydrogen attracts electrons from neighbouring metallic bonds and that those bonds are always weakened as a result. But if the valence band of the metal is more than half full then taking electrons from metallic bonds will *strengthen* them because electrons are removed from *anti*-bonding states. For example, crossing the 4d and 5d transition metal series from left to right the valence d-band is filled progressively (the 3d-series is complicated by magnetism). Taking electrons from bonds in the first half of the 4d and 5d series weakens them, whereas taking electrons from bonds in the second half of the series strengthens them.

[55]Donald G Westlake, US metallurgist

brittle[56]. A mechanism arises in which hydrogen is attracted elastically to a loaded crack tip, leading to supersaturation and precipitation of a hydride particle. The hydride particle grows and it becomes brittle. It provides an easy path for crack growth. The crack pauses at the interface between the hydride and matrix until more hydrogen accumulates ahead of the crack and the process repeats. There are many observations of such a mechanism, and there does not appear to be any controversy about it. In the remainder of this section we consider the influence of hydrogen when hydrides are not formed.

Three years later Beachem[57] published a paper[58] that has proved seminal. Before the widespread use of high resolution scanning electron microscopes to image fracture surfaces the use of carbon replicas to image fracture surfaces in the transmission electron microscope was the standard technique. Beachem was a pioneer of the technique, and a brilliant interpreter of the images he obtained. At higher stress intensity factors he observed dimples on the fracture surfaces, indicating that the crack grew through the coalescence of microvoids. As the stress intensity factor was reduced the extent of the plastic deformation decreased gradually and the fracture mode changed to cleavage, and eventually, at the lowest stress intensity, to intergranular fracture. In a real sense, Beachem turned the thinking up to that point about hydrogen embrittlement on its head[59]. He wrote:

The flat, brittle fractures produced at surprisingly low stresses in the laboratory or in service are therefore thought to be caused by severe, localized crack-tip deformation even when the cracks are propagating along prior austenite grain boundaries, and are not believed to be the result of a cessation, restriction, or exhaustion of ductility. Therefore, the term hydrogen-assisted cracking is probably more descriptive than hydrogen embrittlement cracking. [60]

In the late 1970s, Lynch[61] proposed a mechanism of adsorption-induced dislocation emission (AIDE)[62]. He suggested that *hydrogen at crack tips facilitates the emission of*

---

[56]Westlake, D G, ASM Transactions Quarterly **62**, 1000-1006 (1969).

[57]Cedric Duane Beachem 1932-2022. US metallurgist.

[58]Beachem, C D, Metallurgical Transactions **3**, 441-455 (1972).

[59]As discussed in section 11.6, Beachem extended these ideas in 1976, when he argued that 'brittle' fracture of metals in inert environments is also accomplished by localised plasticity, with support from experiments then and since.

[60]Reprinted by permission from Springer Nature Customer Service Centre GmbH. Published by Springer Nature in Metallurgical Transactions **3**, 441-455, *A New Model for Hydrogen-Assisted Cracking (Hydrogen "Embrittlement")*, by C D Beachem. Copyright 1972. Journal homepage: `https://www.tms.org/pubs/journals/mt/`.

[61]Stanley Peter Lynch 1945–. British and Australian metallurgist.

[62]Lynch, S P, Scr. Metall. **13**, 1051-1056 (1979). The mechanism was first proposed by Lynch, S P and Ryan, N E, in Proceedings of the 2nd International Congress on Hydrogen in Metals, Paper 3D12,

dislocations, enabling the crack to advance into microvoids thereby limiting the extent of plasticity. This was based in part on his observations of similarities between the fracture surfaces seen in hydrogen-embrittled materials, such as steels, iron-silicon alloys and nickel, and those seen in liquid metal embrittlement. In liquid metal embrittlement there is usually limited mutual solubility of the liquid and host metals, e.g. mercury and steels, and cracks can grow at speeds up to several hundred $mm \cdot s^{-1}$. Therefore, the liquid metal must be confined to the crack surface, so that only adsorption can occur. The liquid metal appears to facilitate the emission of dislocations at the crack, enabling the crack to advance into microvoids, resulting in fracture surfaces displaying microscopic dimples. The same features are seen on the crack faces in Beachem's experiments. Lynch's observations further established that hydrogen embrittlement and liquid metal embrittlement of fcc and bcc metallic systems both occurred on $\{100\}$ cleavage planes in $\langle 110 \rangle$ directions. He explained these crystallographic features in terms of dislocation emission from the crack tip on alternate inclined slip planes. He argued it is adsorption of hydrogen at the crack tip, not hydrogen in solution ahead of the crack tip, that leads to plasticity being more localised than during fracture in inert environments. This occurs because dislocations are emitted from crack tips advancing the cracks into voids formed just ahead of the cracks, thereby limiting growth of the voids[63]. In inert environments little or no dislocation emission occurs from crack tips, and crack growth occurs by voids growing and linking up with the crack. There is much more extensive dislocation activity required for the observed void growth and blunting of the crack, which result in the observed much larger dimples on fracture surfaces. It is dislocation emission from crack tips enabling the crack to advance and link up with microvoids before they grow into larger voids through much more extensive plasticity, that determines the degree of embrittlement. His papers contain many examples in fcc, bcc and hcp metals. The generality of his observations is compelling. The question why dislocation emission from crack tips is so difficult in inert environments is discussed in section 11.6.2.

In the late 1980s, Birnbaum[64] and coworkers proposed[65] hydrogen enhanced localised plasticity, or HELP. In this case, hydrogen in solution ahead of the crack tip, not hydrogen adsorbed at the crack tip, is responsible for the localisation of plasticity reported by Beachem. Beachem (1972) had argued on the basis of his torsion tests that hydrogen in solid solution lowered the macroscopic flow stress, and he presumed it

Pergamon Press: Oxford (1977). ISBN 0080221084.

[63]Lynch, S, Corrosion Reviews **37**, 377 (2019).

[64]Howard Kent Birnbaum 1932-2005. US metallurgist.

[65]Shih, D S, Robertson, I M, and Birnbaum, H K, Acta Metallurgica **36**, 111-124 (1988); Birnbaum, H K, and Sofronis, P, Mat. Sci. Eng. A **176**, 191-202 (1994).
Ian M Robertson, US materials scientist. Petros Sofronis, US mechanical engineer.

would also reduce the microscopic flow stress at crack tips. Birnbaum et al. (1994) presented experimental evidence that the activation energies for dislocation motion in Ni and Ni-C alloys decrease in the presence of hydrogen. Birnbaum et al. also observed highly localised ductile failure by microvoid coalescence, and they presumed it was facilitated by enhanced mobility of dislocations (edge, screw and mixed types) in fcc, bcc and hcp metals when hydrogen is in solution.

Following in-situ observations by transmission electron microscopy, Birnbaum and coworkers put forward an explanation of the enhanced dislocation mobility based on the screening of elastic interactions resulting from segregation of hydrogen to dislocations. The degree of segregation is sensitive to strain rate and temperature. There is no reason to doubt that hydrogen is attracted elastically to dislocations forming Cottrell atmospheres. Hydrogen may also be segregated to dislocation cores. Once hydrogen has segregated to dislocations it is likely to reduce the strength and range of elastic interactions between dislocations. However, the connection between reduced elastic interactions and enhanced dislocation mobility is less obvious. Obstacles to dislocation motion such as second phase particles and high angle grain boundaries are short-range and the ability of these obstacles to impede dislocation motion is largely unaffected by elastic interactions. Intrinsic dislocation mobility is determined by properties of the core, such as the formation and migration energies of kinks. Once dislocations are saturated with hydrogen they will be less able to attract other impurities through their elastic fields, but the hydrogen may pin a dislocation rather than enhance its mobility. Evidence for such pinning may be seen in serrated yielding in tensile tests.

There are experimental observations that cannot be explained by the HELP mechanism. As noted above cleavage planes in fcc and bcc metallic systems embrittled by hydrogen are quite specific: they occur on or close to $\{100\}$ planes in $\langle 110 \rangle$ directions. A general increase in dislocation mobility ahead of a crack cannot account for the specific nature of these crystallographic features. Secondly, when hydrogen enters some metals at the crack tip the crack may grow at speeds orders of magnitude faster than hydrogen can diffuse in the metal[66]. But the fracture surfaces may still display dimples indicating localised plasticity. It has also been shown experimentally by a number of researchers[67] that abrupt changes in the environment of a fracture test, such as changing the partial pressure of hydrogen gas or adding oxygen to the hydrogen gas, have an immediate effect on the crack growth rate and the appearance of the fracture surfaces. These observations are consistent with the view that adsorbed hydrogen at the crack tip is responsible for hydrogen embrittlement.

Segregation of hydrogen to dislocations brings about a reduction in the free energy

[66]Lynch, S P, Acta Metallurgica **36**, 2639-2661 (1988).
[67]see references in Lynch (2019).

of the system, as described by the Gibbs adsorption isotherm. As Kirchheim[68] has noted[69] this may be viewed as a reduction in the free energy of *formation* of dislocation kinks. But it may also raise the free energy of *migration* of kinks, since hydrogen atoms would either have to be dragged along with the kinks or kinks would have to break free from them. It is therefore unclear whether hydrogen segregation to dislocation kinks increases dislocation mobility.

In 2001 Nagumo proposed that hydrogen is attracted to vacancies, based on interpretations of thermal desorption experiments[70]. This work suggests that hydrogen lowers the free energy of formation of vacancies, similar to Kirchheim's idea that hydrogen lowers the free energy of formation of dislocation kinks, so that hydrogen enhances the equilibrium concentration of vacancies. DFT calculations by Hickel et al.[71] also showed that the energy of hydrogen in solution in iron is reduced when the hydrogen sits in a vacancy. If the vacancies cluster this may lead to greater ease of formation of nanoscale voids.

5.3 Theory of hydrogen in metals

In 1980 Stott[72] and Zaremba[73] calculated the energy of inserting a hydrogen atom into a homogeneous, paramagnetic electron gas as a function of its density[74]. They found the energy of the hydrogen atom was minimised when the number density of the electron gas was approximately 0.017Å^{-3}. This is a relatively small electron density, comparable to that of potassium metal, and it is consistent with the observation that hydrogen atoms tend to segregate to surfaces and vacancies in metals where they can find such relatively low electron densities. In these calculations they found the s-state of the hydrogen atom is doubly occupied so that it is present as H^-. The negative charge on this ion is screened by a depletion in the electron density surrounding it. The diameter of the screened hydrogen ion must be at least the Fermi wavelength, since this determines the smallest size of any local variation of the electronic charge density.

It has been found in DFT calculations of single hydrogen interstitials that the energy of the 1s-state of the hydrogen atom is less than the bottom of the s-p-d conduction band in iron[75]. When this occurs the 1s-orbital of the hydrogen atom is a localised

[68] Reiner Kirchheim 1943–. German materials scientist.

[69] Kirchheim, R, Scripta Materialia **67**, 767-770 (2012).

[70] Nagumo, M, ISIJ International **41**, 590-598 (2001).

[71] Hickel, T, Nazarov, R, McEniry, E J, Leyson, G, Grabowski, B and Neugebauer, J, JOM **66**, 1399-1405 (2014).

[72] Malcolm J Stott, Canadian physicist.

[73] Eugene Zaremba, Canadian physicist

[74] Stott, M J, and Zaremba, E, Phys. Rev. B **22**, 1564-1583 (1980).

[75] Paxton, A T (2018), private communication. Anthony Thomas Paxton 1953 - , British materials physicist.

state forming a delta-function in the density of electronic states, occupied by two electrons. In this configuration the screened H^- ion cannot form covalent bonds to the metal because it is isoelectronic to helium. The local bonding is then an electrostatic interaction between the imperfectly screened H^- ion and the neighbouring imperfectly screened positive metal ions that have transferred some of their electronic charge to the proton. The screening is imperfect because the screening clouds overlap as their size is comparable to the spacing between the host metal atoms. This picture is consistent with the repulsion that is often observed between hydrogen atoms in transition metals at normal pressures. It is also consistent with hydrogen atoms acting as centres of dilation of the host metal lattice because the screened H^- ion is significantly larger than a neutral hydrogen atom. It is quite different from bonding in a transition metal hydride, where metal atoms are further apart than in the pure metal, enabling s-states on the hydrogen atoms to hybridise with d-states of the metal atoms forming a band of hybridised states. It is conceivable that the increase in the separation of transition metal atoms surrounding a hydrogen interstitial weakens those metal-metal bonds. The bonds may also be weakened electrostatically by the metal atoms surrounding the interstitial becoming slightly positively charged as a result of the transfer of one electron from them to the proton. It would be interesting to test these ideas with accurate electronic structure calculations.

Nazarov et al.[76] carried out DFT calculations for hydrogen atoms in vacancies in twelve fcc metals. They found hydrogen was bound to vacancies in all twelve metals, with binding energies from 0.17 to 0.86 eV. In all cases the hydrogen atom was off-centre in the vacancy. As the chemical potential of hydrogen is raised, for example by exposing the metal to higher partial pressures of hydrogen gas, they found H_2 molecules are more stable in a vacancy than two H-atoms in silver, gold and lead. At extremely high hydrogen partial pressures they found up to eight stable H_2 molecules inside a vacancy in silver. In contrast two H-atoms in nickel and rhodium never bonded to each other in a vacancy.

11.5.4 Suggestions for further research

This brief review suggests the following points for further research:

1. Is there any *direct* experimental evidence for the weakening of metal-metal bonding by hydrogen when the metal is subjected to either tension or shear? Dislocation glide involves bond switching events where bonds crossing the slip plane are broken and reformed. Haydock[77] coined the term *bond mobility*[78] to

[76]Nazarov, R, Hickel, T and Neugebauer, J, Phys. Rev. B **89**, 144108 (2014).

[77]Roger Haydock, US physicist

[78]Haydock, R, J. Phys. C: Solid State Phys. **14** 3807 (1981).

describe the ease with which bonds are broken and reformed in such a process. Strong directional bonds like those in diamond have very limited mobility at room temperature, whereas those in pure copper are much more mobile. Therefore, a sharper question is to ask whether hydrogen increases the mobility of bonds at the surface and within the interior of a metal.

2. Liquid lithium is a potent embrittling element of some transition metals and like hydrogen it has a half-filled valence s-shell. But lithium is a larger atom owing to its full 1s-core. A detailed comparison of the local electronic structures of relaxed configurations of hydrogen, helium and lithium adsorbed on surfaces and grain boundaries of metals would highlight the unique chemistry of hydrogen.
3. How does hydrogen enhance the emission of dislocation loops from cracks? This question is germane to the AIDE mechanism. Hydrogen adsorption at the crack tip reduces the energy of surface steps created when dislocation half loops are nucleated, which reduces the energy barrier to dislocation nucleation. At a more mechanistic level adsorption of a hydrogen atom at a crack tip increases the local coordination number of metal atoms which reduces the strength of bonds between metal atoms neighbouring the adsorbed hydrogen atom. This is a generic feature of metallic bonding that has nothing to do with charge transfers[79]. The greatest need is for a dynamical simulation to explore how adsorbed hydrogen enhances nucleation of half-loops at a loaded crack front. A realistic 3D crack geometry including steps on the crack front where loops can nucleate heterogeneously would probably require multi-million atoms. It would also require a credible model of interatomic forces for hydrogen in the metal. By comparing simulations of the same loaded crack with and without adsorbed hydrogen it should be possible to identify the mechanism by which hydrogen enhances dislocation emission.

11.6 The ductile-brittle transition (DBT)

11.6.1 Introduction

A wide range of materials, from metals and alloys to ceramics and polymers, are ductile at elevated temperatures but become brittle as the temperature is lowered. In some materials the transition occurs gradually over a wide range of temperatures of about 100K or more, and in others it is very abrupt spanning less than 10K. Owing to their importance in safety-critical components, research has been carried out to try to understand the factors affecting the DBT temperature (T_c) in metals and alloys, and the underlying defect-related mechanisms.

[79]See, for example, Sutton, A P, *Electronic structure of materials*, p179-182, Oxford University Press: Oxford (1996). ISBN 0-19-851754-8.

A reduction of the grain size is one of the simplest and most effective ways of reducing T_c even though it also raises the yield stress. This may be because grain boundaries act as barriers to the growth of small cleavage cracks, and/or the increased difficulty of nucleating cracks in smaller grains. In engineering practice there are several factors known to raise T_c and render materials more susceptible to rapid brittle fracture. Increasing the yield stress, in ways other than by reducing the grain size, raises T_c, such as increasing the strain rate, or irradiating the material with high energy particles, or work hardening the material. The size and shape of the component also affect T_c because plastic deformation in large, thick components is more constrained than in small samples of the same material, owing to the larger surface to volume ratio in the latter, which raises T_c. This makes it difficult to evaluate in the laboratory the susceptibility to brittle fracture of large engineering components, such as a pressure vessel or a pipeline. In addition, the conditions to which a component may be subjected in service may be more severe than those that can be readily reproduced in the laboratory. A variety of high strain rate impact tests have been devised on notched specimens, such as the Charpy V-notch impact test, to try to simulate the effects of plastic constraint present in large components and severe service conditions. An example is shown in Fig 11.4.

In this section we briefly review what is known experimentally about the DBT in metals and identify some questions for further research. We shall see that interstitial impurities play a central role in the DBT of bcc metals. Partly for this reason we begin with a summary of fundamental experimental and computational studies carried out at Oxford of the DBT in pre-cracked single crystals of silicon where the impurity content can be controlled to an exceptionally high degree. Through a combination of experiment and modelling this work has provided the most complete understanding of the DBT in any crystalline material, and therefore it is a good place to start. Brittle fracture in silicon is easier to understand than in metals because very little plasticity is involved, if any. In the brittle state a crack in silicon grows by breaking bonds at the crack tip. As we shall see in section 11.6.3, this is usually not the case in metals, even though it has often been assumed in theoretical treatments.

11.6.2 The DBT in silicon

Consider a constant strain rate test on a single crystal of silicon containing a sharp surface crack loaded in mode I. The criterion for brittle fracture in silicon is that the local stress intensity factor along the crack front reaches K_{1c}. The local stress intensity factor includes shielding provided by any pre-existing dislocations. When this criterion is satisfied the crack grows by breaking bonds at the crack front and no dislocations are emitted from the crack. The transition from brittle to ductile fracture is complete when

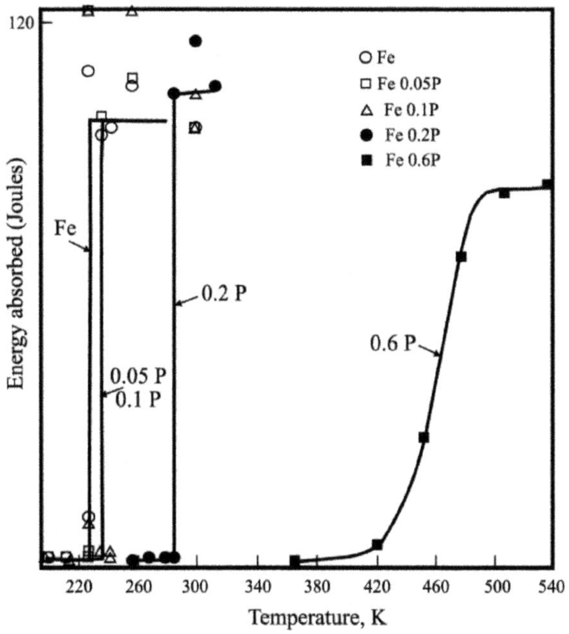

Figure 11.4: Ductile-brittle transitions in high purity iron and iron-phosphorus alloys (0.05, 0.1, 0.2 and 0.6 wt.% P) obtained in Charpy V-notch tests using half standard width specimens, after Spitzig, W A, Metallurgical Transactions A **3**, 1183 (1972); William A Spitzig 1931-2021, US metallurgist. All the alloys had equiaxed ferrite grains of average size 45 μm, with no precipitates. Brittle fractures of the 'pure' iron and 0.05 P and 0.1 P alloys were predominantly transgranular, whereas the 0.2 P and 0.6 P alloys fractures were mostly intergranular due to embrittlement of grain boundaries caused by phosphorus segregation. Notice the abrupt transitions at the lower concentrations of phosphorus and the more gradual transition at 0.6 P.

the local stress intensity factor along the entire crack front becomes less than K_{1c} as a result of shielding dislocations emitted by the crack[80].

For a given strain rate the value of T_c varies by up to 250K for similar silicon samples but different test geometries. However, the transition is always abrupt, occurring over just 5-10K[81]. At temperatures below T_c a pre-cracked silicon crystal, which may or may not contain a small density of pre-existing dislocations, cleaves in an entirely brittle

[80]Samuels, J, and Roberts, S G, Proc. R. Soc. A **421**, 1-23 (1989); Hirsch, P B, Roberts, S G, and Samuels, J, Proc. R. Soc. **421**, 25-53 (1989).
Sir Peter Bernhard Hirsch FRS 1925- , British physicist and materials scientist born in Germany.

[81]Hirsch, P B and Roberts, S G, Philosophical Magazine A **64**, 55 (1991).

manner determined by the Griffith criterion and the low temperature critical stress intensity factor K_{1c} of silicon. At T_c the specimen still fractures in an essentially brittle manner, but sources of dislocations at the crack front operate and emit dislocations. They are shielding dislocations. When the transition is complete the whole sample deforms plastically and behaves in a ductile manner.

The strain rate dependence of T_c was found by equating the rate of increase of the stress intensity factor, $\dot{K}$, in a constant strain rate test, to a constant multiplied by $\exp(-U/k_B T_c)$ where U is the activation energy of the DBT. By performing a series of constant strain rate tests at different strain rates, and by plotting $\ln \dot{K}$ against $1/T_c$, it was shown, to within experimental error, that U is the same as the activation energy for dislocation glide in all specimen geometries. This was done for both intrinsic silicon and n-type silicon where the activation energies for dislocation glide differ owing to the dopant effect[82]. This confirms that the DBT in silicon is controlled by the glide mobility of dislocations. Furthermore, it was shown experimentally that the strain rate dependence of T_c is due to the strain rate dependence of dislocation generation during loading, not during the subsequent fracture. During loading at T_c, pre-existing dislocations in the crystal moved to the crack front where they became sources. These sources emitted shielding dislocations which rapidly reduced the local stress intensity factor along the crack front.

Hirsch and Roberts (1991) addressed the questions of what determines T_c in silicon and why the transition is so abrupt. In dynamical discrete dislocation models they found T_c is determined by the time taken for pre-existing dislocations to reach the crack tip just before the local stress intensity factor reaches K_{1c}. Their modelling showed that once the sources at the crack tip were created they emitted dislocations so rapidly and in such large numbers that the crack was shielded before the applied stress intensity factor reached K_{1c}. The time taken for pre-existing dislocations to reach the crack front depends on the dislocation velocity, which is temperature dependent, and the distance of pre-existing dislocations from the crack front. Thus, T_c depends on the pre-existing dislocation distribution, which accounts for its variability for a given test geometry. To reach the crack tip when the imposed strain rate is greater, pre-existing dislocations have to move faster, which is possible only when the temperature is higher. Hence T_c increases when the strain rate increases.

The abruptness of the transition is determined by how long it takes for the local stress intensity factor at all points along the crack front to be reduced by the emission of shielding dislocations. Shielding dislocations are emitted from the crack front sources as half loops which expand along and normal to the crack front. The degree of shielding at a point along the crack front will be smaller the further it is from sources. Therefore,

[82]Hirsch, P B, Materials Science and Technology **1**, 666 (1985).

the transition is more abrupt the greater the density of sources along the crack front.

This detailed study of the DBT in silicon shows that when emission of dislocations from loaded sharp cracks is difficult, as appears to be the case in silicon, it may be assisted by the ingress of pre-existing crystal dislocations into the crack front that transform into sources. The DBT is then controlled by the mobility of dislocations near the crack as a function of temperature. It is also worth noting that dislocations attracted to the crack front are anti-shielding, that is they increase the local stress intensity factor, which may assist their subsequent transformation into crack front sources. As shielding dislocations are emitted, the crack front remains sharp as it advances along the slip planes and in the directions of the Burgers vectors of the emitted dislocations.

In a later study[83] Roberts and Hirsch analysed experiments by Folk[84] on the DBT in initially dislocation-free, semiconductor-grade single crystals of silicon, which were not pre-cracked and were subjected to three-point bend tests. Their analysis, which was supported by dynamical discrete dislocation modelling, revealed that the DBT was again controlled by the glide mobility of dislocations. But in contrast to the experiments on pre-cracked specimens the fracture stress of samples within $\pm 35K$ of T_c was found experimentally to have a strong dependence on temperature. Roberts and Hirsch argued that this was because the surface flaws which nucleated the cracks in this temperature range were also created by dislocation activity.

The conclusion of these studies is that the DBT in silicon is controlled by the temperature dependence of the glide of dislocations – either to create *new* sources of shielding dislocations at an existing crack, or, in samples that are not pre-cracked, to create *new* sources of dislocations to reduce the load on an incipient surface crack also produced by dislocation activity. The transition from brittle to ductile behaviour is effected by avalanches of dislocations from sources that are created by dislocations gliding either to the surface of a sample that is not pre-cracked, or towards a pre-crack.

One of the most interesting aspects of this research is the experimental observation that a loaded sharp crack in silicon does not emit dislocations until nearby dislocations have entered the crack and transformed into sources of dislocations. This suggests there is an insurmountable barrier to the nucleation of dislocation loops at the loaded crack tip, before nearby dislocations have entered it, despite the large stresses present at the tip. Rice and Thomson[85] postulated that when the barrier to the nucleation of dislocation loops at a loaded atomically sharp crack is high the crack responds in a brittle manner, i.e. by breaking bonds at the crack tip with no dislocation activity involved. Conversely, they argued that when the barrier is low the local stress intensity

[83]Roberts, S G and Hirsch, P B, Philosophical Magazine **86**, 4099 (2006).

[84]Folk, R H, *The brittle to ductile transition in silicon: evidence of a critical yield event*, PhD thesis, University of Pennsylvania (2000).

[85]Rice, J R and Thomson, R, Philosophical Magazine **29**, 73 (1974).

factor is reduced by blunting the crack resulting from the emission of dislocations from the crack tip. The response of the crack is then ductile. We will see in the next section that 'brittle' fracture in metals almost always involves some plasticity. Therefore, the assumed dichotomy between purely brittle fracture, where there is no dislocation activity, and ductile fracture, where there is much dislocation activity, is almost always an oversimplification, at least in metallic systems. Nevertheless, the Rice-Thomson analysis of the forces involved in the emission of a half loop from a crack tip is instructive and useful.

Consider the emission of a dislocation half loop from a crack front. Its Burgers vector will be that of a shielding dislocation, otherwise it will not leave the crack. There are three forces acting on the loop. The Peach-Koehler force, due to the stress field of the loaded crack, tends to expand the loop away from the crack tip. The image interaction force between the dislocation loop and the free surfaces of the crack opposes the Peach-Koehler force. Finally the loop creates a ledge on the crack surface, and this generates a short-range force that tends to constrict the length of the ledge, also opposing the Peach-Koehler force. An additional factor affecting the ease of dislocation nucleation at the crack tip is the width of the dislocation core[86]. A wider dislocation core reduces the nucleation barrier by distributing the atomic displacements associated with its formation over a larger number of atoms.

If the short-range force due to the creation of the surface ledge is neglected the Rice-Thomson analysis shows that crystals with dislocations having wider cores and values of the dimensionless parameter $\gamma/\mu b \gtrsim 0.1$ are able to emit dislocation loops spontaneously from atomically sharp loaded cracks with no activation barrier: this is Rice and Thomson's ductile behaviour. The parameter γ is the energy per unit area of the elastically strained crack surface at the crack tip. There is an activation barrier to the emission of dislocation loops from atomically sharp cracks in crystals with narrower dislocation cores and smaller values of $\gamma/\mu b$. When this barrier is more than a few electron volts the crack is unable to emit dislocation loops spontaneously even with thermal assistance and it is assumed the crack then responds in a brittle manner by breaking bonds at the crack tip with no dislocation activity. Rice and Thomson estimated the activation energy barrier in silicon is 111 eV, and so spontaneous dislocation emission from loaded cracks in silicon is impossible. This conclusion is consistent with the experimental observation discussed in this section that emission of dislocations from a crack in silicon occurred only after sources were created at the crack tip by the ingress of crystal dislocations. The Rice-Thomson analysis is unable to explain the transition to ductile behaviour in silicon as the temperature is raised

[86]The thinking here is that the dislocation core is planar, as in the Peierls-Nabarro model. This is appropriate for fcc crystals and for edge dislocations in bcc crystals.

because the experimentally observed mechanisms of the transition are more complex than the assumed distinction between ductile and brittle behaviours: according to the assumptions of the Rice-Thomson analysis, silicon is brittle at all temperatures.

11.6.3 The DBT in metals

Over the past century an understanding of the DBT in metals has become widely accepted. As in the Rice-Thomson model, the understanding is based on the assumption that there is a competition between dislocation emission from crack tips, leading to ductile fracture, and bond-breaking at crack tips with no dislocation emission, leading to brittle fracture. The occurrence of DBTs observed in bcc metals is attributed to the large increase of the yield stress with decreasing temperature, coupled with a relatively temperature-independent cleavage fracture stress. As the temperature is reduced the yield stress becomes larger than the cleavage stress and the material fractures in a brittle manner before it can yield. The strong temperature dependence of the yield stress in bcc metals is attributed to the exponential dependence on temperature of the creation and movement of kinks on screw dislocations, without which they are unable to overcome the Peierls barrier opposing their glide. A larger strain rate and/or increased plastic constraint raise the effective yield stress, so that the transition to ductile behaviour occurs at a higher temperature. The absence of DBTs in fcc metals is explained by the yield stress having only a small temperature dependence, so that it never rises above the brittle fracture stress as the temperature is reduced. I received this understanding of the DBT in metals as an undergraduate and I continued to believe it when I wrote the first edition of this book. In this section I will argue it is inconsistent with a variety of experimental observations, some of which have existed in the literature for more than half a century.

The previous section showed that experimental observations played a decisive role in the elucidation of the mechanisms of the DBT in silicon. The focus in this section will be on experimental observations of the DBT in metals. *These observations have been largely overlooked in the literature on DBTs in metals, and they cast doubt on the conventional understanding summarised in the previous paragraph.* For example, 'brittle' fracture in metals is often accomplished by localised plasticity. The belief that bcc metals become brittle at low temperatures because screw dislocations become immobile is inconsistent with experimental observations of plasticity in zone-refined bcc metals at liquid helium temperatures. Purified tantalum and potassium are bcc metals with marked temperature-dependences of the yield stress, especially at low temperatures, but there is no evidence that they have a DBT. Iridium is an fcc metal with the same slip systems as other fcc metals and yet it *does* have a DBT.

The classification of fractures of metals and alloys in inert environments under monotonically increasing loads (not oscillatory loads) has become problematic with the recognition that there are areas of overlap between conceptual classes that were formerly distinct. New experimental techniques have revealed significant detail on fracture surfaces and blurred the boundaries of conventional types of fracture. For example, the terms 'ductile' and 'brittle' have become dependent on the length scale of features that can be detected experimentally on fracture surfaces. The terms 'transgranular' and 'intergranular' remain unambiguous because they refer to the path of the crack through a grain or along a grain boundary respectively. On the other hand, 'cleavage' was originally used by mineralogists to describe the brittle fracture of crystals along well-defined crystallographic planes. In the metallurgical literature 'cleavage' is sometimes applied to intergranular fracture. The use of 'cleavage' in this book is confined to transgranular fractures, but unlike its conventional usage it often involves localised plasticity.

It is useful to distinguish *three* modes of fracture under a monotonically increasing load in metallic systems:

- *Ductile fracture*: This always involves extensive plastic deformation. The crack blunts and grows through the absorption of crystal dislocations from sources near the crack. The surfaces of the crack are a sink for these dislocations. Relatively large voids nucleate, grow and coalesce ahead of the crack through intense dislocation activity, resulting in fracture surfaces with large, deep dimples. The work of fracture is several orders of magnitude larger than the energy of the fracture surfaces.
- *Brittle fracture*: The crack is atomically sharp and grows by the tensile breaking of interatomic bonds at the crack tip *with no dislocation activity* at all. The path of the crack may be transgranular or intergranular or a mixture of the two. In single crystals brittle fractures are usually along low index lattice planes ('cleavage planes') and the fracture surfaces are atomically smooth and lustrous. As defined here, brittle fracture in metallic systems is not common, but as an extreme mode of fracture it is useful conceptually. An exceptional case of brittle fracture in a metal was obtained by cleaving a single crystal of chromium at 4.2K in a scanning tunnelling microscope under ultra high vacuum[87]. The (001) cleavage plane of chromium was observed with atomic resolution and showed no evidence of any defects other than occasional atomic steps separating atomically flat terraces. Despite the absence of dislocation activity, brittle fractures are almost always irreversible owing to the excitation of phonons and electrons by the propagating crack. Brittle fracture of non-metals can produce the emission

[87]Kolesnychenko, O Yu, de Kort, R and van Kempen, H, Surface Science Letters **490**, L573 (2001).

of electrons, ions and photons from fracture surfaces, during and for some time after fracture, resulting from the development of high densities of surface charges on the fracture surfaces[88]. Consequently, the work of brittle fracture is more than the energy of creating two free surfaces, but it is still much less than the work of ductile fracture.

- *Quasi-brittle fracture*: The crack grows by the emission and absorption of dislocations and localised plasticity ahead of the crack. Microscopic and/or nanoscopic voids nucleate, grow and coalesce with one another and with the advancing crack leaving correspondingly small dimples on fracture surfaces. The voids may be elongated, forming 'flutes', along slip band intersections when only a limited number of slip systems can be activated, e.g. in hcp metals. Dislocations are created at sources and at voids ahead of the crack and some dislocations enter the crack, contributing to its growth. The crack is both a source and a sink for dislocations. Quasi-brittle cracks may be either transgranular or intergranular or a mixture of the two. Transgranular quasi-brittle cracks are sometimes called 'cleavage-like' cracks. The key point about this mode of relatively low energy 'brittle' fracture is that the crack does not grow by the tensile breaking of bonds at the crack tip, but by emitting and absorbing dislocations and by coalescing with microscopic/nanoscopic voids formed ahead the crack. It was suggested by Beachem[89] in 1976 that 'cleavage' is accomplished by localised plasticity (see also section 11.5.2). He wrote:

  Plasticity precedes cleavage, initiates cracks by dislocation pile-ups or twinning, raising local stresses to produce cleavage, accompanies cleavage propagation, [and] is evident at halted cleavage crack tips... Cleavage is accomplished by plasticity[90].

  Because the plasticity is localised the energy of quasi-brittle fracture is again much less than that of ductile fracture, but it is more than would be consumed by brittle fracture if it were possible. It may be difficult to distinguish experimentally between brittle and quasi-brittle fractures and high resolution techniques may be needed to detect the microscopic dimples on a quasi-brittle fracture surface.

---

[88]Dickinson, J T, Jensen, L C and Jahan-Latibari, A, Journal of Vacuum Science & Technology A **2**, 1112 (1984).

J Thomas Dickinson 1941- . US materials physicist.

[89]Beachem, C D, Proceedings of the 2nd International Conference on Mechanical Behaviour of Materials, p. 322, American Society for Metals: Metals Park Ohio (1976).

[90]Reprinted with permission of ASM International. All rights reserved. `www.asminternational.org`.

Examples of quasi-brittle fractures are given below.

Iron is an example of a bcc metal that displays a DBT in Charpy impact tests, with T_c occurring over a range of temperatures depending on its purity and sample geometry. As indicated above, this has often been attributed to the strong dependence of the mobility of screw dislocations in bcc metals as the temperature is lowered. However, it has been known since the 1950s that polycrystalline bcc iron deforms plastically prior to fracture in tensile tests at temperatures as low as 4.2K if the sample is not pre-cracked[91]. The fracture is 'brittle' when the grain sizes are larger, but becomes ductile with a large reduction of area at a neck when the grain sizes are smaller. Savitskii and Burkhanov[92] note that T_c can be reduced to temperatures close to absolute zero in fine grained, zone-refined polycrystalline iron. At 4.2K the relation between stress and strain in a tensile test becomes highly serrated owing to instabilities in plastic deformation[93]. The instabilities arise from local heating caused by plastic deformation and are expected in all metals and alloys where the flow stress has a strong dependence on temperature. It is a remarkable experimental observation[94] that the same temperature dependences of the flow stress are obtained, between 2K and 293K, in bcc pure iron single crystals and in bcc iron polycrystals containing a small amount of carbon. This indicates that the flow stress *is* an intrinsic property of pure bcc iron, i.e. it is governed by the Peierls stress. Nevertheless, the observations of plasticity at temperatures approaching absolute zero question whether the Peierls stress is the only factor responsible for notched samples of pure iron having a DBT.

According to Savitskii and Burkhanov (1970) the interstitial impurity content has a strong influence on T_c in iron and the refractory bcc metals vanadium, niobium, chromium, molybdenum and tungsten, and even on the existence of a DBT in unnotched samples of these metals. Interstitial impurities are known to pin dislocations by forming Cottrell atmospheres. When the internal stress is sufficiently high dislocations escape their Cottrell atmospheres. This is the origin of the sharp drop of the yield stress observed in a tensile test at a constant strain rate. When the released dislocations reach barriers, such as grain boundaries or second phase particles, they form pile-ups. The stress concentration at the head of a pileup increases as the square root of the grain size.

---

[91] Basinski, Z S and Sleeswyk, A, Acta Metallurgica **5**, 176 (1957).
André Wegener Sleeswijk 1927-2018. Dutch physicist.

[92] Savitskii, E M and Burkhanov, G S, *Physical metallurgy of refractory metals and alloys*, Chapter IV, published in Moscow in 1967 in Russian, translated by Consultants Bureau: New York and London (1970). ISBN: 978-1-4684-1572-8.
Evgenii Mikhailovich Savitskii born 1912, Soviet physical chemist and metallurgist.
Gennadii Sergeevich Burkhanov 1932–, Soviet/Russian metallurgist.

[93] Basinski, Z S, Proceedings of the Royal Society A **240**, 229 (1957).

[94] Altshuler, T L and Christian, J W, Philosophical Transactions of the Royal Society **261**, 253 (1967).

If the stress concentration is not relaxed quickly by further plastic deformation cracks may be nucleated (see problem 10.6). This is one reason why a reduction in the grain size lowers T_c. When the concentration of interstitial and other impurities exceeds their solubility limits in polycrystals they will tend to segregate to grain boundaries. This can lead to the phenomenon of temper embrittlement where a heat treatment enables segregation of impurities to grain boundaries that is subsequently retained by rapid cooling. If grain boundaries become saturated with impurities they may form brittle inclusions such as hydrides, nitrides, oxides and carbides. But grain boundaries may also be embrittled by much smaller concentrations of impurities, where they remain in solution on the boundaries (see the discussion at the end of section 10.5). Brittle inclusions at grain boundaries are not necessarily involved in DBTs.

The group 5 refractory bcc metals, vanadium, niobium and tantalum, have a higher solubility of interstitial impurities than those of group 6 refractory bcc metals, chromium, molybdenum and tungsten. According to Savitskii and Burkhanov (1970), in tensile tests of zone-refined unnotched samples of vanadium and tantalum ductile behaviour is maintained to temperatures as low as 4.2K and as low as 1.4K in zone-refined niobium. The embrittling potency of interstitial impurities in group 5 bcc metals is in the order hydrogen, nitrogen, oxygen and carbon, with hydrogen most potent. For example, in vanadium a carbon content of the order of 0.1 wt.% is permissible, but the hydrogen content must be kept below 0.001 wt.% to retain ductility. In chromium, molybdenum and tungsten the solubility limit of interstitial impurities is less than 0.0001 wt.%. Above these concentrations of interstitial impurities brittle second phases may be precipitated. For example, it is reported in Savitskii and Burkhanov (1970) that tungsten of commercial purity has a T_c between 500 and 700°C. But in zone-refined tungsten T_c is less than -196°C (77K). This was also found[95] in zone-refined single crystals of tungsten tested in tension, if the crystals were not pre-cracked. But when sharp surface cracks were introduced prior to tensile testing, the tungsten crystals failed after smaller strains at all temperatures between 77K and 478K, and at 77K the fracture appeared to be 'brittle'.

More recent experiments on notched and unnotched polycrystalline bars and on single crystals of purified tungsten[96] found the strain rate dependence of T_c has an activation energy of 1.05 eV. This is a remarkable result because it indicates there is a common rate-controlling process of the DBT in these specimens, and therefore the activation energy for the DBT in pure tungsten is a material parameter of pure tungsten. The value of T_c was about 40K lower in the unnotched bar than in the notched bar. The activation energy of the DBT in tungsten is equal to the activation energy of formation

[95]Hull, D, Beardmore, P and Valintine, A P, Philosophical Magazine A **12**, 1021 (1965).
Derek Hull FRS 1931-. British materials scientist.

[96]Giannattasio, A, Yao, Z, Tarleton, E and Roberts, S G, Philosophical Magazine **90**, 3947 (2010).

of a double kink on screw dislocations, if account is taken of the contribution to the activation energy from the stress on dislocations. The significance of this correlation is unclear, because dislocation glide will be involved on both sides of the DBT in tungsten if fracture below T_c involved localised plasticity, as is likely to be the case. Indeed, there are observations of localised plasticity at temperatures below T_c and at low strain rates in molybdenum[97] where it gives rise to stable crack growth prior to rapid fracture. This appears to be similar to the stable crack growth at temperatures below T_c in iridium discussed below.

Contrary to conventional belief, not all bcc metals display a DBT. Pure tantalum is a refractory bcc metal that does not have a DBT even in notched samples at temperatures as low as 22K[98] Potassium has a melting point of 337K and also has the bcc crystal structure. It is ductile in tensile tests in an oxygen-free atmosphere at all temperatures down to 7.5K. The temperature dependences of yielding, flow and the work hardening rate are similar to those of the refractory bcc metals[99], even though potassium is a nearly free electron metal while bonding in the refractory bcc metals is directional and dominated by d-electrons.

Polycrystalline iridium, which is a refractory fcc metal with a melting point of 2,716K, has a DBT at about a third of its melting point[100]. The twelve $\{111\}\langle 110\rangle$ slip systems in iridium are the same as in other fcc metals which do not display a DBT. Panfilov and Yermakov found that single crystals of iridium cleaved under tension at room temperature after elongations of 30-60%[101]. Using Auger electron spectroscopy Hecker et al.[102] found no evidence of impurity segregation at grain boundaries in iridium exposed by intergranular fracture.

In the 1980s Lynch carried out detailed fractographic and metallographic studies of fracture surfaces in iridium, which he reviewed[103] in 2007. Figure 11.5 shows a brittle fracture surface in polycrystalline iridium in which there are intergranular and transgranular areas. Lynch used high purity, notched, cylindrical iridium single crystals with a [100] axis, into which sharp pre-cracks were introduced by fatigue. The samples

97 Hirsch, P B, Booth, A S, Ellis, M and Roberts, S G, Scripta Metallurgica et Materialia **27**, 1723 (1992).

98 Imgram, A G, Mallett, M W, Koehl, B G, Bartlett, E S and Ogden H R, *Notch sensitivity of refractory metals*, Armed Services Technical Information Agency Report AD 275379, p.26, Arlington Virginia (1962).

99 Bernstein, I M and Gensamer, M, Acta Metallugica **16**, 987 (1968).

100 Yang, J, Wang, H, Hu, R, Li, S, Yi, L and Luo, S, Advanced Engineering Materials **20**, 1701114 (2018).

101 Panfilov, P and Yermakov, A, Journal of Materials Science **39**, 4543 (2004).

102 Hecker, S S, Rohr, D L and Stein, D F, Metall Mater Trans A **9**, 481 (1978).
Siegfried S Hecker 1943–, US metallurgist. Dale Franklin Stein 1935–, US metallurgist.

103 Lynch, S P, Scripta Metallurgica **57**, 85 (2007).

were then subjected to monotonic loading in three-point bending at 293K and 77K. Any plastic deformation that occurred during monotonic loading was associated only with crack growth and not any plastic deformation of the crystal before the crack was introduced. He observed stable crack growth preferentially along [011] with crack tip opening angles of $10-15°$ at 293K and with the crack plane approximately parallel to (100). Crack fronts were parallel to $[01\bar{1}]$, which is the line of intersection of (111) and $(1\bar{1}\bar{1})$ slip planes with the (100) crack plane. He found evidence of substantial *localised* plasticity associated with quasi-brittle cleavage fracture of single crystals tested at 293K and 77K, well below T_c. Fig 11.6 shows slip traces on (111) and $(1\bar{1}\bar{1})$ planes at and behind the crack tip. Further evidence of strains beneath fracture surfaces was provided by Laue X-ray back-reflection patterns. The reflections in Fig 11.7(a) are from an electropolished (100) surface and their sharpness indicates the absence of significant strains beneath the surface. In contrast, in Fig 11.7(b) the streaking of reflections indicates large strains just beneath the (100) fracture surface.

Fig 11.8 shows TEM images of carbon replicas of the cleavage fracture surface of iridium shadowed with germanium at small angles (10-15°) and viewed at high tilt angles (30-40°) to enhance contrast and to resolve fine detail. Notice the very fine dimples (shallow depressions) between the serrated steps (thick dark lines). These dimples result from nanoscale voids formed ahead of the crack. In their publications Beachem and Lynch have emphasised that observations by SEM of 'smooth' fracture surfaces may not have sufficient resolution to see nanoscale dimples. However, they may be revealed by using replica techniques in transmission electron microscopy as illustrated in Fig 11.8. Nanoscale dimples on the fracture surfaces provide further evidence of quasi-brittle fracture in iridium through localised plasticity.

Among hcp metals zinc[104] has a DBT where at temperatures below $T_c \approx 350K$ brittle fracture is predominantly transgranular on basal and $\{10\bar{1}0\}$ planes with a significant contribution along $\{10\bar{1}2\}$ twins. Beryllium[105] is similar to zinc with a DBT where the brittle fracture is predominantly transgranular on the basal plane and $\{11\bar{2}0\}$ prism planes, and along $\{10\bar{1}2\}$ twins. Magnesium[106] and cadmium[107] have DBTs where the low temperature fractures are accomplished by localised plasticity. Fig 11.9 shows a scanning electron micrograph of a fracture surface in cadmium obtained at 77K where localised plasticity is evident. It is particularly puzzling that the low temperature fracture modes of zinc (brittle fracture) and cadmium (quasi-brittle fracture) differ,

[104]Hughes, G M and Flewitt, P E J, Materials Science and Technology, **24**, 567 (2008)

[105]Hanes, H D, Porembka, S W, Melehan, J B and Gripshover, P J, *Physical metallurgy of beryllium*, Report to Defense Metals Information Center: Columbus Ohio (1966).

[106]Lynch, S P, Materials Forum **11**, 268 (1988).

[107]Stoloff, N S and Gensamer, M, Transactions of the Metallurgical Society of AIME **227**, 70 (1963). Maxwell Gensamer 1902-1992, US metallurgist. Lynch, S P, unpublished work.

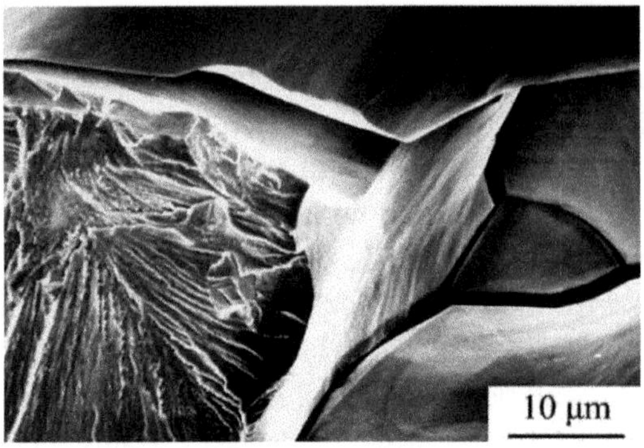

Figure 11.5: Scanning electron micrograph of fracture surface of polycrystalline iridium displaying relatively smooth intergranular surfaces and a rougher transgranular surface. The roughness of the transgranular surface is due to steps on secondary {100} cleavage planes forming 'river lines'. Courtesy of S P Lynch.

Figure 11.6: Optical micrograph of slip around a cleavage crack in a single crystal of iridium at 293K. The crack is the horizontal triangular wedge shape, and its faces are macroscopically parallel to (100) planes. Notice the inclined slip lines at and behind the crack tip. Courtesy of S P Lynch.

considering they are in the same group of the Periodic Table, with the same crystal structure and very similar c/a ratios, 1.856 and 1.886 respectively, and they share the same slip systems.

The tetragonal structure of white tin (β-tin) has a DBT[108] at about half its melting point (505K). Lynch[109] found that a sample of pure white tin subjected to a Charpy impact test at 253K bent without breaking, with extensive stretching at the notch root.

[108] An, Q, An, R, Wang, C and Wang, H, Materialstoday Communications **29**, 102962 (2021).
[109] Lynch, S P, to be published.

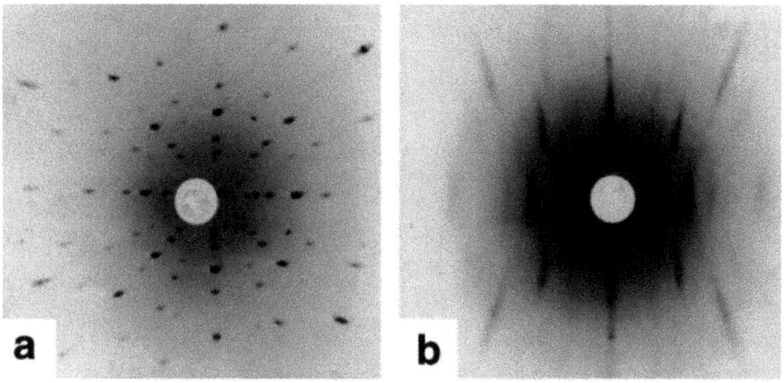

Figure 11.7: Laue X-ray back-reflection patterns from iridium single crystals (a) for an electropolished (100) surface, and (b) for a (100) fracture surface produced by quasi-brittle cleavage at 293K. Courtesy of S P Lynch.

Figure 11.8: Transmission electron micrograph of a carbon replica of a quasi-brittle cleavage surface of iridium produced at 293K showing serrated steps (thick dark lines) and nanoscale dimples. Courtesy of S P Lynch.

Three quarters of the cross-section of the Charpy specimen fractured at 243K in a quasi-brittle manner, with the remainder breaking in a ductile manner owing to the reduced plastic constraint. At 233K the whole cross-section of the Charpy specimen fractured in a quasi-brittle manner. His fractographic observations of the quasi-brittle fracture surfaces revealed a mixture of intergranular and transgranular facets that were dimpled on a fine scale, and exhibited slip lines and twin traces, as seen in Fig 11.10.

Sometimes the DBT in metals is very abrupt with a width of just a few degrees. For example, the DBT in 'pure' iron can be very abrupt when it is determined by Charpy

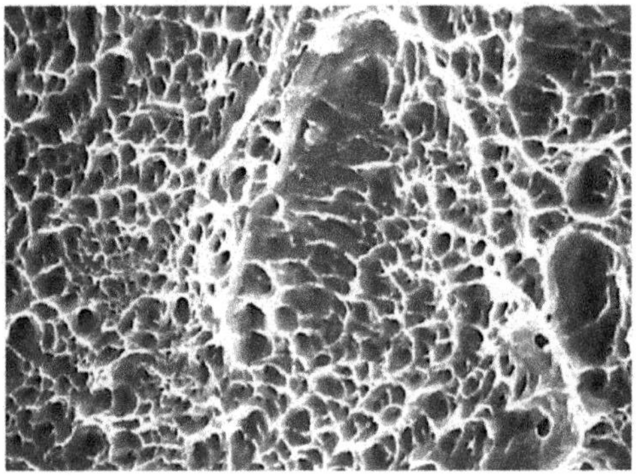

Figure 11.9: Scanning electron micrograph of cleavage fracture surface of cadmium obtained at 77K. The surface is covered with dimples a few microns across providing evidence of localised plasticity during quasi-brittle fracture. Courtesy of S P Lynch.

impact tests, as seen for example in Fig 11.4 and in Syarif et al. (2003)[110]. The width of the DBT in metals can also be hundreds of degrees. It is not clear what determines the width of the DBT in metals.

In section 11.5 we reported that 'brittle' fractures involved in hydrogen assisted cracking and liquid metal embrittlement often proceed in a quasi-brittle manner through *localised plasticity* involving the growth of sharp cracks by dislocation emission and their coalescence with nanoscale voids ahead of the crack, producing fracture surfaces with nanoscopic or microscopic dimples[111]. The similarity with the quasi-brittle fractures in pure metals discussed in this section is striking. Quasi-brittle fractures, which are accomplished by localised plasticity, occur in a wider range of metals, alloys and environments than has been recognised.

It is difficult to explain the persistence of the widely held belief that 'brittle' fracture of metals is accomplished by the tensile breaking of interatomic bonds at a crack tip despite the weight of experimental evidence to the contrary that has accumulated over the past fifty years. In my own case I was unaware of the accumulated experimental evidence until quite recently. Beachem (1976) makes the following observations:

There appears to be no basic principle from which to draw the popular conclusion that the separation of a crystal by cleavage need ever be accomplished by mechanisms

[110]Syarif, J, Tsuchiyama, T and Takaki, S, ISIJ International **43**, 1100 (2003).
[111]Lynch, S P, Acta Metallurgica **36**, 2639 (1988).

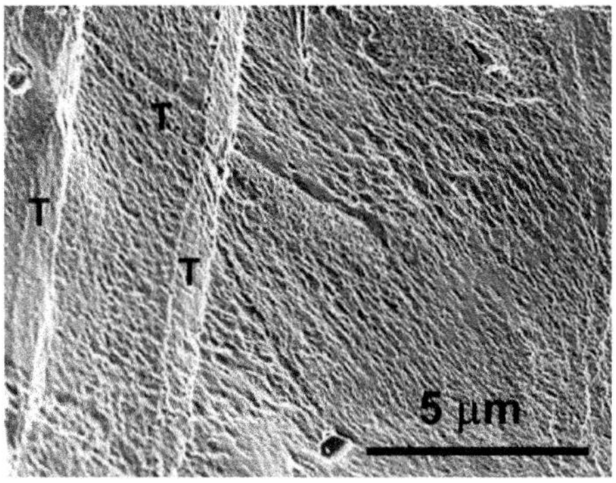

Figure 11.10: Charpy-tested specimen of pure tin at −30°C showing twin traces (T) and nanoscale dimples on an intergranular facet formed by quasi-brittle fracture. Nanoscale dimples were also seen on transgranular fracture surfaces. Courtesy of S P Lynch.

exactly opposite from those mechanisms by which the crystal is held together. The widely held, and almost certainly wrong, concept that cohesion and cleavage are exact opposites, or antitheses, is a product of:

1. *extreme extrapolations of our macroscopic observations of mechanical, physical, and fracture properties down in size scale to the atomic level,*
2. *extreme extrapolations of calculated interatomic binding forces up in size scale to the macroscopic level, and*
3. *extensive efforts to show that interatomic binding forces and fracture forces are the same*[112]*.*

'Brittle' fracture in metals is initiated and propagated by the collective behaviour of defects in a localised region of the material. The volumes of metal in which these defect processes occur are far too large to be treated atomistically. Even if an atomistic approach became possible in the future it would not be trivial to extract the physics from so much redundant information. Atomic interactions are important insofar as they determine the structure and properties of the perfect crystal and individual defects. But, like work-hardening (see section 11.4), 'brittle' fracture in metals involves emergent physics at a larger length-scale. To quote Beachem (1976) again:

Cohesive energies, or cohesive forces, do account for the loads to which tensile or bend specimens can be subjected before they break. Once fracture starts, however, events become localized, and average properties become less important than local stresses, general microstructures become less important than local crack tip microstructures, and

[112]Reprinted with permission of ASM International. All rights reserved. `www.asminternational.org`.

bulk environments become less important than local crack tip environments... Local properties and events become decisive. Once fracture starts, it's a whole new ball game[113].

There is much work to be done to develop the tools needed for a credible simulation of 'brittle' transgranular or intergranular fracture. Only when these tools are available for metals will be able to model the localisation of plasticity that takes place below T_c. That is the central question for theory and simulation to address concerning the DBT in metals.

[113]Reprinted with permission of ASM International. All rights reserved. `www.asminternational.org`.

Recommended books

Elasticity

A treatise on the mathematical theory of elasticity Love, A E H, Dover Publications: New York (1944), reprint of the 4th edition of this book published in 1927. ISBN 0-486-60174-9. A classic, not so easy to read because of the dated notation, but well worth dipping into including the appendices.

Theory of elasticity Landau, L D and Lifshitz, E M, 3rd edn. Pergamon Press: Oxford (1986). Volume 7 of the famous *Course on Theoretical Physics*. ISBN 0-08-033916-6. An advanced treatment with not one superfluous word. Well worth careful study. The chapter on dislocations was written jointly with A M Kosevich and covers a lot of ground in just 24 pages.

Physical properties of crystals Nye, J F, Oxford University Press: Oxford (1985). ISBN 0-19-851165-5. An exceptionally clear account of tensor properties of crystals. Chapter 8 treats the elastic constant tensor in different crystal symmetries.

Dynamical Theory of Crystal Lattices Born, M and Huang, K, Oxford Classic Texts in the Physical Sciences, Oxford University Press: Oxford and New York (1998). This was one of the first books to attempt to derive properties of (non-metallic) crystals from fundamental principles at the atomic scale (it was first published by Oxford University Press in 1954). Chapter 3 deals with elasticity. ISBN 978-0198503699.

Statistical mechanics of elasticity Weiner, J H, Dover Publications: New York (2002). ISBN 0-486-42260-7. First published in 1983, this is another book that treats elasticity at the atomic scale.

Thermodynamics of crystals Wallace, D C, John Wiley & Sons Inc.: New York (1972). ISBN 978-0471918554. This book provides an exceptionally clear treatment of the thermodynamics of crystals including elasticity. It was republished by Dover Publications in 2003 with ISBN 978-0486402123.

Application of elasticity to crystal defects

Introduction to elasticity theory for crystal defects Balluffi, R W, World Scientific: Singapore (2017). ISBN 9814749729. A comprehensive and systematic treatment of defects and their interactions in anisotropic elasticity, with many worked examples. An outstanding pedagogical text.

Imperfections in Crystalline Solids Cai, Wei and Nix, William D, Cambridge University Press: Cambridge (2016). ISBN 9781107123137 hard back. Wide-ranging coverage of mechanics, thermodynamics and kinetics of defects in crystals. Chapters may be downloaded free of charge from Cambridge University Press.

Elastic models of crystal defects Teodosiu, C, Springer-Verlag: Berlin (1982). ISBN 0-387-11226-X. A wonderful book providing a rigorous treatment of elasticity of defects.

Micromechanics of defects in solids Mura, T, Kluwer Academic Publishers: Dordrecht (1991). ISBN 90-247-3256-5. An excellent, wide-ranging mathematical treatment of elasticity and defects.

Collected works of J. D. Eshelby eds. X Markenscoff and A Gupta, Springer: Dordrecht (2006). ISBN 9781402044168 hard back, ISBN 9789401776448 soft cover. Eshelby wrote only 57 papers, but every one is a gem. They are all here.

Anisotropic continuum theory of lattice defects Bacon, D J, Barnett, D M, and Scattergood, R O, Progress in Materials Science **23**, 51-262 (1978). ISBN 0080242472. An excellent introduction to the modern theory of anisotropic elasticity and its application to defects in crystals.

Theory of crystal defects ed. B Gruber, Proceedings of the Summer School held in Hrazany in September 1964, Academia Publishing House of Czechoslovak Academy of Sciences: Prague (1966). Kroupa's chapter on dislocation loops is particularly illuminating.

The physics of metals 2 Defects ed. P B Hirsch, Cambridge University Press: Cambridge (1975). ISBN 0 521 200776. Contains six chapters on different aspects of the physics of defects in metals written by leaders in the field. The absence of heavy mathematical analysis and the focus on physics renders it very readable.

Mechanics of solid materials R J Asaro and V A Lubarda, Cambridge University Press: Cambridge (2006). ISBN 978-0-521-85979-0. A comprehensive treatment from an engineering perspective of continuum mechanics, linear elasticity, micromechanics, thin films and interfaces, plasticity and viscoplasticity, and biomechanics, with solved problems.

Dislocations

Introduction to dislocations Hull, D and Bacon, D J, Pergamon Press: Oxford (1984). ISBN 0-08-028720-4. **This book should be read in conjunction with Chapter 6**. Very readable, covering the basic phenomenology of dislocations.

Elementary dislocation theory Weertman, J and Weertman, J R, Oxford University

Press: Oxford (1992). ISBN 0-19-506900-5. Another very readable introduction to dislocations.

Dislocations and plastic flow in crystals Cottrell, A H, The Clarendon Press: Oxford (1953). The first monograph on dislocations, and still one of the best, written by a pioneer of the field. Few can match Cottrell's clarity and insight.

Dislocations in crystals Read, W T, McGraw-Hill Book Company: New York (1953). Another early monograph on dislocations, written by a pioneer of the field.

Theory of dislocations Anderson, P M, Hirth, J P and Lothe, J, Cambridge University Press: Cambridge England. ISBN 978-0-521-864367. This is the third edition of what for many is the last word on dislocations.

Dislocations Friedel, J, Addison Wesley: Reading Mass. (1964). A classic written by a physicist for metallurgists and engineers in industry as well as researchers in universities. It is an expanded translation of his book *Les dislocations*, which was written in French and published by Gauthier-Villars: Paris in 1956.

Theory of crystal dislocations Nabarro, F R N, Dover Publications: New York (1987). ISBN 0-486-65488-5. A major treatise on dislocations written by one of the central figures in the development of the subject.

Dislocation dynamics and plasticity Suzuki, T, Takeuchi, S and Yoshinaga, H, Springer-Verlag: Heidelberg (1991). ISBN 0-387-52693-5. An excellent introduction to dislocation dynamics in metals, alloys and semiconductors at low and high temperatures.

Anisotropic elasticity theory of dislocations Steeds, J W, Oxford University Press: Oxford (1973). Systematic treatment of dislocations in different crystal symmetries using anisotropic elasticity.

Dislocations in solids eds. F R N Nabarro, M S Duesbery, J P Hirth, L Kubin **1-16**, North-Holland Publishing Company: Amsterdam (1979 - 2009). Each volume contains reviews written by leading researchers on various aspects of dislocations. Eye-wateringly expensive books.

Dislocations, mesoscale simulations and plastic flow Kubin, L P, Oxford University Press: Oxford (2013). ISBN 978-0-19-852501-1. Modern survey of how dislocations control plasticity in elemental metals, from the atomic scale to the microstructural level and the connection to continuum plasticity. Strong on experimental observations with an incisive review of different techniques for modelling dislocation dynamics in 3D.

Computer simulations of dislocations Bulatov, V V and Cai, W, Oxford University Press: Oxford (2006). ISBN 0-19-852614-8. A practical 'how to do it' introduction to computer simulations of dislocation statics and dynamics in 3D.

Strengthening mechanisms of crystal plasticity Argon, A S, Oxford University Press: Oxford (2008). ISBN 978-0-19-851600-2. Systematic treatment of the physics of strengthening mechanisms including the lattice resistance, solid-solution strengthening, precipitation strengthening and strain hardening, all supported by experimental observations.

Cracks

Mathematical theory of dislocations and fracture Lardner, R W, University of Toronto Press: Toronto (1974). ISBN 0-8020-5277-0. Particularly good on continuous distributions of dislocations and cracks.

Physics of fracture Thomson, R, in *Solid State Physics*, **39**, 2-129 (1986). A physicist's view on cracks and dislocations. Particularly good on shielding of cracks by dislocations, and limitations of continuum theories of cracks and fracture.

Dislocation based fracture mechanics Weertman, J, World Scientific: Singapore (1996). ISBN 9810226209. A nice introduction to the treatment of the modelling of cracks by distributions of dislocations. Also very good on shielding of cracks by dislocations.

Solution of crack problems Hills, D A, Kelly, P A, Dai, D N and Korsunsky, A M, Kluwer Academic Publishers: Dordrecht (1996). ISBN 978-90-481-4651-2. An advanced treatment of a wide range of crack problems using distributions of dislocations.

Fracture of brittle solids Lawn, B, Cambridge University Press: Cambridge UK, 2nd edition (1993). ISBN 0-521-40972-1. An outstanding text on fundamental experimental and theoretical aspects of brittle fracture, particularly in ceramics, covering atomistic, microstructural and continuum aspects.

Fundamentals of fracture mechanics Knott, J F, Butterworth & Co.: London (1973). ISBN: 0-408-70529-9. An influential and very readable treatment of cracks and fracture from an engineering perspective.

History of elasticity

History of strength of materials Timoshenko, S P, Dover Publications: New York (1983) reprint of the book published originally in 1953. ISBN 0-486-61187-6. A fascinating account of the historical development of elasticity and the people who worked on it from the seventeenth century to 1950.

A history of the theory of elasticity and of the strength of materials Vol. I Galilei to Saint-Venant 1639-1850, Vol. II parts I and II Saint-Venant to Lord Kelvin, by I Todhunter and K Pearson. Vol. I originally published in 1886. Vol. II parts I and II originally published in 1893. Reissued by Dover Publications in 1960. Delightfully opinionated account of the early history of the subject.

References

Alshits, V I in *Elastic strain fields and dislocation mobility*, eds. V L Indenbom and J Lothe, Chapter 11, p 625-697, North-Holland: Amsterdam (1992). ISBN 0444887733.

Altshuler, T L and Christian, J W, Philosophical Transactions of the Royal Society **261**, 253 (1967).

An, Q, An, R, Wang, C and Wang, H, Materialstoday Communications **29**, 102962 (2021).

Anderson, P M, Hirth, J P and Lothe, J, *Theory of dislocations*, 3rd edn., Cambridge University Press: Cambridge and New York (2017), p.110. ISBN 978-0-521-86436-7.

Anderson, P M, Hirth, J P and Lothe, J, *Theory of dislocations*, 3rd edn., Cambridge University Press: Cambridge and New York (2017), section 15.2. ISBN 978-0-521-86436-7.

Antolovich, S D and Conrad, H, Materials and Manufacturing Processes, **19**, 587-610 (2004).

Ashcroft, N W and Mermin, N D *Solid state physics*, p437-440 (1976). ISBN 978-0030839931.

Atkinson, W and Cabrera, N Phys. Rev. **138**, A763-A766 (1965).

Bacon, D J, Barnett, D M and Scattergood R O, *Anisotropic continuum theory of lattice defects*, Progress in Materials Science, **23**, 51-262 (1979).

Bak, P, *How nature works*, p.1-2, Oxford University Press: Oxford (1997). ISBN 0198501641.

Barenblatt, G I, J. Appl. Math. Mechs. **23**, 622-636 (1959).

Barenblatt, G I, J. Appl. Math. Mechs. **23**, 1009-1029 (1959).

Barenblatt, G I, Adv. Appl. Mech. **7**, 55-129 (1962).

Basinski, Z S and Sleeswyk, A, Acta Metallurgica **5**, 176 (1957).

Basinski, Z S and Basinski, S J, Phil. Mag. **9**, 51-80 (1964).

Beachem, C D, Metallurgical Transactions **3**, 441-455 (1972).

Beachem, C D, Proceedings of the 2nd International Conference on Mechanical

Behaviour of Materials, p. 322, American Society for Metals: Metals Park Ohio (1976).

Bernstein, I M and Gensamer, M, Acta Metallugica **16**, 987 (1968).

Bilby, B A, Proc. Phys. Soc. A **63**, 191 (1950).

Bilby, B A, Cottrell, A H and Swinden, K H, Proc. R. Soc. A **272**, 304-314 (1963).

Bilby, B A, and Eshelby, J D, in *Dislocations and the Theory of Fracture* forming Chapter 2 of *Fracture*, ed. H Liebowitz, volume 1, Academic Press: New York (1968).

Birnbaum, H K, and Sofronis, P, Mat. Sci. Eng. A **176**, 191-202 (1994).

Brown, J M, Abramson, E H and Angel, R J Phys. Chem. Minerals **33**, 256-265 (2006).

Brown, L M, Metallurgical Transactions A, **22**, 1693-1708 (1991).

Brown, L M, Materials Science and Technology **28**, 1209-1232 (2012).

Brown, L M, Philosophical Magazine **96**, 2696-2713 (2016).

Burgers, J M, Koninklijke Nederlandsche Akademie van Wetenschappen, **42**, 293-325 (1939).

Burridge, R and Knopoff, L, Bulletin of the Seismological Society of America **54**, 1875-1888 (1964).

Caro, J A and Glass N, J. Phys. (Paris), Lett. bf 45, 1337-1345 (1984).

Celli, V and Flytzanis, N, J. Appl. Phys. **41**, 4443-4447 (1970).

Chandrasekhar, S, *Ellipsoidal figures of equilibrium*, Dover Publications: New York (1987). ISBN 978-0486652580.

Chang, S-J and Ohr, S M, J. Appl. Phys. **52**, 7174-7181 (1981).

Chang, S-J and Ohr, S M, Int. J. Fracture **23**, R3-R6 (1983).

Chia, K Y and Burns, S J, Scripta Metall., **18**, 467-472 (1984).

Cho, J, `https://www.youtube.com/watch?v=jwK-TF7o2Oo`.

Christian, J W, and Vitek, V, Reports on Progress in Physics **33**, 307 (1970).

Conrad, H and Sprecher, A F in *Dislocations in Solids* **8** Chapter 43, 497-541 (1989). North-Holland: Amsterdam. ISBN 978-0444705150.

Conrad, H, Mat. Sci. Eng. A **287** (2000).

Cottrell, A H, *Dislocations and plastic flow in crystals*, Chapter 10, Clarendon Press: Oxford (1953).

Cottrell, A H, in *Dislocations in Solids* **11**, eds. F R N Nabarro and M S Duesbery, pages vii - xvii, Elsevier: Amsterdam (2002). ISBN 0-444-50966-6.

Dehlinger, U, Ann. Phys. **394**, 749-793 (1929).

Dickinson, J T, Jensen, L C and Jahan-Latibari, A, Journal of Vacuum Science & Technology A **2**, 1112 (1984).

Dimitrov, N K, Liu, Y and Horstemeyer, M F, Mechanics of Advanced Materials and Structures **29**, 705 (2022).

Dudarev, S L and Sutton, A P, Acta Materialia **125**, 425-430 (2017).

Dugdale, D S, J. Mech. Phys. Solids **8**, 100-104 (1960).

Economou, E N, *Green's functions in quantum physics*, Springer-Verlag: Berlin (1983). ISBN 978-3642066917.

Eshelby, J D, Phil. Mag. **40**, 903-912 (1949).

Eshelby, J D, Phil. Trans. R. Soc. London A **244**, 87-111 (1951).

Eshelby, J D, Frank, F C, and Nabarro, F R N, Phil. Mag. **42**, 351-364 (1951).

Eshelby, J D, Solid State Physics **3**, 79-144 (1956).

Eshelby, J D, Proc. R. Soc. A, **241**, 376-396 (1957).

Eshelby, J D, Proc. R. Soc. A **252**, 561-569 (1959).

Eshelby, J D, Progress in Solid Mechanics **2**, 89-140 (1961).

Eshelby, J D, in *Inelastic behaviour of solids*, eds. M F Kanninen, W F Adler, A R Rosenfield and R I Jaffee, p. 77-115, McGraw-Hill: New York (1970).

Eshelby, J D, J. Elasticity **5**, 321-335 (1975).

Eshelby, J D, in *Point defect behaviour and diffusional processes*, eds. R E Smallman and E Harris, pp.3-10, The Metals Society, London (1977).

Eshelby, J D, in *Continuum models of discrete systems*, eds. E Kröner and K-H Anthony, University of Waterloo Press: Waterloo, p.651-665, (1980).

Eshelby, J D, *Collected works of J. D. Eshelby*, eds. X Markenscoff and A Gupta, Springer: Dordrecht (2006). ISBN 9781402044168 hard back, ISBN 9789401776448 soft cover.

Fehlner, W R and Vosko, S H, Can. J. Phys. **54**, 2159-2169 (1976).

Finnis, M W, *Interatomic forces in condensed matter*, (2003), Oxford University Press: Oxford. ISBN 978-0198509776.

Folk, R H, *The brittle to ductile transition in silicon: evidence of a critical yield event*, PhD thesis, University of Pennsylvania (2000).

Frank, F C, and Read, W T, Phys. Rev. **79**, 722 (1950).

Frank, F C, Phil. Mag. **42**, 809 (1951).

Frank, F, C, Proc. R. Soc. Lond. A **371**, 136 (1980).

Frenkel, J, and Kontorova, T, J. Phys. Acad. Sci. USSR **1**, 137-149 (1939).

Friedel, J, *Les dislocations* p.215, Paris: Gauthier-Villars (1956).

Giannattasio, A, Yao, Z, Tarleton, E and Roberts, S G, Philosophical Magazine **90**, 3947 (2010).

Giustino, F, Rev. Mod. Phys. **89** 015003, 63 pages (2017); Erratum Rev. Mod. Phys. **91**, 019901 (2019).

Goldstein, H, *Classical mechanics*, Addison-Wesley: Reading Mass (1980). ISBN 0-201-02969-3.

Green, G, Transactions of the Cambridge Philosophical Society, **7**, 1 (1839).

Griffith, A A, Phil. Trans. R. Soc. A, **221**, 163-198 (1921).

Griffiths, D J, *Introduction to electrodynamics*, International edn., Prentice-Hall (1999), p.351. ISBN 978-9332550445.

Guan, L, Tang, G and Chu, P K, J. Mat. Res. **25**,1215-1224 (2010).

Gurrutxaga-Lerma, B, Verschueren, J, Sutton, A P, and Dini, D International Materials Reviews **66**, 215-255 (2021).

Hall, E O, Proc. Phys. Soc. **B64**, 747-753 (1951).

Hammad, A, Swinburne, T D, Hasan, H, Del Rosso, S, Iannucci, L and Sutton, A P, Proc. R. Soc. A **471** 20150171 (2015).

Haydock, R, J. Phys. C: Solid State Phys. **14** 3807 (1981).

Head, A K, and Louat, N, Australian J. Phys., **8**, 1 (1955).

Hecker, S S, Rohr, D L and Stein, D F, Metall Mater Trans A **9**, 481 (1978).

Hickel, T, Nazarov, R, McEniry, E J, Leyson, G, Grabowski, B and Neugebauer, J, JOM **66**, 1399-1405 (2014).

Hikata, A, Johnson, R A and Elbaum, C, Phys. Rev. B **2**, 4856 (1970).

Hirsch, P B, Materials Science and Technology **1**, 666 (1985).

Hirsch, P B, Roberts, S G, and Samuels, J, Proc. R. Soc. **421**, 25-53 (1989).

Hirsch, P B and Roberts, S G, Philosophical Magazine A **64**, 55 (1991).

Hirsch, P B, Booth, A S, Ellis, M and Roberts, S G, Scripta Metallurgica et Materialia **27**, 1723 (1992).

Hoekstra, J, Sutton, A P, Todorov, T N and Horsfield, A P, Phys. Rev. B **62**, 8568-8571 (2000).

Hoekstra, J, Sutton, A P and Todorov, T N, J. Phys.: Condens. Matter **14**, L137 (2002).

Horton, J A and Ohr, S M, J. Mater. Sci. **17**, 3140-3148 (1982).

Hughes, G M and Flewitt, P E J, Materials Science and Technology, **24**, 567 (2008).

Hull, D, Beardmore, P and Valintine, A P, Philosophical Magazine A **12**, 1021 (1965).

Huu-Duc Nguyen-Tran, Hyun-Seok Oh, Sung-Tae Hong, Heung Nam Han, Jian Cao, Sung-Hoon Ahn and Doo-Man Chun, International Journal of Precision Engineering and Manufacturing Green Technology **2**, 365-376 (2015).

Ishioka, S, J. Phys. Soc. Jpn. **30**, 323-327 (1971).

Imgram, A G, Mallett, M W, Koehl, B G, Bartlett, E S and Ogden H R, *Notch sensitivity of refractory metals*, Armed Services Technical Information Agency Report AD 275379, Arlington Virginia (1962).

Ivory, J, Phil. Trans R. Soc. **99**, 345-372 (1809).

Jackson, P J, Acta Metallurgica **33**, 449-454 (1985).

Jardine, L, *The curious life of Robert Hooke*, Harper Collins: London (2003). ISBN 978-0007151752

Jensen, H J, *Self-organized criticality*, Cambridge University Press: Cambridge (1998). ISBN: 0521483719.

Jinha, Arif E, Learned Publishing **23**, 258-263 (2010).

Johnson, W H, Proc. R. Soc. Lond. **23**, 168-179 (1875).

Kelvin, Lord, Article 37 of Volume 1 of his *Mathematical and Physical Papers* p.97.

Kirchheim, R, Scripta Materialia **67**, 767-770 (2012).

Kolesnychenko, O Yu, de Kort, R and van Kempen, H, Surface Science Letters **490**, L573 (2001).

Kröner, E, *Kontinuumstheorie der Versetzungen und Eigenspannungen*, Springer-Verlag: Berlin (1958). ISBN 978-3540022619.

Kroupa, F in *Theory of crystal defects*, Proceedings of the Summer School held in Hrazany in September 1964, Academia Publishing House of the Czechoslovak Academy of Sciences (1966), pp 275-316.

Kubin, L P, *Dislocations, mesoscale simulations and plastic flow*, section 2.4, Oxford University Press: Oxford (2013). ISBN 978-0-19-852501-1.

Kuhlmann-Wilsdorf, D, Mat. Sci. Eng. **86**, 53-66 (1987).

Landau, L D, and Lifshitz, E M, *Theory of Elasticity*, Pergamon Press: Oxford, 3rd edition (1986), section 27, p.111. ISBN 0-08-033916-6.

Lardner, R W, *Mathematical theory of dislocations and fracture*, University of Toronto Press, Chapter 5 (1974), ISBN 0-8020-5277-0.

Leibfried, G, Z. Phys. **130**, 244 (1951).

Leibfried, G and Breuer, N, *Point defect in Metals I*, Springer-Verlag: Berlin, p.146 (1978). ISBN 978-3662154489.

Love, A E H, *A Treatise on the Mathematical Theory of Elasticity*, Dover: New York (1944), p.4. ISBN 0-486-60174-9.

Lund, F and Scheihing H, B, Phys. Rev. B **99** 214102 (2019).

Lynch, S P and Ryan, N E, in Proceedings of the 2nd International Congress on Hydrogen in Metals, Paper 3D12, Pergamon Press: Oxford (1977). ISBN 0080221084.

Lynch, S P, Scr. Metall. **13**, 1051-1056 (1979).

Lynch, S P, Materials Forum **11**, 268 (1988).

Lynch, S P, Acta Metallurgica **36**, 2639-2661 (1988).

Lynch, S P, Scripta Metallurgica **57**, 85 (2007).

Lynch, S, Corrosion Reviews **37**, 377 (2019).

Lynch, S P, to be published (2023).

Majumdar, B S and Burns, S J, Int. J. Fracture **21**, 229-240 (1983).

Markenscoff, X, J Elasticity **49**, 163-166 (1998).

Mason, D R, Le Page, J, Race, C P, Foulkes, W M C, Finnis, M W and Sutton, A P, J. Phys. Condens. Matter **19**, 436209 (2007).

Maxwell, J C, Philosophical Magazine, **27**, 294-299 (1864).

Mott, N F, Proc. R. Soc. A **220**, 1 (1953).

Mura, T, *Micromechanics of defects in solids* 2nd ed. (1991), section 10 p 65-73, Kluwer: Dordrecht. ISBN 90-247-3256-5.

Mura, T, *Micromechanics of defects in solids*, Chapter 3, Kluwer Academic Publishers: Dordrecht (1991). ISBN 90-247-3256-5.

Muskhelishvili, N I, *Singular integral equations*, Dover Publications: New York (2008), ISBN 9780486462424.

Nabarro, F R N, Proc. Phys. Soc. **59**, 256 (1947).

Nabarro, F R N, Phil. Mag. **42**, 1224 (1951).

Nabarro, F R N, Adv. Phys. **1**, 269-394 (1952), p.360.

Nabarro, F R N, Basinski, Z S and Holt, D B, Adv. Phys. **13**, 193-323(1964).

Nagumo, M, ISIJ International **41**, 590-598 (2001).

Navier, C-L, *Leçons . . . (1833)*

Nazarov, R, Hickel, T and Neugebauer, J, Phys. Rev. B **89**, 144108 (2014).

Neumann, F E, *Vorlesungen über die Theorie der Elastizität der festen Körper und des Lichtäthers*, (1885), ed. O E Meyer. Leipzig, B G Teubner-Verlag.

Nielsen, O H, and Martin, R M, Phys. Rev. B **32**, 3780 (1985)

Noll, W, *The role of the professor* `https://www.math.cmu.edu/~wn0g/RP.pdf`

Nye, J F, *Physical properties of crystals*, Oxford University Press (1957), p.21. ISBN 0-19-851165-5.

Okazaki, K, Kagawa, M and Conrad, H, Scr. Metall. **12**, 1063-1068 (1978).

Orowan, E, Z. Physik **89**, 634 (1934).

Orowan, E, Trans. Inst. Eng. Shipbuilders, Scotland **89**, 165 (1945).

Orowan, E, Rep. Prog. Phys., **12**, 185-232 (1949).

Panfilov, P and Yermakov, A, Journal of Materials Science **39**, 4543 (2004).

Paxton, A T (2018), private communication.

Peach, M O and Koehler, J S, Phys. Rev. **80**, 436 (1950).

Peierls, R, Proc. Phys. Soc. **52**, 34 (1940).

Peierls, R E, in *Dislocation Dynamics*, eds. A R Rosenfield, G T Hahn, A L Bement Jr and R I Jaffee, pages xiii-xiv and xvii-xx, McGraw-Hill: New York (1968).

Peierls, R E, Proc. R. Soc. London A **371**, 28-38 (1980).

Petch, N J, J. Iron and Steel Inst. **174**, 25 (1953).

Pfeil, L B, Proc. R. Soc. A **112**, 182-195 (1926).

Polanyi, M, Z. Physik **89**, 660 (1934).

Prigogine, I, *From being to becoming*, pages 7-8, W.H. Freeman and Co.: San Francisco. ISBN 0-7167-1108-7.

Read, W T, and Shockley, W, Phys. Rev. **78**, 275 (1950).

Ready, A J, Haynes, P D, Rugg, D, and Sutton, A P, Phil. Mag. **97**, 1129-1143 (2017).

Rice, J R, J. Appl. Mech. **35**, 379-386 (1968).

Rice, J R and Thomson, R, Philosophical Magazine **29**, 73 (1974).

Roberts, S G and Hirsch, P B, Philosophical Magazine **86**, 4099 (2006).

Robinson, A, *The last man who knew everything*, One World: Oxford (2006). ISBN 978-0452288058.

Samuels, J, and Roberts, S G, Proc. R. Soc. A **421**, 1-23 (1989).

Savitskii, E M and Burkhanov, G S, *Physical metallurgy of refractory metals and alloys*, Chapter IV, published in Moscow in 1967 in Russian, translated by Consultants Bureau: New York and London (1970). ISBN: 978-1-4684-1572-8.

Shibkov, A A, Denisov, A A, Zheltov, M A, Zolotov, A E, and Gasanov, M F, Materials Science and Engineering A **610**, 338-343,(2014).

Shih, D S, Robertson, I M, and Birnbaum, H K, Acta Metallurgica **36**, 111-124 (1988).

Siems, R Phys. Stat. Sol. **30**, 645-658 (1968).

Sokolnikoff, I S, *Mathematical theory of elasticity*, 2nd edn., section 48, McGraw-Hill Book Company Inc.: New York (1956). ISBN 978-0070596290.

Spitzig, W A, Metallurgical Transactions A **3**, 1183 (1972).

Sprecher, A F, Mannan, S L and Conrad, H, Acta Metallurgica **34**, 1145-1162 (1986).

Stokes, G G, Trans. Camb. Phil. Soc. **9** (1849), 1-62.

Stoloff, N S and Gensamer, M, Transactions of the Metallurgical Society of AIME **227**, 70 (1963).

Stott, M J, and Zaremba, E, Phys. Rev. B **22**, 1564-1583 (1980).

Stroh, A N, Proc. R. Soc. **223**, 404-414 (1954).

Sutton, A P, *Electronic structure of materials*, p179-182, Oxford University Press: Oxford (1996). ISBN 0-19-851754-8.

Sutton, A P and Balluffi, R W, *Interfaces in crystalline materials*, Oxford classic texts in the physical sciences, Clarendon Press: Oxford (2006). ISBN 978-0-19-921106-7.

Sutton, A P and Todorov, T N, Phys. Rev. Materials **5**, 113605 (2021).

Swinburne, T D and Dudarev, S L, Phys. Rev. B **92**, 134302, (2015).

Syarif, J, Tsuchiyama, T and Takaki, S, ISIJ International **43**, 1100 (2003).

Synge, J L *The hypercircle in mathematical physics* (1957) CUP: Cambridge, p.411-413.

Taylor, G I, Proc. R. Soc. A **145**, 388 (1934).

Teodosiu, C, *Elastic models of crystal defects*, section 6.2, Springer Verlag: Berlin (1982). ISBN 0-387-11226-X.

Tewary, V K, Advances in Physics **22**, 757-810 (1973).

Thomson, R, Hsieh, C, and Rana, V, J. Appl. Phys. **42**, 3154-3160 (1971).

Thomson, R, *Solid State Physics*, **39**, 2-129 (1986).

Timoshenko, S P, *History of strength of materials*, Dover Publications: New York (1983), p.108. ISBN 0-486-61187-6.

Ting, T C T, *Anisotropic Elasticity*, p.159-161 Oxford University Press: Oxford and New York (1996). ISBN: 0195074475.

Tittmann, B R and Bömmel, H E, Phys. Rev. **151**, 178-189 (1966).

Todhunter, I *A History of the Mathematical Theories of Attraction and the Figure of the Earth from the time of Newton to that of Laplace*, **2**, paragraph 804, p31 (1873).

Todorov, T N, Dundas, D, Lü, J-T, Brandbyge M, and Hedegård, P, Eur. J. Phys. **35**, 065004 (2014).

Troitskii, O A and Likhtman, V I, Dokl. Akad. Nauk SSSR **148**, 332-334 (1963).

Troitskii, O A, JETP Letters **10**, 11-14 (1969).

Troitskii, O A, Strength of Materials **7** 804 (1975).

Troitskii, O A, Skvortsov, O B and Stashenko, V I, Russian electrical engineering **89**,143-146 (2018).

Valladares, A, White, J A and Sutton, A P, Phys. Rev. Letts., **81**, 4903-4906 (1998).

Verschueren, J, Gurrutxaga-Lerma, B, Balint, D S, Sutton, A P and Dini D, Phys. Rev. Lett. **121**, 145502 (2018).

Vitek, V, Phil. Mag. **18**, 773-786 (1968).

Vitek, V and Paidar, V, *Dislocations in solids* ed. F R N Nabarro, **14**, 439-514 (2008). Elsevier: Amsterdam. ISBN 9780444531667.

Vitek, V, Progress in Materials Science **56**, 577-585 (2011).

Vitek, V, private communication (2019).

Volterra, V, Annales scientifiques de l'École Normale Supérieure **24** 401-517, (1907).

Wallace, D C, *Thermodynamics of crystals*, section 2, John Wiley & Sons Inc.: New York (1972). ISBN 9780471918554.

Weertman, J, Lin, I-H and Thomson, R, Acta Metallurgica **31**, 473-482 (1983).

Westlake, D G, ASM Transactions Quarterly **62**, 1000-1006 (1969).

Willis, J R, J. Mech. Phys. Solids **15**, 151-162 (1967).

Yang, J, Wang, H, Hu, R, Li, S, Yi, L and Luo, S, Advanced Engineering Materials **20**, 1701114 (2018).

Zener, C M in *Fracturing of metals* Symposium, American Society for Metals: Cleveland, Ohio, p.3 (1948).

Zhao, S, Zhang, R, Chong, Y, Li, X, Abu-Odeh, A, Rothchild, E, Chrzan, D C, Asta, M, Morris, J W and Minor, A M, Nature Materials **20**, 468-472 (2021).

Subject Index

Author Index